# Aquatic Insects

Kleber Del-Claro • Rhainer Guillermo
Editors

# Aquatic Insects

## Behavior and Ecology

Springer

*Editors*
Kleber Del-Claro
Institute of Biology
Federal University of Uberlandia, UFU
Uberlândia, Minas Gerais, Brazil

Rhainer Guillermo
Federal University of São Carlos, UFSCar
São Carlos, São Paulo, Brazil

ISBN 978-3-030-16326-6 ISBN 978-3-030-16327-3 (eBook)
https://doi.org/10.1007/978-3-030-16327-3



This Springer imprint is published by the registered company Springer Nature Switzerland AG
The registered company address is: Gewerbestrasse 11, 6330 Cham, Switzerland

# Preface

In our current environmental scenario, it is a matter of the utmost importance to highlight the loss of freshwater habitats. Notwithstanding, aquatic insects have taken the role of ambassadors with the mission to show us how anthropogenic impacts have been depleting freshwater reserves worldwide. Although several aquatic insect orders have been intensively studied for decades, most groups lack crucial biological information required for species conservation and for the maintenance of ecological communities and processes they are part of. For instance, their behavior, natural history, general biology, and ecology are underexplored, mostly in the tropics.

This book offers us the opportunity to look to the future and address important gaps in knowledge and research. The aim here was to provide an overview and appraise the behavior and ecology of aquatic insects under a multidisciplinary approach, as a way to address the neglected dimensions of their complex biodiversity. We expect that each chapter serves not only as a tool for current research but also as an inspiration for undergraduate and graduate students. We urge young scientists to pursue their career researching the myriad of themes and areas of interest we outline here.

This extensive set of work produced by several renowned and young scientists would not have been possible without the priceless support of universities, research institutes, state financial agencies for science, and a long list of teachers, advisors, students, and support staff.

We are most grateful to our families, friends, and colleagues that made this work possible and embarked with us in this endeavor.

This book is dedicated to every person who believes in science as a way to have a better world.

Uberlândia, Minas Gerais, Brazil — Kleber Del-Claro
São Carlos, São Paulo, Brazil — Rhainer Guillermo

# Contents

1 **Aquatic Insects: Why It Is Important to Dedicate Our Time on Their Study?** . . . . . 1
Kleber Del-Claro

2 **Distribution and Habitats of Aquatic Insects** . . . . . 11
Ricardo Koroiva and Mateus Pepinelli

3 **Dispersal by Aquatic Insects** . . . . . 35
Michael L. May

4 **Effects of Abiotic Factors and Ecogeographic Patterns on the Ecology, Distribution, and Behavior of Aquatic Insects** . . . . . 75
Vinicius Marques Lopez, Aurélio Fajar Tonetto, Ricardo Cardoso Leite, and Rhainer Guillermo

5 **The Biotic Environment: Multiple Interactions in an Aquatic World** . . . . . 95
Gabriela C. Mendes, Guilherme Gonzaga da Silva, Leonardo Samuel Ricioli, and Rhainer Guillermo

6 **Drumming for Love: Mating Behavior in Stoneflies** . . . . . 117
J. Manuel Tierno de Figueroa, Julio M. Luzón-Ortega, and Manuel Jesús López-Rodríguez

7 **Aquatic Insect Sensilla: Morphology and Function** . . . . . 139
Manuela Rebora, Gianandrea Salerno, and Silvana Piersanti

8 **Territoriality in Aquatic Insects** . . . . . 167
Gregory F. Grether

9 **Defenses of Water Insects** . . . . . 191
Konrad Dettner

10 **Reproductive Behavior and Sexual Selection** . . . . . 263
Diogo Silva Vilela and Iago Sanmartín-Villar

**11 Sexual Conflict in Water Striders, Dragonflies and Diving Beetles** . . . . . . . . . . . . . . . . . . . . . . . . . . . . . . . . . . . . . . . . . . . 295
Adolfo Cordero-Rivera and Anais Rivas-Torres

**12 Paternal Care in Giant Water Bugs** . . . . . . . . . . . . . . . . . . . . . . . . . . . 321
Shin-ya Ohba, José Ricardo I. Ribeiro, and Melania Santer

**13 The Aquatic Lepidopterans: A Mysterious and Unknown Fauna** . . . . . . . . . . . . . . . . . . . . . . . . . . . . . . . . . . . . . . . . . . . 341
Iasmim De-Freitas, Julia De Agostini, and Vanessa Stefani

**14 Urban Aquatic Insects** . . . . . . . . . . . . . . . . . . . . . . . . . . . . . . . . . . . . . . . 349
Javier Muzón, Lía S. Ramos, and Alejandro del Palacio

**15 The Tinbergen Shortfall: Developments on Aquatic Insect Behavior that Are Critical for Freshwater Conservation** . . . . . . . . . . 365
Fabio de Oliveira Roque, Francisco Valente-Neto, Marciel Elio Rodrigues, Francine Novais Souza, Davidson Gomes Nogueira, Ana Cláudia Piovezan-Borges, Alan P. Covich, and Michael J. Samways

**16 Genetic Connectivity in Conservation of Freshwater Insects** . . . . . . . 381
Drielly da Silveira Queiroga, Renan Fernandes Moura, and Jessica Ware

**17 Aquatic Insects as a Source for Biomimetics** . . . . . . . . . . . . . . . . . . . . 401
Stanislav N. Gorb and Elena V. Gorb

**Index** . . . . . . . . . . . . . . . . . . . . . . . . . . . . . . . . . . . . . . . . . . . . . . . . . . . . . . . . . 427

# Contributors

**Adolfo Cordero-Rivera** ECOEVO Lab, EE Forestal, Campus Universitario, Pontevedra, Spain

**Alan P. Covich** Odum School of Ecology, University of Georgia, Athens, GA, USA

**Guilherme Gonzaga da Silva** Laboratory of Ecological Studies on Ethology and Evolution (LESTES), Federal University of São Carlos, São Carlos, SP, Brazil

**Drielly da Silveira Queiroga** Department of Biology, University of São Paulo (USP), Ribeirão Preto, Brazil

**Julia De Agostini** Laboratório de História Natural e Reprodutiva de Artrópodes (LHINRA), Instituto de Biologia, Universidade Federal de Uberlândia, Uberlândia, MG, Brazil

**J. Manuel Tierno de Figueroa** Facultad de Ciencias, Departamento de Zoología, Universidad de Granada, Granada, Spain

**Fabio de Oliveira Roque** Instituto de Biociências, Universidade Federal de Mato Grosso do Sul, Campo Grande, MS, Brazil

Centre for Tropical Environmental and Sustainability Science (TESS) and College of Science and Engineering, James Cook University, Cairns, QLD, Australia

**Alejandro del Palacio** Laboratorio de Biodiversidad y Genética Ambiental (BioGeA), UNDAV, Buenos Aires, Argentina

**Iasmim De-Freitas** Laboratório de História Natural e Reprodutiva de Artrópodes (LHINRA), Instituto de Biologia, Universidade Federal de Uberlândia, Uberlândia, MG, Brazil

**Kleber Del-Claro** Laboratório de Ecologia Comportamental e de Interações, Instituto de Biologia, Universidade Federal de Uberlândia, Uberlândia, MG, Brazil

**Konrad Dettner** Department of Animal Ecology II, University of Bayreuth, Bayreuth, Germany

**Elena V. Gorb** Department Functional Morphology and Biomechanics, Zoological Institute of the Kiel University, Kiel, Germany

**Stanislav N. Gorb** Department Functional Morphology and Biomechanics, Zoological Institute of the Kiel University, Kiel, Germany

**Gregory F. Grether** Department of Ecology and Evolutionary Biology, University of California, Los Angeles, Los Angeles, CA, USA

**Rhainer Guillermo** Federal University of São Carlos, UFSCar, São Carlos, São Paulo, Brazil

**Ricardo Koroiva** Mapinguari Lab, Federal University of Mato Grosso do Sul, Campo Grande, MS, Brazil

Departamento de Hidrobiologia (DHb), Universidade Federal de São Carlos, São Carlos, SP, Brazil

**Ricardo Cardoso Leite** Instituto Federal de São Paulo, Piracicaba, Brazil

**Manuel Jesús López-Rodríguez** Facultad de Ciencias, Departamento de Ecología, Universidad de Granada, Granada, Spain

**Vinicius Marques Lopez** Universidade de São Paulo, Ribeirão Preto, Brazil

**Julio M. Luzón-Ortega** Facultad de Ciencias, Departamento de Zoología, Universidad de Granada, Granada, Spain

**Michael L. May** Department of Entomology, Rutgers University, New Brunswick, NJ, USA

**Gabriela C. Mendes** Laboratory of Ecological Studies on Ethology and Evolution (LESTES), Federal University of São Carlos, São Carlos, SP, Brazil

**Renan Fernandes Moura** Universidade Federal de Uberlândia – UFU – Centro de Estudos do Cerrado/Laboratório de Ecologia Comportamental e de Interações (LECI), Uberlândia, MG, Brazil

**Javier Muzón** Laboratorio de Biodiversidad y Genética Ambiental (BioGeA), UNDAV, Buenos Aires, Argentina

**Davidson Gomes Nogueira** Instituto de Biociências, Universidade Federal de Mato Grosso do Sul, Campo Grande, MS, Brazil

**Shin-ya Ohba** Biological Laboratory, Faculty of Education, Nagasaki University, Nagasaki, Japan

**Mateus Pepinelli** Department of Natural History, Royal Ontario Museum, Toronto, ON, Canada

**Silvana Piersanti** Dipartimento di Chimica, Biologia e Biotecnologie, University of Perugia, Perugia, Italy

**Ana Cláudia Piovezan-Borges** Instituto de Biociências, Universidade Federal de Mato Grosso do Sul, Campo Grande, MS, Brazil

**Lía S. Ramos** Laboratorio de Biodiversidad y Genética Ambiental (BioGeA), UNDAV, Buenos Aires, Argentina

**Manuela Rebora** Dipartimento di Chimica, Biologia e Biotecnologie, University of Perugia, Perugia, Italy

**José Ricardo I. Ribeiro** Laboratório de Estudos da Biodiversidade do Pampa (LEBIP), Universidade Federal do Pampa (UNIPAMPA), São Gabriel, Rio Grande do Sul, Brazil

**Leonardo Samuel Ricioli** Laboratory of Ecological Studies on Ethology and Evolution (LESTES), Federal University of São Carlos, São Carlos, SP, Brazil

**Anais Rivas-Torres** ECOEVO Lab, EE Forestal, Campus Universitario, Pontevedra, Spain

**Marciel Elio Rodrigues** Departamento de Ciências Biológicas, Universidade Estadual de Santa Cruz (UESC), Ilhéus, BA, Brazil

**Gianandrea Salerno** Dipartimento di Scienze Agrarie, Alimentari e Ambientali, University of Perugia, Perugia, Italy

**Michael J. Samways** Department of Conservation Ecology and Entomology, Stellenbosch University, Matieland, South Africa

**Iago Sanmartín-Villar** IVEP Team, CRCA—CNRS, Université Toulouse III—Paul Sabatier, Toulouse, France

**Melania Santer** Laboratório de Estudos da Biodiversidade do Pampa (LEBIP), Universidade Federal do Pampa (UNIPAMPA), São Gabriel, Rio Grande do Sul, Brazil

**Francine Novais Souza** Departamento de Ciências Biológicas, Universidade Estadual de Santa Cruz (UESC), Ilhéus, BA, Brazil

**Vanessa Stefani** Laboratório de História Natural e Reprodutiva de Artrópodes (LHINRA), Instituto de Biologia, Universidade Federal de Uberlândia, Uberlândia, MG, Brazil

**Aurélio Fajar Tonetto** Universidade Paulista, Jundiaí, Brazil

**Francisco Valente-Neto** Instituto de Biociências, Universidade Federal de Mato Grosso do Sul, Campo Grande, MS, Brazil

**Diogo Silva Vilela** Graduate Program in Entomology, Department of Biology, University of São Paulo (USP), Ribeirão Preto, Brazil

**Jessica Ware** Biology Department, Rutgers University, Newark, NJ, USA

# Chapter 1
# Aquatic Insects: Why It Is Important to Dedicate Our Time on Their Study?

**Kleber Del-Claro**

**Abstract** Evolution begets diversity, and insects are the most diverse organisms in the history of life, so insects should provide profound insight into evolution. Insects are considered keystone species because loss of their critical ecological functions (e.g., pollinators; seed dispersers; predators; and parasitoids) could lead to collapse of the wider ecosystem. Indeed, insects dominate freshwater aquatic systems, but in general we know very few about the biology and natural history of these organisms. Here, we present a good set of basic data and knowledge in ecology, natural history, and behavior of this unknown fauna. Our hope is that it encourages young researchers to investigate basic aspects of the life history of aquatic insects, mainly in tropical systems.

**Keywords** Biology · Behavior · Behavioral ecology · Ecological networks · Wetlands

The diversity of life is one of the most astonishing aspects of our planet. Organisms are spread from the highest mountains to the deepest oceans of the Earth. They are under the bark of trees and inside flowers, in deep soil, in air, and even in frozen lakes. How many are they? How long will they stay as they are, or will evolve? Or disappear like so many others?

In a more recent and conservative perspective we suspect that there are more than 8.7 million (±1.3 million SE) eukaryotic species globally (Mora et al. 2011). But these numbers can be higher, as some researchers suggest that only to arthropods we could have between 5 and 10 million species (Ødegarrd 2000), perhaps 30 million (Erwin 1991). Regarding insects, the estimations vary from a minimum of 1 to possibly more than 5 million species (Scheffers et al. 2012), in a conservative approach. The unique real consensus is that we do not know the majority of species; 86% of existing organisms on continental environments and 91% of species in the ocean

---

K. Del-Claro (✉)
Laboratório de Ecologia Comportamental e de Interações, Instituto de Biologia, Universidade Federal de Uberlândia, Uberlândia, Minas Gerais, Brazil
e-mail: delclaro@ufu.br



K. Del-Claro, R. Guillermo (eds.), *Aquatic Insects*,
https://doi.org/10.1007/978-3-030-16327-3_1

still await description (Mora et al. 2011). If they need description and basic taxonomic studies, then what we know about their behavior, ecology, and natural history? Indeed, very little.

I was a naturalist for my whole life, roaming around tropical forests, savannas, high mountains, and margins of rivers and lakes. When asked by a student about "what that insect does for a living?" my most common answer, after some general comment, is "but … indeed, in this specific case, it is a very interesting question! Why you do not study it?" This is particularly real in relation to aquatic insects, mainly in the tropics, where their diversity is absurdly high and most species are unknown. A book like this has as objective to stimulate young biologists to go deep in the water exploring these amazing creatures' life histories.

Evolution begets diversity, and insects are the most diverse organisms in the history of life, so insects should provide profound insight into evolution (Grimaldi and Engel 2005). The vast majority of known insects are among the most conspicuous creatures, like the flying butterflies, beetles, wasps, and bees, that call our attention for beauty and also for agronomic, economic, or sanitary reasons. Insects may dominate food chains and food webs in a wide aspect being essentials in (a) nutrient recycling and soil turnover, like ecosystem engineers; (b) plant propagation, including pollination and seed dispersal; (c) control of plant community composition and structure, via herbivory; (d) base of food chain, serving as food for insectivorous animals; and (e) regulation of animal community structure, via predation, parasitism, or like a vector of several diseases (Fig. 1.1). Thus "Some insects are considered *keystone species* because loss of their critical ecological functions (e.g. pollinators; seed dispersers; predators and parasitoids) could lead to collapse of the wider ecosystem" (Gullan and Cranston 2014). Despite this clear importance of insects, even in the twenty-first century we still need to harp on the same string: the vast majority of our knowledge on insects is based on the most visible, beautiful, or important animals to man's health, economy, or their ecological services. What can we say about the immense hidden insect fauna? The "Darwin's entangled bank" was buried in the ground or dipped in lakes, ponds, streams, and rivers.

All inland water bodies (rivers, streams, or lakes) support a biological community. Invertebrates (insects, crustaceans, and nematodes) provide the highest levels of biomass, production, and number of individuals in freshwater communities. Indeed, insects dominate freshwater aquatic systems. However, in general we know more about the biology and natural history of the few vertebrate species (mainly fishes and birds) in any inland aquatic system than of the whole community of invertebrates that structure all the bottom-up process in these complex ecological networks. We cannot neglect the immense and extremely important set of knowledge we have produced on aquatic insects' biology, physiology, taxonomy, environmental monitoring, and general zoology (e.g., Grimaldi and Engel 2005; Gullan and Cranston 2014). But to the major orders of insects that are almost exclusively aquatic in their immature stages, the Ephemeroptera (mayflies), Odonata (damselflies and dragonflies), Plecoptera (stoneflies), and Trichoptera (caddisflies), we know in fact little about their life histories, ecology, and behavior. It is surprising

**Fig. 1.1** Insects are the most diverse organisms in the history of life, so insects can act as ecosystem engineers (**a**) modifying the environment like the ants *Pogonomyrmex naegeli* that revolve soil and also prey on or disperse seeds. Their activities, like pollination by bees (**b**) and herbivory by beetles (**c**), shape plant community composition and structure. Serving as food for insectivorous animals, like *Epicadus heterogaster* (**d**), a crab spider that prey on pollinators while hidden in flowers, insects also interfere in the whole animal community directly or indirectly

that it also occurs to adults of the most species of Odonata, animals commonly conspicuous, very attractive for their beauty (Fig. 1.2).

I am an ethologist, and to me it is incomprehensible to how to think about the preservation of these amazing species, without thinking about how they relate each other, and how each behavioral decision of one impacts reproduction, life, and death of others. We need to know details of life histories and behavior, to understand the ecological interactions and the network behind the Darwin's entangled bank.

The study of ecological networks is derived from graph theory to investigate the structure and shape of ecological interactions around the world. Nowadays, it is incontestable that graph theory brings enormous advances to the knowledge of ecological interactions modifying the manner we look to the tree of life (Del-Claro et al. 2018). Thus, to indeed understand the dynamic of aquatic ecological networks, we need to know deeply the set of interconnected biotic interactions. The sum of

**Fig. 1.2** *Argia angelae* (Odonata: Coenagrionidae), example of a very flashy group of aquatic insects on which we have relatively little knowledge

this knowledge, that conducts us to understand like conserve and/or recover the severely endangered freshwater systems, has its basis in basic biology, natural history, and behavior.

Thus, the aim of our book is to present a good set of basic data and knowledge in ecology, natural history, and behavior of this unknown fauna. Our hope is that it encourages young researchers to investigate basic aspects of the life history of these insects, mainly in tropical systems. It is imperative, especially in endangered biomes like the tropical savanna of South America and Africa, where species are quickly disappearing due to the expansion of human activities on natural areas. See what WWF says about Cerrado, the Brazilian savanna, in a recent publication titled "The Big Five of the Cerrado":

"*Imagine a place where there are over 11,000 species of plants and the fauna is as diverse as the flora. The Cerrado harbours 837 species of birds, 120 reptiles, 150 amphibians, 1,200 fish, 90,000 insects and 199 mammals. Together, they account for 5% of the world's species and 30% of Brazil's biodiversity. ... Less than 10% of the Cerrado is covered by protected areas, and less than 3% of its area is strictly protected, putting various animals at risk of extinction.*"

(https://www.wwf.org.br/informacoes/english/?50242/The-Big-Five-of-the-Cerrado).

This text is about the five big mammals of Cerrado, the maned wolf, the giant anteater, the jaguar, the tapir, and the giant armadillo. Nevertheless, in just one small pound in central Brazil, a Vereda (palm swamp communities, Fig. 1.3) surrounded by tropical savanna, only 5 km far from the borders of a big city (population of ~750,000), we recorded at least 31 damselflies and dragonflies species of 5 different

**Fig. 1.3** In a single Vereda, a palm swamp environment, in central Brazil we can find an enormous and unknown diversity of aquatic insects. Of damselflies, like *Mnesarete pudica* (**a**), *Ischnura capreolus* (**b**), and *Acanthagrion truncatum* (**c**), we do not know anything about the life underwater and know very little about their adult life

families and 21 genera (Vilela et al. 2016), some possibly endemic, and some new species. Our knowledge on this fauna is still poor, and areas like that are in general located in private country clubs and small farms. Being so, these areas are subject to real estate speculation, or risk being used for agricultural purposes; in any case they are in evident risk of extinction.

The study of behavior, or in simple words "what an animal does or does not do" (Del-Claro 2010), is not only a basic tool to better explore and know the ecology and the outcomes of interactions involving animals, like aquatic insects. It is also a tool to present amazing and surprising strategies of animals in their constant fight for survival and reproduction, in a captivating manner to the general public. Behavior is not only a source of knowledge; it's also a way to make people fall in love with nature. For example, Vilela et al. (2017) studying the tropical tiger damselfly, *Tigriagrion aurantinigrum*, combining techniques of computer sciences with basic observations of animal behavior and behavioral ecology, showed that the species replaced direct and physical conflict with the use of visual signals to solve territorial contests. The authors demonstrated that winner males have more orange-yellow patches in their frontal head than the losers, and that threat displays are more efficient to win contests than physical conflict. This study confirms the importance of coloration patters in Odonata and the importance of these insects as models in the study of animal visual communication. There are several other studies, just as much or even more interesting to the general public, that need better scientific divulgation.

Then, in our search of a way to stimulate more people, mainly young biologists, to dedicate their time on the study of aquatic insects, we edited this book that presents in few chapters a general vision of ecology and behavior of aquatic insects, beginning with basic knowledge and finishing with possible future applications. Thus, after this introduction, the second chapter, headed by Ricardo Koroiva and Mateus Pepinelli, will show how aquatic insects have successfully achieved a global distribution, with an extraordinary capacity of survival in the most distinct conditions. Considering the breadth of the subject, this chapter aims to show some of the factors related to the distribution of aquatic insects, illustrating the main habitats used by the main orders. But not only distribution, but also dispersal, is one of the fundamental processes that influence the ability of organisms to reach suitable habitats, to find mates, and to avoid potentially disastrous disturbances. In the third chapter, Michael L. May brightly discusses fascinating questions related to aquatic insects' dispersal capabilities and its implications. In the two following chapters, headed by Rhainer Guillermo and an enthusiastic group of young researchers, the abiotic (Vinicius Marques Lopez, Aurélio Fajar Tonetto) and biotic factors (Gabriela C. Mendes, Guilherme Gonzaga da Silva, Leonardo Samuel Ricioli) that influence the interactions and success of aquatic insects, depending on its biology and behavior, are explored. These chapters shape an interesting background for beginners in this research field.

Communication is imperative to the evolutionary success of any organism. In the sixth chapter, J. Manuel Tierno de Figueroa, Julio M. Luzón-Ortega, and Manuel Jesús López-Rodríguez use the mating behavior of stoneflies to show us how vibrational signals can be produced by drumming, rubbing, tremulation, or a combination of these methods, and are species specific; therefore, they can be used as a method to confirm/infer phylogenies or define species. Following the line of communication and visual perception, our colleagues Manuela Rebora, Gianandrea Salerno, and Silvana Piersanti give us an excellent contribution. In their chapter titled "Aquatic Insect Sensilla: Morphology and Function," they teach us how insect success is directly related to their ability to respond quickly to external and internal cues, thanks to a great variety of sensory systems, able to evolve over a relatively short time.

The sensory and communication abilities open a new window to aquatic insects, the habitat contestation, and the defense of territories. Gregory F. Grether after providing a brief introduction to territoriality in general in his chapter reviewed the taxonomic distribution and types of territoriality reported in aquatic insects, before delving into more theoretical topics. This was extremely well done, offering some general advice for studying territoriality and concluding by identifying areas where more research is needed. For example, what is the importance of communication to the sex of aquatic insects? Bright colors associated to face-to-face displays exhibited by males can communicate strong, ability, dominance to females and result in success? Our colleagues, Diogo Silva Vilela and Iago Sanmartín-Villar, revised the communication aspects involved in the reproductive behavior and sexual selection among aquatic insects. In one of the best illustrated chapters, the authors lecture about it in a so clear manner that any college student will be able to comprehend the text completely. This presentation opens space for a deeper analysis, which was done in sequence by Adolfo Cordero-Rivera and Anais Rivas-Torres. According to them, "the field of sexual selection has been historically dominated by a stereotyped view of the sexual roles, with competing males and selective females, but in recent decades there has been a paradigm switch, with the emergence and dominance of the concept of sexual conflict." In an amazing chapter these colleagues stressed the hypothesis that "the dimensional structure of the habitat will affect the intensity of sexual conflict over mating rate, copulation duration and postcopulatory guarding." To achieve their objective, authors illustrate the text with wonderful examples in which behavior, basic biology, and ecology are fundamental to our understanding of what is going on.

Animals live to survive, grow up, and reproduce; they evolved distinct mechanisms and features to have success in this process, which in the endgame may also be needed to protect their best chance for eternity: the progeny. In the book sequence, parental care is discussed among aquatic insects, having as model the giant water bugs (Belostomatidae). Our colleagues Shin-ya Ohba, José Ricardo I. Ribeiro, and Melania Santer revel how behavioral ecology may be important providing basic tools to enable us to study parental care as it deserves, with experimentation. This chapter is finalized in a very stimulating manner, presenting plausible directions that must be followed. One of these directions calls attention to unknown fauna underwater, unexpected life forms that we can find especially in tropical systems. Vanessa Stefani and her scientific initiation students Iasmim De-Freitas and Julia De Agostini presented us an amazing case. Although most Lepidoptera are exclusively terrestrial there are some lineages associated with the aquatic environment. Little is known about aspects of the biology and life history of its representatives, leading to a gap in scientific knowledge. In this chapter, possible evolutionary paths explored which aquatic lepidopteran larvae had to live in aquatic environments, and the authors examine the incipient studies that address life aspects of these extraordinary animals.

The three final chapters of the book are more applied and point to us future directions. Not only the very preserved and untouched natural areas are under our sights and deserve our attention, but also study and preservation of aquatic insects. In the urban environment, our colleagues Javier Muzón, Lía S. Ramos, and Alejandro del Palacio show that "aquatic insects, as key components of urban wetlands, are critically impacted by human environmental changes and practices. The main threats are those derived from the loss, replacement or fragmentation of natural habitats, ecosystem homogenization, and the modification of hydrological, sedimentological and thermal wetland characteristics, due to the surrounding urban matrix." Therefore, this chapter will focus on the importance of insect knowledge as indicators of environmental health in urban wetlands, and the promotion of the citizen science to improve their conservation. Thinking in the same direction, Drielly da Silveira Queiroga, Renan Fernandes Moura, and Jessica Ware bring us an excellent new look over aquatic environments in their chapter "Genetic Connectivity in Conservation of Freshwater Insects." The authors will show us in this amazing chapter that "rich biodiversity is fundamental for the balance and stability of ecosystems, providing vital ecosystem services such as climate regulation. Although aquatic environments are essential to human health, they are often the most exploited environments. Many aquatic insects may increase or decrease their abundance in response to environmental disturbances; these species are known bioindicators and may be used to evaluate environmental quality of an aquatic ecosystem. Thus, it is imperative to preserve bioindicators as they can alert us before degradation becomes large enough to extinguish ecologically important species. To best determine which species require the most attention, many scientists have turned to genetic tools to identify genetically diverse populations and their relative genetic connectivity: the exchange of genetic material among populations of the same species, enabling the maintenance of genetic diversity."

We decided to close the book, with a futuristic chapter that very competently Stanislav N. Gorb and Elena V. Gorb produced to us. In this awesome chapter the authors demonstrate how the knowledge of basic biology, morphology, and behavior can help humans solve contemporary problems. "Aquatic insects, due to their aquatic or amphibiotic habit, evolved many particularly interesting functional solutions, which are of special interest for biomimetics. Understanding functional principles of insect materials, structures, sensors, actuators, locomotion patterns, control systems, and behavior is of major scientific interest, since we can learn about their functional principles and biological role. On the other hand, this knowledge is also highly relevant for technical applications. One of the greatest challenges for today's engineering science is miniaturization." I am sure that this chapter will call attention of a broad and diverse audience.

I am a beginner in the study of aquatic insects and this book was a challenge to me: to produce a piece of art supported in the shoulders of others, lovers of aquatic insects: people devoted to give part of their time in this wonderful planet to protect, to preserve, and to recover the unknown fauna that can maintain our survival as a species.

## References

Del-Claro K (2010) Introdução à Ecologia Comportamental: um manual para o estudo do comportamento animal. Technical Books, Rio de Janeiro

Del-Claro K, Lange D, Torezan-Silingardi HM, Anjos DV, Calixto ES, Dáttilo W, Rico-Gray V (2018) The complex ant–plant relationship within tropical ecological networks. In: Dáttilo W, Rico-Gray V (eds) Ecological networks in the tropics. Springer, Cham. https://doi.org/10.1007/978-3-319-68228-0_5

Erwin TL (1991) How many species are there? Rev Conserv Biol 5:1–4

Grimaldi DA, Engel MS (2005) Evolution of the insects. Cambridge University Press, Cambridge

Gullan PJ, Cranston PS (2014) The insects: an outline of entomology, 5th edn. Wiley-Blackwell, Oxford, p 624

Mora C, Tittensor DP, Adl S, Simpson AGB, Worm B (2011) How many species are there on earth and in the ocean? PLoS Biol 9(8):e1001127. https://doi.org/10.1371/journal.pbio.1001127

Ødegarrd F (2000) How many species of arthropods? Erwin's estimate revised. Biol J Linn Soc 71(4):583–597. https://doi.org/10.1111/j.1095-8312.2000.tb01279.x

Scheffers BR, Joppa LN, Pimm SL, Laurance WF (2012) What we know and don't know about Earth's missing biodiversity. Trends Ecol Evol 27(12):712–713

Vilela DS, Ferreira RG, Del-Claro K (2016) The odonata community of a Brazilian Vereda: seasonal patterns, species diversity and rarity in a palm swamp environment. Biosci J 32(2): 486–495. https://doi.org/10.14393/BJ-v32n2a2016-30491

Vilela DS, Tosta TAA, Rodrigues RR, Del-Claro K, Guillermo-Ferreira R (2017) Colours of war: visual signals may influence the outcome of territorial contests in the tiger damselfly, Tigriagrion aurantinigrum. Biol J Linn Soc 121(4):786–795. https://doi.org/10.1093/biolinnean/blx024

# Chapter 2
# Distribution and Habitats of Aquatic Insects

**Ricardo Koroiva and Mateus Pepinelli**

**Abstract** Understanding how communities, populations, and species are organized and distributed is one of the main objectives of ecology. Aquatic insects are composed of several orders of diverse groups found in many different types of aquatic (and semiaquatic) habitats. From the highest water trickles to the largest rivers and deepest lakes, aquatic insects have successfully achieved a global distribution, with an extraordinary capacity of survival in the most distinct conditions. Considering the breadth of the subject, this chapter aims to show some of the factors related to the distribution of aquatic insects, illustrating the main habitats used by the main orders.

**Keywords** Biodiversity · Spatial factors · Lentic-lotic · Saltwater · Environments inhabited

## 2.1 Introduction

Any place that has freshwater is likely to harbor aquatic insects. Although they represent about 6% of all insect diversity (Dijkstra et al. 2014), they are the dominant organisms in aquatic ecosystems and are defined by species that spend part of their life cycle or all development in the water. Such organisms show a varied range of metabolic and morphological adaptations, allowing them to occupy virtually any kind of aquatic freshwater habitat, including high-salinity water bodies and streams and lakes that freeze during winter. The distribution of aquatic insects is driven by several factors acting in different scales, beyond the lotic versus lentic or permanent versus temporary freshwater binary views. Nonetheless, there is one aquatic

R. Koroiva (✉)
Mapinguari Lab, Federal University of Mato Grosso do Sul, Campo Grande, MS, Brazil

Departamento de Hidrobiologia (DHb), Universidade Federal de São Carlos, São Carlos, SP, Brazil

M. Pepinelli
Department of Natural History, Royal Ontario Museum, Toronto, ON, Canada



K. Del-Claro, R. Guillermo (eds.), *Aquatic Insects*,
https://doi.org/10.1007/978-3-030-16327-3_2

ecosystem that insects didn't successfully colonize, the oceans, having only relatively few species adapted to live in marine water (Springer 2009).

Several factors influence the distribution of aquatic insects. On a large scale, historical processes and climate are considered the main factors. Over the last decades, studies have shown that, at local scales, variables such as substrate heterogeneity, land use, abiotic components, aquatic flora, and water current velocity are important factors determining communities. Some of these variables are related to the structure of the watershed, ecosystem differences, and microhabitats. Although recent studies have shown that variables at multiple scales can substantially affect aquatic insects' community structuring, local scale variables have been suggested as the most important drivers of aquatic insects' distribution and diversity, due to the action of filters on trait composition of communities (Bae et al. 2011; Tolonen et al. 2016).

The close association between different orders of aquatic insects and aquatic environments is reflected in several adaptations developed for life on water, including a wide range of physiological and structural solutions (physiological and morphological traits) used by the many aquatic groups to colonize different water systems. The results are dramatic new ways to breathe, move, find mates, acquire food, etc.

Considering the complexity of factors related to the distribution of aquatic insects, in this chapter, we provide an overview of the main aspects that affect their distribution and some of the most common habitats used by them. First, we indicate how large and local scale factors play important roles in determining their distribution. Then, we present some freshwater systems (water bodies) with examples of groups with peculiar adaptations to the various habitats explored. In this way, we intend to illustrate the variety of environments inhabited by aquatic insects and how they overcome the challenges of living in water.

## 2.2 Question of Scale

The processes that determine diversity are scale dependent and, therefore, it is crucial to examine the relative roles of factors at multiple scales (Poff 1997). As stated by Tolonen et al. (2016), for some life-history traits of benthic macroinvertebrates, the importance of environmental factors that vary at large spatial scales could be more pronounced than the importance of local environmental variables. In a large-scale view, historical factors, such as speciation, extinction, dispersion, and climatic factors, such as temperature and precipitation, are important in structuring species pools, acting over traits and filtering species. Zooming into the spatial scale of environmental filters, factors such as river basin features, and more localized abiotic and biotic conditions, are some of the main factors influencing the assemblages. Altogether, these filters determine the diversity, composition, and functional taxonomic variability of the assemblages through the characteristics of each species. Thus, only species with adequate characteristics are able to overcome the challenges presented by filters at each scale (Heino 2009).

### 2.2.1 Large Scale

The influence of historical effects in determining the distribution of aquatic insects on a large scale is directly related to the historical and current development of natural environments. Geological processes such as the formation of mountains and the continental drift have been some of the factors associated with large-scale distributions of aquatic insects (Grimaldi and Engel 2005). Probably, the earliest example of how large-scale continental processes (Wegener 1922) shaped the distribution of aquatic insects came from a monograph published in 1966 by the Swedish entomologist Lars Brundin, who studied the transantarctic relationships of chironomid midges (Brundin 1966).

More recent examples are exemplified from a study on the spring brook headwater-specific caddisflies (Trichoptera), of the genus *Palaeagapetus* Ulmer. Using a molecular approach, the authors showed how the genetic structure of *Palaeagapetus* clearly reflects the separation of eastern and western Japan in the Fossa Magna region (Tojo et al. 2017). All nine species of *Palaeagapetus* caddisflies belong to two major clades, whose geographic boundaries are in the center of the Japanese Islands (i.e., center of Honshu), corresponding to the westernmost side of Fossa Magna, the Itoigawa-Shizuoka Tectonic Line.

While continental drift was responsible to separate several clades of animals and plants, it was the climatic conditions that shaped and are still a strong mechanism shaping distribution. Temperature and precipitation, for example, strongly influence river ecosystems. As evidenced by Poff et al. (2010), water temperature, annual thermal regime, and precipitation regulate the growth of aquatic insects and the structure and function of the flow in aquatic environments, which directly and indirectly regulate many aspects of the aquatic community, including its composition.

Critical changes introduced by climate events promote impacts on the water quality and on the physical habitats in water ecosystems. The increase in air temperature, for example, leads to other changes such as alterations in rainfall cycles, increase of water temperature, shifting habitat dynamics, affecting species with weak adaptability (Sáinz-Bariáin et al. 2016). These changes influence the phenology, distribution, composition, and dynamics of aquatic communities, providing a broad subject for ecological studies.

Recent studies suggest that, as a result of global warming caused by human activities, the freshwater biodiversity is declining at much higher rates than the most affected terrestrial ecosystems, because the increase of temperature can promote habitat reduction and physiological changes for several aquatic insects (Fenoglio et al. 2010). Even long-term climate cycles of natural conditions such as El Niño and La Niña can dramatically alter aquatic communities. These cycles are characterized by complex weather patterns, such as the change in temperature and the amount of precipitation. For aquatic insects, studying the effects of this type of phenomenon has served as an indicator of how directional changes in global temperature will affect species distribution, and studies have predicted that many aquatic insects are vulnerable to extreme changes over different scenarios (Kingsolver et al. 2011; Shah et al. 2017).

Finally, large-scale variables have their share influencing aquatic insect distribution; however the local scale ones have been suggested as the most important drivers of aquatic insect distribution and diversity, both in terms of explaining variability in community composition and those that act directly on the organisms (Tornwall et al. 2017).

## 2.2.2 *Local Scale*

Decreasing the spatial scale, factors such as watershed abiotic and biotic conditions are the environmental filters determining the assemblages, by filtering the organisms based on their functional attributes (Poff 1997). According to recent metacommunity studies, dispersal is an important mechanism structuring aquatic insect communities and it varies according to the type of water body and the dispersal capacity of the insects in question; however, environmental factors usually overlap with spatial factors (Heino et al. 2015). Several authors have shown broad categorizations of local factors related to community structuring, and here we list five of the most important, substrate heterogeneity, land use, abiotic factors, aquatic flora, and water current velocity.

### 2.2.2.1 Substrate Heterogeneity

The characteristic and composition of the bed substrate are of extreme importance especially in lotic environments, since they are the primary components determining which groups of insects are able to live and assess the resources. The colonization of insects on aquatic substrates is influenced by substrate heterogeneity and complexity, such as interstitial spaces, surface complexity, different types of sediment, and presence of organic matter from the riparian vegetation. Heterogeneous and complex substrates harbor higher density and richness of aquatic insects than homogenous do, because it provides more hiding places, creates different current flow speed, and provides substrates for egg deposition, protection from predators, and greater availability of food (e.g., Brown 2003, Barnes et al. 2013 but see Palmer et al. 2010). For example, aquatic detritivores can be affected when their food supply is either buried under sediments or diluted by increased inorganic sediment load. For the family Chironomidae, higher substrate heterogeneity promotes higher diversity of microhabitats and usually harbors greater number of species and guilds (Gurski et al. 2014).

### 2.2.2.2 Land Use

Closely associated with bed composition, land use has recently been considered one of the main factors driving environmental changes in aquatic systems, because the impacts almost always result in structurally simplified habitats. Changes of land cover may increase the surface runoff rates, as well as the amount of sediment and nutrient inflow (Kasangki et al. 2008). The conversion of forest areas reduces the entry of allochthonous material into headstreams, reducing the abundance of species associated with leaves and woods (Valente-Neto et al. 2015).

In forested streams, terrestrial detritus is an important resource for the aquatic food web; therefore the dependence on their input directly influences the composition of aquatic insects (Minshall 1978; Power et al. 2013). Summarily, species composition can be considered the best predictor for severe alterations of land use in the insect community (see Siegloch et al. 2014; Gimenez et al. 2015). In lentic environments, the effects of forest removal are expected to be of a lower magnitude, considering that these systems are already open to sunlight and the influence of allochthonous material is low (Richardson 2008). However, studies have shown that concentrations of nutrients and other components are positively correlated with the nature of the environment modification. For example, agropastoral activities have been related to nutrient enrichment (Abell et al. 2011) and increasing the abundance of some aquatic insects such as chironomids (Blumenshine et al. 1997).

#### 2.2.2.3 Aquatic Vegetation

The presence of vegetation inside a water course also greatly affects the environment and the fauna. The so-called aquatic macrophytes are plants that require a water environment to complete all or most of their life cycle. They can be classified as emergent, floating, or submerged plants; are present in both lotic and lentic environments; and play a key role in structuring communities not only of aquatic insects but also of fish and other organisms. In temperate shallow lakes, for example, these plant communities are probably the most important group that affect whole lake ecosystems (Scheffer 2004). Also, in the situation of degradation of riparian vegetation, the presence of macrophytes can provide an important alternative food source, and sustain in-stream productivity and, consequently, faunal biodiversity (Paice et al. 2017). In northern lakes, macrophytes play an important role in structuring littoral macroinvertebrate communities (Heino 2013), because they alter the dynamics of nutrients, dissolved organic and inorganic carbon, oxygen, pH, sediment quality, etc., having been recognized as "biological engineers" to aid in restoring water quality (Byers et al. 2006).

For many aquatic insect groups, the food availability, both directly in the form of fresh and detrital material, structural habitat, and refuge against predation (Heck and Crowder 1991; Bell et al. 2013) increase the density and richness of aquatic insect species on macrophyte vegetation, when compared to areas without this resource. Also, studies such as that of Gonçalves Jr. et al. (2004) demonstrate the ability of different species of macrophytes to support different aquatic insects, considering that the chemical composition of the macrophyte detritus strongly influences the decay rate, while degradative ecological succession is the driving force of the invertebrate community.

#### 2.2.2.4 Abiotic and Environmental Variables

Other important geomorphological factors related to the ecology of aquatic insects are the abiotic and environmental variables of the water body, such as water temperature, pH, salinity, and oxygen concentration. They directly affect the tolerance limits of organisms and consequently the distributional range of species (Death and Winterbourn 1995). Water temperature is considered one of the most important environmental factors, because if affects the aquatic insect growth. The "optimum thermal regime" (Vannote and Sweeney 1980), the temperature in which optimal growth, reproduction, and other measures of fitness are greatest (Dallas and Day 2004), and "thermal tolerance range," the range over which a species may survive, allow us to understand diverse biological aspects of the species as well as their fundamental niches, geographical distribution, and evolutionary dynamics.

The relation between thermal limits and evolution led to the development of different biogeographic hypotheses such as the climate variability hypothesis proposed by Janzen (1967), which predicts that more variable climates select for organisms with broader thermal tolerances, whereas less variable (stable) climates select for narrower thermal tolerance. Recent studies (e.g., Shah et al. 2017) have corroborated this hypothesis for aquatic insects, but also need to take into consideration altitudinal factors. Like terrestrial insects, many aquatic species have adapted to extreme temperature conditions, such as a *Tanytarsus* van der Wulp species present in hot springs (<34 ° C) in California (Lamberti and Resh 1985), as well as the survival of long periods of freezing as for some chironomids capable of maintaining activities even at −16 °C (Paterson 1971; Kohshima 1984).

Directly related to water temperature, the concentration of oxygen is another important and limiting factor. The sensitivity of aquatic insects to oxygen depletion is characterized by both the oxygen concentration in the water and the respiration capacity. The concentration of oxygen in natural environments varies with other biotic factors, such as effects of microbial respiration in eutrophic environments, and abiotic factors, like water temperature and current velocity. Anthropic influence can affect both the biotic, with the addition of nutrients, and abiotic equilibrium, changing water temperatures, reducing the concentration of dissolved oxygen, etc. For example, some cold-water lotic mayflies and stoneflies cannot tolerate dissolved oxygen concentrations below 8.0 mg/L (e.g., *Epeorus sylvicola* (Pictet) and *Baetis alpinus* (Pictet); Barton and Taylor 1994), as they possess gill membranes, require high concentrations of oxygen, and display specialized behaviors for greater diffusion of oxygen (Genkai-Kato et al. 2000). On the other hand, some aquatic insect species such as *Chironomus* Meigen species, are practically not dependent on the dissolved oxygen content of the water and can live where the oxygen is nearly zero at all times (Shobanov 2001) and other dipterans, such as mosquito larvae (Culicidae), have respiratory syphons that enable them to use atmospheric oxygen, and are thus adapted to live in poor water quality conditions (Chipps et al. 2006).

Besides the function of gas diffusion, gill membranes present in some orders, like Odonata and Ephemeroptera, have an important role in ion exchange. For insects without gills, other epithelial membranes on the surface of the body, or spe-

**Fig. 2.1** Lotic and lentic environments. (**a**) Stream with predominance of roots. (**b**) Stream flowing over rock slab (in the background), sandy sediment (center), and foliage (foreground). (**c**) Rapid creek water. (**d**) Lake densely covered by hydrophytes. (**e**) Small lake. Modified from Salles and Ferreira-Júnior (2014). Reprinted with permission from Insetos aquáticos na Amazônia Brasileira: taxonomia, biologia e ecologia, edited by Hamada, N.; Nessimian J.L.; Querino, R. © 2014. Published by Editora do INPA

cialized structures, are responsible for the ion transport. The salinity of the environment, especially Na and Cl ions, is essential for maintaining hormone-signaling pathways, generating electrical cell potentials, and regulating bodily fluids. However, the physiological limitations of osmotic control in high concentrations of salinity (>35 ppt) limit the presence of aquatic insects in this condition. A broad review about ion diffusion such as sodium, potassium, calcium, magnesium, chlorine, bicarbonate, and sulfate in aquatic insects is revealed by Griffith (2017).

Finally, current velocity is considered the most important factor in structuring the community of aquatic environments. The division into standing (lentic) and fluvial (lotic) waters represents the two major ways used to classify water bodies, despite the wide range of aquatic systems for both environments (see Fig. 2.1). Water flow has a wide influence acting on both biotic (e.g., vegetation composition) and abiotic (e.g., pH, water oxygenation) elements, changing key conditions for community structuring. Although many groups occur in both environments (for instance the presence of pool areas in lotic environments), the current velocity strongly determines the ecological distribution of the species, especially because of its relation with morphological and behavioral adaptations.

## 2.3 Environments Inhabited

Considering the differences between lotic and lentic environments, the analysis of the composition and distribution of the species allows us to investigate how each environment acts in the characterization of its community. However, far beyond rivers and lakes, aquatic insects can be found in virtually any freshwater habitat worldwide. In this topic, we present different types of water bodies and peculiarities about morphological traits related to the success of aquatic insects in colonizing these environments.

### 2.3.1 *Rivers and Streams*

Defined by limnologists as lotic environments, rivers and streams share the same function of carrying flowing water back to the oceans, differing only on their sizes. Classification systems such as the Strahler stream order (Strahler 1952), where numbers are used to define stream size based on a hierarchy of tributaries, or how the tributaries coalesce together to form a bigger stream, allow an efficient and visual classification. Order numbers may range from 1 (a stream without any tributaries) to 12 (Amazon River in its mouth), being the vast majority of streams from first through third order, also called headwater streams.

Considering this classification, Vannote et al. (1980) in his River Continuum Concept (RCC) used flow order (Fig. 2.2), energy sources and nutrients, and food webs to characterize the community of aquatic insects. In a river system flowing through a forested region, the headwaters (orders 1–3) are heavily shaded where algae growth is limited by light. According to Vannote et al. (1980), in this region it is expected the predominance of shredders, a functional feeding group, or organisms that feed of large-sized organic matter such as animal carcasses, leaves, and woody debris, also called coarse particulate organic matter (CPOM) (Cummins 2016). In spite of remembering that functional feeding groups are not an immutable classification, common shredding invertebrate taxa are found among Ephemeroptera (mayflies), Plecoptera (stoneflies), Diptera (true flies and midges), and Trichoptera (caddisflies) (Merritt and Cummins 1996).

Increasing the order flows (orders 4–6), an increase in the diversity of both vegetation and insect species is expected since the tributaries are wider and less shaded, generating a high diversity of macrohabitats to be explored. Primary production levels increase in response to the increase of light, along with the production of periphyton and the consumption of fine particulate organic matter (FPOM). Functional groups such as grazers and collectors increase in abundance and this section is characterized by the greater diversity of groups such as predators, shredders, collectors, and grazers with predominant occurrence of Ephemeroptera (mayflies), Plecoptera (stoneflies), and Diptera (true flies and midges) (Allan 1996).

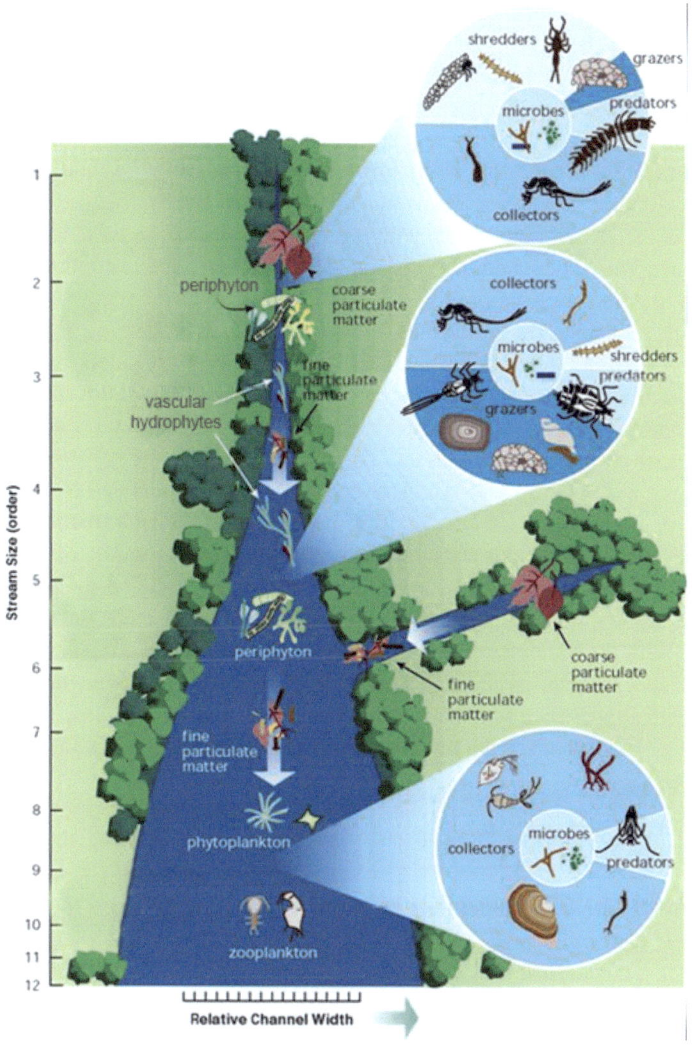


**Fig. 2.2** The river continuum concept by Vannote et al. (1980). The proportion of invertebrate feed groups corresponds to changes of environmental conditions in the structure of the river basin (Figure taken from USDA 2001). Reprinted with permission from Stream Corridor Restoration: Principles, Processes, and Practices, edited by Federal Interagency Stream Restoration Working Group. © 1998. Published by USDA. In addition, reprinted with permission from Canadian Science Publishing (Vannote et al. 1980), © 1980

Finally, the rivers of higher orders are too wide to be dependent on allochthonous material and too deep so that the production of algae in the bed is not preponderant. Thus, organic inputs of the floodplain with both phytoplankton and zooplankton production play an important role in structuring the community, with the domain of collectors, such as chironomids and mayflies (Battle et al. 2007; Masson et al. 2010).

Despite the broad characterization, the applicability of the River Continuum Concept has limitations because it is a model that considers a perennial flow and absence of changes in riparian vegetation. As it is well known, streams and rivers from the different biogeographic realms have different structures and face different seasonal climatic variations (Vinson and Hawkins 2003), factors that are not included in the RCC model. Even within these limitations, the RCC has served as a useful conceptual model and has stimulated research on aquatic community characterization to date (see Xenopoulos et al. 2017).

In recent decades, the traditional linear conceptual models of aquatic ecosystems, such as RCC, have been replaced by a view that considers the role of the spatial structure of river networks in determining diversity patterns (Siqueira et al. 2014). Dendritic networks (so-called river networks) are structural and functional backbones of river basins, which can be viewed as rooted trees formed by fluvial erosion over a drainage basin (Balister et al. 2018). As evidenced by Tonkin et al. (2018), the organization of the dendritic network and the unidirectional physical flows can exert strong controls on the consequent distribution of biodiversity, depending on species traits which influences the dispersion and the ability to survival (see more in next subtopic). In this sense, studies of metacommunities have deepened with the understanding of how local and regional factors act in the distribution of species and, as already mentioned, local environmental factors tend to overlap with spatial factors in structuring insect communities in lotic environments (Heino et al. 2017, but see about very small spatial scales in Grönroos et al. 2013).

#### 2.3.1.1 Life in Running Waters

Several morphological adaptations have evolved for living under the influence of streamflow, especially ones related to breathing, acquiring food, and locomoting underwater. The dorsoventral flattening is a classic example of adaptation for living under fast currents, because it is a hydrodynamic shape that provides the animals the perfect balance of surface contact with the substratum, water drag, and lift that they need to be able to live close to the surface layer. Dorsoventral flattening is well expressed in a number of mayflies (e.g., Heptageniidae, Ephemerellidae) and stonefly nymphs (e.g., Perlidae), reaching its highest degree in water-penny larvae (Psephenidae). Other animals developed innovative mechanisms and behaviors for anchoring in rapid waters, such as suction cups (Blephariceridae), hooks and grapples (Simuliidae, Fig. 2.3), and architectural cases and shelters (different families of Trichoptera and some Chironomidae).

Despite the fact that most of rivers and streams have well-oxygenated water, stream-dwelling insects are at least as well provided with respiratory apparatus as are the inhabitants of ponds (Hynes 1970).

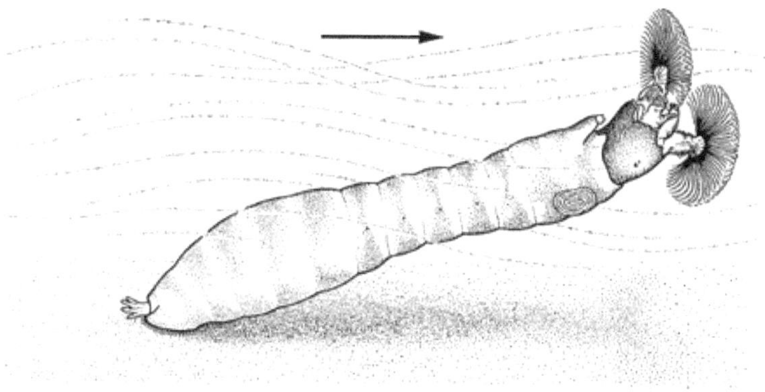

**Fig. 2.3** Black fly larvae live exclusively in running waters. Larvae anchor themselves to the substrate by means of a posterior circlet of hooks embedded in a dab of silk. Food particles borne in the water current are captured by the cephalic fans (Gullan and Cranston 2014). Arrow indicates the direction of flow. Reprinted with permission from John Wiley and Sons: The Insects: An Outline of Entomology, edited by P. J. Gullan and P. S. Cranston. © 2014. Published by Springer Publishing Company. In addition, modified and reprinted with permission from Cambridge University Press (Currie 1986), © 1986

### 2.3.2 Intermittent Streams

In many regions of the world, streams can dry out seasonally, occasionally, or even for several years. In these environments, aquatic insects can take refuge in moist cracks within the bed, under leaves, barks, or stones, to wait for the water to return (Boulton and Lake 2008; Chester and Robson 2011). Despite the great diversity, most species can be classified as "facultative" because they are present in both perennial and intermittent streams. One of the few species that have been preferentially related to non-perennial streams are larvae of arrowhead spiketail (*Cordulegaster obliqua* (Say); Odonata: Cordulegastridae) (Santos and Stevenson 2011). Other species go rapidly throughout their life cycles and disperse as adults before drying is completed, such as have been speculated for the pupae of *Rhyacophila grandis* Banks (Trichoptera: Rhyacophilidae) (London 2017), and there are cases where there is a substantial exchange of individuals between temporary or ephemeral flows and near-permanent waters (Godoy et al. 2016).

### 2.3.3 Lakes and Ponds

Also called as lentic environments, lakes and ponds are widely distributed in all continents. In broad characterization, most lakes have at least one natural outflow in the form of a river or stream, which maintain the average level by allowing the drainage of excess water. However, some remain as closed bodies of water, such as lakes formed in volcano craters, and lose water solely by evaporation or underground seepage or both. The difference between lakes and ponds is mostly related to their sizes, but the thresholds are yet debatable. There are several distinct types of lakes, and a landmark study is provided by Hutchinson (1957).

Lakes provide habitats for many aquatic insect groups, some specialized in standing water such as the Chaoboridae and Corethrellidae flies, in which their larvae are able to move and migrate in the water column. Other groups such as some coleopterans and heteropterans can also move in the water column. However, because they need to obtain air from the atmosphere for breathing, they are always visiting the water surface or living in regions with aquatic vegetation. In general, insects in lentic environments are concentrated mainly near the shores of the lake (the littoral zone), where the water is relatively shallow and lush vegetation may be present.

The vertical component (depth) of the lake creates several effects. Deep lakes typically have strong temperature gradients associated with reduction on the concentrations of dissolved oxygen as you dive. The deep zone of a lake is typically a homogeneous habitat where only a handful of taxa are able to thrive, as some very characteristic genera of chironomids such as *Chironomus*. These genera have evolved the presence of high hemoglobin concentration in the hemolymph, and can obtain carbon derived from methane, both examples of adaptations for living in this type of habitat (van der Valk 2012).

Other important factor associated with spatial structuring of aquatic insects in lakes and ponds is the dispersal ability of terrestrial forms. Recent studies (e.g., Hill et al. 2017) have shown that for lentic and temporary environments, weak dispersers generally showed less environmental control and more spatial structuring, while strong dispersers, such as most aquatic insects, showed greater environmental control and no significant spatial structuring (Heino et al. 2015). However, other studies have shown different results, without any significant spatial and environmental structuring in aquatic small organisms (e.g., Nabout et al. 2009).

### 2.3.4 Wetland, Temporary Ponds, Madicolous Habitats, and Human-Made Environments

Despite being a single word, wetlands represent a complex of ecosystems that are related to flooded lands or areas that have soil saturated with moisture (Fig. 2.4). This definition includes lands covered by water, either salty or fresh, marshes, rock pools, edge of a lake or ocean, delta at the mouth of the river, peatlands, freshwater swamps, and mangroves.

Among the diverse of wetland environments, extreme conditions of low oxygen concentration, low nutrient richness, and high salinity affect the structuring of communities. However, most of the main orders are present, such as Diptera—Chironomidae (Armitage et al. 1995; Batzer and Sharitz 2007), Ephydridae, etc.; Heteroptera—Corixidae, etc.; Coleoptera—Dytiscidae, Hydrophilidae, etc.; Odonata; and Ephemeroptera. Batzer and Ruhí (2013) found that the Chironomidae

**Fig. 2.4** Flooded areas in Bolivia (**a**) and rock pools in Brazil (**b**)

(midges) and Dytiscidae (predaceous diving beetles) were the only families that were virtually ubiquitous across the 447 wetlands they studied. Corixidae (water boatmen) and Hydrophilidae (water scavenger beetles) occurred in most (>50%) of the wetlands. A large revision of over 25 families of aquatic insects can be found in Batzer and Boix (2016).

Another type of freshwater habitat that has been shown to have a great diversity of aquatic insects is the madicolous habitats. They are characterized by a thin layer of water temporarily or permanently flowing over rocky surfaces. Such habitats harbor a rich fauna of aquatic, semiaquatic, and terrestrial organisms, some of which are rare or even endemic, living exclusively in this biotope, and termed eumadicolous fauna (Vaillant 1956). For example, Sinclair and Marshall (1987) listed 13 eumadicolous species from southern Ontario, including 12 dipteran (Tipulidae, Psychodidae, Thaumaleidae, Ceratopogonidae, Stratiomyidae, and Dolichopodidae species) and 1 Trichoptera species (Hydroptilidae). Shimabukuro et al. (2015) listed 20 higher taxa (families or order) of insects from madicolous habitats in Brazilian mountains.

Despite these records, it is a fact that those habitats have remained historically understudied. As evidenced for rock pools (Jocque et al. 2010), large wetlands (see Keleher and Sada 2012), and madicolous habitats in mountainous regions (Shimabukuro et al. 2015), these environments are unique habitats housing a high diversity of specialist and endemic species and therefore contribute substantially to regional diversity.

Under the pressure of evaporation, temporary pools are another important environment explored by aquatic insects. They include any habitat that has intermittently standing water and which, once flooded, retains water long enough for some species to complete the aquatic phases of their life cycle ignoring the size and time of resilience of these pools. It might include temporary lakes, temporary ponds, and environments created by humans.

As evidenced by Wiggins (2004), many species have adapted to life in temporary freshwater ponds, and at least two advantages can be recognized by the colonization of this environment: firstly, the rare presence of fish species, considered the main predators of aquatic insects. Other predatory insects such as water bugs, beetles, and dragonfly nymphs that are also less diverse and abundant in this type of habitat. The second advantage is due to the nutrition and growth of the species. In general, temporary pools have their food web based mainly on detritus which, differently from submerged environments, tends to present a phase of degradation in the air. This period causes the debris to be degraded faster and to have higher protein concentration, especially by the colonization of fungi.

A classic example of organisms present in temporary pools in the northern hemisphere is certain species of the caddisfly family Limnephilidae (Wiggins 2004). These species, such as *Limnephilus indivisus* Walker, present several adaptations for survival such as suspension of ovarian development by late summer and eggs wrapped with thick gelatinous mucopolysaccharide matrix, which allows resistance to dissection for months. In this way, oviposition has become independent of water, allowing these species to lay eggs in temporary pools after surface water has been depleted.

Finally, human-made environments, artificial containers (Fig. 2.5), should not be overlooked. Broadly, urbanization has led to an increase in the number of man-made containers, including water storage systems for human use (e.g., water boxes) and urban waste (e.g., tires and unloaded bottles). In forest environments, artificial containers often attract the same fauna as natural systems, while in open environments many species may not tolerate the abiotic conditions found (e.g., high temperature) being common the presence of generalist species like *Culex* Linnaeus mosquitoes (Kling et al. 2007; Yee and Yee 2007).

Species of medical importance often exhibit a wide survival capacity and are well adapted to extreme conditions. As summarized by Manguin and Boëte (2011), three major biological characteristics are found among the most invasive mosquito species: (1) close association with humans, (2) egg resistance to desiccation (genus *Aedes* Meigen), and (3) small larval habitats such as a wide variety of water-holding containers among which man-made ones are suitable. Thus, the proximity to humans and the use of artificial containers have allowed an assistance to transport of most of mosquito invasions as well as their maintenance in urbanized areas. Lloyd (2003) brought to attention the need of eliminating artificial containers in urban environments to tackle diseases associate with its inhabitants.

### 2.3.5 *Phytotelma and Water-Filled Tree Holes*

Phytotelma and water-filled tree holes also provide an environment for the development of a diverse community (Fig. 2.6). Tanks of bromeliads, carnivorous plants, tree holes filled with water, bamboo internodes, and axillary waters collected by leaves, bracts, or petals are water reservoirs formed from parts of living plants capable of being inhabited by aquatic insects, while water collection cavities are formed in the fallen parts of plants.

**Fig. 2.5** *Aedes albopictus* (Skuse) breeding sites. Examples include (**a**) metal containers, (**b**) terrarium, (**c**) stone holes, (**d**) ceramic vessels, (**e**) plastic containers, (**f**) gutters, (**g**) used tire dumps, (**h**) surface accumulated water, (**i**) disposable containers, (**j**) parking poundings, (**k**) flowerpot trays, and (**l**) metal containers. Reprinted with permission from Elsevier (Bonizzoni et al. 2013), © 2013

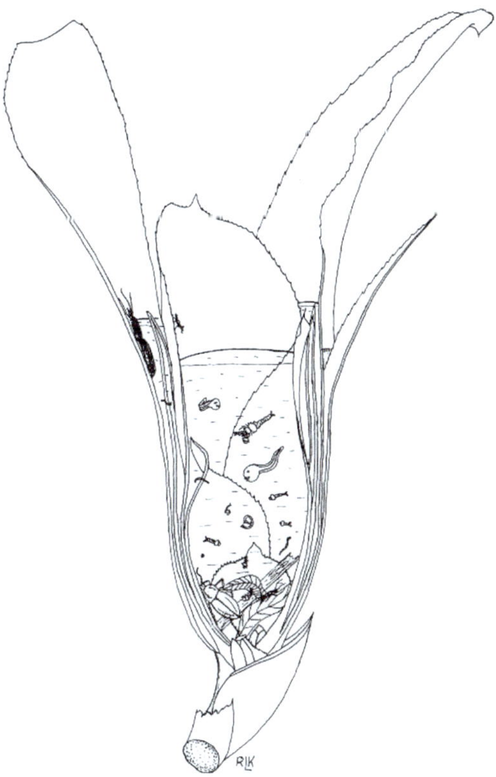

**Fig. 2.6** Cut-away schematic of a community in a tank bromeliad (composite) (figure taken from Kitching 2009). Reprinted with permission from Food Webs and Container Habitats. The Natural History and Ecology of Phytotelmata, edited by R. L. Kitching, © 2009. Published by Cambridge University Press

There are at least 1500 species of plants, belonging to 29 families, which have been recorded as phytotelmic environments. There are records of over 70 families and 11 orders of insects inhabiting those habitats, with Diptera being the most abundant order (Greeney 2001) with over 20 families occuring in these habitats (Williams 2006). Like artificial containers, natural containers such as phytotelma may play an important role in the presence of insects of medical importance, but factors such as competition and predation affect the abundance of their inhabitants (Zahouli et al. 2017).

Among symbolic species that have their development associated with phytothelmic habitats, we highlight the dragonflies of the family Pseudostigmatidae. Hypoxia conditions and pH variation are some of the factors that their larvae had to overcome to colonize this type of environment. Behavioral differences among larvae provide additional evidence that species of Pseudostigmatidae have adapted and specialized for either tree holes or bromeliads, but not both. For example, larvae of the *Megaloprepus* Rambur species and some *Mecistogaster* Rambur species rarely leave its tree hole breeding site (Fig. 2.7), whereas larvae of *Mecistogaster modesta* Selys, a species of bromeliad odonate, usually crawl out of the plant seeking better and bigger habitats as they grow (Fincke 1998).

### 2.3.6 Marine Environments

Undoubtedly, the greatest diversity of aquatic insects occurs in continental freshwaters; however, there are a few groups that have also successfully colonized marine environments. They are mostly considered semiaquatic or live in the region between tides (intertidal). Among the main representatives are Diptera, Hemiptera, and Coleoptera. One of the most iconic examples of intertidal insects is the caddisflies of the Chathamiidae family. Present in Oceania, the larvae live in intertidal pools and build cases from algae (usually *Corallina* Linnaeus spp) and shell or sand (Cowley 1978). In this peculiar family, the species *Philanisus plebeius* Walker lay the eggs within the coelom of the starfish *Patiriella regularis* (Verrill), where they develop until they are released as first instar larvae (Winterbourn and Anderson 1980).

Few species are truly marine. *Pontomyia* Edwards, an exclusive genus of flightless chironomid insect, are rarely observed on the ocean surface (Fig. 2.8). As a form of adaptation, the females had extreme body reduction, having lost the wings and the legs, and reduced the head and the thorax, becoming animals similar to worms. Their males have highly reduced and modified wings, and have lost the power to fly, but they walk on the surface of the sea for displacement. Another iconic group in marine environments is the water striders in the genus *Halobates* Eschscholtz. Although most *Halobates* species are coastal and associated with mangroves or other vegetation, at least one species is completely oceanic (Andersen and Cheng 2004).

**Fig. 2.7** Artistic view of a female *Mecistogaster* sp. throwing eggs in the water contained in a tree hole. The proportions of the insect to the fallen trunk and its hole were purposely modified. Reprinted with permission from Osmylus Scientific Publishers (Machado and Martinez 1982), © 1982

Here we presented an introduction to the main factors influencing the distribution of aquatic insects and its most common water habitats. Considering that the topics covered in this chapter are vast, from spatial spread to diversity, from habitats to aquatic insect adaptations, they will percolate through all the following chapters, with particular emphasis in the next two chapters looking into dispersal and the role of abiotic factors (Chap. 3 and Chap. 4).

**Fig. 2.8** (**a**) *Pontomyia oceana* Tokunaga masses at Turquoise Bay, Ningaloo Reef, in the Indian Ocean. (**b**) Male and (**c**) female specimen of *P. oceana*. Reprinted with permission from Springer Nature (Bessey and Cresswell 2016), © 2016

## References

Abell JM, Özkundakci D, Hamilton D, Miller S (2011) Relationships between land use and nitrogen and phosphorus in New Zealand lakes. Mar Freshw Res 62:162–175

Allan JD (1996) Stream ecology: structure and function of running waters. Chapman & Hall, New York

Andersen NM, Cheng L (2004) The marine insect Halobates (Heteroptera: Gerridae): biology, adaptations, distribution and phylogeny. Oceanogr Mar Biol Annu Rev 42:119–180

Armitage PD, Pinder LC, Cranston P (1995) The chironomidae: biology and ecology of non-biting midges. Springer, Dordrecht

Bae M-J, Kwon Y, Hwang S-J et al (2011) Relationships between three major stream assemblages and their environmental factors in multiple spatial scales. Ann Limnol Int J Limnol 47:S91–S105. https://doi.org/10.1051/limn/2011022

Balister P, Balogh J, Bertuzzo E et al (2018) River landscapes and optimal channel networks. Proc Natl Acad Sci 115:6548–6553. https://doi.org/10.1073/pnas.1804484115

Barnes JB, Vaughan IP, Ormerod SJ (2013) Reappraising the effects of habitat structure on river macroinvertebrates. Freshw Biol 58:2154–2167. https://doi.org/10.1111/fwb.12198

Barton BA, Taylor BR (1994) Dissolved oxygen requirements for fish of the Peace, Athabasca, and Slave River Basins

Battle JM, Jackson JK, Sweeney BW (2007) Mesh size affects macroinvertebrate descriptions in large rivers: examples from the Savannah and Mississippi Rivers. Hydrobiologia 592:329–343. https://doi.org/10.1007/s10750-007-0771-x

Batzer D, Boix D (2016) Invertebrates in Freshwater Wetlands. Springer, Cham

Batzer DP, Ruhí A (2013) Is there a core set of organisms that structure macroinvertebrate assemblages in freshwater wetlands? Freshw Biol 58:1647–1659. https://doi.org/10.1111/fwb.12156

Batzer DP, Sharitz RR (2007) Ecology of freshwater and estuarine wetlands. University of California Press, Berkeley, CA

Bell N, Riis T, Suren AM, Baattrup Pedersen A (2013) Distribution of invertebrates within beds of two morphologically contrasting stream macrophyte species. Fundam Appl Limnol/Arch für Hydrobiol 183:309–321. https://doi.org/10.1127/1863-9135/2013/0517

Bessey C, Cresswell AK (2016) Masses of the marine insect *Pontomyia oceana* at Ningaloo Reef, Western Australia. Coral Reefs 35:1225. https://doi.org/10.1007/s00338-016-1488-y

Blumenshine SC, Vadeboncoeur Y, Lodge DM et al (1997) Benthic-pelagic links: responses of benthos to water-column nutrient enrichment. J N Am Benthol Soc 16:466–479. https://doi.org/10.2307/1468138

Bonizzoni M, Gasperi G, Chen X, James AA (2013) The invasive mosquito species *Aedes albopictus*: current knowledge and future perspectives. Trends Parasitol 29:460–468. https://doi.org/10.1016/j.pt.2013.07.003

Boulton AJ, Lake PS (2008) Effects of drought on stream insects and its ecological consequences. In: Lancaster J, Briers RA (eds) Aquatic insects: challenges to populations. CABI, Wallingford, pp 81–102

Brown BL (2003) Spatial heterogeneity reduces temporal variability in stream insect communities. Ecol Lett 6:316–325. https://doi.org/10.1046/j.1461-0248.2003.00431.x

Brundin L (1966) Transantarctic relationships and their significance, as evidenced by chironomid midges. With a monograph of the subfamilies Podonominae and Aphroteniinae and the Austral Heptagyiae. K Sven Vetenskapakadamiens Handl 11:1–474

Byers JE, Cuddington K, Jones CG et al (2006) Using ecosystem engineers to restore ecological systems. Trends Ecol Evol 21:493–500. https://doi.org/10.1016/j.tree.2006.06.002

Chester ET, Robson BJ (2011) Drought refuges, spatial scale and recolonisation by invertebrates in non-perennial streams. Freshw Biol 56:2094–2104. https://doi.org/10.1111/j.1365-2427.2011.02644.x

Chipps SR, Hubbard DE, Werlin KB et al (2006) Association between wetland disturbance and biological attributes in floodplain wetlands. Wetlands 26:497–508

Cowley DR (1978) Studies on the larvae of New Zealand trichoptera. N Z J Zool 5:639–750. https://doi.org/10.1080/03014223.1978.10423816

Cummins KW (2016) Combining taxonomy and function in the study of stream macroinvertebrates. J Limnol 75. https://doi.org/10.4081/jlimnol.2016.1373

Currie DC (1986) Morphology and systematics of primitive Simuliidae (Diptera: Culicomorpha). University of Alberta, Edmonton

Dallas H, Day J (2004) The effect of water quality variables on aquatic ecosystems: a review. Water Research Commission, Pretoria

Death RG, Winterbourn MJ (1995) Diversity patterns in stream benthic invertebrate communities: the influence of habitat stability. Ecology 76:1446–1460. https://doi.org/10.2307/1938147

van der Valk AG (2012) Microorganisms and invertebrates. In: The Biology of Freshwater Wetlands. Oxford University Press, Oxford, pp 45–68

Dijkstra K-DB, Monaghan MT, Pauls SU (2014) Freshwater biodiversity and aquatic insect diversification. Annu Rev Entomol 59:143–163. https://doi.org/10.1146/annurev-ento-011613-161958

Fenoglio S, Bo T, Cucco M et al (2010) Effects of global climate change on freshwater biota: a review with special emphasis on the Italian situation. Ital J Zool 77:374–383. https://doi.org/10.1080/11250000903176497

Fincke OM (1998) The population ecology of Megaloprepus caerulatus and its effect on species assemblages in water-filled tree holes. In: Insect populations in theory and in practice. Springer Netherlands, Dordrecht, pp 391–416

Genkai-Kato M, Nozaki K, Mitsuhashi H et al (2000) Push-up response of stonefly larvae in low-oxygen conditions. Ecol Res 15:175–179. https://doi.org/10.1046/j.1440-1703.2000.00336.x

Gimenez BCG, Lansac-Tôha FA, Higuti J (2015) Effect of land use on the composition, diversity and abundance of insects drifting in neotropical streams. Braz J Biol 75:52–59. https://doi.org/10.1590/1519-6984.03914

Godoy BS, Queiroz LL, Lodi S et al (2016) Successional colonization of temporary streams: an experimental approach using aquatic insects. Acta Oecol 77:43–49. https://doi.org/10.1016/j.actao.2016.08.010

Gonçalves JF Jr, Santos AM, Esteves FA (2004) The influence of the chemical composition of Typha Domingensis and *Nymphaea ampla* detritus on invertebrate colonization during decomposition in a Brazilian coastal lagoon. Hydrobiologia 527:125–137. https://doi.org/10.1023/B:HYDR.0000043190.49651.dc

Greeney HF (2001) The insects of plant-held waters: a review and bibliography. J Trop Ecol 17:241–260. https://doi.org/10.1017/S026646740100116X

Griffith MB (2017) Toxicological perspective on the osmoregulation and ionoregulation physiology of major ions by freshwater animals: teleost fish, crustacea, aquatic insects, and Mollusca. Environ Toxicol Chem 36:576–600. https://doi.org/10.1002/etc.3676

Grimaldi D, Engel MS (2005) Evolution of the insects. Cambridge University Press, Cambridge

Grönroos M, Heino J, Siqueira T et al (2013) Metacommunity structuring in stream networks: roles of dispersal mode, distance type, and regional environmental context. Ecol Evol 3:4473–4487. https://doi.org/10.1002/ece3.834

Gullan PJ, Cranston PS (2014) The insects: an outline of entomology, 5th edn. Wiley & Sons, Chichester

Gurski FDA, Pinha GD, Moretto Y et al (2014) Effect of habitat heterogeneity in the composition and distribution of Chironomidae (Diptera) assemblage in different microhabitats of preserved streams in the Brazilian Atlantic Forest. Acta Limnol Bras 26:163–175. https://doi.org/10.1590/S2179-975X2014000200007

Heck KL, Crowder LB (1991) Habitat structure and predator—prey interactions in vegetated aquatic systems. In: Habitat structure. Springer Netherlands, Dordrecht, pp 281–299

Heino J (2009) Biodiversity of aquatic insects: spatial gradients and environmental correlates of assemblage-level measures at large scales. Freshwater Rev 2:1–29. https://doi.org/10.1608/FRJ-2.1.1

Heino J (2013) Does dispersal ability affect the relative importance of environmental control and spatial structuring of littoral macroinvertebrate communities? Oecologia 171:971–980. https://doi.org/10.1007/s00442-012-2451-4

Heino J, Melo AS, Bini LM (2015) Reconceptualising the beta diversity-environmental heterogeneity relationship in running water systems. Freshw Biol 60:223–235. https://doi.org/10.1111/fwb.12502

Heino J, Soininen J, Alahuhta J et al (2017) Metacommunity ecology meets biogeography: effects of geographical region, spatial dynamics and environmental filtering on community structure in aquatic organisms. Oecologia 183:121–137. https://doi.org/10.1007/s00442-016-3750-y

Hill MJ, Heino J, Thornhill I, et al (2017) Effects of dispersal mode on the environmental and spatial correlates of nestedness and species turnover in pond communities. Oikos 126:1575–1585. https://doi.org/10.1111/oik.04266

Hutchinson G (1957) A treatise on limnology, vol 1. John Wiley & Sons, New York

Hynes HBN (1970) The ecology of running waters. Toronto Press, Toronto, ON

Janzen DH (1967) Why mountain passes are higher in the tropics. Am Nat 101:233–249. https://doi.org/10.1086/282487

Jocque M, Vanschoenwinkel B, Brendonck L (2010) Freshwater rock pools: a review of habitat characteristics, faunal diversity and conservation value. Freshw Biol 55:1587–1602. https://doi.org/10.1111/j.1365-2427.2010.02402.x
Kasangki A, Chapman LJ, Balirwa J (2008) Land use and the ecology of benthic macroinvertebrate assemblages of high-altitude rainforest streams in Uganda. Freshw Biol 53:681–697. https://doi.org/10.1111/j.1365-2427.2007.01925.x
Keleher MJ, Sada DW (2012) Desert spring wetlands of the Great Basin. In: Batzer DP, Baldwin AH (eds) Wetland habitats of North America. University of California Press, Berkeley, CA, pp 329–341
Kingsolver JG, Arthur Woods H, Buckley LB et al (2011) Complex life cycles and the responses of insects to climate change. Integr Comp Biol 51:719–732. https://doi.org/10.1093/icb/icr015
Kitching RL (2009) Food webs and container habitats: the natural history and ecology of phytotelma. Cambridge University Press, Cambridge
Kling LJ, Juliano SA, Yee DA (2007) Larval mosquito communities in discarded vehicle tires in a forested and unforested site: detritus type, amount, and water nutrient differences. J Vector Ecol 32:207–217. https://doi.org/10.3376/1081-1710(2007)32[207:LMCIDV]2.0.CO;2
Kohshima S (1984) A novel cold-tolerant insect found in a Himalayan glacier. Nature 310:225–227. https://doi.org/10.1038/310225a0
Lamberti GA, Resh VH (1985) Distribution of benthic algae and macroinvertebrates along a thermal stream gradient. Hydrobiologia 128:13–21. https://doi.org/10.1007/BF00008935
Lloyd LS (2003) Best practices for dengue prevention and control in the Americas. Environ Heal Proj:1–105
London MS (2017) A comparison of benthic macroinvertebrate assemblages between perennial and intermittent headwater streams of the Mattole river in Northern California, USA. Faculty of Humboldt State University
Machado ABM, Martinez A (1982) Oviposition by egg-throwing in a zygopteran, *Mecistogaster jocaste* Hagen, 1869 (Pseudostigmatidae). Odonatologica 11:15–22
Manguin S, Boëte C (2011) Global impact of mosquito biodiversity, human vector-borne diseases and environmental change. Importance Biol Interact Study Biodivers:27–50. https://doi.org/10.5772/22970
Masson S, Desrosiers M, Pinel-Alloul B, Martel L (2010) Relating macroinvertebrate community structure to environmental characteristics and sediment contamination at the scale of the St. Lawrence. River Hydrobiol 647:35–50. https://doi.org/10.1007/s10750-009-9915-5
Merritt RW, Cummins KW (1996) An introduction to the aquatic insects of North America. Kendall/Hunt Publishing Company, Dubuque
Minshall GW (1978) Autotrophy in stream ecosystems. BioScience 28:767–771. https://doi.org/10.2307/1307250
Nabout JC, Siqueira T, Bini LM, Nogueira IS (2009) No evidence for environmental and spatial processes in structuring phytoplankton communities. Acta Oecol 35:720–726. https://doi.org/10.1016/j.actao.2009.07.002
Paice RL, Chambers JM, Robson BJ (2017) Potential of submerged macrophytes to support food webs in lowland agricultural streams. Mar Freshw Res 68:549–562. https://doi.org/10.1071/MF15391
Palmer MA, Menninger HL, Bernhardt E (2010) River restoration, habitat heterogeneity and biodiversity: a failure of theory or practice? Freshw Biol 55:205–222. https://doi.org/10.1111/j.1365-2427.2009.02372.x
Paterson CG (1971) Overwintering ecology of the aquatic fauna associated with the pitcher plant *Sarracenia purpurea* L. Can J Zool 49:1455–1459
Poff NL (1997) Landscape filters and species traits: towards mechanistic understanding and prediction in stream ecology. J N Am Benthol Soc 16:391–409. https://doi.org/10.2307/1468026
Poff NL, Pyne MI, Bledsoe BP et al (2010) Developing linkages between species traits and multiscaled environmental variation to explore vulnerability of stream benthic communities to climate change. J N Am Benthol Soc 29:1441–1458. https://doi.org/10.1899/10-030.1

Power ME, Holomuzki JR, Lowe RL (2013) Food webs in Mediterranean rivers. Hydrobiologia 719:119–136. https://doi.org/10.1007/s10750-013-1510-0
Richardson JS (2008) Aquatic arthropods and forestry: effects of large-scale land use on aquatic systems in Nearctic temperate regions. Can Entomol 140:495–509. https://doi.org/10.4039/N07-Ls04
Sáinz-Bariáin M, Zamora-Muñoz C, Soler JJ et al (2016) Changes in Mediterranean high mountain Trichoptera communities after a 20-year period. Aquat Sci 78:669–682. https://doi.org/10.1007/s00027-015-0457-9
Salles FF, Ferreira-Júnior N (2014) Hábitat e hábitos. In: Hamada N, Nessimian JL, Querino R (eds) Insetos aquáticos na Amazônia Brasileira: taxonomia, biologia e ecologia. Editora do INPA, Manaus, pp 39–49
Santos AN, Stevenson RD (2011) Comparison of macroinvertebrate diversity and community structure among perennial and non-perennial headwater streams. Northeast Nat 18:7–26. https://doi.org/10.1656/045.018.0102
Scheffer M (2004) Ecology of Shallow Lakes. Springer Netherlands, Dordrecht
Shah AA, Gill BA, Encalada AC et al (2017) Climate variability predicts thermal limits of aquatic insects across elevation and latitude. Funct Ecol 31:2118–2127. https://doi.org/10.1111/1365-2435.12906
Shimabukuro EM, Pepinelli M, Perbiche-Neves G, Trivinho-Strixino S (2015) A new trap for collecting aquatic and semi-aquatic insects from madicolous habitats. Insect Conserv Divers 8:578–583. https://doi.org/10.1111/icad.12135
Shobanov NA (2001) Function of ventral and lateral processes in larvae of chironomus (Diptera, Chironomidae). J Evol Biochem Physiol 37:384–387. https://doi.org/10.1023/A:1012970711312
Siegloch AE, Suriano M, Spies M, Fonseca-Gessner A (2014) Effect of land use on mayfly assemblages structure in neotropical headwater streams. An Acad Bras Cienc 86:1735–1747. https://doi.org/10.1590/0001-3765201420130516
Sinclair BJ, Marshall SA (1987) The madicolous fauna in southern Ontario. Proc Entomol Soc Ont 117:9–14
Siqueira T, Durães LD, de Oliveira RF (2014) Predictive modelling of insect metacommunities in biomonitoring of aquatic networks. In: Ecological modelling applied to entomology. Springer, Cham, pp 109–126
Springer M (2009) Marine insects. In: Wehrtmann IS, Cortés J (eds) Marine Biodiversity of Costa Rica, Central America. Springer Netherlands, Dordrecht, pp 313–322
Strahler A (1952) Dynamic basis of geomorphology. Geol Soc Am Bull 63:923–938
Tojo K, Sekiné K, Suzuki T et al (2017) The species and genetic diversities of insects in Japan, with special reference to the aquatic insects. In: Motokawa M, Kajihara H (eds) Species diversity of animals in Japan. Springer, Japan, pp 229–247
Tolonen KE, Tokola L, Grönroos M et al (2016) Hierarchical decomposition of trait patterns of macroinvertebrate communities in subarctic streams. Freshw Sci 35:1032–1048. https://doi.org/10.1086/687966
Tonkin JD, Heino J, Altermatt F (2018) Metacommunities in river networks: the importance of network structure and connectivity on patterns and processes. Freshw Biol 63:1–5. https://doi.org/10.1111/fwb.13045
Tornwall BM, Swan CM, Brown BL (2017) Manipulation of local environment produces different diversity outcomes depending on location within a river network. Oecologia 184:663–674. https://doi.org/10.1007/s00442-017-3891-7
USDA (2001) Stream corridor restoration: principles, processes, and practices. USDA
Vaillant F (1956) Recherches sur la faune madicole de France, de Corse et d'Afrique du Nord. Mem du Mus d'Histoire Nat Paris 11:1–258
Valente-Neto F, Koroiva R, Fonseca-Gessner AA, Roque F de O (2015) The effect of riparian deforestation on macroinvertebrates associated with submerged woody debris. Aquat Ecol 49:115–125. doi: https://doi.org/10.1007/s10452-015-9510-y

Vannote R, Sweeney BW (1980) Geographic analysis of thermal equilibria: a conceptual model for evaluating the effect of natural and modified thermal regimes on aquatic insect communities. Am Nat 115:667–695

Vannote RL, Minshall GW, Cummins KW et al (1980) The river continuum concept. Can J Fish Aquat Sci 37:130–137. https://doi.org/10.1139/f80-017

Vinson MR, Hawkins CP (2003) Broad-scale geographical patterns in local stream insect genera richness. Ecography (Cop) 26:751–767. https://doi.org/10.1111/j.0906-7590.2003.03397.x

Wegener A (1922) Die Entstehung der Kontinente und Ozeane. Vieweg, Braunschweig

Wiggins GB (2004) Caddisflies: the underwater architects. University of Toronto Press, Toronto, ON

Williams DD (2006) The biology of temporary waters. Oxford University Press, Oxford

Winterbourn MJ, Anderson NH (1980) The life history of *Philanisus plebeius* Walker (Trichoptera: Chathamiidae), a caddisfly whose eggs were found in a starfish. Ecol Entomol 5:293–304. https://doi.org/10.1111/j.1365-2311.1980.tb01151.x

Xenopoulos MA, Downing JA, Kumar MD et al (2017) Headwaters to oceans: ecological and biogeochemical contrasts across the aquatic continuum. Limnol Oceanogr 62:S3–S14. https://doi.org/10.1002/lno.10721

Yee DA, Yee SH (2007) Nestedness patterns of container-dwelling mosquitoes: effects of larval habitat within variable terrestrial matrices. Hydrobiologia 592:373–385. https://doi.org/10.1007/s10750-007-0775-6

Zahouli JBZ, Koudou BG, Müller P et al (2017) Urbanization is a main driver for the larval ecology of Aedes mosquitoes in arbovirus-endemic settings in south-eastern Côte d'Ivoire. PLoS Negl Trop Dis 11:1–23. https://doi.org/10.1371/journal.pntd.0005751

# Chapter 3
# Dispersal by Aquatic Insects

**Michael L. May**

**Abstract** All habitats change over time, so most organisms must resist unfavorable conditions or disperse to more favorable localities. Typically, aquatic habitats are relatively short lived because of drying or infilling. Aquatic insects, then, often have adaptations for effective dispersal, sometimes over long distances and most often by flight. This chapter examines some of the environmental drivers and organismal responses that affect the nature of dispersal. These include consideration of how different habitats affect dispersal, especially some differences between lentic and lotic habitats. Dispersal characteristics may also have major effects on genetic structure of populations. Both selective forces and proximate cues affect when insects disperse and when and where they colonize new habitats; availability of space, presence of predators, and availability of food may all play a role, depending on species and circumstances. Adaptations for dispersal include, in addition to active flight, behaviors that promote passive movement by wind, dispersal polymorphism (i.e., changes in body structure, such as wing development, that enhance dispersal, usually hormonally controlled and incurring some cost in fecundity), increased body size, and timing of diapause and reproduction. In a few species dispersal extends to migrations of hundreds of kilometers and may have important seasonal effects on habitats of origin and of destination. Dispersal is also integral to the concept of metapopulations and in fact may be a major driver of community composition and dynamics. Simultaneous dispersal of very large insect populations can have an important effect on nutrient and energy flow to and from communities. Finally, dispersal may be a critical determinant of whether and how aquatic insects respond to climate warming.

**Keywords** Geographic range · Use of habitat · Habitat template · Community dynamics · Climate warming

M. L. May (✉)
Department of Entomology, Rutgers University, New Brunswick, NJ, USA
e-mail: mimay@scarletmail.rutgers.edu



K. Del-Claro, R. Guillermo (eds.), *Aquatic Insects*,
https://doi.org/10.1007/978-3-030-16327-3_3

## 3.1 Introduction

Dispersal is one of the fundamental processes that influences the ability of organisms to reach suitable habitats, to find mates, and to avoid potentially disastrous disturbances. It can also have major effects on geographic range, genetic structure of populations, population persistence, and movement of energy and nutrients. Because the habitats of aquatic species are physically relatively well defined and cover broad ranges of size, permanence, and faunal complexity and because it is often relatively easy to maintain small populations and communities in the laboratory or under seminatural conditions they lend themselves to many sorts of observational and manipulative studies that are more difficult in many other animals. My hope is that this chapter will make clear some of the fascinating questions that have been and can be addressed using aquatic and semiaquatic insects as subjects.

## 3.2 The Evolution of Dispersal by Flight

The evolution of wings was the sine qua non of most insect dispersal, and one of the great morphological innovations of Metazoa—the first wings on earth. A brief discussion of that landmark therefore seems appropriate, especially since aquatic insects may have been the first to evolve proto-wings. For many years, most entomologists subscribed to the idea that insect wings developed from paranotal lobes, flattened extensions of the thoracic terga, and present on many Paleozoic insect fossils (Snodgrass 1935). The work of Kukalová-Peck (1978, 1983), however, based on detailed analysis of fossil morphology, turned the focus to aquatic taxa. She argued that wings originated as outgrowths of the insect pleura, possibly as modifications of movable gill covers that were found on the thoracic and abdominal segments of aquatic Paleozoic insect nymphs and that have a pattern similar to the inferred plesiomorphic pattern of veins in the wings. Although modifications of her scheme have been proposed (Trueman 1990; Elias-Neto and Belles 2016; Linz and Tomoyasu 2018), the main features seem likely to endure. Marden and Kramer (1994) suggested that the behavior like that of brachypterous stoneflies skimming across the water surface might have been a predecessor of flight, implying that flight may actually have originated in aquatic insects (but see Dudley et al. 2007). Wings then enabled insects to reach and exploit almost every nonmarine habitat on earth.

## 3.3 The Habitat Template

No habitat is perfectly stable, although change may occur at timescales ranging from hours to millennia. All eventually change owing to physical and/or biological processes, including human disturbance. Consequently, most organisms must be able to evade or resist unfavorable conditions. Aquatic insects, especially those inhabiting temporary ponds or pools, may resist periodic drought by, e.g., burrowing into mud or sheltering under vegetation or stones as adults or (rarely) larvae, diapausing in a drought-resistant form, or entering a terrestrial adult life stage (Batzer and Wissinger 1996). Most, however, have at some time in their life cycle the capacity to move from their area of birth or current residence to another, more suitable area. Many species that inhabit seasonal ponds and wetlands develop rapidly, with the resulting adults flying to permanent waters where they pass the dry season, in some cases reproducing there and in others returning to temporary waters for reproduction (Batzer and Resh 1992; Batzer and Wissinger 1996; Saijo 2001). Other insects may pass the dry season as adults in reproductive diapause (Corbet 1999). This movement from their birthplace or place of initial reproduction to other locations is dispersal, the subject of this chapter.

Dispersal of aquatic insects is not usually observed directly, and almost never from departure to arrival at a new habitat patch. Most often it is inferred from changes in populations in response to seasonal or other environmental changes. Sometimes direct observation of important portions of the dispersal process is possible and very useful, however. This is usually most feasible with large, diurnal species such as dragonflies (e.g., Russell et al. 1998; Wikelski et al. 2006; Anderson 2009; May 2013) or in cases of unusual dispersing swarms of smaller insects (e.g., Stevens et al. 2007). Mark-recapture methods have sometimes been useful, notably by using fluorescent powders to mark mosquitoes (e.g., Service 1993; Epopa et al. 2017; this technique does not allow discrimination of individuals) or paint marks or numbers on wings of Odonata (e.g., Michiels and Dhondt 1991) and on several species of water striders (Gerridae; Ditrich 2016), corixid bugs (water boatmen; Boda and Csabai 2009), and dytiscids (predaceous diving beetles, Davy-Bowker 2002). Medeiros et al. (2017) marked large numbers of the mosquitoes, *Culex quinquefasciatus and Aedes albopictus*, by adding small amounts of $^{15}KNO_3$ or $^{14}C$-glucose to water in plastic tubs colonized by wild mosquitoes. Using light traps for recapture, they determined that *C. quinquefasciatus* females seeking oviposition sites often dispersed 1–2 km from the marking site while females of *A. albopictus* seldom went further than 300 m. Thus, despite the limitations of all mark-recapture techniques, a great deal has been learned through careful observation and experiment with both undisturbed and manipulated populations.

In general, dispersal increases in frequency and importance as habitat stability declines and habitat isolation increases (Southwood 1962). Some aquatic habitats are effectively permanent, including large, deep lakes, e.g., L. Baikal, estimated to be 20–25 million years old (Wikipedia 2018a), and a few river channels (Wikipedia 2018b; even if these estimates are correct, however, the habitats in these rivers have probably changed radically over the millennia). The vast majority of aquatic environments are of relatively short duration, and many still waters, such as shallow lakes, ponds, and wetlands, are seasonal or sometimes dry for several years at a time. Datry et al. (2014) estimated that ca. 30% of the total length of rivers globally are intermittent. Thus many aquatic habitats are relatively unstable and isolated from similar habitats by environments that are unsuitable for aquatic insects. Some aquatic species can survive desiccation in a state of diapause, commonly as eggs (e.g., Williams 1996; Corbet 1999), but for many dispersal is an essential, often urgent, adaptation that affects individual fitness, population and metapopulation dynamics, and community composition and persistence.

Intermittent rivers notwithstanding, still, or lentic, waters are much more likely to desiccate or suffer from other catastrophic events at relatively frequent intervals and will eventually be obliterated by sediment infilling. Moreover, lentic habitats are generally less interconnected than are lotic (running) waters, which normally are continuous with other streams and river systems. For that reason, insects of lentic waters are expected to have greater powers of dispersal than lotic species. Despite the difficulty of observing dispersal directly because of the small size of most insects and the difficulty of marking and recapturing large numbers, this expectation is supported indirectly by numerous studies.

For example, Arribas et al. (2012) studied sister species pairs of hydrophilids (water scavenger beetles) and showed that the member of each pair that lives in ponds had a substantially larger geographic ranges and longer wings than their stream-dwelling relative, both of which suggested greater powers of dispersal. Ribera and Vogler (2000) found that other aquatic beetles from lentic habitats had on average considerably larger ranges than lotic species in the Iberian Peninsula. On a wider geographic scale, Ribera et al. (2003) showed a similar difference, as well as higher species turnover, among lotic species throughout Europe. These differences were also present in each of the four independent beetle clades. Hof et al. (2008) presented evidence that in freshwater animals generally, not only insects, lotic species richness in Europe declines steadily with latitude, suggesting poor dispersal from southern Pleistocene refugia from glaciation, while lentic species richness showed a maximum at intermediate latitudes, and β-diversity (a measure of species turnover) was greater at any latitude for lotic than for lentic species, suggesting less mixing of populations among the former. Hof et al. (2012), using more detailed information on the European ranges of Odonata, found that lentic species usually occupy more of the projected suitable range, based on climatic characteristics, than do lotic species, again suggesting that the former disperse more readily into areas made suitable by a warming climate. Finally, many species of southern European Odonata have recently expanded their range northward, probably in response to climate warming, but this effect has been much stronger in lentic than in lotic species (Grewe et al. 2013). All these studies, however, measured presumed indicators of dispersal rather than dispersal directly, and comparable data apparently are not available from other continents.

Perhaps the most persuasive evidence of the correlation of range size with dispersal capacity is that of McCauley et al. (2014) for North American Libellulidae. These authors used data from McCauley (2007) on observed dispersal distance and rate from natural source populations to initially uninhabited artificial ponds. This was used along with two measures of niche breadth and several other variables as possible factors that might explain range in multiple regressions; only dispersal and niche breadth had significant effects, positive in both cases, on range size, which was extracted for each species from the web site, OdonataCentral (https://www.odonatacentral.org/) as of 2007.

Another indicator of the relative degree of dispersal is the genetic structure of populations. In species with high rates of dispersal, populations should experience a high level of genetic exchange, whereas if dispersal rates are low populations are more isolated and genetic differentiation among populations should be higher. This is commonly expressed in terms of $F_{ST}$ (Wright 1951), which is a measure of the relative extent of genetic variation among subpopulations; the value ranges from 0, if the entire population is randomly mixed, to 1 if all variation is between, rather than within, subpopulations. Since the advent of allozyme frequency analysis, and especially of DNA sequence analysis, population structure of many organisms has been measured in this way. For example, Marten et al. (2006) compiled $F_{ST}$ values for European aquatic molluscs, crustaceans, and insects. In all three cases, lotic species had, on average, significantly higher $F_{ST}$s. As might be expected, average $F_{ST}$ values were lower in insects generally than in the other organisms, since among these only insects can fly.

On the other hand, such broad generalizations may obscure a great deal of variation within and overlap among habitat types. Short and Caterina (2009), e.g., demonstrated that three largely sympatric, lotic beetles in southern California diverge markedly in genetic and phylogeographic population structure and corresponding variation in inferred population connectivity. They varied from practically unstructured, implying almost complete mixing throughout the entire population, to having an extremely close correlation of haplotype with geographic origin, thus presumably very little movement away from their place of origin. Likewise, Phillipsen et al. (2015) compared population genetic structure in three lotic species, *Abedus herberti* (a flightless bug), a stonefly, *Mesocapnia arizonensis* (males brachypterous, females fly), and *Boreonectes aequinoctialis* (a strong-flying beetle). All three showed precisely the genetic pattern that was predicted: *A. herberti* had high mean and variance in genetic distance among populations, and no indication that this was affected by geographic distance (i.e., almost no genetic mixing among its isolated populations); *Boreonectes* had very low mean and variance of genetic distance, and again no effect of geographic distance (populations all very similar because of great dispersal ability and intermixing); and in *Mesocapnia* the mean and variance increased smoothly with distance (i.e., so-called isolation by distance, consistent with intermediate dispersal; Fig. 3.1). Thus, while the broad patterns found by Marten et al. (2006) are probably correct, many individual exceptions surely exist.

Although streams are typically more stable than lentic waters, organisms in streams that dry up partly or entirely face many of the same problems as those of still waters. In some cases residual pools remain throughout the dry season and can provide refuges for at least some stream inhabitants, and a few species may shelter in the hyporheic zone (Stubbington 2012), but others must either disperse to running streams or diapause in a drought-resistant form (e.g., Corbet 1999; Garcia and Hagen 1987). A form of disturbance peculiar to streams is the occurrence of spates, or sudden floods may displace organisms far downstream. These may be avoided by movement into the hyporheic region, which seems to be a more common and

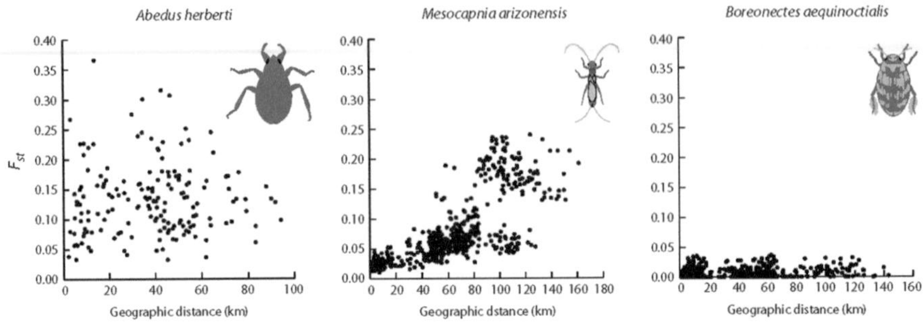


**Fig. 3.1** Relationships between genetic distance (Fst) and geographic distance in three species of desert insects from the southwestern USA. The pattern for *Abedus herberti* shows highly variable genetic distance with virtually no relationship to distances between populations, suggesting that there is very little genetic exchange even between nearby populations. In *Mesocapnia arizonensis* nearby populations are more similar genetically than are distant populations, i.e., isolation by distance. *Boreonectes aequinoctialis* populations are all closely similar in genetic characteristics irrespective of distance between them, indicating that the regional population is essentially panmictic. Figure from Phillipsen et al. (2015), used with permission

successful strategy in the face of spates than in drought (Stubbington 2012), by sheltering under rocks or other in-stream objects, or, especially in univoltine species, by timing adult eclosion for the end of the dry season, so that most larvae have exited the stream just before the onset of flooding (Corbet 1999; Lytle 2008; pers. obs. 1994). In addition, many very small larvae may disperse short distances into the hyporheic zone, possibly to avoid predators as well as physical stresses (Mugnai et al. 2014; Williams 1981).

Lotic organisms also face another challenge that has received a great deal of attention over the years. Because of the persistent downhill flow of water in streams, their insect inhabitants, as well as other aquatic organisms, will tend to be displaced downstream; this is known as drift and has been documented in a number of lotic systems (e.g., Waters 1972). Thus headwater streams apparently would be depleted of aquatic fauna in the absence of substantial compensatory upstream movement. This process was first described explicitly by Müller (1954, 1982), who called it the "colonization cycle" (also known as "drift compensation" or "the drift paradox"). Müller presented data in support of his ideas, but subsequent tests of the phenomenon have been inconclusive; one study using stable isotope labeling with $^{15}N$ strongly supported the pattern of drift and return in arctic mayflies (Hershey et al. 1993), and Coutant (2008) used a similar technique to recover caddisflies up to 16 km upstream from a source of effluent of $^{65}Zn$, but Briers et al. (2004), using $^{15}N$ as a label, showed little indication of preferential upstream movement in stoneflies in Wales. Several studies using malaise traps or sticky traps along a stream have obtained equivocal results (e.g., Svensson 1974; Bird and Hynes 1981; Williams and Williams 1993; Macneale et al. 2004; Winterbourn et al. 2007).

Anholt (1995) was probably the first to suggest that "drift compensation" by mass upstream movements of adults may be unnecessary if repopulation is density dependent; that is, under reasonable assumptions, if drift depletes larval populations upstream it is likely that the offspring of only a few adult pairs or gravid females can replenish the population, since few existing larvae will be present to compete with the newly colonizing individuals. This basic idea has been refined (Pachepsky et al. 2005; Mazzucco et al. 2015) and appears to provide a reasonable mechanism for maintenance of upstream populations, but without accurate measurements of insect movements it is difficult to verify in detail. It is also possible in some cases that movement of aquatic larvae, by swimming or crawling upstream, might partly reverse the effects of drift, and significant upstream crawling has been reported (Elliot 1971; Hershey et al. 1993; Williams and Williams 1993), although most investigators have found this to be of much less importance than adult flight (e.g., Graham et al. 2017). Humphries and Ruxton (2002), however, modified Anholt's model to show that, under some reasonable assumptions about population size and persistence, upstream crawling of less than 1 m might be sufficient to compensate for drift without reliance on adult flight. Substantial upstream crawling by larvae can occur (Graham et al. 2017) and may even facilitate independent habitat selection by larvae. Galatowitsch and Batzer (2011) observed large numbers of mayfly larvae moving upstream against strong currents into floodplain wetland, where they enjoyed faster growth than in the adjacent river channel.

Another fraught issue in the study of dispersal of insects in lotic habitats is the extent of overland movement away from natal streams that might indicate dispersal to separate streams or watersheds. It is clear that many insects do move along streams or stream valleys (e.g., Bogan and Boersma 2012). Most studies using passive traps have concluded that lateral movement away from streams is small, often less than 100 m (e.g., Collier and Smith 1998; Griffith et al. 1998; Briers et al. 2002; Petersen et al. 1999). In contrast, some genetic studies suggest substantial gene flow among separate watersheds at distances of up to tens or even hundreds of kilometers (Wilcock et al. 2003, 2007; Chaput-Bardy et al. 2008; Hughes et al. 2011). A possible reconciliation of these results is suggested by Didham et al. (2012), who placed arrays of passive interception traps at standardized distances from several streams, both at ground level and in the canopy of adjacent forest fragments approximately 15 m above ground. They collected more than three times as many adult aquatic insects in the canopy as at ground level, including at more than 1 km (the greatest distance of trap placement) from the nearest stream, although with some differences among taxa. Thus the apparent discrepancy between trapping and genetic results may stem, at least in part, from trapping schemes that failed to collect a majority of dispersing adults because traps were only placed at ground level.

Somewhat counterintuitively, several studies have demonstrated appreciably greater genetic uniformity across catchment basins than among reaches within streams (Bunn and Hughes 1997; Schmidt et al. 1995; Miller et al. 2002; Hughes et al. 2011). This could be explained by the so-called "patchy recruitment hypothesis" (Bunn and Hughes 1997). If each section of stream is colonized by offspring of only a few individual insects, then the allele complement in different reaches may differ strongly simply by the accident of which females oviposited there. If, at the same time, there is ample mixing of populations across catchments, the genetic profile of each catchment, which would include many individual reaches, should be similar to others in the region. This pattern may be common in many situations, although it seems less likely in taxa that emerge synchronously and oviposit quickly thereafter (Bunn and Hughes 1997) or that maintain large populations in very stable habitats (e.g., Krosch et al. 2011). Another pattern shown by some populations of insects confined to headwater streams, especially in mountains, is that populations in different catchments are more closely related to others on the same mountain than to lower portions of the same catchment. This suggests that such species are limited, possibly by temperature and stream gradient, to the uppermost reaches of streams but can disperse overland to nearby headwaters of other systems originating on the same mountaintop (Finn et al. 2007). Hughes et al. (2009) summarize the characteristics and predictions of these and several other models of genetic exchange.

## 3.4 Cues for Dispersal and Colonization

Given that dispersal allows organisms to anticipate and avoid deterioration of their habitat, what are the indicators of decline, currently or in the near future? Besides leaving a deteriorating habitat, dispersing individuals must locate and, often, assess the suitability of a new locale. This section examines some of the known solutions to these exigencies.

### *3.4.1 Initiation of Dispersal*

Few experiments have addressed the direct effects of drying of water bodies on insect dispersal. Boersma and Lytle (2014) placed individuals of the giant water bug, *Abedus herberti*, which occurs in intermittent streams in the southwestern USA, into small tanks either with or without water and lined with screening so the insects could escape. These were placed into larger outer tanks that contained water. After 24 h, ten times as many bugs had left the dry small tanks as the small tanks with water. Velasco and Millan (1998) studied one hydrophilid and two dytiscid beetles and a corixid bug (water boatman), also from intermittent streams, and found that two of the beetles and the corixid left their aquaria if the water temperature was raised to about 40 °C. In addition, they found that, when they placed individuals of the same species in aquaria with different depths of water, from 10 cm deep to entirely dry, all of the beetles flew or crawled out of the aquaria when the depth was 1 cm or less; the corixid did not respond to water depth. In each replicate, 5–10 individuals of a single species were placed in the same aquarium, irrespective of depth, so it is possible that the beetles were responding to crowding rather than to water depth per se (Boersma et al. 2014). *Rhantus* spp. (Dytiscidae), which live primarily in temporary seepage pools, also respond to crowding with escape behavior (which would lead to dispersal), and to high temperature and to unsuitable substrate (smooth glass bottom in experimental containers; Smith 1973). Drummond et al. (2015) observed assemblages of insects in natural scour pools during progressive drying; emergence rates of most, but not all, taxa increased as drying progressed. Some larval mosquitoes from temporary pools accelerate development in response to drying (Chodorowski 1969; Schäfer and Lundström 2006); Chodorowski (1969) also observed slowing of development when pools were diluted with rainwater. On the other hand, De Block and Stoks (2005) found that damselfly larvae (*Lestes viridis*) from temporary ponds actually developed more slowly and emerged later than controls as water volume was reduced in outdoor tubs, although individuals that hatched late in the season did accelerate development (and see Johansson and Rowe 1999). The authors suggested that in this case crowding increased competition for food, which made rapid growth impossible. Jannot (2009) showed that development time in the caddisfly, *Limnephilus indivisus,* was unaffected by water depth but female size at eclosion was reduced. Some *Asynarchus* caddisfly larvae, which live in high-altitude snowmelt basins, become aggressive toward and may cannibalize conspecifics in response to high temperature, crowding, and lowered water level; this behavior may accelerate pupation and allow for successful dispersal by the resulting adults (Lund et al. 2016). Based on statistical analysis of field trap catches, Klečka (2008) concluded that dytiscids, hydrophilids, and scirtids (marsh beetles) in the Czech Republic made more dispersal flights when water in occupied pools was low. Finally, Lytle et al. (2008) observed large numbers of adult water beetles and dragonfly larvae displaying uncharacteristic positive rheotaxis and crawling upstream in a desert stream just ahead of the receding waterline of a drying stream reach.

An excess of water, at least in lotic habitats, can be nearly as harmful as the absence of water. Particularly in arid regions subject to occasional heavy rains, spates may represent a substantial danger to aquatic organisms, and several taxa, especially Hemiptera and Coleoptera, in the arid American southwest climb out the stream bed in response to rainfall and escape the danger of flash floods (Lytle 1999, 2001; Lytle and Smith 2004). Alternatively, some insects may time adult eclosion to just precede the onset of rains, thus avoiding spates (Lytle and Smith 2004). A number of other stimuli can induce dispersal, and their effect may depend in part on the phenotype of the dispersing insect.

One might imagine that the presence of predators would promote dispersal, but this has not often been demonstrated. The clearest evidence of a direct effect on dispersal behavior is probably the work of McCauley and Rowe (2010) and Baines et al. (2015, 2018) on the backswimmer, *Notonecta undulata*, in mesocosms (water-filled cattle tanks). In the first of these studies, the authors showed that in the presence of predators (caged fish that could not actually reach the insects), *Notonecta* dispersal increased dramatically in the presence of one fish, but there was no significant additional effect of adding a second fish. If one or more *Notonecta* were deliberately fed to the fish, however, there was a sharp increase in dispersal that increased further as additional *Notonecta* were eaten; therefore, these insects are sensitive not only to the presence of predators but also to some signal that conspecifics were consumed. The experiments of Baines et al. examined the effects of predators in combination with body condition adjusted by pre-experimental feeding regime (Baines et al. 2015), or conspecific density and predation (Baines et al. 2018). In the first experiment, emigration was higher in the presence of caged fish, irrespective of body condition. Better-conditioned bugs initially emigrated more rapidly than those in poorer condition, but the effect declined over time so the difference was marginal for the entire experiment. In the later study emigration increased across three levels of increasing density. Predation significantly increased emigration only at medium density for reasons that were not clear. Dispersal was also positively correlated with beetle density in dytiscids (Yee et al. 2009). Plant density was also manipulated and had a significant but small negative influence on dispersal.

The effect of predators on dispersal of mayfly (*Baetis* spp.) larvae has been investigated extensively by Peckarsky and her colleagues. *Baetis* usually coexist with insectivorous trout in mountain streams and in fact maintain higher densities in streams with trout than without (Peckarsky et al. 2011). They feed by drifting with the current until they encounter rocks with good patches of benthic algae, so they drift more in streams with less food available (Hernandez and Peckarsky 2014). Peckarsky and McIntosh (1998) placed larvae in one of the four mesocosms with flowing water and stones with roughly uniform growths of algae: (1) exposed to water dripping from a tank containing trout, (2) trout odor and predaceous stonefly larvae with mandibles glued so they could not prey on the mayflies, (3) stonefly larvae and odorless water, and (4) odorless water and no stoneflies. All *Baetis* were predominantly active at night, but fish odor reduced drifting sharply during the night. The presence of stoneflies increased crawling during the day in the absence of fish odor and drifting at night with or without fish odor. Thus fish tended to reduce dispersal while the presence of stoneflies increased dispersal even in the absence of actual predation (Peckarsky et al. 2008; Peckarsky et al. 2011). Wooster and Sih (1995) review earlier work on the effects of predators on drift in stream insects.

Because dispersal is a crucial part of the adaptation of an animal's life cycle to its environment, it must be coordinated with other facets of the life cycle, e.g., adult eclosion and timing of reproduction. Photoperiod supplies a cue that triggers many events in the lives of insects. Apparently no evidence has demonstrated a direct effect of photoperiod on dispersal, but certainly events that are closely coordinated with dispersal are strongly influenced by photoperiod. Lytle (2002, 2008) showed how photoperiodically controlled metamorphosis allows insects to time their adult eclosion, and thus their dispersal from the stream, before flooding is likely to occur. The trichopteran, *Phylloicus aeneus*, probably uses just such a strategy (Lytle 2002). In addition, photoperiod controls development of either winged or wingless forms in at least some wing-polymorphic species (Harrison 1980; Zera and Denno 1997), which clearly has profound effects on dispersal ability. Among related species, aptery is often more common in those that occupy more permanent habitats (e.g., Ditrich et al. 2008).

### 3.4.2 Colonization of New Habitats

Dispersal, i.e., moving away from a now or soon-to-be unsuitable place, also implies colonization, the occupation of a new and presumably more suitable environmental patch. How, then, do insects find appropriate habitats? At least part of the answer for aquatic species is by detection of polarized light reflected from water surfaces, a phenomenon demonstrated by Horváth and his associates (Bernath, et al. 2001; Horváth 1995a, b; Horváth and Kriska 2008). Several researchers have taken advantage of this phenomenon by using other polarizing surfaces, such as black plastic, as a means of attracting and collecting insects for study (e.g., Boda and Csabai 2009); this also explains the strange propensity for dragonflies to attempt oviposition on black gravestones and on the surface of cars (Horváth et al. 2007; Wildermuth and Horváth 2005).

More than likely some insects use other visual, tactile, or chemical cues to locate appropriate habitat, but few of these have been identified. Some aquatic insects can detect the presence of underwater predators without direct contact. Blaustein et al. (2004) showed that mosquitoes, *Culiseta longiareolata*, whose larvae are highly vulnerable to predation by notonectids (backswimmers), avoid ovipositing in artificial or natural pools containing *Notonecta*, apparently responding to chemical cues; midges, *Chironomus riparius*, have much less vulnerable larvae and do not avoid *Notonecta*. *Culiseta longiareolata* females also avoid laying eggs in pools containing a larva of the dragonfly, *Anax imperator* (Stav et al. 1999). Mosquitoes also apparently respond to other chemical and physical cues. Among tree hole-breeding mosquitoes in Florida, USA, drought-susceptible species chose more permanent holes that are larger and lower, with larger, more vertical openings, and contain darker water with higher conductivity, pH, and tannin-lignin content (Bradshaw and Holzapfel 1988). In coastal NSW, Australia, Mokany and Mokany (2006) studied larvae of *Ochlerotatus notoscriptus*, a relatively drought-resistant species, and the more drought-susceptible *Anopheles annulipes*, in small mesocosms with filtered pond water concentrated by evaporation by fivefold; mesocosms were filled with either 2 L or 4 L of either dilute or concentrated water. Larvae of *A. annulipes* were consistently recruited to higher volume containers, regardless of concentration, while *O. notoscriptus* favored higher concentration containers, regardless of volume. Thus *O. notoscriptus*, at least, responded to water chemistry.

Resetarits and Silberbush (2016) found that two species of *Culex* mosquitoes also avoid fish predators in arrays, each consisting of nine mesocosms constructed from plastic pools. In each array, the pools ("patches") were arranged into three groups of three ("localities"). One locality contained no patches with fish, one a single patch with fish, and one two patches with fish. In this configuration, female mosquitoes exhibited a phenomenon called "compression" by these authors (Fig. 3.2). Mosquitoes laid fewer eggs in patches containing fish than in fish-free patches, but also laid fewer eggs in fish-free patches in localities also containing one and still fewer in those containing two patches with fish. That is, female mosquitoes were sensitive not only to the condition of individual patches but also to their

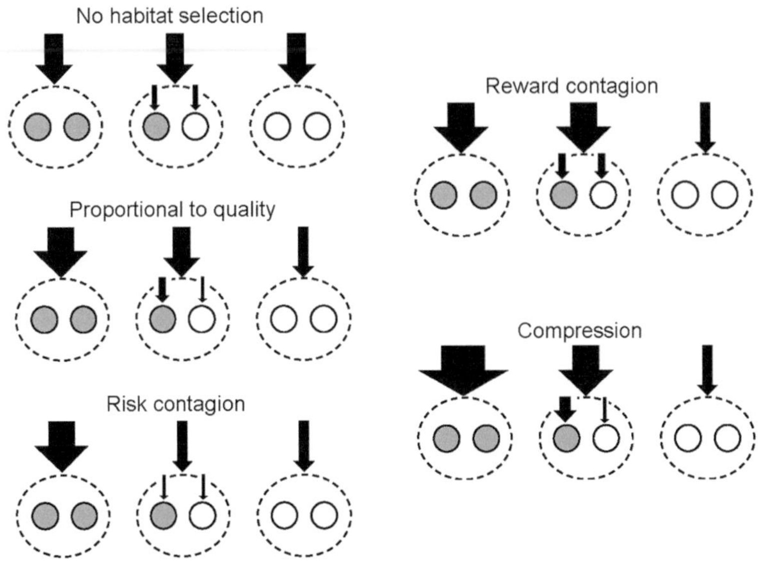


**Fig. 3.2** Schematic illustration of effects of locality (dashed ovals) and patch (small circles) quality on colonization by dytiscid and hydrophilid beetles. Filled circles are patches of higher quality, e.g., lacking predators or with higher food quality, than open circles. Widths of vertical arrows are proportional to the number of beetles colonizing each locality or patch. Modified from Pintar and Resetarits (2017a), used with permission

context, i.e., the condition of each locality, so that more eggs were "compressed" into localities entirely without fish.

Similar experiments on dytiscid and hydrophilid beetles and some aquatic Hemiptera have examined several different contrasts among patches. Binkley and Resetarits (2007), e.g., found that these taxa were dramatically more likely to colonize mesocosms under an open tree canopy than under a closed canopy. The beetles also detected and avoided fish in mesocosms and even discriminated among fish species in a pattern that reflected the degree of threat they posed (Resetarits and Pintar 2016). Resetarits and Binkley (2009, 2013, 2014) arranged localities with variable numbers of patches (cattle tanks) with and without fish, resulting in three patch types: (1) fish free and distant from any fish-containing patches, (2) with fish, and (3) without fish but adjacent to patches with fish ("fish-associated" patches). In fish-free localities the number of colonists did not differ significantly among patches. Patches without fish always had more beetles than patches with fish, and the difference increased as the number of patches per locality increased, although the fish could not reach the beetles. Fish-associated patches had intermediate numbers. The difference in colonists between fish-free localities and localities with fish increased with locality size, even though the ratio of patches with and without fish was always 1:1 in mixed localities; beetles apparently perceived increasing rather than constant risk as the number of patches with fish increased ("spatial risk contagion"; Resetarits et al. 2005; Fig. 3.2). This response might lead to underutilization of some suitable natural habitats.

Vonesh et al. (2009) also reported that caged fish reduced the abundance and altered the composition of colonizing insects (mostly Diptera in their case). Subsequent exposure of the colonized containers to uncaged fish reduced abundance further due to predation. However, tanks that were fish free during the colonization phase still had more insects after actual predation than did the fish-associated tanks. Thus, differences in dispersal behavior can have long-lasting effects on community composition despite exposure to predators. Kraus and Vonesh (2010) showed that in their system the avoidance of fish by dytiscids was not strong if other colonizing species were allowed to accumulate but was much stronger if earlier colonizers were removed periodically. Hydrophylids and mosquitoes strongly avoided fish under all conditions, but, surprisingly, chironomid midges preferred to oviposit with fish, possibly because fish usually eliminate many invertebrate predators. Thus the responses of insect colonists may depend on subtle differences in patch characteristics. The presence of predators may also have other effects on life history and morphology, such as causing earlier metamorphosis, and, consequently, smaller size at metamorphosis, or development of defensive spines (Benard 2004).

The presence of food may also influence colonization of new habitat patches. Pintar and Resetarits (2017a, b, c) studied colonization by dytiscids (predators) or hydrophilids (predators as larvae, scavengers as adults). Hydrophilids preferred hardwood litter initially but later favored pools with pine needles, presumably due to changes in zooplankton composition. Dytiscids showed no significant preference. Individual species of both families mostly followed the same patterns, so the change among hydrophilids was not primarily due to species turnover. In separate experiments, dytiscids, but not hydrophilids, responded markedly to removal or addition of zooplankton to half the mesocosm containers (Pintar and Resetarits 2017b).

Aquatic beetles are subject to spatial contagion and compression in response to variation in resource level (Pintar and Resetarits 2017a). Beetles were as strongly attracted to localities with one high and one low resource patch as they were to localities with two high patches, and in the former case they compressed into the high-quality patch. As expected, localities with two low patches were less attractive. Finally, Pintar et al. (2018) compared the effects of three levels of predatory fish combined with three nutrient levels (i.e., nine possible combinations). In all cases they found that the presence of predators depressed colonization but, in these circumstances, the effect of nutrient level was not significant for any one species of beetle, although it was significant for all beetles taken together. They also counted egg rafts of *Culex* mosquitoes, which responded to both nutrients and presence of predators. Overall, they concluded that predators had a stronger effect than nutrient level on immigration. The giant water bug, *Lethocerus deyrollei*, however, emigrates more readily if deprived of food (Ohba and Takagi 2005).

### 3.4.3 Timing of Dispersal

The daily and seasonal peaks of dispersal may be responses to specific environmental cues, although this has received less attention than the questions of whether and where to disperse. As seen in the next section, seasonal timing is often a crucial adaptation for dispersal, especially when the latter is necessary to avoid seasonally unfavorable conditions. The timing of dispersal, however, may also serve more quotidian ends and may respond to several immediate cues. In some cases, as with many Odonata, adult emergence occurs at night or near dawn and is followed by dispersal away from the natal site as soon as hardening of the exoskeleton allows for flight (e.g., Corbet 1957; Michiels and Dhondt 1991; Geenen et al. 2000). This probably reduces predation and also spares the still rather fragile new adults from attack and injury by mating attempts from mature conspecific males at the water before their muscles are fully developed and their exoskeleton is fully hardened.

In other insects various physical cues may stimulate and assist dispersal. These are summarized exhaustively by Johnson (1961). More recently Boda, Csabai, and colleagues in Hungary have investigated conditions favoring or inhibiting dispersal by aquatic beetles and bugs. One important factor is wind. Dispersal flights were most common at wind speeds of $<6$ km $h^{-1}$ and was completely inhibited by wind speeds of $>12$ km $h^{-1}$ (Bota and Csabai 2005). Csabai et al. (2012) found that most of species of Coleoptera and Hemiptera flew during one or more of three distinguishable daily intervals: mid-morning, near noon, or at nightfall. In the spring, however, a few species flew throughout the day, and a number of others had different activity patterns in different seasons. Nearly all had a distinct seasonal peak, although relatively few were confined strictly to a single season. The authors concluded that the seasonal variation was driven by changes in air temperature in relation to the minimum temperature for flight and the (higher) minimum for mass dispersal although they did not report measurements of minimum or routine body temperatures. Certainly, however, wind and temperature are likely to have marked effects on dispersal in these and other insects.

## 3.5 Adaptations for Dispersal

The division between this section and the last is somewhat artificial. Clearly the responses to the various "cues" that stimulate dispersal and colonization ordinarily are adaptive, so I have made some arbitrary decisions about how to allocate information to one or the other topic. Here we will be concerned with characteristics that have evolved to favor successful dispersal rather than the circumstances that dictate when and where dispersal begins or ends. However, the sources of many cues for dispersal and colonization, such as food sources or presence of predators in a habitat patch, are the immediate result of features that make avoidance or exploitation of those patches adaptive, as shown clearly in the preceding discussion.

### 3.5.1 Passive and Active Dispersal

An important distinction is that between passive and active dispersers, i.e., between those that are propelled entirely by wind or water or that attach themselves to other animals, and those that move primarily under their own power. Most insects, because they can fly, fall into the latter category. Nevertheless, a number of insects, especially small Diptera, including mosquitoes, ceratopogonid midges, and blackflies, are dispersed by wind (Elbers et al. 2015; Jones et al. 1999; Service 1980). Japanese encephalitis is thought to have been introduced to Australia by wind-blown *Culex* mosquitoes from Papua New Guinea (Chapman et al. 2003), and genetic analysis suggested a panmictic population of these mosquitoes throughout western PNG and the Australian Cape York Peninsula. So-called aerial plankton—small organisms that are transported passively by winds, usually at hundreds to thousands of meters altitude—includes small insects of many orders, including a number of aquatic groups (Compton 2002). *Ischnura aurora*, a tiny zygopteran, also probably migrates passively on the wind while teneral (Rowe 1978). All these insects have little control of their direction except during takeoff and landing, and their movements may not always be either intentional or adaptive.

Of course not all insects are airborne dispersers. Many aquatic insects can walk or swim for some distance to reach emergence sites, escape drying habitats (Lytle et al. 2008), or compensate for downstream drift (as discussed above). Among terrestrial species nymphs of some locusts (Acrididae) and Mormon crickets (Tettigoniidae), e.g., walk for long distances to find food (Reynolds et al. 2014). In many Lepidoptera first instar caterpillars "balloon" by spinning out long silk threads that catch the wind. Some small Hemiptera nymphs can position themselves at the edge of leaves to launch themselves into the wind (Washburn and Washburn 1984). These methods are mostly unavailable to aquatic insects, however. Several semi-aquatic insects and Collembola are also propelled by wind or currents on the surface of water, and this may effect successful dispersal but probably is often maladaptive (Reynolds et al. 2014). One known case of adaptive dispersal in this manner is that of the aphid, *Pemphigus trehernei*, for which the sea aster, a plant of tidal salt marshes, is secondary host. First instar larvae are initially photopositive and crawl to and walk on the soil surface until they encounter the incoming tide. They are small enough to remain on the surface due to surface tension and often are blown along the surface by wind until they reach their host, at which time they become photonegative and descend through cracks in the soil to the aster roots where they feed (Foster and Treherne 1978).

Darwin (1859) famously observed that newly hatched larval snails readily attached themselves to a duck's feet. He conclude that small aquatic animals might easily be dispersed on the feet and legs of water birds, and this has been confirmed in a number of cases for some small crustaceans and other zooplankton (Figuerola and Green 2002). Desiccation is a danger in such circumstances, but it should be possible for desiccation-resistant resting stages, including diapausing insect eggs. Apparently, though, this has never been definitely confirmed for insects.

### 3.5.2 Dispersal Polymorphism

A dramatic example of adaptive morphology is that of dispersal polymorphism, in which some individuals are adapted for dispersal and have fully developed wings while others do not disperse and usually have reduced or absent wings (brachyptery or aptery). For non-dispersing individuals there can be a considerable advantage to reduced or absent wings and flight muscles because in most insects these structures represent a considerable investment of energy and material, which may slow development and reduce fecundity compared to individuals without them (Harrison 1980; Zera 1984; Zera and Denno 1997).

Among aquatic taxa, this has been studied extensively in Gerridae, or water striders (summarized by Vepsäläinen 1978). Wing development varies among species, often geographically within species, and often between diapausing (usually wintering as inactive adults) and non-diapausing (summer, actively breeding) generations. Some species of *Gerris* are not dimorphic at all, while in others the diapause generation is long-winged, since they need to find sheltered sites for diapause and then return to habitats suitable for reproduction. The non-diapause generation generally does not disperse, sacrificing the ability for higher fecundity. In some species, populations in certain areas retain functional wings in summer, perhaps to avoid drying of their habitat. In some species dimorphism is seasonal, as described above, and probably is controlled by photoperiod, while in others dimorphism is genetically determined. Several other combinations of diapause and wing development have been identified. Zera et al. (1983) found in the North American genus *Limnoporus* that both genetic factors and photoperiod affect wing development, with short and declining photoperiod favoring production of long-winged morphs. Harada and Nishimoto (2007) reported that in *Aquarius paludum*, adults held under a long-day photoperiod regime and fed daily quickly broke reproductive diapause and matured ovaries while also histolyzing their flight muscle; others on the same photoperiod but fed only once every 3 days remained in diapause and retained functional flight muscles and their ability to disperse by flight. Thus the underfed individuals could disperse and possibly colonize a more favorable patch, while those with ample food diverted resources to gamete production and were ready to reproduce in situ (see Dingle and Arora 1973 for a more detailed study of a similar phenomenon in terrestrial bugs).

In the closely related and ecologically similar family Veliidae (riffle bugs), somewhat similar patterns are found (Ditrich 2016; Ditrich et al. 2008). As an interesting aside, at least one species of *Velia* is mostly flightless and is dispersed overland by walking (Ditrich and Papáček 2009).

A diverse group of other aquatic taxa also exhibits flight polymorphism. These include at least one mayfly (Ruffieux et al. 1998), a number of stoneflies (Brittain 1990; Lillehammer 1985; Zwick 2000), several Hemiptera in addition to gerrids and veliids especially corixids and notonectids (Young 1965; Baines et al. 2018), a few Trichoptera (Giudicelli and Botosaneanu 1999), and a few Diptera (Byers 1969;

Berendonk and Bonsall 2002). Whether reduced wings and/or flight muscles are genetically determined or induced by environmental circumstances is mostly unknown. The selective pressures that might lead to this condition may include the trade-off between flight and rate of development, fecundity, or longevity, or they could be in response to the danger of being permanently transported away from suitable habitat (Bilton 1994), as might be the case for species living on oceanic islands (e.g., Short and Liebherr 2007); these authors questioned the "island effect" as an explanation in this case, however, as did Roff (1990) more generally.

One other taxon deserves brief mention here. Although not truly aquatic, several species of planthoppers in the family Delphacidae are restricted to intertidal salt marshes. These were extensively studied by Denno and his collaborators (Denno and Roderick 1990) and subsequently reviewed in a broader context by Zera and Denno (1997). This is one of the most thorough studies of flight polymorphism to date, including ecological, physiological, and evolutionary aspects such as the close correlation of brachyptery with stable habitats. This correlation is also supported by Roff (1990) based on data from a very wide array of aquatic and terrestrial insects.

Aside from physical polymorphism, more subtle changes in the flight system affect dispersal. Dispersal behavior is usually accompanied by hormonal and other physiological changes. Modulation of juvenile hormone (JH) is probably the most pervasive of these. JH probably plays a major role in determining whether individuals develop normal or reduced wings, as well as muscle histolysis, although ecdysone, the molting hormone, is also important. The interplay of these hormones also largely controls the ultimate body size of individuals. In addition, a suite of hormones, most prominently adipokinetic hormone (AkH), but also including octopamine, regulate the deposition and mobilization of fat and the relative utilization of fat, carbohydrate, and, in some taxa, amino acids as fuel for energy metabolism. Fat is the normal fuel for all but very-short-range dispersers because it contains the largest amount of metabolic energy per unit mass (Dingle 2014).

Body size is influenced by many selective and life history factors (e.g., Peters 1986), but requirements for dispersal are often among these. Although not thoroughly investigated, a number of instances are known in which dispersers are larger on average than non-dispersers (reviewed by Benard and McCauley 2008). These include several cases among Odonata (e.g., Angelibert and Giani 2003; Conrad et al. 2002; Anholt 1990; and Michiels and Dhondt 1991). Roff (1991) found similar relationships in intraspecific comparisons. Larger body size may benefit active migrants because the mass-specific metabolic cost of flight tends to be smaller at larger body size. Flight speed and hence the ability to control flight direction at higher wind velocity are usually greater. Larger size might also tend to increase fecundity and make successful colonization of new habitats more likely. On the other hand, Thompson (1991) found no size difference among resident and dispersing individuals of a coenagrionid damselfly. McCauley (2005, 2010) also very carefully investigated males of two species of libellulid Anisoptera, *Leucorrhinia intacta* and *Pachydiplax longipennis*, both of which are highly territorial and aggressive, dispersing from permanent to recently refilled temporary ponds. In both cases, dispersing males were smaller than non-dispersers. The large males dominated smaller ones in territorial interactions, and the latter may disperse because of their greater difficulty in establishing new territories at the occupied ponds.

Since variation in body size does not necessarily prevent dispersal, it is often said to influence the propensity to disperse, i.e., a behavioral effect rather than a physical barrier to dispersal like aptery. Another such adaptation is subtle variation in wing size or shape. For example, Rundle et al. (2007) found that in *Enallagma* damselflies in North America wing length was positively correlated with range size, even when controlling for body length. In three species of *Libellula* dragonflies McCauley (2013) showed that two species with a broader habitat range and that dispersed further and more frequently had larger wings with a higher aspect ratio than a third species that had a narrow habitat range and was less dispersive. Suarez-Tovar and Sarmiento (2016) found that, when corrected for phylogenetic relatedness, migratory South American Libellulidae had larger and more deeply corrugated wings and a more expanded hindwing anal lobe than in nonmigratory species. In Swedish mayflies and stoneflies wing length was positively correlated with range size and in mayflies also with occupancy (the proportion of sampled sites actually occupied by a species) (Malmqvist 2000), and a similar, although not entirely consistent, correlation with habitat predictability occurs in chironomids (McLachlan 1985). Adaptive changes in dispersal propensity need not depend only on morphological features—physiology and behavior are clearly involved and may be of greater importance. Iversen et al. (2017), e.g., argued that the difference in flight tendency and distance in two genera of dytiscids are purely behavioral, since the wings and flight muscles remained intact in both.

McPeek (1989) found that larvae of two species of *Enallagma* damselflies only inhabit permanent lakes with predatory fish (which usually extirpate large invertebrates) and two others occur only in fishless lakes, dominated by invertebrate predators and generally less permanent than fish lakes. One species, *E. ebrium*, mostly inhabited "winterkill" lakes, which normally have fish but where winter oxygen levels may drop below lethal levels for fish, allowing large invertebrate predators to take over. *Enallagma ebrium* can coexist with fish but not with invertebrate predators, so adults disperse temporarily to lakes with fish only. The four species that coexist only with fish or with predatory invertebrates but not both are very philopatric. *Enallagma ebrium*, however, is adapted to lakes in which predator populations change radically and thus is much less philopatric. McCauley (2006), in another mesocosm experiment, showed that connectivity to other tanks explained a high proportion of the species richness of Odonata adults, and a still higher proportion of larvae, in the focal tank. The identities of adults seen nearby natural habitats and fields without cattle tanks were also extremely similar to those of adults and larvae at or within the tanks, suggesting that no physical barriers to dispersal existed. McCauley concluded that dispersal behavior alone limited colonization and largely determined community composition of these odonates.

### 3.5.3 Diapause, Reproduction, and Dispersal

As suggested in several examples already described, diapause, reproduction, and dispersal, with or without morphological changes, are intimately coordinated (Southwood 1962). The case of *Aquarius paludum* females (Harada and Nishimoto 2007) was described above. Other examples include several Odonata. In Japan, most *Sympetrum frequens* disperse from emergence sites in lowlands, often rice paddies, into highland areas during the summer, where they undergo partial reproductive diapause, although gametes slowly mature and mature coloration develops (Corbet 1999, pp. 390–394; Ueda 1988), and a very similar phenomenon occurs in *Sympetrum meridionale*, *S. striolatum*, and *Aeshna mixta* in Algeria (Samraoui et al. 1998). In many tropical dragonflies in areas with a distinct dry season, prereproductive adults may move from bodies of water where they reproduce into forest where they remain in reproductive diapause, often until the onset of rains, then quickly mature gametes, and develop mature coloration (Corbet 1999, pp. 261–262; pers. obs. 1974, 1994).

Because dispersal and diapause are intimately related, it may be difficult to distinguish independent environmental conditions that induce the two states, and in fact the cues may be the same. It is commonly assumed that photoperiod usually affects both, and Goehring and Oberhauser (2002) showed that in Monarch butterflies diapause induction and readiness to migrate are closely parallel. In other cases, it has been implicitly assumed that one effect of diapause induction (and sometimes also termination) is to initiate migration (Saunders 2010), but I know of no comparable studies on aquatic insects.

The close connection among dispersal, diapause, and reproduction led Johnson (1969) to propose the "oogenesis-flight syndrome," that is, female insects have hypertrophied fat bodies and maintain their gametes in an immature state until dispersal is accomplished, after which JH titer increases, mature eggs develop, and oviposition ensues. This should allow them to begin dispersal with maximum energy stores and possibly reduced flight costs, and arrive at their destination with maximal reproductive value. This syndrome does apply to many dispersing insects. Rankin (1978) showed that in the large milkweed bug, *Oncopeltus fasciatus*, migration is stimulated by an intermediate level of JH and maturation of gametes by a high level. Instances of oogenesis-flight syndrome have not been studied in detail in aquatic insects but seem likely in the hydrophilid, *Helophorus brevipalpis* (Landin 1980), and dytiscids in the genus *Graphoderus* (Iversen et al. 2017). There is also good evidence for the syndrome in belostomatid bugs (Cullen 1969; Lytle and Smith 2004), and it is likely in Gerridae and probably numerous other taxa, although its hormonal control has not been shown definitively in either aquatic Hemiptera or Coleoptera. On the other hand, it certainly is absent, or present in abbreviated form, in some migrating dragonflies. These normally begin migration when sexually immature (Corbet 1999, pp. 395–396), but they mature gametes and may mate and oviposit long before reaching their final destinations (May and Matthews 2008; May 2013; May et al. 2017).

## 3.6 Long-Distance Migration

Many authors have used the terms migration and dispersal interchangeably or ambiguously (Johnson 1969, pp. 3–8). Here, I consider migration to be a category of dispersal, more or less equivalent to Corbet's (1999, p. 394) interhabitat displacement, that is, usually a movement over multiple tens to hundreds of kilometers. Typically it entails a later return to the original habitat, although, in insects, not usually by the same individuals that migrated initially. Among terrestrial insects, this would include the famous migrations of monarch (e.g., Agrawal 2017) and painted lady butterflies (Stefanescu et al. 2012), the bug, *Oncopeltus fasciatus* (e.g., Dingle 1996), or plague locusts such as *Schistocerca gregaria* (Rainey 1951, 1976). Only a few aquatic insects, notably Anisoptera, undertake similar migrations under their own power, although a number of taxa may be dispersed passively for long distances, as noted above. Kennedy (1985) defined migration in behavioral terms as "... persistent and straightened-out movement effected by the animal's own locomotory exertions or by its active embarkation on a vehicle. It depends on some temporary inhibition of station-keeping responses, but promotes their eventual disinhibition and recurrence." This is still a useful and insightful definition and is widely used, although it does not include purely passive dispersal (but does include, e.g., cases in which insects actively fly high into the air so as to become entrained in air currents—a "vehicle"—which thereafter carry them passively). It also does not depend on any particular adaptive advantage accruing or even require that dispersal be adaptive, although Kennedy certainly recognized the advantages and trade-offs involved.

Although butterflies, locusts, and dragonflies perform the most visually spectacular migrations, recent attention has been focused on vast migrations at hundreds of meters aloft by a variety of smaller insects at night across Britain and northern Europe (Hu et al. 2016). These fly well above their flight boundary layer (Taylor 1958), so they depend largely on the wind for propulsion and have been shown to actively seek out altitudes where wind direction matches their intended flight direction. They have been studied partly by aerial sampling e.g., (Chapman et al. 2004) but mainly by upward-looking radar that reveals patterns, velocities, and even headings, of insects as small as 10 mg. Upwards of $3 \times 10^{12}$ insects (including many <10 mg, estimated from high-altitude net captures) may pass over southern Britain during the autumn, always on northerly winds. The smallest species are carried passively, but even slightly larger insects can modify their course by shifting their heading if wind is not blowing within 20° of the preferred flight direction (Chapman et al. 2015). The vast majority of these are terrestrial species, although a small number of aquatic insects—Corixidae, Dytiscidae, and Hydrophilidae—are included. These flights illustrate, however, the extent and importance of long-distance insect migration.

Odonata are probably the most flight-worthy of aquatic or semiaquatic insects and rank high among all insects. Several species are known to make long migrations, in at least two cases over thousands of kilometers. Although they do fly under their own power, they also depend, at least to some extent, on favorable wind for propulsion. One species that has been studied extensively, *Pantala flavescens* (the "Wandering Glider" or "Globetrotter"), has been shown to fly some 500–1000 km from India to the Maldives archipelago and thence on to east and southeast Africa, a total distance of around 3500 km (Anderson 2009). Based on observations in the Maldives and literature reports from southern India and east to southeast Africa, Anderson inferred that *P. flavescens* adults emerging in India during late summer fly southwest at high altitude on winds associated with the Intertropical Convergence Zone (ITCZ) to Africa. Corbet (1962) noted that adults are at various latitudes in Africa in synchrony with the ITCZ, and they may breed continuously within the continent. During the northern spring, however, a strong westerly upper air current, the Somali Jet, develops off the Horn of Africa and blows across the Arabian Sea and northern Indian Ocean, also bringing moisture that helps initiate the southwest monsoon in south Asia. This jet probably enables a reverse movement of *P. flavescens* back to northern India. Hobson et al. (2012), using stable hydrogen isotope ratios in wing samples of *P. flavescens*, showed that the most probable origin of specimens collected on the Maldives in October was in northeasternmost India, Nepal, and Bangladesh. This confirms Anderson's proposed route from India to the Maldives and is strongly supportive of a circuit very much as he suggested.

*Pantala* mostly breed in ephemeral pools created by rains along the ITCZ and consequently have very rapid larval development (Corbet 1999, p. 227). It is very likely that they never diapause as larvae and are obligate migrants in most places where they occur. Some other migrant odonates may diapause as larvae and thus could have two options for surviving conditions unsuitable for development. The common North American migrant, *Anax junius* (Fig. 3.3), is one such species. It has

**Fig. 3.3** Photograph of male *Anax junius* in flight, as during migration. Used with permission of the photographer, Dennis R. Paulson

attracted attention since the nineteenth century because of its sometimes spectacular mass migrations (Russell et al. 1998). It was long thought (Trottier 1971) that two temporally separate cohorts exist, one emerging in early summer and breeding in the neighborhood of its natal site, and the other emerging in late summer and migrating south. Freeland et al. (2003) showed that migrant and nonmigrant individuals could not be distinguished genetically. Thereafter Matthews (2007a) and Hobson et al. (2012), again using stable hydrogen isotopes, confirmed that late summer-autumn migrants collected in Texas and Mexico had very likely emerged in the northern USA or southernmost Canada. Matthews (2007a) also found that *A. junius* population show very little genetic differentiation across their entire eastern North American range. Stable isotope studies indicate, as expected, that individuals collected in the northern USA in early spring emerged far to the south (Matthews 2007a; Macfarland K, pers. comm. 2015). Most recently May et al. (2017) examined developmental phenology of *A. junius* at several sites in the eastern USA. Two emergence groups of larvae could usually be distinguished, but they overlapped broadly both in larval size and adult emergence, making genetic differentiation unlikely. Larval growth and timing of eclosion suggest that the early emerging group are largely individuals that overwintered as late instar diapausing larvae and probably will lay eggs locally, most of which will hatch as larvae that again overwinter in diapause. Late emerging adults are thought to be the offspring of adults that migrated from the south and laid eggs in early spring. The resulting larvae develop directly, emerging in late summer or autumn, and migrate south. However, because of the genetic evidence already cited and the broad overlap of larval size groups and adult emergence, it is probable that the timing of emergence and hence the likelihood of migration are determined by the effect of some environmental cues such as photoperiod on early larval instars (Matthews 2007b). Other known but less studied migrants among Odonata include *Aeshna mixta, Anax ephippiger, A. parthenope, Libellula quadrimaculata, Pantala hymenaea, Rhionaeschna bonariensis, Tramea lacerata*, and several species of *Sympetrum*, among others (Corbet 1999).

Long-distance migrants are likely to face some challenges that are absent or much less acute for insects that disperse short distances. One of these may be the necessity for navigation over hundreds of kilometers—the mere fact that it is far away makes a remote habitat patch harder to find than a comparable nearby patch. Presumably these and other long-distance migrants recognize suitable habitat using cues similar to those already described after arrival in the appropriate region, although it is not always clear how they recognize the latter. In the case of *Anax junius*, and probably some other migrants, migrating adults may mate and oviposit several times before reaching their final destination (May and Matthews 2008; May 2013).

Monarch butterflies have been studied extensively and have a complex, multifaceted navigation system (Reppert et al. 2016; Shlizerman et al. 2016) that allows individuals from all over the eastern USA to find a few patches of forest in a limited area in the mountains of Mexico. Comparable information is not available for any aquatic insect. On the other hand, most migrants are much less constrained than Monarchs as to their destination. May and Matthews (2008) suggested that *A. junius* might find suitable habitats, because the latter are quite widespread, simply by orienting very generally southward and avoiding dangerous areas such as open sea. It seems unlikely that *Anax* navigation is quite that simple, since they do preferentially fly when tailwinds are available (Wikelski et al. 2006) and may follow landmarks such as roadways (pers. obs. 1992). Other migrants have also been observed flying along river courses (Dumont and Hinnekint 1973) and probably dry wadis in the northern Sahara (Dumont and Desmet 1990).

In the case of ITCZ migrants or others that rely mostly on flying downwind to more favorable regions, usually where sustained rainy periods are starting, navigation problems are considerably less, since they need, for the most part, simply to maintain a flight heading that takes best advantage of the wind. Recognition of suitable habitat may also be simplified, because the downwind flight automatically brings them to areas where suitable pools are available, although they must still, of course, recognize specific sites for reproduction. Also, most, and probably all, animals that migrate by flight depend to a significant degree on wind to assist and guide their flight. For example, visible migration of *Anax junius* usually takes advantage of northerly winds associated with cold fronts that appear to stimulate flight and typically also strongly influence its direction (e.g., Russell et al. 1998; Wikelski et al. 2006). During fall migration, dragonflies, like many birds, take advantage of updrafts over mountain ridges to reduce flight cost and gain altitude.

The stimuli that prompt Odonata to begin or end migration are not well known but are more likely to be related primarily to seasonal changes and weather than immediate responses to predation, competition, or food supply because predation probably is often more intense during migration than during other periods (Nicoletti 1997), and long-distance migration is generally initiated before food availability deteriorates. Some other migrants such as *Oncopeltus* may respond to restriction of their food supply (Dingle 1968). Among so-called obligate migrants (Corbet 1999, pp. 408–418), migration may be initiated endogenously shortly after the teneral period (i.e., as soon as the cuticle has become sufficiently hardened); this appears to be the case in many Odonata, in which dispersal begins shortly after adult emergence with the so-called maiden flight (Corbet 1957). Alternatively, in ITCZ migrants it could be triggered by the onset or the cessation of rains or in response to habitat drying. In North America southward migration often is associated with cold fronts (Russell et al. 1998; Wikelski et al. 2006). As suggested above for *A. junius*, photoperiod might trigger developmental processes that determine whether eclosing individuals are prone to migrate or not. In principle, photoperiod might directly inhibit or stimulate migratory behavior, as it does in birds (Dingle 1996, 2014). In addition, conditions that prompt shorter dispersal episodes, described above (e.g., drying of natal ponds, crowding, high or low temperature, food supply, or availability of local oviposition sites), may stimulate or modulate long-distance migration in some species (Johnson 1969, pp. 203–239). Regardless of the specifics, it is likely that bouts of migration alternate with intervals of maintenance behavior such as feeding, probably at least in part because of reciprocal inhibition of flight and maintenance; for example, at the beginning of migratory flight, stimuli that would ordinarily elicit feeding are inhibited, but as flight is prolonged, feeding stimuli become more salient and eventually inhibit flight, and vice versa. This general pattern was first described by Kennedy (1961, 1985) in "migrating" aphids, but it appears to apply to many migratory organisms (e.g., Wikelski et al. 2006). The separation of migration and maintenance activity is not absolute, however, as I have watched migrating dragonflies divert from their flight path to grasp prey, albeit much less frequently than during nonmigratory feeding aggregations (Russell et al. 1998).

## 3.7 Population, Community, and Ecosystem Consequences

### *3.7.1 Metapopulations*

We often think of populations and communities as existing as isolated and self-contained entities (Forbes 1887). Many organisms, however, exist not in isolated populations but rather in groups of subpopulations, each within a restricted area (a "patch"), so that population dynamics of these subpopulations may strongly influence one another. They may, e.g., experience fairly frequent stochastic extinctions within a patch but with the possibility of "rescue" by dispersal from other nearby patches. This led Levins (1969) to propose the concept of a metapopulation, i.e., a group of these interacting populations, and to analyze how their dynamics might differ from those of larger but more remote populations. These ideas have been tested, modified, and expanded by a number of authors (e.g., Hanski 1998; Bohonak and Jenkins 2003), for instance by explicitly incorporating variation in patch characteristics and isolation; by allowing for priority effects, including local adaptation, that may give initial colonizers an advantage in exploiting the patch; or by considering assemblages of species within patches (a metacommunity; Wilson 1992; Hanski 1998). These metapopulation models seem particularly appropriate for understanding the dynamics of small ponds scattered through a landscape, isolated from one another by variable stretches of terrestrial habitat. The general concept has also been modified (Hanski 2001) to include circumstances similar to those envisaged by MacArthur and Wilson (1967), with a single large and relatively stable population, e.g., in a lake, with a number of smaller ponds nearby containing populations that might show local adaptation or even become extinct but could be rescued by dispersal from the lake; the large, stable population would presumably be little changed. Metapopulation analyses have also been extended to include population along different reaches in a single stream catchment (Fagan 2002; Downes and Reich 2008). There is general agreement that the model is applicable to many organisms and that interchange among subpopulations can stabilize the wider population of a species, depending on the size, number, and connectivity (i.e., the ease of moving among subpopulations).

A number of studies described above, including most mesocosm experiments in which initial conditions are very similar among widely separated patches, probably fit the assumptions of metapopulation models reasonably well, although few, if any, have continued long enough to observe extinction and rescue. Nevertheless, they do illustrate that dispersal ability can have an important influence on population and community dynamics. For example, McCauley (2006) showed that dispersal was probably the primary determinant of species richness of Odonata larvae in tanks placed at varying differences from source populations of adult odonates, and in fact the more distant tanks contained, for the most part, a nested subset of the species found in the closer tanks. Most of the experiments of Resetarits and colleagues, cited above (pp. 11–13), directly show the effects of dispersal and colonization but not subsequent population and community development, because colonizing insects were completely removed each week. Recruitment varied among treatments, often among taxa, over time, and in response to conditions in neighboring patches, e.g., presence of predators. All these effects were dependent on dispersal followed by acceptance or rejection of patches at close range and made large differences in the sizes of populations and community composition. A frequent assumption in modeling dispersal in metacommunities is either that (1) dispersal is random and selection of dispersing organisms occurs after entering a habitat patch, e.g., by competition or predation, or (2) populations are philopatric, i.e., they rarely disperse from their natal patch (McPeek 1989). Resetarits et al. (2005) argued that nonrandom dispersal and subsequent individual habitat selection (IHS) based on perceived patch quality are likely to be frequent, and they showed experimentally that this, rather than random dispersal or philopatry, is the rule among water beetles (Binckley and Resetarits 2005; Resetarits and Binckley 2013). In turn, habitat selection makes possible the phenomena of compression and contagion, discussed above. These authors argue that IHS is likely to provide a better description than truly random dispersal of population changes and species assembly for many active dispersers. Philopatry may be common principally among species inhabiting extremely stable habitats, perhaps including large, deep lakes, or organisms that cannot detect potential predators in a new habitat, as may be the case for many adult Odonata in habitats with larval predators (McPeek 1989, pers. obs.).

### 3.7.2 Dispersal and Genetic Change

Dispersal may also influence species characteristics or speciation since it strongly affects the extent of genetic interchange among (sub)populations. In general it is assumed that greater dispersal results in more gene exchange and thus increases the genetic similarity among populations and reduces the opportunity for speciation. Bohonak (1999) found a general, albeit weak, tendency for this to be true in a wide variety of animals. It is particularly evident in very mobile and widespread taxa like the dragonflies, *Anax junius* (Matthews 2007a) and *Pantala flavescens*, which Troast et al. (2016) suggested may be panmictic throughout its nearly global range (but see Pfeiler and Markow 2017). Several possible mechanisms, however, could alter this outcome. Rapid population growth and local adaptation (De Meester et al. 2002) or low dispersal rates (Bohonak and Jenkins 2003) might make it difficult for genes from individuals dispersing from outside to invade an existing population, thus increasing the likelihood of divergence of populations, and possibly speciation. On the other hand, if dispersal leads to a suitable patch with few or no conspecifics or allows the disperser to successfully establish itself in an occupied patch, new alleles will be introduced that may, if they spread initially, result in an altered genetic landscape within that patch. Finally, if dispersal propensity is contingent on properties of dispersing individuals (e.g., body size) or of their natal patch (e.g., crowding), the chances are increased of the same contingent behavior being spread (Arendt 2015).

### *3.7.3 Dispersal and Community Dynamics*

Dispersal might be important in determining community composition and dynamics even in the absence of niche-based interactions such as competition or predation. Hubbell (2001) introduced the so-called neutral theory of biodiversity, which suggests that similar organisms' coexistence does not depend on differences in niche, but rather that they enter a community via dispersal until resource availability is limiting and then are extirpated or fail to invade purely by chance. In such a case, the existing assemblage of species depends largely on their probability of immigrating into the community. A likely example applying to aquatic insects is a study by Siepielski et al. (2010), who investigated *Enallagma* damselfly larvae that coexist with fish. These clearly occupy a different niche than their congeners that cannot coexist with fish, as well as non-congeneric Odonata, but among themselves they are ecologically extremely similar. Siepielski et al. compared growth and mortality of seven species of *Enallagma* larvae, with special emphasis on two species, across a range of primary production, macrophyte abundance and diversity (as refugia for the larvae), and abundance and diversity of potential predators. They found that the species appeared to be ecologically equivalent, i.e., to share a single niche. This led to the conclusion that coexistence of these species depended on stochastic immigration and local extinction. It should be borne in mind, however, that dispersal is assumed to be stochastic here, but this is rarely known to be the case. Other studies have suggested that adaptations to particular aquatic niches are, in some habitats, the principal determinants of insect distribution (e.g., Heino and Mykrä 2008).

Immigration may increase intraspecific competition for resources and for mates (if the sex ratio of the newcomers is not 1:1), while emigration could have the reverse effects, and might also ameliorate the effects of kin selection, if this exists. Profound effects on other organisms occupying the patch could occur. In fish-free bodies of water, the top predators are often dragonfly or dytiscid and hydrophilid larvae. If large numbers of these insects invade a habitat where none or only a few lived before, it is conceivable that they could seriously compete with native species or decimate their prey, with effects throughout the food web. This has never been investigated, but would surely be worth examining. Conversely, if their progeny later leave the habitat, as probably occurs in long-distance migrants, a substantial relaxation of competition or predator pressure, or a reduction of food resources, might occur. The potential for nutrient and energy transfer by migrating insects in general is quite large. Hu et al. (2016) estimated that the nocturnal fall migration of insect over England transports as much as $5.78 \times 10^{12}$ Joules of energy, 100,000 kg of nitrogen, and 10,000 kg of phosphorus annually. Over the long term quantities transferred southward are very close to those transferred northward in spring, but in some years amounts are markedly unbalanced, resulting in substantial net movement of nutrients and energy in one direction or the other.

### *3.7.4 Dispersal and Nutrient Subsidies*

Although little is known about the effects of mass long-distance migration of aquatic insects on the ecosystems they enter or leave, the effects of nutrient and energy flux from aquatic to terrestrial habitats have been studied rather extensively on a smaller scale and can be impressive (reviewed by Polis et al. 1997; also more briefly by Schindler and Smits 2016). It must be said at the outset that energy and nutrient subsidies are nearly always greater from terrestrial to aquatic systems than vice versa, simply because of gravity—mineral nutrients either in solution or as suspended material, and carbon and energy from detritus and dissolved organic solids, are all brought in by runoff from the surrounding land. In many headwater streams energy from this source substantially outstrips in-stream primary production (Fisher and Likens 1973). Nevertheless, the energy and nutrients from the bodies of dispersing insects can provide a substantial subsidy downstream or to nearby terrestrial habitats. Among the more impressive reported instances are at Lake Mỳvatn (Lake of Midges) in Iceland, where huge numbers of midges emerge in summer. Dreyer et al. (2015) estimated that some 290 kg $ha^{-1}$ $yr^{-1}$ of midges were deposited within 450 m of the lake edge (mostly within the nearest 100 m) during the most prolific of 4 years of study, although the amount was less than 10% of this during the least prolific year. This was equivalent to about 10 kg N $ha^{-1}$ $yr^{-1}$ and 1 kg P $ha^{-1}$ $yr^{-1}$ during the best year. Thus the nutrient subsidy was very significant, at least during good midge years.

Jackson and Fisher (1986) measured production of aquatic insects in Sycamore Creek, Arizona, during April to November, spread over part of two calendar years. In-stream secondary productivity was about 1200 kg $ha^{-1}$ $yr^{-1}$ and biomass of emerged insects was 230 kg $ha^{-1}$ $yr^{-1}$, i.e., similar to the Lake Mỳvatn values, although in a starkly different habitat. Both of these results are unusually high for productivity of aquatic insects, but they illustrate the potential for nutrient transfer as a result of dispersal away from the immediate site of emergence. Paetzold et al. (2006) artificially subsidized small plots along a stream with about 75 mg $day^{-1}$ of dried aquatic insects, i.e., approximately 3 kg N $ha^{-1}$ $yr^{-1}$, assuming that the insects were ca. 10% N (Fagan et al. 2002), and recorded slightly more than twice the abundance of aquatic arthropods in the enriched plots as in control plots. Thus the natural subsidies at Lake Mỳvatn and Sycamore Creek cold have quite a significant effect on the later abundance of aquatic insects.

Export of insect biomass from aquatic habitats may also have marked effects on predators of flying adults of aquatic insects. For example, Fukui et al. (2006) used netting to prevent emergence of insects along a 1.2 km reach of a forest stream in Japan and measured bat foraging activity along that reach and an undisturbed adjacent control reach. They found that bat activity was dramatically reduced along the experimental reach compared to the control reach in May and June; thereafter, the abundance of aquatic adults decreased in both reaches while flying terrestrial insects increased throughout, and bat foraging increased, apparently because bats then depended on terrestrial prey.

### *3.7.5 Dispersal and Climate Warming*

As the most recent report of the IPCC (2018) makes clear, climate warming is proceeding rapidly and is likely to continue for some time even under the most optimistic scenarios. These changes already are affecting aquatic insects. For many this means that to avoid extinction, they must quickly evolve resistance to the rapidly changing conditions or be able to disperse to more a favorable climate; in other cases, northward expansion may enable species to exploit otherwise suitable habitats from which they had been excluded by low temperature. Indications that this has happened during and since postglacial warming (Grewe et al. 2013) and probably still is occurring (Hof et al. 2012) have already been cited. Additional studies, especially in the UK, suggest that the pace of northward expansion has accelerated in recent decades. Hickling et al. (2006) showed that the northern limit of the ranges of both aquatic bugs and Odonata on average moved northwards about 75 km from 1970 to 2000 and 1960 to 1996, respectively. Hassall and Thompson (2008), in a general review of effects of climate change in UK Odonata, reported that emergence periods have shifted to earlier dates in many species, an ability that might improve their success if moving north but is complicated by its interaction with photoperiod as well as temperature. Considerable anecdotal evidence suggests that some tropical and subtropical odonate species have expanded their ranges northward (e.g., Paulson 2011), although this has not yet been supported by formal studies. Odonata are, among aquatic taxa, among the best adapted for dispersal, but they are also relatively easily detected and identified. It is likely that members of other orders have also moved northward.

Hering et al. (2009) estimated the sensitivity of European Trichoptera to climate change, based on current habitat requirements. Perhaps counterintuitively, they predicted that southern taxa would probably be more vulnerable to adverse effects of warming. Southern species lived in more restricted habitats and had smaller ranges, probably because they occupied areas south of the limit of Pleistocene glaciation and had been present in their current ranges for longer and become more specialized. These authors also, as expected (e.g., Laurance et al. 2011), found that high-altitude species are at rather high risk. Similarly, De Knijf et al. (2011) estimated that the range of two boreo-alpine *Somatochlora* dragonflies in the mountains of

Romania were predicted to be forced upward by at least 200 m and thus to lose ca. 40% of their current range by a temperature increase of 1.5 C. *Epiophlebia laidlawi*, a relict odonate from the Himalayas, is predicted to lose around 60% of its current potential range by 2050 under even a moderate warming regime (Shah et al. 2012).

## References


Agrawal A (2017) Monarchs and milkweed: a migrating butterfly, a poisonous plant, and their remarkable story of coevolution. Princeton University Press, Princeton, NJ, p 296

Anderson RC (2009) Do dragonflies migrate across the western Indian Ocean? J Trop Ecol 25:347–334

Angelibert S, Giani N (2003) Dispersal characteristics of three odonate species in a patchy habitat. Ecography 26:13–20

Anholt BR (1990) Size-biased dispersal prior to breeding in a damselfly. Oecologia 83:385–387

Anholt BR (1995) Density dependence resolves the stream drift paradox. Ecology 76:2235–2239

Arendt JD (2015) Effects of dispersal plasticity on population divergence and speciation. Heredity 115:306–311

Arribas P, Velasco J, Abellán P, Sánchez-Fernandez D, Andújar C, Calosi P, Millán A, Ribera I, Bilton DT (2012) Dispersal ability rather than ecological tolerance drives differences in range size between lentic and lotic water beetles (Coleoptera: Hydrophilidae). J Biogeogr 39:984–994

Baines CB, McCauley SJ, Rowe L (2015) Dispersal depends on body condition and predation risk in the semi-aquatic insect, *Notonecta undulata*. Ecol Evol 5:2307–2316

Baines CB, McCauley SJ, Rowe L (2018) The interactive effects of competition and predation risk on dispersal in an insect. Biol Lett 10:20140287

Batzer DP, Resh VH (1992) Macroinvertebrates of a California seasonal wetland and responses to experimental habitat manipulation. Wetlands 12:1–7

Batzer DP, Wissinger SA (1996) Ecology of insect communities in nontidal wetlands. Annu Rev Entomol 41:75–100

Benard MF (2004) Predator-induced phenotypic plasticity in organisms with complex life histories. Ann Rev Ecol Evol Syst 35:651–673

Benard MF, McCauley SJ (2008) Integrating across life history stages: consequences of natal habitat effects on dispersal. Am Nat 171:553–567

Berendonk TU, Bonsall MB (2002) The phantom midge and a comparison of metapopulation structures. Ecology 83:116–128

Bernath B, Szedenics G, Molnar G, Kriska G, Horváth G (2001) Visual ecological impact of 'shiny black anthropogenic products' on aquatic insects: oil reservoirs and plastic sheets as polarized traps for insects associated with water. Arch Nat Cons Landsc Res 40:89–109

Bilton DT (1994) The flight apparatus and flying ability of *Hydroporus glabriusculus* (Coleoptera, Dytiscidae) with a brief review of structural modifications in flightless beetles. Entomol Tidskr 114:23–32

Binckley CA, Resetarits WJ (2005) Habitat selection determines abundance, richness and species composition of beetles in aquatic communities. Biol Lett 1:370–374

Binkley CA, Resetarits WJ (2007) Effects of forest canopy on habitat selection in tree frogs and aquatic insects: implications for communities and metacommunities. Oecologia 153:951–958

Bird, BA, Hynes HBN (1981) Movements of adult aquatic insects near streams in southern Ontario. Hydrobiologia 77:65–69

Blaustein L, Kiflawi M, Eitam A, Mangel M, Cohen JE (2004) Oviposition habitat selection in response to risk of predation in temporary pools: mode of detection and consistency across experimental venue. Oecologia 138:300–305

Boda P, Csabai Z (2009) Seasonal and diel dispersal activity characteristics of Sigara lateralis (Leach, 1817) (Heteroptera: Corixidae) with special emphasis on possible environmental factors and breeding state. Aquat Insects 31:301–314
Boda P, Csabai Z (2013) When do beetles and bugs fly? A unified scheme for describing seasonal flight behaviour of highly dispersing primary aquatic insects. Hydrobiologia 703:133–147
Boersma KS, Lytle DA (2014) Overland dispersal and drought-escape behavior in a flightless aquatic insect, *Abedus herberti* (Hemiptera: Belostomatidae). Southwest Nat 59:301–302
Boersma KS, Bogan MT, Henrichs BA, Lytle DA (2014) Invertebrate assemblages of pools in arid-land streams have high functional redundancy and are resistant to severe drying. Freshw Biol 59:491–501
Bogan MT, Boersma KS (2012) Aerial dispersal of aquatic invertebrates along and away from arid-land streams. Freshw Sci 31:1131–1144
Bohonak AJ (1999) Dispersal, gene flow, and population structure. Q Rev Biol 74:21–44
Bohonak AJ, Jenkins DG (2003) Ecological and evolutionary significance of dispersal by freshwater invertebrates. Ecol Lett 6:783–796
Bradshaw WE, Holzapfel C (1988) Drought and the organization of tree-hole mosquito communities. Oecologia 74:507–514
Briers, RA, Cariss, HM, Gee, JHR (2002) Dispersal of adult stoneflies (*Plecoptera*) from upland streams draining catchments with contrasting land-use. Arch.Hvdrobiol. 155: 627–644
Briers RA, Gee JHR, Cariss HM, Geoghegan (2004) Inter-population dispersal by adult stoneflies detected by stable isotope enrichment. Freshw Biol 49: 425–431
Brittain JE (1990) Life history strategies in Ephemeroptera and Plecoptera. In: Campbell IC (ed) Mayflies and stoneflies: life histories and biology. Springer, Dordrecht
Bunn ES, Hughes JM (1997) Dispersal and recruitment in streams: evidence from genetic studies. J N Am Benthol Soc 16:338–346
Byers GW (1969) Evolution of Wing Reduction in Crane Flies (Diptera: Tipulidae). Evolution 23: 346–354
Chapman HF, Hughes JM, Ritchie SA, Kay BH (2003) Population structure and dispersal of the freshwater mosquitoes *Culex annulirostris* and *Culex palpalis* (Diptera: Culicidae) in Papua New Guinea and northern Australia. J Med Entomol 40:165–169
Chapman JW, Reynolds DR, Smith AD, Smith ET, Woiwod IP (2004) An aerial netting study of insects migrating at high altitude over England. Bull Entomol Res 94:123–136
Chapman JW, Reynolds DR, Wilson K (2015) Long-range seasonal migration in insects: mechanisms, evolutionary drivers and ecological consequences. Ecol Lett 18:287–302
Chaput-Bardy A, Lemaire C, Picard D, Secondi J (2008) In-stream and overland dispersal across a river network influences gene flow in a freshwater insect, *Calopteryx splendens*. Mol Ecol 17:3496–3505
Chodorowski A (1969) The desiccation of ephemeral pools and the rate of development of *Aedes communis* larvae. Pol Arch Hydrobiol 16:79–91
Collier KJ, Smith BR (1998) Dispersal of adult caddisflies (Trichoptera) into forests alongside three New Zealand streams. Hydrobiologia 361:53–65
Compton SG (2002) Sailing with the wind: dispersal by small flying insects. In: Bullock JM, Kenward RE, Hails RS (eds) Dispersal ecology. Blackwell Science, Oxford, pp 113–133
Conrad KF, Willson KH, Whitfield K, Harvey IF, Thomas CJ, Sherratt YN (2002) Characteristics of dispersing Ischnura elegans and Coenagrion puella (Odonata): age, sex, size, morph and ectoparasitism. Ecography 25:439–445
Corbet PS (1957) The life-history of the emperor dragonfly Anax imperator leach (Odonata: Aeshnidae). J Anim Ecol 26:1–69
Corbet PS (1962) A biology of dragonflies. E.W. Classey, Faringdon, p 247
Corbet PS (1999) Dragonflies: behavior and ecology of Odonata. Cornell University Press, Ithaca, NY, p 829
Coutant CC (2008) Evidence for upstream dispersion of adult caddisflies (Trichoptera: Hydropsychidae) in the Columbia River. Aquat Insects 4:61–66

Csabai Z, Kálmán Z, Szivák I, Boda P (2012) Diel flight behaviour and dispersal patterns of aquatic Coleoptera and Heteroptera species with special emphasis on the importance of seasons. Naturwissenschaften 99: 751–765

Cullen MJ (1969) The biology of giant water bugs (Hemiptera: Belostomatidae) in Trinidad. Proc R Entomol Soc Lond A 44:123–136

Darwin C (1859) On the origin of species by means of natural selection. J. Murray, London

Datry T, Larned ST, Tockner K (2014) Intermittent rivers: a challenge for freshwater ecology. Bioscience 64: 229–235

Davy-Bowker J (2002) A mark and recapture study of water beetles (Coleoptera: Dytiscidae) in a group of semi-permanent and temporary ponds. Aquat Ecol 36:435–446

De Block M, Stoks R (2005) Pond drying and hatching date shape the tradeoff between age and size at emergence in a damselfly. Oikos 108:485–494

De Knijf G, Flenker U, Vanappelghem C, Manci CO, Kalkman VJ, Demolder H (2011) The status of two boreo-alpine species, *Somatochlora alpestris* and *S. arctica*, in Romania and their vulnerability to the impact of climate change (Odonata: Corduliidae) Int J Odonatology 14: 111–126

De Meester L, Gómez A, Okamura B, Schwenk K (2002) The monopolization hypothesis and the dispersal–gene flow paradox in aquatic organisms. Acta Oecol 23:121–135

Denno RF, Roderick GK (1990) Population biology of planthoppers. Annu Rev Entomol 35:489–520

Didham RK, Blakely TJ, Ewers RM, Hitchings TR, Ward JB, Winterbourn MJ (2012) Horizontal and vertical structuring in the dispersal of adult aquatic insects in a fragmented landscape. Fundam Appl Limnol 180: 27–40

Dingle H (1968) The influence of environment and heredity on flight activity in the milkweed bug *Oncopeltus*. J Exp Biol 48:175–184

Dingle H (1996) Migration: the biology of life on the move, 1st edn. Oxford University Press, Oxford, p 480

Dingle H (2014) Migration: the biology of life on the move, 2nd edn. Oxford University Press, Oxford, p 326

Dingle H, Arora G (1973) Experimental studies of migration in bugs of the genus *Dysdercus*. Oecologia 12:119–140

Ditrich T (2016) Migration and dispersal patterns of semiaquatic bugs from lentic and lotic freshwater habitats. In: Abstracts of the 25th International Congress of Entomology, Orlando, FL, pp 25–30

Ditrich T, Papáček M (2009) Correlated traits for dispersal pattern: terrestrial movement of the water cricket Velia caprai (Heteroptera: Gerromorpha: Veliidae). Eur J Entomol 106:551–555

Ditrich T, Papáček M, Broum T (2008) Spatial distribution of semiaquatic bugs (Heteroptera: Gerromorpha) and their wing morphs in a small scale of the Pohořský Potok stream spring area (Novohradské Hory Mts.). Silva Gabreta 14:173–178

Downes BJ, Reich P (2008) What is the spatial structure of stream insect populations? Dispersal behaviour at different life-history stages. In: Lancaster J, Briers RA (eds) Aquatic insects challenges to populations. CABI, Wallingford, pp 184–203

Dreyer J, Townsend PA, Hook JC III, Hoekman D, Vander Zanden MJ, Gratton C (2015) Quantifying aquatic insect deposition from lake to land. Ecology 96:499–509

Drummond LR, McIntosh AR, Larned ST (2015) Invertebrate community dynamics and insect emergence in response to pool drying in a temporary river Freshw Biol 60: 1596–1612

Dudley R, Byrnes G, Yanoviak SP, Borrell B, Brown RM, McGuire JA (2007) Gliding and the functional origins of flight: biomechanical novelty or necessity? Ann rev Ecol. Evol Syst 38:179–201

Dumont HJ, Desmet K (1990) Transsahara and transmediterranean migratory activity of *Hemianax ephippiger* (Burmeister) in 1988 and 1989 (Anisoptera: Aeshnidae). Odonatologica 19:181–185

Dumont HJ, Hinnekint BON (1973) Mass migration in dragonflies, especially in *Libellula quadrimaculata* L. a review, a new ecological approach and a new hypothesis. Odonatologica 2:1–20

Elbers ARW, Koenraadt CJM, Meiswinkel R (2015) Mosquitoes and *Culicoides* biting midges: vector range and the influence of climate change. Rev Sci Tech 4:123–137

Elias-Neto M, Belles X (2016) Tergal and pleural structures contribute to the formation of ectopic prothoracic wings in cockroaches. R Soc Open Sci 3:160347

Elliot JM (1971) Upstream movements of benthic invertebrates in a Lake District stream. J Anim Ecol 40:235–252

Epopa PS, Millogo AA, Collins CM, North A, Tripet F, Benedict MQ, Diabate A (2017) The use of sequential mark-release recapture experiments to estimate population size, survival and dispersal of male mosquitoes of the *Anopheles gambiae* complex in Bana, a west African humid savannah village. Parasit Vectors 10:376–380

Fagan WF (2002) Connectivity, fragmentation, and extinction risk in dendritic metapopulations. Ecology 83:3243–3249

Fagan WF, Siemann E, Mitter C, Denno RF, Huberty AF, Woods HA, Elser JJ (2002) Nitrogen in insects: implications for trophic complexity and species diversification. Am Nat 160:784–802

Figuerola J, Green AJ (2002) Dispersal of aquatic organisms by water birds: a review of past research and priorities for future studies. Freshw Biol 47:483–494

Finn DS, Blouin MS, Lytle DA (2007) Population genetic structure reveals terrestrial affinities for a headwater stream insect. Freshw Biol 52:1891–1897

Fisher SG, Likens GE (1973) Energy flow in bear brook, New Hampshire: an integrative approach to stream ecosystem metabolism. Ecology 43:421–439

Forbes SA (1887) The lake as a microcosm. Bull Peoria Sci Assoc:77–87

Foster WA, Treherne JE (1978) Dispersal mechanisms in an intertidal aphid. J Anim Ecol 47:205–217

Freeland JR, May M, Lodge R, Conrad KF (2003) Genetic diversity and widespread haplotypes in a migratory dragonfly, the common green darner *Anax junius*. Ecol Entomol 28:413–421

Fukui D, Murakami M, Nakano S, Aoi T (2006) Effect of emergent aquatic insects on bat foraging in a riparian forest. J Anim Ecol 75:1252–1258

Galatowitsch ML, Batzer DP (2011) Benefits and costs of Leptophlebia (Ephemeroptera) mayfly movements between river channels and floodplain wetlands. Can J Zool 89:714–723

Garcia R, Hagen KS (1987) Summer dormancy in adult *Agabus disintegratus* (Crotch) (Coleoptera: Dytiscidae) in dried ponds in California. Ann Entomol Soc Am 80:267–271

Geenen S, Jordaens K, De Block MJ (2000) Genetic differentiation and dispersal among populations of the damselfly *Lestes viridis* (Odonata). J N Am Benthol Soc 19:321–328

Giudicelli J, Botosaneanu L (1999) The remarkable case of a water striding, mainly brachypterous new species of Rhyacophila Pictet (Trichoptera) from the River Sorgue (southern France). Ann Soc Entomol France 35:265–273

Goehring L, Oberhauser KS (2002) Effects of photoperiod, temperature, and host plant age on induction of reproductive diapause and development time in *Danaus plexippus*. Ecol Entomol 27:674–685. https://doi.org/10.1046/j.1365-2311.2002.00454.x

Graham SE, Storey R, Smith B (2017) Dispersal distances of aquatic insects: upstream crawling by benthic EPT larvae and flight of adult Trichoptera along valley floors. N Z J Mar Freshwater Res 51:146–164

Grewe Y, Hof C, Dehling DM, Brandl R, Brändle M (2013) Recent range shifts of European dragonflies provide support for an inverse relationship between habitat predictability and dispersal. Glob Ecol Biogeogr 22:403–409

Griffith MB, Barrows EM, Perry SA (1998) Lateral dispersal of adult aquatic insects (Plecoptera, Trichoptera) following emergence from headwater streams in forested Appalachian catchments. Ann Entomol Soc Am 91:195–201

Hanski I (1998) Metapopulation dynamics. Nature 396:41–49

Hanski I (2001) Spatially realistic theory of metapopulation ecology. Naturwissenschaften 88:372–381

Harada T, Nishimoto T (2007) Feeding conditions modify the photoperiodically induced dispersal of the water strider, Aquarius paludum (Heteroptera: Gerridae). Eur J Entomol 104:33–37. https://doi.org/10.14411/eje.2007.005
Harrison RG (1980) Dispersal polymorphisms in insects. Annu Rev Ecol Syst 11:95–118
Hassall C, Thompson DJ (2008) The impacts of environmental warming on Odonata: a review. Int J Odonatol 11:131–153
Heino J, Mykrä H (2008) Control of stream insect assemblages: roles of spatial configuration and local environmental factors. Ecol Entomol 33:614–622
Hering D, Schmidt-Kloiber A, Murphy JF, Wolfram G (2009) Potential impact of climate change on aquatic insects: a sensitivity analysis for European caddisflies (Trichoptera) based on distribution patterns and ecological preferences. Aquat Sci 71(1):3–14
Hernandez SA, Pecarsky BL (2014) Do stream mayflies exhibit trade-offs between food acquisition and predator avoidance behaviors? Freshw Sci 33:124–133
Hershey AE, Pastor J, Peterson BJ, Kling GW (1993) Stable isotopes resolve the drift paradox for *Baetis* mayflies in an Arctic river. Ecology 74:2315–2325
Hickling R, Roy DB, Hill JK, Fox R, Thomas CD (2006) The distributions of a wide range of taxonomic groups are expanding polewards. Glob Chang Biol 12:1–6
Hobson KA, Anderson RC, Soto DX, Wassenaar LI (2012) Isotopic evidence that dragonflies (*Pantala flavescens*) migrating through the Maldives come from the Northern Indian Subcontinent. PLoS One 7:e52594
Hof C, Brändle M, Brandl R (2008) Latitudinal variation of diversity in European freshwater animals is not concordant across habitat types. Glob Ecol Biogeogr 17:539–546
Hof C, Brändle M, Dehling DM, Mungúıa M, Brandl R, Araujo MB, Rahbek C (2012) Biol Lett 8:639–643
Horváth G (1995a) How do water insects find their aquatic habitat? Természet Világa 125:44–49
Horváth G (1995b) Reflection-polarization patterns at flat water surfaces and their relevance for insect polarization vision. J Theoret Biol 175:27–37
Horváth G, Kriska G (2008) Polarization vision in aquatic insects and ecological traps for polarotactic insects. In: Lancaster J, Briers RA (eds) Aquatic insects: challenges to populations. CABI, Wallingford, pp 204–229
Horváth G, Malik P, Kriska G, Wildermuth H (2007) Ecological traps for dragonflies in a cemetery: the attraction of Sympetrum species (Odonata: Libellulidae) by horizontally polarizing black gravestones. Freshw Biol 52:1700–1709
Hu G, Lim KS, Horvitz N, Reynolds DR CSJ, Sapir N, Chapman JW (2016) Mass seasonal bioflows of high-flying insect migrants. Science 354:1584–1587
Hubbell SP (2001) The unified neutral theory of biodiversity and biogeography. Princeton University Press, Princeton, NJ
Hughes JM, Schmidt DJ, Finn DS (2009) Genes in streams: using DNA to understand the movement of freshwater fauna and their riverine habitat. BioScience 59:573–583
Hughes JM, Huey JA, McLean AJ, Baggiano O (2011) Aquatic insects in eastern Australia: a window on ecology and evolution of dispersal in stream. Insects 2:447–461
Humphries S, Ruxton GD (2002) Is there really a drift paradox? J Anim Ecol 71:151–154
IPCC (2018) Special Report on Global Warming of 1.5°C. http://www.ipcc.ch/report/sr15/. Accessed 8 Oct 2010
Iversen LL, Rannap R, Briggs L, Sand-Jensen K (2017) Timerestricted flight ability influences dispersal and colonization rates in a group of freshwater beetles. Ecol Evol 7:824–830
Jackson JK, Fisher SG (1986) Secondary production, emergence, and export of aquatic insects of a Sonoran Desert stream. Ecology 67:629–638
Jannot JE (2009) Life history plasticity and fitness in a caddisfly in response to proximate cues of pond-drying. Oecologia 161:267–277
Johansson F, Rowe L (1999) Life history and behavioral responses to time constraints in a damselfly. Ecology 80:1242–1252
Johnson CG (1969) Migration and dispersal of insects by flight. Methuen, London, 766 pp

Jones CJ, Isard SA, Cortinas MR (1999) Dispersal of synanthropic Diptera: lessons from the past and technology for the future. Ann Entomol Soc Am 92:829–839

Kennedy JS (1985) Migration: behavioral and ecological. In Rankin MA, ed., Migration: mechanisms and adaptive significance. Contrib Mar Sci 27(Suppl):5–26

Klečka J (2008) The structure and dynamics of a water beetle community in a semipermanent wetland (Vrbenské rybníky Nature Reserve, South Bohemia). BSc. Thesis, Faculty of Science, University of South Bohemia, České Budějovice, Czech Republic, p 63

Kraus JM, Vonesh JR (2010) Feedbacks between community assembly and habitat selection shape variation in local colonization. J Anim Ecol 79:795–802

Krosch MN, Baker AM, Mather PB, Cranston PS (2011) Spatial population genetic structure reveals strong natal site fidelity in Echinocladius martini (Diptera: Chironomidae) in northeast Queensland, Australia. Freshw Biol 56:1328–1341

Kukalová-Peck J (1978) Origin and evolution of insect wings and their relation to metamorphosis, as documented by the fossil record. J Morphol 156:53–125

Kukalová-Peck J (1983) Origin of the insect wing and wing articulation from the arthropodan leg. Can J Zool 61:1618–1669

Landin J (1980) Habitats, life histories, migration and dispersal by flight of two water-beetles Helophorus brevipalpis and H. strigifrons (Hydrophilidae). Ecography 3:190–201

Laurance WF, Useche DC, Shoo LP, Herzog SK, Kessler M, Escobar F, Brehm G, Axmacher JC, Chen I-C, Arellano Gamez L, Hietz P, Fiedler K, Pyrcz T, Wolf J, Merkord CL, Cardelus C, Marshall AR, Ah-Peng C, Aplet GH, del Coro Arizmendi M, Baker WJ, Barone J, Bruehl CA, Bussmann RW, Cicuzza D, Eilu G, Favila ME, Hemp A, Hemp C, Homeier J, Hurtado J, Jankowski J, Kattan G, Kluge J, Kroemer T, Lees DC, Lehner M, Longino JT, Lovett J, Martin PH, Patterson BD, Pearson RG, Peh KS-H, Richardson B, Richardson M, Samways MJ, Senbeta F, Smith TB, Utteridge TMA, Watkins JE, Wilson R, Williams SE, Thomas CD (2011) Global warming, elevational ranges and the vulnerability of tropical biota. Biol Conserv 144:548–557

Levins R (1969) Some demographic and genetic consequences of environmental heterogeneity for biological control. Bull Entomol Soc Am 15:237–240

Lillehammer A (1985) Studies in short-wingedness in stoneflies (Plecoptera). Fauna Norvegica Ser B 32:58–61

Linz DM, Tomoyasu Y (2018) Dual evolutionary origin of insect wings supported by an investigation of the abdominal wing serial homologs in *Tribolium*. Proc Natl Acad Sci U S A 115:E658–E667

Lund JO, Wissinger SA, Peckarsky BL (2016) Caddisfly behavioral responses to drying cues in temporary ponds: implications for effects of climate change. Freshwater Sci 35(2):619–630

Lytle DA (1999) Use of rainfall cues by *Abedus herberti* (Hemiptera: Belostomatidae): a mechanism for avoiding flash floods. J Insect Behav 12:1–12

Lytle DA (2001) Disturbance regimes and life history evolution. Am Nat 157:525–536

Lytle DA (2002) Flash floods and aquatic insect life-history evolution: evaluation of multiple models. Ecology 83:370–385

Lytle DA (2008) Life-history and behavioural adaptations to flow regime in aquatic insects. In: Rankin J, Briers RA (eds) Aquatic insects: challenges to populations. CAB International, Wallingford, pp 122–138

Lytle DA, Smith RL (2004) Exaptation and flash flood escape in the giant water bugs. J Insect Behav 17:169–178

Lytle DA, Olden JD, McMullen LE (2008) Drought-escape behaviors of aquatic insects may be adaptations to highly variable flow regimes characteristic of desert rivers. Southwest Nat 53:399–402

MacArthur RH, Wilson EO (1967) The theory of island biogeography. Princeton University Press, Princeton, NJ

Macneale KH, Peckarsky BL, Likens GE (2004) Contradictory results from different methods for measuring direction of insect flight. Freshw Biol 49:1260–1268

Malmqvist B (2000) How does wing length relate to distribution patterns of stoneflies (Plecoptera) and mayflies (Ephemeroptera)? Biol Conserv 93:271–276
Marden JH, Kramer MG (1994) Surface-skimming stoneflies: a possible intermediate stage in insect flight evolution. Science 266:427–430
Marten A, Brändle M, Brandl R (2006) Habitat type predicts genetic population differentiation in freshwater invertebrates. Mol Ecol 15:2643–2651
Matthews JH (2007a) Large-scale dragonfly migration by *Anax junius* (Odonata: Aeshnidae) results in genetic homogeneity despite temporally and spatially discrete populations. In: Research in motion: patterns of large-scale migration in dragonflies and birds. PhD Dissertation, University of Texas, pp 33–61
Matthews JH (2007b) Phenotypic plasticity determines migration strategy across large spatial scales. In Research in motion: patterns of large-scale migration in dragonflies and birds. PhD Dissertation, University of Texas, pp 62–91
May ML (2013) A critical overview of progress in studies of migration of dragonflies (Odonata: Anisoptera), with emphasis on North America. J Insect Conserv 17:1–15. https://doi.org/10.1007/s10841-012-9540-x
May ML, Matthews JH (2008) Migration in Odonata: an overview with special focus on Anax junius. In: Córdoba-Aguilar A (ed) Dragonflies, Model organisms for ecological and evolutionary research. Oxford University Press, Oxford, pp 63–77
May ML, Gregoire JS, Gregoire SM, Aliberti Lubertazzi M, Matthews JH (2017) Emergence phenology, uncertainty, and the evolution of migratory behavior in *Anax junius* (Odonata: Aeshnidae). PLoS One 12:e0183508. https://doi.org/10.1371/journal.pone.0183508
Mazzucco R, Nguyen TV, Kim D-H, Chon T-S , Dieckmann (2015) Adaptation of aquatic insects to the current flow in streams. Ecol Model 309–310:143–152
McCauley SJ (2005) Differential dispersal propensities between individualsin male *Leucorrhinia intacta* (Odonata: Libellulidae). Int J Odonatology 8:223–232
McCauley SJ (2006) The effects of dispersal and recruitment limitation on community structure of odonates in artificial ponds. Ecography 29:585–595
McCauley SJ (2007) The role of local and regional processes in structuring larval dragonfly distributions across habitat gradients. Oikos 116:121–133
McCauley SJ (2010) Body size and social dominance influence breeding dispersal in male *Pachydiplax longipennis* (Odonata). Ecol Entomol 35:377–385
McCauley SJ (2013) Relationship between morphology, dispersal, and habitat distribution in threes species of Libellula (Odonata, Anisoptera). Aquat Insects 35:195–204
McCauley SJ, Davis J, Werner EE, Robeson MS (2014) Dispersal, niche breadth and population extinction colonization ratios predict range size in North American dragonflies. J Anim Ecol 83: 858–865
McCauley SJ, Rowe L (2010) *Notonecta* exhibit threat-sensitive, predator-induced dispersal. Biol Lett 6:449–452
McLachlan A (1985) The relationship between habitat predictability and wing length in midges (Chironomidae). Oikos 44:391–397
McPeek MA (1989) Differential dispersal tendencies among *Enallagm*a damselflies (Odonata) inhabiting different habitats. Oikos 56:187–195
Medeiros MCI, Boothe EC, Roarke EB, Hamer GL (2017) Dispersal of male and female *Culex quinquefasciatus* and *Aedes albopictus* mosquitoes using stable isotope enrichment. PLoS Negl Trop Dis 11:e0005347
Michiels NK, Dhondt AA (1991) Characteristics of dispersal in sexually mature dragonflies. Ecol Entomol 16:449–459
Miller MP, Blinn DW, Keim P (2002) Correlations between observed dispersal capabilities and patterns of genetic differentiation in populations of four aquatic insect species from the Arizona White Mountains, USA. Freshw Biol 47:1660–1673
Mokany A, Mokany K (2006) Effects of habitat permanence cues on larval abundance of two mosquito species. Hydrobiologia 563:269–276

Mugnai R, Filho AS, Nessimian JL, Kury AB (2014) Spatial distribution of hyporheic chironomid larvae in a neotropical stream, Atlantic forest, Rio de Janeiro, Brazil. J Insect Sci Photon 115:195–206
Müller K (1954) Investigations on the organic drift in North Swedish streams. Inst Freshwater Res Drottningholm Rep 52:133–148
Müller K (1982) The colonization cycle of freshwater insects. Oecologia 52:202–207
Nicoletti FJ (1997) American Kestrel and Merlin migration correlated with Green Darner movements at Hawk Ridge. Loon 68:216–221
Ohba S, Takagi H (2005) Food shortage affects flight migration of the giant water bug *Lethocerus deyrollei* in the prewintering season. Limnology 6:85–90
Pachepsky E, Lutscher F, Nisbet RM, Lewis MA (2005) Persistence, spread and the drift paradox. Theor Pop Biol 67:61–73
Paetzold A, Bernet JF, Tockner K (2006) Consumer-specific responses to riverine subsidy pulses in a riparian arthropod assemblage. Freshw Biol 51:1103–1115
Paulson DR (2011) Dragonflies of the East. Princeton University Press. 544 pp
Peckarsky BL, McIntosh AR (1998) Fitness and community consequences of avoiding multiple predators. Oecologia 113:565–576
Peckarsky BL, Kerans BL, Taylor BW, McIntosh AR (2008) Predator effects on prey population dynamics in open systems. Oecologia 156:431–440
Peckarsky BL, Encalada AC, AR MI (2011) Why do vulnerable mayflies thrive in trout streams? Am Entomol 57:152–164
Peters RH (1986) The ecological implications of body size. Cambridge University Press, Cambridge, p 329
Petersen I, Winterbottom JH, Orton S, Friberg N, Hildrew AG, Spiers DC, Gurney WSC (1999) Emergence and lateral dispersal of adult Plecoptera and Trichoptera from Broadstone Stream, UK. Freshw Biol 42:401–416
Pfeiler E, Markow TA (2017) Population connectivity and genetic diversity in long distance migrating insects: divergent patterns in representative butterflies and dragonflies. Biol J Linnean Soc 20:1–8
Phillipsen IC, Kirk EH, Bogan MT, Mims MC, Olden JD, Lytle DA (2015) Dispersal ability and habitat requirements determine landscape-level genetic patterns in desert aquatic insects. Mol Ecol 24:54–69
Pintar MR, Resetarits WJ (2017a) Context-dependent colonization dynamics: regional reward contagion drives local compression in aquatic beetles. J Anim Ecol 86:1124–1135
Pintar MR, Resetarits WJ (2017b) Prey-driven control of predator assemblages: zooplankton abundance drives aquatic beetle colonization. Ecology 98:2201–2215
Pintar MR, Resetarits WJ (2017c) Tree leaf litter composition drives temporal variation in aquatic beetle colonization and assemblage structure in lentic systems. Oecologia 183:797–807
Pintar MR, Bohenek JR, Eveland LL, Resetarits WJ (2018) Colonization across gradients of risk and reward: nutrients and predators generate species specific responses among aquatic insects. Funct Ecol 32(6):1589–1598. https://www.matthewpintar.net/publications.html. Accessed 22 May 2018. https://doi.org/10.1111/1365-2435.13086
Polis GA, Anderson WB, Holt RD (1997) Toward an integration of landscape ecology and food web ecology: the dynamics of spatially subsidized food webs. Annu Rev Ecol Syst 28:289–316
Rainey RC (1951) Weather and the movement of locust swarms: a new hypothesis. Nature 168:1057–1060
Rainey RC (1976) Flight behaviour and aspects of the atmospheric environment. In: Rainey RC (ed) Insect flight. Blackwell Scientific Publishers, Oxford, pp 75–112
Rankin MA (1978) Hormonal control of insect migratory behavior. In: Dingle H (ed) Evolution of insect migration and diapause, Proceedings in life sciences. Springer, New York, NY, pp 5–32
Reppert SM, Guerra PA, Merlin C (2016) Neurobiology of monarch butterfly migration. Annu Rev Entomol 61:25–42

Resetarits WJ, Binkley CA (2009) Spatial contagion of predation risk affects colonization dynamics in experimental aquatic landscapes. Ecology 90: 869–876
Resetarits WJ, Binckley CA (2013) Patch quality and context, but not patch number, drive multiscale colonization dynamics in experimental aquatic landscapes. Oecologia 173:933–946
Resetarits WJ, Binckley CA, Chalcraft DR (2005) Habitat selection, species interactions, and processes of community assembly in complex landscapes: a metacommunity perspective. In: Holyoak M, Leibold MA, Holt RD (eds) Metacommunities: spatial dynamics and ecological communities. University of Chicago Press, Chicago, IL, pp 374–398
Resetarits WJ, Pintar MR (2016) Functional diversity of non-lethal effects, chemical camouflage, and variation in fish avoidance in colonizing beetles. Ecology 97:3517–3529
Resetarits WJ, Silberbush A (2016) Local contagion and regional compression: habitat selection drives spatially explicit, multiscale dynamics of colonisation in experimental metacommunities. Ecol Lett 19:191–200
Reynolds AM, Reynolds AR, Chapman JR (2014) Non-volant modes of migration in terrestrial arthropods. Anim Migr 2:8–28
Ribera I, Foster GN, Vogler AP (2003) Does habitat use explain large scale species richness patterns of aquatic beetles in Europe? Ecography 26:145–152
Ribera I, Vogler AP (2000) Habitat type as a determinant of species range sizes: the example of lotic-lentic differences in aquatic Coleoptera. Biol J Linn Soc 71:33–52
Roff DA (1990) The evolution of flightlessness in insects. Ecol Monogr 60:389–421
Roff DA (1991) Life history consequences of bioenergetic and biomechanical constraints on migration. Am Zool 31:243–251
Rowe RJ (1978) *Ischnura aurora* (Brauer), a dragonfly with unusual mating behaviour (Zygoptera: Coenagrionidae). Odonatologica 7:375–383
Ruffieux L, Elouard J-L, Sartori M (1998) Flightlessness in mayflies and its relevance to hypotheses on the origin of insect flight. Proc R Acad Lond B 265:2135–2140
Rundle SD, Bilton DT, Abbott JC, Foggo A (2007) Range size in North American Enallagma damselflies correlates with wing size. Freshw Biol 52:471–477
Russell RW, May ML, Soltesz KL, Fitzpatrick JW (1998) Massive swarm migrations of dragonflies (Odonata) in eastern North America. Am Midl Nat 140:325–342
Saijo H (2001) Seasonal prevalence and migration of aquatic insects in paddies and an irrigation pond in Shimane Prefecture. Jpn J Ecol 51:1–11
Samraoui B, Bouzid S, Boulahbal R, Corbet PS (1998) Postponed reproductive maturation in upland refuges maintains life-cycle continuity during the hot, dry season in Algerian dragonflies (Anisoptera). Int J Odonatol 1(2):119–135
Saunders DS (2010) Photoperiodism in insects: migration and diapause responses. In: Nelson RJ et al (eds) Photoperiodism: the biological calendar. Oxford University Press, Oxford, pp 218–258
Schäfer ML, Lundström JO (2006) Different responses of two floodwater mosquito species, Aedes vexans and Ochlerotatus sticticus (Diptera: Culicidae), to larval habitat drying. J Vector Ecol 31:123–128
Schindler DE, Smits AP (2016) Subsidies of aquatic resources in terrestrial ecosystems. Ecosystems 20:78–93
Schmidt SK, Hughes JM, Bunn SE (1995) Gene flow among conspecific populations of *Baetis* sp. (Ephemeroptera): adult flight and larval drift. J N Am Benthol Soc 14:147–157
Service MW (1980) Effects of wind on the behaviour and distribution of mosquitoes and blackflies. Int J Biometeorol 24:347–353
Service MW (1993) Mosquito ecology field sampling methods. Elsevier, London, p 988
Shah RDT, Shah DN, Domisch S (2012) Range shifts of a relict Himalayan dragonfly in the Hindu Kush Himalayan region under climate change scenarios. Int J Odonatol 15:209–222
Shlizerman E, Phillips-Portillo J, Forger DB, Reppert SM (2016) Neural integration underlying a time-compensated sun compass in the migratory monarch butterfly. Cell Rep 15:683–69

Short AEZ, Caterino MS (2009) On the validity of habitat as a predictor of genetic structure in aquatic systems: a comparative study using California water beetles. Mol Ecol 18:403–414
Short AEZ, Liebherr JK (2007) Systematics and biology of the endemic water scavenger beetles of Hawaii (Coleoptera, Hydrophilidae, Hydrophilini). Syst Entomol 32:601–624
Siepielski AM, Hung K-L, Bein EEB, McPeek MA (2010) Experimental evidence for neutral community dynamics governing an insect assemblage. Ecology 91:847–857
Smith RL (1973) Aspects of the biology of three species of the genus *Rhantus* (Coleoptera: Dytiscidae) with special reference to the acoustical behavior of two. Can Entomol 105:909–919
Snodgrass RE (1935) Principles of insect morphology. McGraw-Hill, New York, p 667
Southwood TRE (1962) Migration of terrestrial arthropods in relation to habitat. Biol Rev 37:171–214
Stav G, Blaustein L, Margalit J (1999) Experimental evidence for predation sensitive oviposition by a mosquito, *Culiseta longiareolata*. Ecol Entomol 24:202–207
Stefanescu C, Páramo F, Åkesson S et al (2012) Multi-generational long-distance migration of insects: studying the painted lady butterfly in the Western Palaearctic. Ecography 35:1–14
Stevens LE, Polhemus JT, Durfee RS, Olson CA (2007) Large mixed-species dispersal flights of predatory and scavenging aquatic Heteroptera and Coleoptera, northern Arizona, USA. West North Am Nat 67:587–592
Stubbington R (2012) The hyporheic zone as an invertebrate refuge: a review of variability in space, time, taxa and behaviour. Mar Freshw Res 63:293–311
Suarez-Tovar CM, Sarmiento CE (2016) Beyond the wing planform: morphological differentiation between migratory and nonmigratory dragonfly species. J Evol Biol 29:690–703
Svensson, BW (1974) Population movements of adult Trichoptera in a south Swedish stream. Oikos 25:175–175
Taylor LR (1958) Aphid dispersal and diurnal periodicity. Proc Linn Soc Lond 169:67–73
Thompson DJ (1991) Size-biased dispersal prior to breeding in a damselfly: conflicting evidence from a natural population. Oecologia 87:600–601
Troast D, Suhling F, Jinguji H, Sahlén G, Ware J (2016) A global population genetic study of *Pantala flavescens*. PLoS One 11:e0148949. https://doi.org/10.1371/journal.pone.0148949
Trottier R (1971) Effect of temperature on the life-cycle of *Anax junius* (Odonata: Aeshnidae) in Canada. Can Entomol 103:1671–1683
Trueman JWH (1990) Evolution of insect wings a leg exite plus endite model. Can J Zool 68:1333–1335
Ueda T (1988) Diversity in the life history of the dragonfly Sympetrum frequens (Odonata: Insecta). Bull Ishikawa Agric Coll 18:98–110
Velasco J, Millan A (1998) Insect dispersal in a drying desert stream: effects of temperature and water loss. Southwest Nat 43:80–87
Vepsäläinen K (1978) Wing dimorphism and diapause in *Gerris*: determination and adaptive significance. In: Dingle H (ed) Evolution of insect migration and diapause, Proceedings in Life Sciences. Springer, New York, NY, pp 218–253
Vonesh JR, Kraus JM, Rosenberg JS, Chase JM (2009) Predator effects on aquatic community assembly: disentangling the roles of habitat selection and post-colonization processes. Oikos 118:1219–1229
Washburn JO, Washburn L (1984) Active aerial dispersal of minute wingless arthropods: exploitation of boundary-layer velocity gradients. Science 223:1088–1089
Waters TF (1972) The drift of stream insects. Annu Rev Entomol 17:253–272
Wikelski M, Moskowitz D, Adelman J, Cochran J, Wilcove DS, May ML (2006) Simple rules guide dragonfly migration. Biol Lett 2:325–329
Wikipedia (2018a) Lake Baikal. https://en.wikipedia.org/wiki/Lake_Baikal. Accessed 6 June 2018
Wikipedia (2018b) List of rivers by age. https://en.wikipedia.org/wiki/List_of_rivers_by_age. Accessed 6 June 2018
Wilcock HR, Bruford MW, Nichols RA., Hildrew AG (2007) Landscape, habitat characteristics and the genetic population structure of two caddisflies. Freshw Biol 52:1907–1929

Wilcock HR., Nichols RA, Hildrew AG (2003) Genetic population structure and neighbourhood population size estimates of the caddisfly Plectrocnemia conspersa. 48: 1813–1824
Wildermuth H, Horváth G (2005) Visual deception of a male *Libellula depressa* by the shiny surface of a parked car (Odonata: Libellulidae). Int J Odonatol 8:97–105
Williams DD (1981) Migrations and distribution of stream benthos. In: Lock MA, Williams DD (eds) Perspectives in running water ecology. Plenum, New York, pp 155–207
Williams DD (1996) Environmental constraints in temporary fresh waters and their consequences for the insect fauna. J N Am Benthol Soc 15:634–650
Williams DD, Williams NE (1993) The upstream/downstream movement paradox of lotic invertebrates: quantitative evidence from a Welsh mountain stream Freshw Biol 30:199–218
Wilson DS (1992) Complex interactions in metacommunities, with implications for biodiversity and higher levels of selection. Ecology 73:1984–2000
Winterbourn MJ, Chadderton WL, Entrekin SA, Tank JL, Harding JS (2007) Distribution and dispersal of adult stream insects in a heterogeneous montane environment. Archiv für Hydrobiologie 168: 127 135
Wooster D, Sih A (1995) A review of the drift and activity responses of stream prey to predator presence. Oikos 73:3–8
Wright S (1951) The genetical structure of populations. Ann Eugen 15:323–354
Yee DA, Taylor S, Vamosi SM (2009) Beetle and plant density as cues initiating dispersal in two species of adult predaceous diving beetles. Oecologia 160:25–36
Young EC (1965) Flight muscle polymorphism in British Corixidae: ecological observations. J Anim Ecol 34:353–390
Zera AJ (1984) Differences in survivorship, development rate and fertility between long-winged and wingless morphs of the water strider, *Limnoporus canaliculatus*. Evolution 38:1023–1032
Zera AJ, Denno RF (1997) Physiology and ecology of dispersal polymorphism in insects. Annu Rev Entomol 42:207–230
Zwick P (2000) Phylogenetic system and zoogeography of the Plecoptera. Annu Rev Entomol 45:709–746

# Chapter 4
# Effects of Abiotic Factors and Ecogeographic Patterns on the Ecology, Distribution, and Behavior of Aquatic Insects

**Vinicius Marques Lopez, Aurélio Fajar Tonetto, Ricardo Cardoso Leite, and Rhainer Guillermo**

**Abstract** Aquatic insects are inserted in a physical and chemical world with a wide range of challenges that selected a myriad of adaptations and strategies throughout their evolutionary history. Mayflies, stoneflies, dragonflies, and many other insects inhabit both still and running waters, freshwater, and brackish waters; hence, they must deal with temperature regimes, variations on habitat complexity, and water chemical composition on a daily basis. All these environmental features determine not only how aquatic insects behave and occupy microhabitats in a stream or pond, but also define species distribution at macroscale. Here, in this chapter we attempt to show how this fascinating water world influences the lives of aquatic insects and their distribution across space.

**Keywords** Environment · Habitat · Microhabitat · Heterogeneity · Complexity · Hydraulics · Temperature

V. M. Lopez
Universidade de São Paulo, Ribeirão Preto, Brazil
e-mail: viniciuslopez@usp.br

A. F. Tonetto
Universidade Paulista, Jundiaí, Brazil

R. C. Leite
Instituto Federal de São Paulo, Piracicaba, Brazil

R. Guillermo (✉)
Federal University of São Carlos, UFSCar, São Carlos, São Paulo, Brazil



K. Del-Claro, R. Guillermo (eds.), *Aquatic Insects*,
https://doi.org/10.1007/978-3-030-16327-3_4

## 4.1 Introduction

Freshwater environments are highly heterogeneous and widely distributed across the land. Water bodies are characterized by not only extremely variable chemical composition, but also physical components. For instance, rocks, sand grains, plants, and other types of substrates altogether make part of the physical environment that insects must interact with in freshwater systems. Hydrodynamics, habitat complexity, and visual background are then strong forces that aquatic insects are well adapted to. Here, we attempt to show how abiotic factors may play a remarkable role on the ecology, behavior, and distribution of aquatic insects. First, we present a brief revision of ecogeographic patterns that encompass a variety of climatic variables to predict species occurrence and distribution at macroscales. Further, we explore how water temperature, chemical composition, altitude, hydrodynamics, and other factors affect behavioral and physiological adaptations of aquatic insects at meso and microscales. To better illustrate the role of such abiotic factors, we present a selection of examples that we hope to instigate interest in this fascinating aquatic world of insects.

### *4.1.1 Ecogeographic Patterns and World Climate*

Over the last few centuries, one of the major goals in biological sciences is to understand macroecological patterns. In this sense, scientists established ecogeographic rules, which attempt to link phenotypic variations in populations/communities to climatic and biogeographic gradients. For example, several factors can drive the variations in body coloration exhibited by taxa of the same clade such as temperature, humidity, latitude, and altitude (Chown and Gaston 2010; Cooper 2010; Shelomi 2012; Zeuss et al. 2014). Body size and coloration are usually specific traits that influence aquatic insects' behavior and ecology, predicting sexual selection, predation regimes, and population dynamics. Therefore, ecogeographic rules are crucial for the comprehension of how climatic and geographical features influence aquatic insects' ecology and behavior, habitat occupancy, niches, and species distribution.

There are many ecogeographic rules (Gaston et al. 2008), however, here we propose a brief review regarding aquatic insects. It is important that readers know that biological rules do not apply to all groups of organisms. Biological rules have been proposed by noted biologists such as Lamarck, Humboldt, Darwin and Haeckel, and they must consider the probabilistic factor of the question addressed. This exemplifies the difficulty of creating universal laws in the biological sciences, because of the probabilistic and causal characteristic of biological phenomena, being one of the primordial differences of Biology in relation to Exact Sciences (Mayr 1996).

#### 4.1.1.1 Thermal Melanism Hypothesis

The thermal melanism hypothesis (TMH) usually applies to ectothermic animals and predicts that individuals with darker coat or tegument are more effective in converting solar radiation to heat (Watt 1968). Thus, darker animals would have greater thermoregulatory advantages in cold environments, because they may become active before lighter morphotypes. For example, in some groups of European dragonflies, the predominant selective pressure for the evolution of body coloration is thermoregulation (Fig. 4.1, Zeuss et al. 2014).

Comparing the North American dragonfly assemblies with the European ones, Pinkert et al. (2017) have reported that darker colorations occur in dragonflies from regions with cold climates, and species with brighter body coloration are more common in regions with warmer climates. These studies corroborate others that show the relationship between coloration and distribution in odonates on all continents (see Hassall and Thompson 2008). Although data suggest a general pattern, supporting the TMH as a central predictor of coloration, idiosyncratic features in the

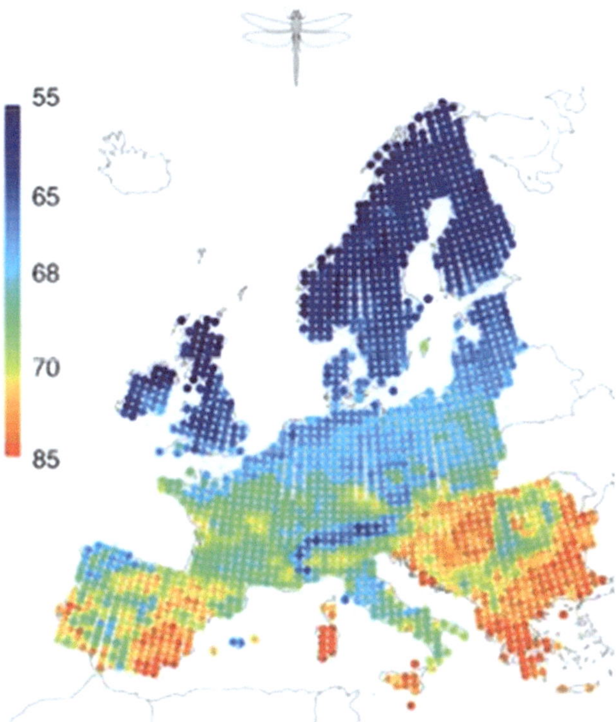


**Fig. 4.1** Average color values across species on a scale from 0 (black) to 255 (white). Red indicates light-colored Odonata assemblages and blue indicates dark-colored assemblages ($n = 1845$). There is a decrease from assemblages dominated by light-colored species in the south to assemblages dominated by dark-colored species in the north. Adapted from Zeuss et al. (2014)

evolutionary history of taxonomic clades, such as sexual selection, resistance to diseases, or even pleiotropic genes, may be the main drivers for these phenotypes (Roulin 2016). Studies on TMH could have implications for predicting the effects of climatic change and can assist in species conservation, since environmental changes can alter species distribution and assembly structure (Deutsch et al. 2008; Dillon et al. 2010; Pateman et al. 2012; Zeuss et al. 2014).

#### 4.1.1.2 Bergmann's Rule and the Temperature–Size Rule

Bergmann (1847) described a relationship between latitude and body size and proposed an intraspecific phenomenon where larger endothermic species can be found at higher latitudes (Bergmann 1847; Mayr 1956). However, some changes have been proposed to describe the observed patterns (mainly in ectotherms). For example, there is a strong positive correlation between latitude and body size in the damselfly *Erythromma viridulum* Charpentier (Odonata: Coenagrionidae), which is consistent with Bergmann's rule (Hassall et al. 2014).

Body size is a very important feature indeed, influencing a wide variety of important parameters to organisms (Roff et al. 2006; De Biee et al. 2012). Although this characteristic can be influenced by a variety of factors (e.g., genetic and environmental; see also Shelomi 2012), there may be a significant relationship between temperature and body size in aquatic arthropods—the "temperature–size rule" (TSR) (Horne et al. 2015). The TSR proposes that individuals that developed at colder temperatures exhibit larger body sizes than individuals that developed at warmer temperatures (Atkinson 1994). The TSR is diffused among ectotherms (Horne et al. 2015) and is often confused with Bergmann's rule, because there is a strong correlation between latitude and temperature.

The aquatic taxa may have a stronger negative TS response than terrestrial taxa, and they follow the "size-temperature" rule in over 90% of cases. In Diptera, species with aquatic larval phases had a TS negative response more expressive than other taxa. In *Calopteryx maculata* (Palisot de Beauvois) (Odonata: Calopterygidae), the TSR also explains body size range in part of the species distribution (Hassall 2013). Forster et al. (2012), in a meta-analysis, proposed that oxygen supply plays a central role in explaining the "size-temperature" responses in ectotherms, with the response of aquatic organisms being ten times higher than terrestrial organisms. Hence, oxygen gradients in space are probably one of the main factors of response to TSR in aquatic insects (Forster et al. 2012).

#### 4.1.1.3 Other Possibilities

Some other hypotheses are proposed to explain the distribution of aquatic insect species. For instance, eco-morphology (Wainwright and Reilly 1994) may explain the distribution of an endemic fauna of New Zealand stoneflies (106 species/21 genera). They are distributed over a wide altitudinal gradient (0–2000 m), with a great variety of phenotypes, such as the reduced-wing (26 spp.) and the winged (80 spp.) forms. In this group, there is a significant correlation between altitude and body size for brachypterous taxa, but the same correlation could not be found for winged taxa (McCulloch and Waters 2018). These results suggest that morphology and altitude may interfere in the predictions made by ecogeographic rules, since the maintenance of traits related to flight represents an evolutionary constraint on body size (McCulloch and Waters 2018).

Finally, the ecophysiological hypothesis (EH) may also explain the distribution of aquatic insects at local and regional scales (De Marco Júnior et al. 2015). The EH suggests that the organization of the metacommunities of dragonflies and damselflies of small-to-medium tropical streams is related to the degree of sunlight availability, thermoregulatory characteristics, and body size of each species. In other words, environmental filters (e.g., narrow streams) can differentially select certain odonate taxonomic groups (Zygoptera and Anisoptera). For example, anisopteran abundance and species richness may increase with channel width, while zygopteran abundance usually decreases (De Marco Júnior et al. 2015). This effect is theoretically based on the thermoregulatory strategies of each taxon. Ectothermic species are expected to exhibit heliothermic behavior and perch on sunny spots to achieve optimal body temperature, while endothermic species adopt a "flier" strategy and keep constant flight activity (De Marco Júnior et al. 2015).

### *4.1.2 Temperature*

Temperature regimes may determine aquatic insect distribution at macroscale, but it also varies spatially according to position in water bodies, mostly regarding depth and distance to banks. Several studies have shown the importance of diel temperature regimes for aquatic insects. For instance, hyporheic waters provide more constant thermal conditions than surface waters. Such variation in water temperature is an abiotic factor that has a predominant effect, mainly, on aquatic eggs and larvae.

Water temperature plays a crucial role in larval development and growth. For example, 90% hatching success was recorded for *Lestagella penicillata* (Barnard) (Ephemeroptera: Teloganodidae) eggs in the 10, 15, and 20 °C (Ross-Gillespie et al. 2018). Larval development rates in odonates are progressively faster at higher temperatures (Lutz 1968; Mendonça et al. 2018). Adult mayflies of *Cloeon dipterum* (L) (Ephemeroptera: Baetidae) become smaller as temperature warms, because warming temperature causes development rates to increase proportionately faster than growth rate (Sweeney et al. 2018).

Aquatic insects may encounter a wide range of thermal conditions when moving in a water body, when facing seasonal changes and throughout a single day; hence, they must develop mechanisms to adapt to these fluctuations and to extreme climatic changes. As an example, a recently discovered stonefly species was found living under ice in Patagonia, Argentina (Pablo Pessacq personal communication). In winter, some aquatic insects (e.g., many chironomids) build cocoons to suspend metabolism and development (Lencioni et al. 2015). Some other aquatic insects have striking adaptations to cope with heat and cold. Temperature may impact oxygen concentration in plastrons and ultimately affect the metabolism and activity of aquatic insects (Jones et al. 2018). Diving beetles, for example, have hollow setae on the elytra that interconnect with the tracheal system. This way, beetles are able to extract oxygen directly from the water. In this case, hypoxia may reduce heat tolerance. Consequently, diving beetle species with higher densities of structures have high cold tolerance, but reduced heat tolerance (Verberk et al. 2018).

As other examples, caddisfly shredders' leaf-breakdown and survival rates are affected by warming temperature (Martins et al. 2017). Crawling water beetles (Coleoptera: Haliplidae) stay submerged for shorter durations in response to both increasing water temperature and decreasing dissolved oxygen (Banks et al. 2018). Temperature may also influence predator-prey dynamics. Temperature influenced larvae of *Anisops sardea* Herrich-Schäffer (Hemiptera: Notonectidae) in the predation of *Anopheles stephensi* Liston (Diptera: Culicidae), the most common malarial vector in Southeast Asia (Mondal et al. 2017). Through a laboratory bioassay, Mondal et al. (2017) suggest that the rate of prey consumption was positively related to the increase in temperature. Finally, the timing and duration of emergence in aquatic insects involve responses to temperature, often interacting with photoperiod (e.g., Norling 2018).

### *4.1.3 Light Environment and Turbidity*

Solar irradiance may be a key factor determining habitat occupancy, thermoregulation, voltinism, and sexual behavior in some aquatic insects. Characteristics of the thermal environment, as discussed in this chapter, and selection of locations for behavioral displays and territory defense may present conflicting interests, generating trade-offs. For example, male calopterygid damselflies defend sunny spots along streams, which females visit to mate and oviposit. In this case, the light environment may increase recognition by potential mates and increase detection by predators (e.g., Endler 1991; Kemp et al. 2018). Understanding patterns of habitat/microhabitat use can provide good insights into their ecological drivers and, in this context, top predators such as dragonflies and damselflies are good tools for these studies.

Many studies report the importance of the thermal and light environments for Odonata, as aspects of microhabitat luminosity have been reported by researchers as factors predictive of species occurrence. In forestall environments, for example, light may become a limiting factor. Riparian shading has a crucial role in tropical and subtropical streams, refraining the explosive growths of aquatic and semiaquatic plants, and algae, while creating a high allochthonous influx of organic matter, mainly from decaying leaves. In the tropics, these habitats support diverse and well-adapted endemic aquatic insects. For instance, mayflies and other larvae seek shelter and hide under leaves, which also act as a foraging site (Fig. 4.2a), and there are some shade-seeking adult Odonata that prefer to inhabit dark habitats in forests

**Fig. 4.2** Mayflies may hide under leaves, seeking shelter (**a**). The abiotic factors can affect the development and emergence of aquatic insects; for example, this unsuccessful emergence attempt by libellulid dragonfly in the Neotropical savannah, being eaten by ants (white arrowheads) (**b**). *Erythrodiplax berenice* (Drury) (Odonata: Libellulidae) is known to oviposit in brackish waters (**c**). Net-spinning caddisflies may act as environmental engineers and create complex habitats (**d**). Photos by Adolfo Cordero (**a**, **b**, **d**) and Dennis Paulson (**c**)

when adults (Henry et al. 2017) or larval forms that hide in darker substrates (Tavares et al. 2018), showing the importance of absence of light or dim light for some taxa.

Some other taxa exhibit positive phototropism. Stoneflies, caddisflies, aquatic Lepidoptera, Coleoptera, Hemiptera, and others exhibit this behavior that, in the natural environment, favors adults during mate searching. However, the same insects are attracted to artificial light and light traps. Consequently, the tendency of some aquatic insects to confuse certain colored or polarizing structures with the water surface is quite common. For instance, it is known that dragonflies find water bodies due to polarization of light. Many aquatic insects that are attracted to horizontally polarized light sources are also attracted to unpolarized UV blacklight (Nowinszky 2003). Cars, for example, are visual deceptions to aquatic insects, solely explained by the reflection–polarizational characteristics of the car paintwork (Kriska et al. 2006). In this context, light act as an evolutionary and sensorial trap. This can cause high mortality rates and select insects that are not attracted to light, affecting normal sexual behavior.

Finally, photoperiod has a remarkable effect on larval development and adult behavior. For instance, odonate larvae may exhibit synchronous development that depends on photoperiod (Norling 2018). In mayflies, photoperiod may affect size at emergence and survival (Corkum and Hanes 1992). Turbidity, which relates to the aquatic light environment, is also an important regulatory factor for aquatic insects. Particularly, turbidity may be one major factor determining the occurrence of dragonflies in water bodies (Klein et al. 2018). Nevertheless, turbidity may be related to water flow, suspended sediment, and temperature, which might be more important for aquatic larvae than the light environment. For example, turbidity due to suspended sediments may affect leaf shredding by caddisflies, and hence suppress processing of organic matter (Hornig and Brusven 1986).

### *4.1.4 Water Chemical Composition*

Water is an effective medium of dissolved chemicals and thus generates many different gradients in freshwater habitats, e.g., pH, dissolved oxygen, and particles from headwaters to river mouths, or salinity degrees from freshwater to marine. The chemical composition of water bodies is another environmental trait that predicts the occurrence and behavior of aquatic insects. Salt, ion concentration, and conductivity may have an impact on osmotic regulation and select for species that tolerate certain ranges of water chemical composition.

Several chemical components and abiotic factors are widely recognized by their influence on the composition of aquatic insect communities. Even small variations in these environmental characteristics can exert direct changes in microhabitat conditions and, consequently, influence habitat occupancy and colonization. For instance, abiotic factors such as pH, electrical conductivity, dissolved oxygen, total dissolved solids, and oxidation reduction potential influence richness, abundance, and composition of Odonata species (see Table 4.1, Mendes et al. 2018). All these features of water composition may affect larval development and growth, for example, resulting in death during emergence (Fig. 4.2b).

In this context, salinization of water bodies is currently one of the problems that deeply concerns scientists (Cañedo-Argüelles et al. 2013; Kefford et al. 2016). Salinization has been reported in all inhabited continents and is mainly associated with human activities. In the short term, the results of this imbalance may be a reduction in the number of species and changes in community composition and ecosystem processes (see Kefford et al. 2016 for review). In a climate change scenario and growing demand for water, studies on behavioral adaptations and on how aquatic insects respond to changing conductivity regimes may help to elucidate the real impact of this phenomenon worldwide.

In fact, salinity is important for metabolic functions, and may even increase the richness of tolerant insect species (Kefford et al. 2011; Kefford et al. 2016). However, the richness of groups such as Trichoptera, Ephemeroptera, and Plecoptera diminishes with increased salinity (Kefford et al. 2011). Other taxa exhibit variable sensitivities, such as Diptera, Odonata, Coleoptera, and Hemiptera (Kay et al. 2001; Kefford et al. 2012; Cañedo-Argüelles et al. 2013). Nevertheless, in naturally saline streams, Diptera, Heteroptera, and Coleoptera seem to be the most diverse groups (Millán et al. 2011).

**Table 4.1** Table illustrating the relationship between assemblages of Odonata, Anisoptera, and Zygoptera and abiotic factors. Results from Mendes et al. (2018) through the MDRM and PCA analyses

| Assemblage | Variable | Df | Sum of squares | Mean squares | *F* model | $R^2$ | Pr (>*F*) |
|---|---|---|---|---|---|---|---|
| Odonata | pH | 1 | 0.378 | 0.378 | 1.183 | 0.032 | 0.265 |
| | DO | 1 | 0.349 | 0.349 | 1.091 | 0.030 | 0.339 |
| | HII | 1 | 0.558 | 0.558 | 1.746 | 0.047 | 0.058 |
| | ORP | 1 | 0.574 | 0.574 | 1.795 | 0.049 | 0.049[a] |
| | Residuals | 31 | 9.905 | 0.320 | – | 0.842 | – |
| | Total | 35 | 11.763 | – | – | 1.000 | – |
| Zygoptera | pH | 1 | 0.160 | 0.160 | 15 | 0.014 | 0.852 |
| | DO | 1 | 0.466 | 0.465 | – | 0.039 | 0.209 |
| | IIII | 1 | 0.506 | 0.506 | – | 0.043 | 0.165 |
| | ORP | 1 | 0.331 | 0.331 | – | 0.028 | 0.444 |
| | Residuals | 31 | 10.323 | 0.333 | 2 | 0.876 | – |
| | Total | 35 | 11.786 | – | 2 | 1.000 | – |
| Anisoptera | pH | 1 | 0.008 | 0.008 | 20.902 | 0.054 | <0.01[a] |
| | DO | 1 | 0.120 | 0.120 | 303.442 | 0.777 | <0.01[b] |
| | HII | 1 | 0.012 | 0.012 | 29.178 | 0.075 | <0.01[a] |
| | ORP | 1 | 0.002 | 0.002 | 5.890 | 0.015 | 0.01[a] |
| | Residuals | 31 | 0.012 | 0.000 | – | 0.079 | – |
| | Total | 35 | 0.155 | – | – | 1.000 | – |

Abbreviations: *DO* dissolved oxygen, *ORP* oxidation reduction potential, *HII* habitat integrity index

[a]The test was significant ($p < 0.05$), but with a very low power of explanation

[b]The test was significant ($p < 0.05$), with a high power of explanation ($R^2 = 0.777$). Adapted from Mendes et al. 2018

It is axiomatic that insects are the most dominant group of invertebrates in lotic habitats, but the evolutionary history of each group may limit their adaptive capacity to environmental changes, such as increased salinity. Several taxa inhabit saline or marine habitats. For example, marine thysanurans comprise an abundant group in the ocean shore; some caddisflies tolerate high salt concentrations, such as *Philanisus plebeius* Walker (Trichoptera: Chathamiidae), a sea algae eater that builds its case with silk and sand grains (Leader and Bedford 1979); *Halobates* (Hemiptera: Gerridae) are wingless hemipterans that skim the water surface of the sea (Cheng 1985); *Erythrodiplax berenice* (Drury) (Odonata: Libellulidae) oviposits on the sea shore and marshes and larvae are adapted to a wide range of salinity (Fig. 4.2c, Dunson 1980); *Hexagenia limbata* (Serville) (Ephemeroptera: Ephemeridae) nymphs can survive elevated salinity (Chadwick and Feminella 2001). Adaptations include, for example, folding of osmoregulatory cell membranes in the tracheal gills of stonefly nymphs (Kapoor 1978).

The diversity of Ephemeroptera is also negatively affected by pH (Sutcliffe and Carrick 1973; O'Halloran et al. 2008). Despite apparently unfavorable conditions, some specialists do thrive in acidic waters (e.g., a few Odonata, Juen and De Marco

Jr 2011). Furthermore, aquatic insect biology is influenced by other factors such as organic matter input (in autochthonous and allochthonous environments); concentration of phosphorus, nitrogen, and carbon; and dissolved oxygen and conductivity, which may constrain development and growth (e.g., Johnson et al. 2015).

### 4.1.5 Habitat Complexity

Environments are heterogeneous and vary in complexity and visual and structural characteristics of the habitat that affect biological and behavioral aspects of organisms and interfere with predator-prey dynamics (Xiao et al. 2016). Complex habitats may provide shelter and conditions for camouflage, making it difficult to locate prey (Merilaita 2003) and reduce the risk of predation. However, structural complexity also provides structures such as perches and shelters that increase predation rates of sit-and-wait predators, such as odonate larvae (Klecka and Boukal 2014). Some aquatic insects, such as web-spinning caddisflies, act as ecosystem engineers, creating a complex habitat for themselves and other organisms (Fig. 4.2d). Background visual complexity can determine the distribution of predators and prey; for example, larvae of *Acanthagrion lancea* Selys (Odonata: Coenagrionidae) have preference for more complex backgrounds (Tavares et al. 2018).

### 4.1.6 Hydraulics and Substrate Roughness

Hydraulics and hydrodynamics influence the spatial and temporal distribution of insects in stream habitats, where features of flow physically constrain habitat occupancy by benthic organisms. Hence, insects must deal with substrate heterogeneity to achieve the best suitable places for growth and development. All these features, mostly related to unidirectional flow, affect whole communities by displacing organisms, resources, and sediments. Indeed, the disturbance of water flow is a dominant organizing factor in stream ecology (Resh et al. 1988; Gordon et al. 2004). Water movement comprises a very complex system that combines fluid transportation from upstream to downstream (longitudinally), from bank to bank (laterally), and from bottom to surface (vertically), while the whole system varies temporally (Ward 1989).

These would be all the dimensions of water flux to be considered if streams ran in smooth and regular channels. However, real water bodies present substrate heterogeneity and variances in the channel topography. There are changes in depth and in water velocity in each narrowing or widening of a creek to adapt the flow to the streambed shape. As the water accelerates or decelerates by diverting from solid structures of the substrate, it creates the turbulent flow that is the remarkable characteristic of lotic systems. Given this physical complexity of the fluid medium, the hydraulic features are predictors for the aquatic assemblage's establishment (Brooks et al. 2005; Biggs et al. 2005; Graba et al. 2013) because successful attachment and colonization in this environment are determined by the relationship between water flow and substrate structure (Ditsche-Kuru et al. 2010; Ditsche et al. 2014; Biggs and Hickey 1994). To better understand how hydrodynamics mold the aquatic insects' world, it is important to look at different spatial scales.

Starting from the mesohabitats, which are represented by the riffles and pools, it is easy to recognize that a very-high-speed flow between two rocks represents a physical barrier to insects that live on each stone. Now, looking to the microscale level, it is possible for insects to survive in these environments positioning themselves in shelters. These small zones of slow water created by the interaction between the substrate roughness and the turbulent flow itself are the key factors to keep the diversity in fast waters. In fact, hydrodynamic laws state that the layer of water just above a rough substrate has a speed close to zero, and the water surface has the maximum speed. The rougher the substrate, the thicker is this slow water layer. Hence, the aquatic insects exhibit numerous types of behaviors regarding their feeding, dispersal, or reproduction modes (Merritt and Cummins 1996), as a way to deal with substrate heterogeneity and water flow to achieve the best suitable places for the best ecological fitness. These complex and dynamic variations of water flow influence larval behavior and, thus, drive the variation of community structure in freshwater habitats.

Historically, the literature focuses on physical factors to explain the variation of fauna distribution in a range of spatial scales. Variations in the macrohabitat level occur mainly due to hydrological and geomorphological factors (Sheldon and Walker 1998). In mesohabitats, the highlight goes for the longitudinal gradients in the water flow, particle size, and hydraulic variables (Growns and Davis 1994; Downes and Hindell 2000). In smaller scales, species microdistribution is correlated with hydrodynamics, depth, substrate features, or combinations of these elements (Davis and Barmuta 1989). Today, we have a better comprehension that all these scales are important for studies on aquatic insects, however, in a recent past, most studies addressed the effect of abiotic factors at macroscales only. A common point of view in the study of stream aquatic macroinvertebrates, illustrated by the review made by Cooper et al. (1998), considered that scales smaller than the section did not need to be considered for faunistic and ecological studies in different water bodies or different drainage basins because they have a low interference at the larger scales. While the scientific community was focused on the larger scales, significant variability was observed in the fauna of aquatic insects between stretches of streams of the same order, between sections of the same stretch, between riffles of the same section, and between groups of rocks of the same riffles (Downes et al. 1998).

Larger scales studies are important for obvious reasons that we showed previously in this chapter; nevertheless, we also highlight the importance of studying aquatic insect communities at smaller scales. In mesohabitat scale, the contrast between riffles and pools caught the attention of curious minds that asked how these organisms could handle with these two different worlds that coexist at such a small distance. Several studies have shown that insects exhibit a great variability in habitat occupancy regarding faster waters in riffles and sluggish flows in ponds (e.g., Roque and Trivinho-Strixino 2001; Crisci-Bispo et al. 2007). Indeed, there is a high number of species and even higher taxonomical categories recognized as riffle inhabitants, such as dipterans *Simulium incrustatum* Lutz (Diptera: Simuliidae), *Leptohyphes* spp. (Ephemeroptera: Leptohyphidae), *Parametriocnemus* sp. (Diptera: Chironomidae), *Lopescladius* sp. (Chironomidae), *Farrodes* spp. (Leptophlebiidae), *Rheotanytarsus* sp. (Chironomidae), diamphipnoid and eustheniid stoneflies, and polythorid and calopterygid damselflies.

Pools dominated by sand as the main substrate provide no shelter for predators and are mainly inhabited by small dipterans (e.g., Chironomidae) and larger insects that can burrow (e.g., some Gomphidae, Odonata). Furthermore, higher richness and abundance in riffles seem to be a general pattern in tropical streams (Logan and Brooker 1983; Brown and Brussock 1991; Baptista et al. 2001; Silveira et al. 2006), and the substrate composition is related to this. Riffles are typically composed by rocky streambed while pools exhibit leaves, sediment, and organic matter in addition to stones. These substrate features are often perceived as a major factor contributing to the distribution of many invertebrates (Merritt and Cummins 1996; Taniguchi and Tokeshi 2004).

Mesohabitat studies have improved our knowledge on stream ecology and suggest that some questions on insect distribution can be answered only at microscale level. The variation of substrate shape exerts a great influence on stream benthic communities. Different surface irregularities such as roughness, pits, crevices, and macrophytes have been considered determinant habitat components that may explain richness and abundance of freshwater invertebrates (Downes et al. 1995, 1998). In order to evaluate substrate heterogeneity that occurs in both mesoscale and microscale, the concept of microhabitat architecture can be very useful. This concept was approached unpretentiously in the work of Robson (1996), who suggested its use to avoid possible confusion in experimental descriptions, such as using the terms complexity and heterogeneity to characterize the substrate. According to the author, habitat architecture can be represented by different dimensions such as substrate type, texture, and peculiar characteristics such as cracks and presence of vegetation.

Here, we approach this concept to the niche theory (Hutchinson 1957), using the multiple dimensions that can be attributed to habitat architecture to characterize the conditions and resources available in each microhabitat. Considering the lotic environment and insect behavior and ecology, we can define three basic dimensions of microhabitat architecture: (a) instability of the environment—representing the hydraulic forces acting on individuals; (b) potential refugia—a three-dimensional composition of the substrate; and (c) food availability in the environment. To represent these dimensions and to work at microscale, we have a challenge: we must adopt metrics consistent with the microscale, and the collection of aquatic organisms must also be adjusted to the reality of the microhabitat.

Taniguchi and Tokeshi (2004) developed a creative and elegant method to assess substrate complexity. In terms of fractal geometry, they modelled the structural pattern of microhabitat architecture focusing on surface features. The authors found that the number of species and invertebrate density were scaled positively with complexity levels. For instance, more complex surfaces provide more refuge areas because the irregularities may change the flow pattern. Furthermore, insect body size tends to decrease in function of increasing complexity, since shelters tend to become smaller. In other words, the size of refuges matches the organism's size (Bergey 1999; Tokeshi 1994).

In a way, the use of fractal geometry allowed us to assess the complexity of the substrate, but other complex factors, such as microhabitat hydrodynamics and biodiversity itself, cannot be represented satisfactorily by simplified measures. In this regard, water flow in a streambed can be considered a complex variable. Heterogeneity of surface architectures can change flow profiles and associated drag forces, turbulence, gravity, and viscosity. Hence, one small patch may exhibit several different hydraulic conditions. Therefore, there are so many dimensions of water flow to describe the stream dwellers that there is a shortfall on measures that represent all water flow complexity. Nevertheless, there are a few classic measures that together help with the description of microhabitat architecture.

One of them is the Reynolds number: a dimensionless coefficient used to characterize different flow regimes over a specific surface and describe whether the mean flow is laminar or turbulent (Davis and Barmuta 1989). Brooks et al. (2005) found that Reynolds number explained more the spatial variation of richness and abundance of invertebrates than other hydraulic variables. They found that most species preferred the areas of riffles with the lowest turbulence close to the streambed, highlighting the microhabitat architecture importance. Probably, this condition provides a friendlier boundary layer (distance between free water flow and substratum surface), where metabolic requirements are more easily accomplished (e.g., acquisition of oxygen). Additionally, insects also seek a place to stay without constantly fighting the current; in fact, most riffle species live under the rocks most of the time and go to the upper part to feed. In this context, a flattened and streamline body seems to be typical for many insects. For instance, many mayflies and stoneflies have legs positioned laterally reducing drag and providing better attachment (Merritt and Cummins 1996).

As streams are highly heterogenous, it is highly expected to find a very complex gradient of hydraulic conditions even in a small scale. Insects deal with these hydraulic factors all the time to experience the best ecological fitness possible. However, all these conditions change through time (Gordon et al. 2004). Specially in tropical regions, it is possible to distinguish rainy and drought periods. Seasonal variation of flow affects invertebrates through hydrological alterations influencing dispersal, attachment, and resource acquisition, and also changing abiotic factors like temperature (increasing with low flow) and dissolved oxygen (reducing with low flow) (Sharpe and Downes 2006; Brooks and Haeusler 2016). Brooks and Haeusler (2016) suggested that during a low-flow period faster waters can act as a refuge habitat for filter feeders. Thus, during droughts these insects track local velocity conditions to obtain enough resources and resist extended low-flow periods. On the other hand, gill-respiring invertebrates (e.g., Ephemeroptera, Plecoptera, Zygoptera) seem to be less influenced by low flows since they can rhythmically beat their gills to control oxygen uptake.

However, the relationship between stream invertebrates and hydraulics is also species specific. There are many obligates, facultative, or flow avoiders among macroinvertebrates, and it seems to be related to functional groups. For example, filterers prefer more turbulence and shear velocities where they can constantly receive supplies from the rapid flow, while shredders occur on slower waters with lower shear velocities and Froude number (related to depth and gravitational forces) containing detrital material (Growns and Davis 1994). Moreover, the morphology of the insect gills seems to be associated to water velocity. Species with higher gill morphological complexity tend to be pond dwellers, while aquatic insects with simpler gill structures inhabit faster waters at microscale (Tonetto et al. 2018). Nevertheless, this relationship seems to vanish when water velocities are analyzed at mesoscale (Fig. 4.3). Thus, the attributes of a species are expected to respond consistently along specific environmental gradients (Brooks and Haeusler 2016), and it is reasonable to note that a combination of body plan, behavior to obtain resources, and hydraulic conditions strongly influence the spatial distribution of the species in stream habitats.

All these environmental features of water bodies can be condensed in the River Continuum Concept (RCC, Vannote et al. 1980). The RCC is a model that relates all the physical parameters described above, e.g., channel width, water depth, slope angle, flow velocity, and sediment load, to explain the variation of biological communities and their distribution along stream and river patches. For instance, headwaters and low-order streams have a large input of allochthonous plant material, which favors the occurrence of shredders, such as caddisflies. Middle patches of a stream have a larger concentration of autochthonous material and particulate matter that favors collectors and gazers, while further stream patches have more fine particulate matter that favors collector-filterers. Therefore, here we suggest that macroecological processes that are inherent to the RCC are also intrinsically related to meso and mesohabitat characteristics, in a wider scenario that predicts the behavioral (e.g., habitat selection, resource exploration), functional (e.g., shredders, grazers, collectors, filterers, and predators), and ecological (e.g., population dynamics and assemblage composition) traits of insect communities.

### *4.1.7 Drought*

Temporary water bodies are home to many aquatic insects. These species tend to have fast development and explosive synchronic mating events. For instance, when water levels decline, caddisflies may use conspecific density as a cue to adjust behavior and development in the face of habitat drying. Since early drying is a constraint for population stability in temporary habitats (Lund et al. 2016), aquatic insects must be well adapted to hydric stress. The larvae of *Neohermes filicornis* (Banks) (Megaloptera: Corydalidae) actively burrow into the dry substrates after draught (Cover et al. 2015).

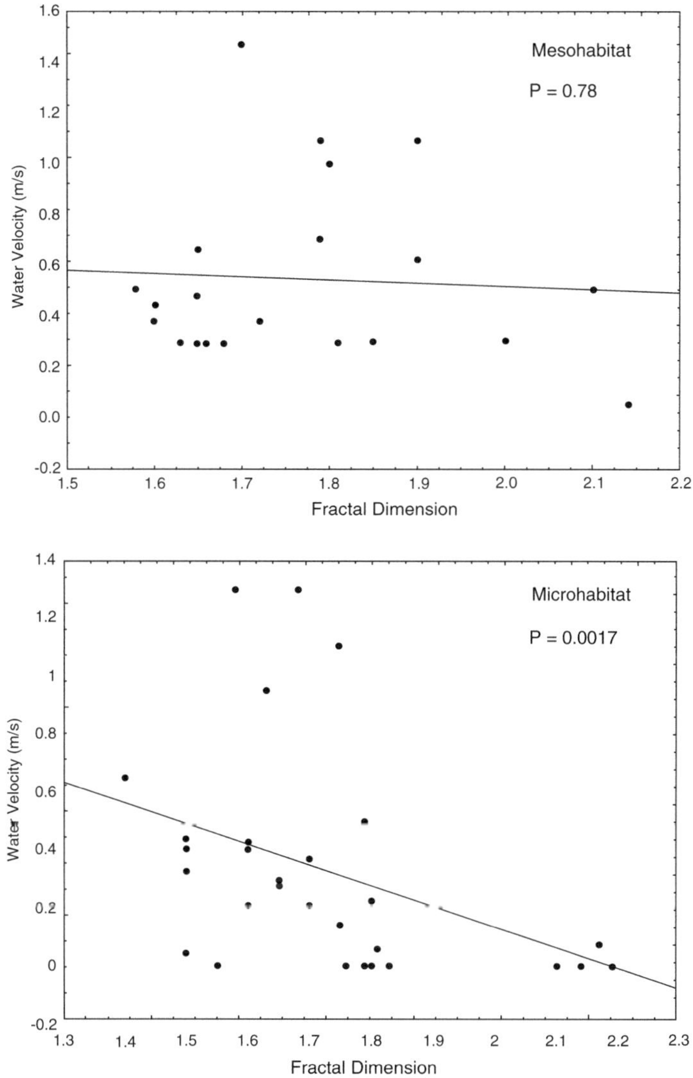


**Fig. 4.3** The relationship between the fractal complexity of aquatic insect gills and water velocity at meso- and microscales (from Tonetto et al. 2018)

### 4.1.8 Altitude

As previously shown in this chapter, the aquatic insect community may be under selective pressures of temperature and hydraulics. In water bodies located at high altitudes, there is a great number of endemic lineages and species constrained by adaptations to alpine environments. Historically, the negative correlation between diversity along increasing altitudinal gradients is widely recognized (Vinson and

Hawkins 2003), mainly its influence on aquatic fauna (Godoy et al. 2017). However, no general rule was observed in aquatic insects. The richness differences along the altitudinal gradients appear to be singularities of the taxa. In other words, diversity is highest in higher latitudes for: Odonata; lower for Plecoptera-Ephemeroptera; and not significant for Simullidae and Trichoptera (Vinson and Hawkins 2003; McCreadie et al. 2018). The main constraints are lower temperature and higher slopes, besides the autochthonous food resources that form the basis for high-altitude trophic webs, while there is a shift towards allochthonous aquatic systems at lower elevations (Atkinson et al. 2018).

### *4.1.9 Concluding Remarks*

Here, we have described and explored several ecogeographical rules and abiotic factors that interact to regulate ecological processes at macro-, meso-, and microscales. Understanding how such factors function to shape the diversity of aquatic insects and to determine their occurrence and distribution is crucial to conserve species and freshwater resources. Which ecogeographical model is most appropriate or which abiotic factor is more important in a given situation depends on the particular taxon and objectives of the study in question. We finally suggest that future studies should focus on understanding how these rules and factors select different behavioral and physiological adaptations. These approaches have been widely neglected or vilipended in studies on aquatic insects; however, this kind of study is strikingly necessary since they involve dimensions that create and maintain biodiversity.

## References

Atkinson D (1994) Temperature and organism size: a biological law for ectotherms? Adv Ecol Res 25:1–58

Atkinson CL, Encalada AC, Rugenski AT, Thomas SA, Landeira-Dabarca A, Poff NL, Flecker AS (2018) Determinants of food resource assimilation by stream insects along a tropical elevation gradient. Oecologia 187(3):1–14

Banks TB, Kincaid RM, Boersma KS (2018) Temperature and dissolved oxygen determine submersion time in aquatic beetle *Peltodytes callosus* (Coleoptera: Haliplidae). J Insect Behav 31(4):427–435

Baptista DF, Dorvillé LFM, Buss DF, Nessiamian JL (2001) Spatial and temporal organization of aquatic insects assemblages in the longitudinal gradient of a tropical river. Rev Bras Biol 61(2):295–304

Bergey EA (1999) Crevices as refugia for stream diatoms: effect of crevice size on abraded substrates. Limnol Oceanogr 44(6):1522–1529

Bergmann C (1847) Über die Verhältnisse der Wärmeökonomie der Thiere zu ihrer Grösse. Göttingen Studien 1:595–708

Biggs BJ, Hickey CW (1994) Periphyton responses to a hydraulic gradient in a regulated river in New Zealand. Freshw Biol 32(1):49–59

Biggs BJ, Nikora VI, Snelder TH (2005) Linking scales of flow variability to lotic ecosystem structure and function. River Res Appl 21(2-3):283–298
Brooks AJ, Haeusler T (2016) Invertebrate responses to flow: trait-velocity relationships during low and moderate flows. Hydrobiologia 773(1):23–34
Brooks AJ, Haeusler TIM, Reinfelds I, Williams S (2005) Hydraulic microhabitats and the distribution of macroinvertebrate assemblages in riffles. Freshw Biol 50(2):331–344
Brown AV, Brussock PP (1991) Comparisons of benthic invertebrates between riffles and pools. Hydrobiologia 220(2):99–108
Cañedo-Argüelles M, Kefford BJ, Piscart C, Prat N, Schäfer RB, Schulz CJ (2013) Salinisation of rivers: an urgent ecological issue. Environ Pollut 173:157–167
Chadwick MA, Feminella JW (2001) Influence of salinity and temperature on the growth and production of a freshwater mayfly in the Lower Mobile River, Alabama. Limnol Oceanogr 46(3):532–542
Cheng L (1985) Biology of halobates (Heteroptera: Gerridae). Annu Rev Entomol 30(1):111–135
Chown SL, Gaston KJ (2010) Body size variation in insects: a macroecological perspective. Biol Rev 85(1):139–169
Cooper IA (2010) Ecology of sexual dimorphism and clinal variation of coloration in a damselfly. Am Nat 176(5):566–572
Cooper SD, Diehl S, Kratz KIM, Sarnelle O (1998) Implications of scale for patterns and processes in stream ecology. Aust J Ecol 23:27–40
Corkum LD, Hanes EC (1992) Effects of temperature and photoperiod on larval size and survivorship of a burrowing mayfly (Ephemeroptera, Ephemeridae). Can J Zool 70(2):256–263
Cover MR, Seo JH, Resh VH (2015) Life history, burrowing behavior, and distribution of *Neohermes filicornis* (Megaloptera: Corydalidae), a long-lived aquatic insect in intermittent streams. West N Am Nat 75(4):474–490
Crisci-Bispo VL, Bispo PC, Froehlich CG (2007) Ephemeroptera, Plecoptera and Trichoptera assemblages in litter in a mountain stream of the Atlantic Rainforest from Southeastern Brazil. Rev Bras Zool 24(3):545–551
Davis JA, Barmuta LA (1989) An ecologically useful classification of mean and near-bed flows in streams and rivers. Freshw Biol 21(2):271–282
De Bie T, De Meester L, Brendonck L, Martens K, Goddeeris B, Ercken D, Hampel H, Denys L, Vanhecke L, Van Der Gucht K, Van Wichelen J (2012) Body size and dispersal mode as key traits determining metacommunity structure of aquatic organisms. Ecol Lett 15(7):740–747
De Marco Júnior P, Batista JD, Cabette HSR (2015) Community assembly of adult odonates in tropical streams: an ecophysiological hypothesis. PLoS One 10(4):e0123023
Deutsch CA, Tewksbury JJ, Huey RB, Sheldon KS, Ghalambor CK, Haak DC, Martin PR (2008) Impacts of climate warming on terrestrial ectotherms across latitude. Proc Natl Acad Sci 105(18):6668–6672
Dillon ME, Wang G, Huey RB (2010) Global metabolic impacts of recent climate warming. Nature 467(7316):704
Ditsche-Kuru P, Koop JHE, Gorb SN (2010) Underwater attachment in current: the role of setose attachment structures on the gills of the mayfly larvae Epeorus assimilis (Ephemeroptera, Heptageniidae). J Exp Biol 213(11):1950–1959
Ditsche P, Michels J, Kovalev A, Koop J, Gorb S (2014) More than just slippery: the impact of biofilm on the attachment of non-sessile freshwater mayfly larvae. J R Soc Interface 11(92):20130989.
Downes BJ, Hindell JS (2000) What's in a site? Variation in lotic macroinvertebrate density and diversity in a spatially replicated experiment. Austral Ecol 25(2):128–139
Downes BJ, Lake PS, Schreiber ESG (1995) Habitat structure and invertebrate assemblages on stream stones: a multivariate view from the riffle. Aust J Ecol 20(4):502–514
Downes BJ, Lake PS, Schreiber ESG, Glaister A (1998) Habitat structure and regulation of local species diversity in a stony, upland stream. Ecol Monogr 68(2):237–257
Dunson WA (1980) Adaptations of nymphs of a marine dragonfly, *Erythrodiplax berenice*, to wide variations in salinity. Physiol Zool 53(4):445–452

Endler JA (1991) Variation in the appearance of guppy color patterns to guppies and their predators under different visual conditions. Vision Res 31(3):587–608

Forster J, Hirst AG, Atkinson D (2012) Warming-induced reductions in body size are greater in aquatic than terrestrial species. Proc Natl Acad Sci 109(47):19310–19314

Gaston KJ, Chown SL, Evans KL (2008) Ecogeographical rules: elements of a synthesis. J Biogeogr 35(3):483–500

Godoy BS, Queiroz LL, Lodi S, Oliveira LG (2017) Environment and spatial influences on aquatic insect communities in Cerrado streams: the relative importance of conductivity, altitude, and conservation areas. Neotrop Entomol 46(2):151–158

Gordon ND, McMahon TA, Finlayson BL, Gippel CJ (2004) Stream hydrology: an introduction for ecologists. John Wiley and Sons, Chichester

Graba M, Sauvage S, Moulin FY, Urrea G, Sabater S, Sanchez-Pérez JM (2013) Interaction between local hydrodynamics and algal community in epilithic biofilm. Water Res 47(7):2153–2163

Growns IO, Davis JA (1994) Longitudinal changes in near-bed flows and macroinvertebrate communities in a Western Australian stream. J N Am Benthol Soc 13(4):417–438

Hassall C (2013) Time stress and temperature explain continental variation in damselfly body size. Ecography 36(8):894–903

Hassall C, Thompson DJ (2008) The effects of environmental warming on Odonata: a review. Int J Odonatol 11(2):131–153

Hassall C, Keat S, Thompson DJ, Watts PC (2014) Bergmann's rule is maintained during a rapid range expansion in a damselfly. Glob Chang Biol 20(2):475–482

Henry ER, Rivera JA, Linkem CN, Scales JA, Butler MA (2017) Damselflies that prefer dark habitats illustrate the importance of light as an ecological resource. Biol J Linn Soc 123(1):144–154

Horne CR, Hirst AG, Atkinson D (2015) Temperature-size responses match latitudinal size clines in arthropods, revealing critical differences between aquatic and terrestrial species. Ecol Lett 18(4):327–335

Hornig CE, Brusven MA (1986) Effects of suspended sediment on leaf processing by *Hesperophylax occidentalis* (Trichoptera: Limnephilidae) and *Pteronarcys californica* (Plecoptera: Pteronarcidae). Great Basin Nat 46(1):33–38

Hutchinson GE (1957) Concluding remarks. Cold Spring Harb Symp Quant Biol 22:415–427

Johnson BR, Weaver PC, Nietch CT, Lazorchak JM, Struewing KA, Funk DH (2015) Elevated major ion concentrations inhibit larval mayfly growth and development. Environ Toxicol Chem 34(1):167–172

Jones KK, Hetz SK, Seymour RS (2018) The effects of temperature, activity and convection on the plastron PO2 of the aquatic bug Aphelocheirus aestivalis (Hemiptera; Aphelocheiridae). J Insect Physiol 106(Pt 3):155–162

Juen L, De Marco P Jr (2011) Odonate biodiversity in terra firme forest streamlets in Central Amazonia: on the relative effects of neutral and niche drivers at small geographical extents. Insect Conserv Div 4(4):265–274

Kapoor NN (1978) Effect of salinity on the osmoregulatory cells in the tracheal gills of the stonefly nymph, *Paragnetina media* (Plecoptera: Perlidae). Can J Zool 56(12):2608–2613

Kay WR, Halse SA, Scanlon MD, Smith MJ (2001) Distribution and environmental tolerances of aquatic macroinvertebrate families in the agricultural zone of southwestern Australia. J N Am Benthol Soc 20(2):182–199

Kefford BJ, Marchant R, Schäfer RB, Metzeling L, Dunlop JE, Choy SC, Goonan P (2011) The definition of species richness used by species sensitivity distributions approximates observed effects of salinity on stream macroinvertebrates. Environ Pollut 159(1):302–310

Kefford BJ, Piscart C, Hickey HL, Gasith A, Ben-David E, Dunlop JE, Palmer CG, Allan K, Choy SC (2012) Global scale variation in the salinity sensitivity of riverine macroinvertebrates: Eastern Australia, France, Israel and South Africa. PLoS One 7(5):e35224

Kefford BJ, Buchwalter D, Canedo-Argüelles M, Davis J, Duncan RP, Hoffmann A, Thompson R (2016) Salinized rivers: de graded systems or new habitats for salt-tolerant faunas? Biol Lett 12(3):20151072

Kemp DJ, Batistic FK, Reznick DN (2018) Predictable adaptive trajectories of sexual coloration in the wild: evidence from replicate experimental guppy populations. Evolution 72(11):2462–2477

Klecka J, Boukal DS (2014) The effect of habitat structure on prey mortality depends on predator and prey microhabitat use. Oecologia 176:183–191
Klein CE, Pinto NS, Spigoloni ZAV, Bergamini FM, de Melo FR, De Marco JP, Juen L (2018) The influence of small hydroelectric power plants on the richness and composition of Odonata species in the Brazilian Savanna. Int J Odonatol 21(1):33–44
Kriska G, Malik P, Csabai Z, Horváth G (2006) Why do highly polarizing black burnt-up stubble-fields not attract aquatic insects? An exception proving the rule. Vision Res 46(26):4382–4386
Leader JP, Bedford JJ (1979) Oviposition by the marine caddisfly, *Philanisus plebeius* (Walk.). Search 10:275–276
Lencioni V, Jousson O, Guella G, Bernabò P (2015) Cold adaptive potential of chironomids over-wintering in a glacial stream. Physiol Entomol 40(1):43–53
Logan P, Brooker MP (1983) The macroinvertebrate faunas of riffle and pools. Water Res 17(3):263–270
Lund JO, Wissinger SA, Peckarsky BL (2016) Caddisfly behavioral responses to drying cues in temporary ponds: implications for effects of climate change. Freshw Sci 35(2):619–630
Lutz PE (1968) Effects of temperature and photoperiod on larval development in *Lestes eurinus* (Odonata: Lestidae). Ecology 49(4):637–644
Martins RT, Melo AS, Gonçalves JF, Campos CM, Hamada N (2017) Effects of climate change on leaf breakdown by microorganisms and the shredder Phylloicus elektoros (Trichoptera: Calamoceratidae). Hydrobiologia 789(1):31–44
Mayr E (1956) Geographical character gradients and climatic adaptation. Evolution 10(1):105–108
Mayr E (1996) The autonomy of biology: the position of biology among the sciences. Q Rev Biol 71(1):97–106
McCreadie JW, Williams RH, Stutsman S, Finn DS, Adler PH (2018) The influence of habitat heterogeneity and latitude on gamma diversity of the Nearctic Simuliidae, a ubiquitous group of stream-dwelling insects. Insect Sci 25(4):712–720
McCulloch GA, Waters JM (2018) Does wing reduction influence the relationship between altitude and insect body size? A case study using New Zealand's diverse stonefly fauna. Ecol Evol 8(2):953–960
Mendes TP, Luiza-Andrade A, Cabette HSR, Juen L (2018) How Does Environmental Variation Affect the Distribution of Dragonfly Larvae (Odonata) in the Amazon-Cerrado Transition Zone in Central Brazil? Neotrop Entomol 47(1):37–45
Mendonça FZ, Bernardy JV, Oliveira CEK, Oliveira PBG, De Marco P (2018) Temperature effect on the development of tropical dragonfly eggs. Neotrop Entomol 47(4):484–491
Merilaita S (2003) Visual background complexity facilitates the evolution of camouflage. Evolution 57:1248–1254
Merritt RW, Cummins KW (eds) (1996) An introduction to the aquatic insects of North America. Kendall Hunt, Dubuque
Millán A, Velasco J, Gutiérrez-Cánovas C, Arribas P, Picazo F, Sánchez-Fernández D, Abellán P (2011) Mediterranean saline streams in southeast Spain: what do we know? J Arid Environ 75(12):1352–1359
Mondal RP, Chandra G, Bandyopadhyay S, Ghosh A (2017) Effect of temperature and search area on the functional response of Anisops sardea (Hemiptera: Notonectidae) against Anopheles stephensi in laboratory bioassay. Acta Trop 166:262–267
Norling U (2018) Constant and shifting photoperiods as seasonal cues during larval development of the univoltine damselfly Lestes sponsa (Odonata: Lestidae). Int J Odonatol 21(2):1–22
Nowinszky L (2003) The orientation of insects by light—major theories. The handbook of light trapping. Savaria University Press, Szombathely, pp 15–18
O'Halloran K, Cavanagh JA, Harding JS (2008) Response of a New Zealand mayfly (Deleatidium spp.) to acid mine drainage: implications for mine remediation. Environ Toxicol Chem 27(5):1135–1140
Pateman RM, Hill JK, Roy DB, Fox R, Thomas CD (2012) Temperature-dependent alterations in host use drive rapid range expansion in a butterfly. Science 336(6084):1028–1030
Pinkert S, Brandl R, Zeuss D (2017) Colour lightness of dragonfly assemblages across North America and Europe. Ecography 40(9):1110–1117

Resh VH, Brown AV, Covich AP, Gurtz ME, Li HW, Minshall GW, Reice SR, Sheldon AL, Wallace JB, Wissmar RC (1988) The role of disturbance in stream ecology. J N Am Benthol Soc 7(4):433–455

Robson BJ (1996) Habitat architecture and trophic interaction strength in a river: riffle-scale effects. Oecologia 107(3):411–420

Roff DA, Heibo E, Vøllestad LA (2006) The importance of growth and mortality costs in the evolution of the optimal life history. J Evol Biol 19(6):1920–1930

Roque FO, Trivinho-Strixino S (2001) Benthic macroinvertebrates in mesohabitats of different spatial dimensions in a first order stream (São Carlos, SP). Acta Limnol Bras 13(2):69–77

Ross-Gillespie V, Picker MD, Dallas HF, Day JA (2018) The role of temperature in egg development of three aquatic insects Lestagella penicillata (Ephemeroptera), Aphanicercella scutata (Plecoptera), *Chimarra ambulans* (Trichoptera) from South Africa. J Therm Biol 71:158–170

Roulin A (2016) Condition-dependence, pleiotropy and the handicap principle of sexual selection in melanin based colouration. Biol Rev 91(2):328–348

Sharpe AK, Downes BJ (2006) The effects of potential larval supply, settlement and post-settlement processes on the distribution of two species of filter-feeding caddisflies. Freshw Biol 51(4):717–729

Sheldon F, Walker KF (1998) Spatial distribution of littoral invertebrates in the lower Murray–Darling River system, Australia. Mar Freshw Res 49(2):171–182

Shelomi M (2012) Where are we now? Bergmann's rule sensu lato in insects. Am Nat 180(4):511–519

Silveira MP, Buss DF, Nessimian JL, Baptista DF (2006) Spatial and temporal distribution of benthic macroinvertebrates in a Southeastern Brazilian river. Braz J Biol 66(2B):623–632

Sutcliffe DW, Carrick TR (1973) Studies on mountain streams in the English Lake District: I. pH, calcium and the distribution of invertebrates in the River Duddon. Freshw Biol 3(5):437–462

Sweeney BW, Funk DH, Camp AA, Buchwalter DB, Jackson JK (2018) Why adult mayflies of *Cloeon dipterum* (Ephemeroptera: Baetidae) become smaller as temperature warms. Freshw Sci 37(1):64–81

Taniguchi H, Tokeshi M (2004) Effects of habitat complexity on benthic assemblages in a variable environment. Freshw Biol 49(9):1164–1178

Tavares RI, Pestana GC, Rocha AD, Schiavone DC, Guillermo-Ferreira R (2018) Come to the dark side: habitat selection of larval odonates depends on background visual patterns. Ecol Entomol 46(5):640–646

Tokeshi M (1994) Community ecology and patchy freshwater habitats. In: Giller PS, Hildrew AG, Raffaelli DG (eds) Aquatic ecology: scale, pattern and process. Blackwell Scientific Publications, Oxford, pp 63–91

Tonetto AF, Guillermo-Ferreira R, Cardoso-Leite R, Novaes MC, Peres CK (2018) The relationship between water velocity and morphological complexity of stream dwellers. Limnologica 72:22–27

Vannote RL, Minshall GW, Cummins KW, Sedell JR, Cushing CE (1980) The river continuum concept. Can J Fish Aquat Sci 37(1):130–137

Verberk WCEP, Calosi P, Spicer JI, Kehl S, Bilton DT (2018) Does plasticity in thermal tolerance trade-off with inherent tolerance? The influence of setal tracheal gills on thermal tolerance and its plasticity in a group of European diving beetles. J Insect Physiol 106(Pt 3):163–171

Vinson MR, Hawkins CP (2003) Broad-scale geographical patterns in local stream insect genera richness. Ecography 26(6):751–767

Wainwright PC, Reilly SM (eds) (1994) Ecological morphology: integrative organismal biology. University of Chicago Press, Chicago, IL

Ward JV (1989) The four-dimensional nature of lotic ecosystems. J N Am Benthol Soc 8(1):2–8

Watt WB (1968) Adaptive significance of pigment polymorphisms in Colias butterflies. I Variation of melanin pigment in relation to thermoregulation. Evolution 22(3):437–458

Xiao F, Yang C, Shi H, Wang J, Sun L, Lin L (2016) Background matching and camouflage efficiency predict population density in four-eyed turtle (*Sacalia quadriocellata*). Behav Processes 131:40–46

Zeuss D, Brandl R, Brändle M, Rahbek C, Brunzel S (2014) Global warming favours light-coloured insects in Europe. Nat Commun 5:3874

# Chapter 5
# The Biotic Environment: Multiple Interactions in an Aquatic World

**Gabriela C. Mendes, Guilherme Gonzaga da Silva, Leonardo Samuel Ricioli, and Rhainer Guillermo**

**Abstract** The distribution and diversity of aquatic insects is a result of their interactions with the environment and other organisms. Right from the egg to larval and adult stages, insects must deal with a great biodiversity of natural enemies and mutualists. Such relationships evolved for millions of years in such a way that aquatic insects developed behavioral, ecological, and evolutionary strategies to cope with predation, parasitism, and competition. In the same way, they have joined forces with other organisms to solve problems, such as the interaction with gut bacteria to digest cellulose. These strategies and selective forces not only predict behavior and occurrence of aquatic insects, but also shape their diversity, community structures, and population dynamics. In this chapter, we sought to bring to the reader a useful source of information and a background for future studies. Although our current knowledge on species interactions in freshwater habitats allows us to discuss several topics, there is an open field of possibilities and gaps to be addressed in future research.

**Keywords** Ecological interactions · Biotic factors · Aquatic animals · Trophic webs · Networks · Trophic cascades

## 5.1 Introduction

The freshwater environment is a rich amalgam of functional groups. Several taxa are parasites, predators, parasitoids, or prey of aquatic insects, and aquatic insects themselves make part of an extensive interaction network that harbors terrestrial, aquatic, and semiaquatic organisms. Here, in this chapter, we attempt to cover most of this biodiversity and give an insight into the evolution of behavioral strategies

G. C. Mendes · G. G. da Silva · L. S. Ricioli
Laboratory of Ecological Studies on Ethology and Evolution (LESTES), Federal University of São Carlos, São Carlos, SP, Brazil

R. Guillermo (✉)
Federal University of São Carlos, UFSCar, São Carlos, São Paulo, Brazil



K. Del-Claro, R. Guillermo (eds.), *Aquatic Insects*,
https://doi.org/10.1007/978-3-030-16327-3_5

adopted by aquatic insects in their interactions with the biotic environment. This issue is extremely relevant, mainly considering the current perspective in which biodiversity must embrace the enormous richness present in biotic interactions, encompassing life histories, biology, and behavior of related species (Del-Claro and Torezan-Silingardi 2012). Therefore, we must consider that the outcomes of each interaction vary depending on proximal (internal factors) and distal (environmental, external factors) biotic changes that may modify the character of each interaction (e.g., Del-Claro et al. 2016).

Interactions between aquatic insects and other organisms will vary, depending on the context, like any other type of interaction. Hence, the outcomes of each interaction can change from positive to neutral or negative according to variations that may depend (e.g., prey abundance) or not on density (e.g., seasonal plant phenology). Thus, aquatic insects may present harmonic (e.g., commensalism, facilitation, and mutualisms) or disharmonic interactions (e.g., predation, herbivory, competition, exploitation, and parasitism), several of them subject to spatial and/or temporal variation. Predation and parasitism (including herbivory) are deleterious relationships that act as selective pressures on aquatic insects, which have developed strategies and behaviors to prey and to avoid predation, and to parasitize and to avoid parasitism. Competition and exploitation may be intra- and interspecific, where individuals compete for sites, mates, and resources. These negative relationships often force selection to favor strategies that increase fitness of those who are better competitors in specific habitats. Finally, commensals, facilitation between sympatric species, and mutualisms are rarely documented when researchers focus on aquatic insects, but here we show some interesting cases and suggest future approaches that will bring light towards such interactions in freshwater trophic cascades.

### *5.1.1 Predation*

Selective pressures from predation may directly affect population dynamics, species distribution, and occurrence in natural environments. The presence of predators in the habitat interferes with prey survival in a direct and lethal way, through consumption, or indirectly by causing behavioral modifications on prey due to predation risk (Hoverman et al. 2005), the nonlethal effects that reduce reproductive success. Studies on invertebrates, fish, and amphibians have shown that nonlethal effects may be even larger than lethal effects in determining behavior, condition, density, and distribution of animals over a range of trophic levels (Luttbeg and Kerby 2005). Ultimately, predator presence can alter the distribution of threat-sensitive aquatic insects, interfering with the choice of a patch, its size, or the connectivity between two or more occupied patches (Resetarits Jr and Binckley 2009). For instance, in heterogeneous habitats, prey select patches with the highest expected fitness in order to maximize growth and minimize predation risk (Lima and Dill 1990). As an example, stonefly predators may force changes in behavior and habitat occupation by mayflies (Peckarsky 1980; Peckarsky and

**Fig. 5.1** A pisaurid spider preying on a male *Boyeria irene (Fonscolombe)* (Odonata: Aeshnidae) (**a**); corixids, dragonflies, and mayflies caught on spider webs (**b–d**). Photographs by Adolfo Cordero Rivera

Dodson 1980) and affect benthic communities and ecological processes through consuming and scaring prey (Majdi et al. 2015).

Aquatic insects may constitute a great portion of a predator's diet. For instance, the presence of aquatic insects in riparian environments can contribute to a great amount of all the energy requirements of spiders (Fig. 5.1, Sanzone et al. 2003). This contribution may be significant for *Bembidion* Latreille and *Nebria* Latr. carabid beetles that depend entirely on the consumption of such prey (Paetzold et al. 2005). A significant part of fish diet is also composed by aquatic insects—mayflies are among the preferred food items of many species, along with caddisflies and stoneflies (Allan 1995; Nair et al. 2015). Other predators rely on aquatic insects to complement their diet, such as amphibians, birds, lizards, bats, and other arthropods (Nakano and Murakami 2001; Sabo et al. 2002; Hering and Plachter 1997). Humans also consume aquatic insects, mainly in China, Mexico, Colombia, Thailand, Indonesia, Japan, Australia, and Africa (Williams and Williams 2017). Aquatic insects must even face predation from plants. The plant *Utricularia australis* R. Brown (Lentibulariaceae) has water-filled bladders that function as a suction trap for aquatic prey, such as damselflies, mayflies, and dipterans (Martens and Grabow 2011). In soils especially poor in N, many carnivorous plants use insect bodies as a source of N, and damselflies are commonly found as prey of species like *Drosera* L. (Droseraceae), very common plant in the borders of ponds, for example at the South American savannas.

In response to such a great biodiversity of predators, and consequent multitude of predatory tactics, aquatic insects have developed strategies to avoid predation. Insects with complex life cycles, such as most aquatic insects, can exhibit different antipredatory behaviors at various stages of life. In insects with an aquatic larval stage, the recognition of predator cues can induce ovipositing females to choose habitats with a low risk of predation to increase offspring success (Andrade et al. 2017). For example, female *Culex* L. *(Diptera: Culicidae)* mosquitoes select water bodies that are relatively free of predators (Silberbush and Resetarits Jr 2017). In some cases, the ability to distinguish environments occurs through the recognition of kairomones released by fish that feed on larvae (Eveland et al. 2016), which act as chemical cues of predation risk. Waterborne fish cues may also provoke inducible defense on mayflies. *Drunella* Needham nymphs that developed in streams with fish had significantly longer caudal filaments and lower mortality when exposed to predation by trout than those mayflies that developed in fishless streams (Dahl and Peckarsky 2002). Mayflies often use the cerci to detect disturbances in the water and detect predators; however, they may also be used to deflect predators. For example, mayfly nymphs adopt a "scorpion posture" and avoid predation from stoneflies (Peckarsky 1987).

This recognition of predatory signals decreases the costs of avoiding predators and can occur visually, chemically, or through the recognition of species-specific signals (Sih 1982). A recent study also showed that mosquito larvae can identify water vibrations from predatory dragonfly larvae and reduce foraging activity (Roberts 2017). The presence or absence of antipredatory behavior may also affect prey community composition. For example, the swimming larvae of the libellulid dragonfly *Leucorrhinia dubia* VD Lind. are more vulnerable to attack by predatory fish than the larvae of *Libellula quadrimaculata* L. and *Orthetrum coerulescens* (Fabricius) that feint death when a predator approaches (Henrikson 1988), affecting species composition and distribution in odonate communities.

Predation risk may select and change foraging behavior, induce morphological defenses, and affect emergence time and rates of aquatic insects (Benard 2004). For instance, some aquatic insects have developed behaviors to avoid attacks from predators. When facing a predator, water striders adjust the angulation of their legs as they push the surface of the water to make a great escape jump, increasing their chances of surviving an attack (Yang et al. 2016). Odonate larvae also have behavioral adaptations to increase their chances against predators. Aeshnid larvae use water jet propulsion as way of locomotion, when water is expelled by the anal region. When there is an encounter with a predator, larvae can rapidly burst their way out of danger (Corbet 1999).

The intensity of predator-prey interactions is relatively stable, and defensive strategies can be permanent in these environments. However, there are cases where environments are extremely heterogeneous or unstable and subject to change over short periods of time. In these cases, the prey may develop inducible antipredator responses. For example, the nymphs of some dragonflies have spines as defenses against their predators (Johansson and Samuelsson 1994), sharp protrusions used to pierce those who try to devour them. Larvae of *L. dubia* develop different sizes of spines, depending on the presence or absence of the predators (Johansson 2002). Larvae exposed to the presence of predatory fish have developed longer spines than nymphs under low predation risk—this phenomenon is called "inducible defense." Inducible defense strategies are advantageous due to the energy saved by avoiding unnecessary production of defensive characters in the absence of predators (Auld and Relyea 2011), which could decrease the net gain of individuals (DeWitt 1998).

Caddisfly larvae case design and predator-avoidance behavior may affect the vulnerability to predators (Nislow and Molles Jr 1993; Wissinger et al. 2006). Trichopteran larvae may also exhibit glands that secrete defensive exudates that inhibit predators from approaching (Duffield et al. 1977). Stoneflies may exhibit autohemorrhage at the intersegmental membranes of the coaxial and tibiofemoral joints of the legs when subjected to traumatizing stimuli, a defense against predators, mainly ants and fish (Benfield 1974; Moore and Williams 1990). Moreover, this behavior creates a sound that may delay the attack from vertebrate predators (Benfield 1974). Stonefly larvae may also be aggressive towards each other, causing spacing and decreasing predation risk as well (Moore and Williams 1990).

Antipredatory behavior of *Anax junius* Drury (Odonata: Aeshnidae) larvae varies according to developmental stage, where older larvae are successful in escaping predatory tadpoles, while only 12.5% of the young achieve the same result, showing that ontogeny may represent a vulnerability for young larvae (Hopkins et al. 2011). The larvae of mosquitoes that occupy environments with water accumulated on epiphytic plants (as is the case of bromeliads) are exposed to predation by tadpoles, as demonstrated by Salinas et al. (2018), who evaluated the predatory capacity and rate of predation of tadpoles *Phyllodytes luteolus* (Wied) (Anura: Hylidae) feeding on mosquito larvae. This study also suggests strong evidence of the relationship between body size and stage of development and predatory efficacy of tadpoles, demonstrating that the risk of predation may also change according to the level of predator development. Although aquatic insects are small in size and serve as prey to many taxa, a great part of freshwater biodiversity is composed of predatory insects.

#### 5.1.1.1 Predatory Aquatic Insects

Many aquatic insects are important predators in trophic networks that connect water and land. This is the case of dragonflies that are top predators when aquatic larvae, and predators of remarkable aerial abilities that prey on other insects when adults. Dragonflies are often generalists, attacking every prey with certain size that moves. Giant water bugs, *Belostoma* Latreille (Heteroptera: Belostomatidae), are sit-and-wait predators that feed on fish, amphibians, snails, and other aquatic insects (Menke 1979). These animals use their raptorial legs to capture their prey and pierce it to inject toxins and digestive enzymes that cause prey paralysis and necrosis (Swart and Felgenhauer 2003). However, other taxa are specialists, such as diving beetles (Coleoptera: Dytiscidae) that are often specialized in certain prey. For instance, many larval insects are adapted to prey on mosquito larvae. In some cases, they can be even more specialized. *Hyphydrus japonicus* Sharp (Dytiscidae: Hydroporinae), have adapted mandibles and a projection on the head to open the valves of seed shrimps and gain access to the prey within (Hayashi and Ohba 2018). Helicopter damselflies (Odonata: Pseudostigmatinae) glean spiders from their webs or steal wrapped prey from the web (Fincke 1984; Ingley et al. 2012). Some

Heptageniidae (Ephemeroptera) are collectors (gathers) and some species are even predators (engulfers) (e.g., Soluk and Clifford 1985).

Top predators in freshwater habitats, such as dragonflies, diving beetles, water bugs, water striders, some stoneflies, and even chironomids, compose one of the most striking selective forces acting on communities' structuring, affecting prey behavior, life history, and population dynamics (Chase 1999; Babbitt and Jordan 1996; Kehr and Schnack 1991). For example, trophic cascades that occur when predators reduce prey abundance, indirectly impacting lower trophic levels, may regulate species interactions that include aquatic insects. In an elegant study, Knight et al. (2005) showed that ponds with fish may reduce larval dragonfly abundances, hence fewer adult dragonflies hunting for prey in the air. Consequently, pollinators were benefited and plants near ponds with fish received more pollinator visits than plants near ponds without fish. Finally, freshwater predators may be a driving force for community species composition by acting as selective filters. Dragonfly larvae prey on water boatman (Hemiptera: Corixidae), especially on small species like *Trichocorixa* Kirkaldy; hence, larger species occupy water bodies with odonates, while *Trichocorixa* colonizes dragonfly-free brackish waters (Coccia et al. 2014).

### *5.1.2 Parasitism*

Parasites and parasitoids may also have striking impacts on the ecology of aquatic insects, at individual, population, and community levels. Such interactions may affect insects directly, by decreasing individual fitness, or indirectly, by changing population densities, species interactions, and behavior (Kohler 2008). An impressive variety of parasites and parasitoids inhabit riparian environments and rely on aquatic insects as hosts. Consequently, aquatic insects have their fitness impacted, which may decrease interspecific competition and ultimately benefit species coexistence. Nevertheless, intraspecific competition may favor those individuals that are able to defend themselves against parasites, by adopting strategies or exhibiting good immunocompetence. With few exceptions, aquatic insects regularly need to face attacks from Microsporidia, Nematoda, Nematomorpha, Gregarina, Trematoda, Diptera, Hymenoptera, and Acari (Poulin and Morand 2000).

Trematodes search for intermediate insect hosts after development in a mollusk, until infecting the definitive host, usually a vertebrate (Chae et al. 2000; Esch et al. 2002). These parasites can infect aquatic insects and alter their behavior. For instance, larvae of *Aedes aegypti (Linnaeus in Hasselquist)* that are heavily infected with metacercariae are less active and spend more time at the water surface. As a result, host behavior manipulation facilitates predation or accidental ingestion by aquatic vertebrates, the final host (Poulin 1995).

Females of *Forcipomyia* Meigen (Diptera: Ceratopogonidae) are ectoparasites of flying insects (Fig. 5.2a). In odonates, for example, female midges attach to the host exoskeleton to feed on their hemolymph. *Forcipomyia paludis* (Macfie) firmly attaches to the main veins at the wing base of odonates (Wildermuth and Martens 2007). Another midge species, *F.* (*Pterobosca*) *incubans Macfie*, was recorded parasitizing

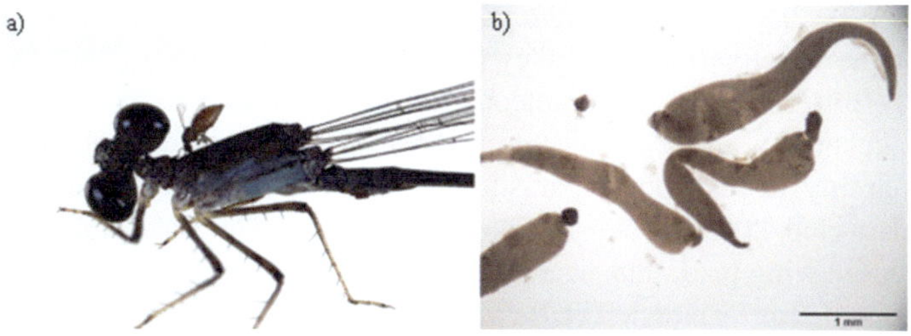


**Fig. 5.2** A female *Forcipomyia* midge biting the thorax of a male *Epipleoneura tariana* Machado (Odonata: Protoneurniae) in the Brazilian Amazon (**a**) and gregarines infecting the gut of a *Polythore gigantea* (Selys) (Odonata: Polythoridae) (**b**). Photographs by Adolfo Cordero Rivera

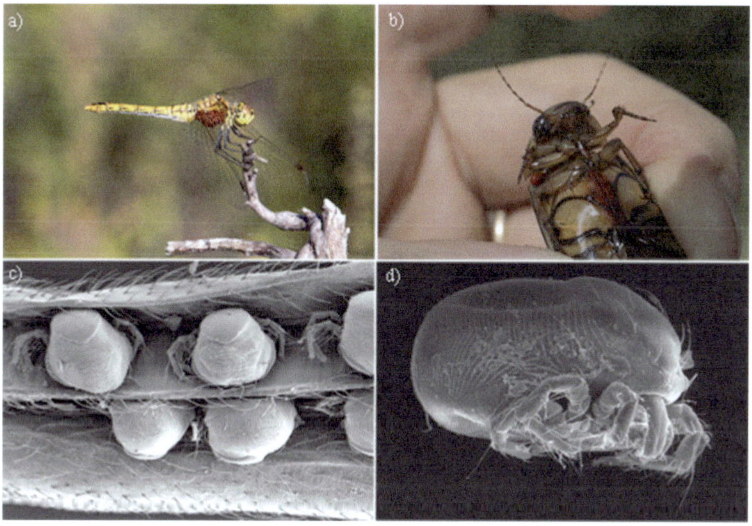


**Fig. 5.3** Water mites infecting a dragonfly *Sympetrum sanguineum* (Müller) (Odonata: Libellulidae) (**a**) and a diving beetle (*Dytiscus* L.) (**b**). Scanning electron microscopy micrographs showing the mite larvae attached to a damselfly's exoskeleton (*Ischnura graellsii* Rambur) (**c**) and a detached one (**d**). Photographs by Adolfo Cordero Rivera

wings of several odonate species throughout the neotropics (Guillermo-Ferreira and Vilela 2013). While some parasites may infect hosts in specific developmental stages, others do not have this restriction. This is the case of gregarines (Fig. 5.2b), which may infect aquatic insects at any stage of their development (González-Tokman et al. 2011). When infecting dragonflies, gregarines may cause metabolic syndromes and obesity in the host, decreasing activity and causing alterations in performance of flight muscles (Schilder and Marden 2007).

Water mites can also infect adults of aquatic insects (Fig. 5.3), usually attaching to the ventral part of the body (Böttger 1976). Most water mites act as predators of chironomid larvae in lotic habitats. In lentic environments, it is common to find water mites acting as predators of eggs and as ectoparasites, attached to aquatic insects (Proctor and Pritchard 1989), such as Coleoptera, Odonata, Plecoptera,

Diptera and Trichoptera. *Hydrachna globosa* (De Geer) larvae, for example, swim around on the water surface until they find a host, *Nepa rubra* L. and *Ranatra linearis* (L.) (Böttger 1972). As an outcome of the infection, insects may have lower longevity, fecundity, and survival (Smith 1988), because the infection may compromise the ability to respond to parasites and pathogens in subsequent infections (Honkavaara et al. 2009). When odonates are parasitized by water mites, they suffer with increased wing load, thus decreasing flight distance and dispersal, resulting in increased predation (Marden 1989). Moreover, mites can cause damage to flight muscles, result in less symmetric wings, and reduce fat content (i.e., energy reserves), which potentially affects odonate dispersal, foraging rates, and mate acquisition (Stechmann 1978; Smith 1988; Stoks and Córdoba-Aguilar 2012).

Nematodes and horsehair worms (Nematomorpha) (Fig. 5.4a) infect some species of mayflies, caddisflies, stoneflies, corixids, black flies, and beetle larvae (White 1969; Grabner 2017). In nematodes that infect multiple hosts, the parasite frequently changes the intermediate host behavior to facilitate transference to the final host. For example, odonates may act as paratenic hosts of nematodes that parasitize freshwater turtles (Wiles and Bolek 2015). Even humans may be part of complex parasitic nematode cycles, where human consumption of odonate larvae may lead to infection (Corbet 1999). Consequently, the aquatic insect may exhibit altered behaviors that facilitate transmission. When mayflies *Baetis bicaudatus* Dodds (Ephemeroptera: Baetidae) are infected by parasites, they are preferentially attacked by their stonefly predator (Vance and Peckarsky 1997), as a result of pathological side effects.

Parasite transmission is utterly linked to parasite fitness; hence, in order to get to its final host, a parasite needs to adopt strategies to facilitate infection. Some parasites of hematophagous insects do it by enhancing sensibility of their hosts to odor cues (Rossignol et al. 1986). When infected, *Anopheles gambiae Giles*—a common vector of *Plasmodium falciparum*, a parasite that causes human malaria—were significantly more attracted to a substrate emanating human odor than noninfected ones. This occurs because the parasite enhanced mosquito olfactory sensibility to humans, augmenting infection chances (Smallegange et al. 2013). Females of *A. gambiae* also increase the feeding frequency when parasitized and take larger blood

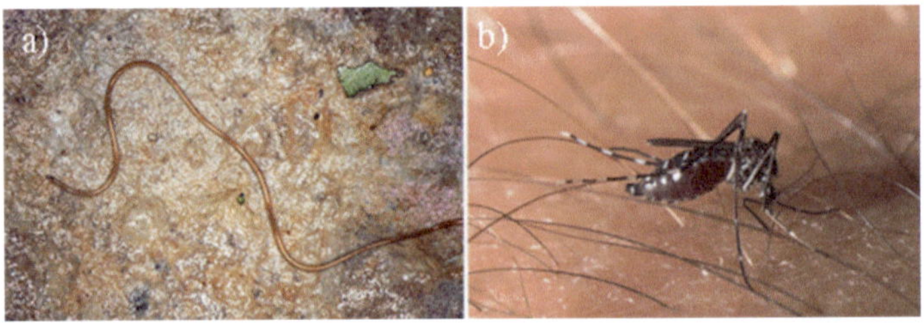


**Fig. 5.4** A gordiid horsehair worm (Nematomorpha) (**a**). Some larval forms of these worms infect aquatic insects. *Aedes albopictus* (Skuse) (Diptera: Culicidae), vector of the yellow fever virus, dengue fever, and Chikungunya fever, having a blood meal (**b**). Photographs by Adolfo Cordero Rivera

meals than uninfected females. Malaria parasites may decrease risky behaviors of their hosts during pre-infectious phase, such as blood feeding, but this behavior resumes when the parasite is ready to be transferred to its final host (Schwartz and Koella 2001; Anderson et al. 1999; Koella et al. 2002).

*Leishmania* also manipulates its host behavior. Sand flies infected with the parasite present altered behavior, depending on the parasite developmental stage. When the parasite is infectious, it manipulates sand fly behavior, enhancing biting persistence, thus enhancing host infection (Rogers and Bates 2007). Parasitized sand flies fed on multiple hosts much more frequently than non-parasitized sand flies, augmenting infection chances (Ready 2008). *Leishmania* also produces a mucin-like gel in the sand fly gut that blocks its mouthparts, causing regurgitation during blood feeding. This effect enhances parasite transmission during feeding (Rogers 2012).

Viruses can also manipulate the host behavior. La Crosse virus (LACV) depressed blood meal size in *Aedes albopictus* (Skuse) (Fig. 5.4b) and *Ae. triseriatus (Say)*, making them more avid to feed again. In this way, infected mosquitoes feed more frequently, enhancing horizontal transmission (Jackson et al. 2012). In females, LACV increased sexual activity, and infected females mated significantly more than noninfected ones (Gabitzsch et al. 2006), although it is still not clear if this is a host response to infection, or if it benefits the parasite fitness. Viruses may also alter the host locomotion behavior: in *Ae. aegypti*, parasitized mosquitoes presented higher activity levels (Lima-Camara et al. 2011).

All aquatic Hymenoptera are parasitoids of aquatic invertebrates. They search for insect immature stages, mostly eggs and larvae of Coleoptera, Diptera, Hemiptera, Lepidoptera, Odonata, and Megaloptera. At least one-third of the hymenopteran parasitoid species can be found in lotic environments, while all the other species are associated with lentic habitats (Bennett 2007). Some pompilid wasps may also hunt adult semiaquatic spiders (Shimizu 1992). Most hosts are herbivore insects associated with aquatic plants (Feitosa et al. 2016; Kula et al. 2017). While most terrestrial plants produce volatile organic compounds in response to herbivory to attract natural enemies of herbivores (Turlings and Erb 2018), some aquatic parasitoids search for hosts in a random pattern, apparently not using chemical cues produced by the host insect or plant. The female of *Apsilops japonicus* Yoshida et al. (Hymenoptera: Ichneumonidae), for example, needs to probe underwater petioles of the yellow water lily *Nuphar* Sibth. & Sm. one by one to find the host pupae *Neoschoenobia testacealis Hampson* (Lepidoptera: Crambidae) (Hirayama et al. 2014). *Caraphractus cinctus* Walker (Hymenoptera: Mymaridae) parasitizes eggs of diving beetles (Dytiscidae). Males usually emerge first, and search for host eggs and wait for females to emerge and mate. This wasp is able to swim using their wings and live for about 15 days, sometimes completely submerged (Jackson 1966). The Trichogrammatidae egg parasitoid *Pseudoligosita longifrangiata* (Viggiani) finds the host plant *Tonina fluviatilis* Aubl. (Eriocaulaceae) and searches for *Argia insipida* Hagen in Selys (Odonata, Coenagrionidae) eggs underwater (Querino and Hamada 2009). *Anagrus obscurus* (Hymenoptera: Mymaridae) parasitizes eggs of *Calopteryx haemorrhoidalis* (Vander Linden) and *Platycnemis pennipes* (Pallas) (Odonata: Calopterygidae: Platycnemididae)

(Santolamazza et al. 2011). Many aquatic Hymenoptera have adaptations like dense pubescence and elongated strongly curved claws that seem to be adaptive to live underwater (Bennett 2007). The ectoparasitoid genus *Agriotypus* Haliday exhibits a unique pupal respiratory filament to breathe after death of its hosts, caddisfly pupae (Elliott 1982).

#### 5.1.2.1 Response to Parasitism

Since pathogens and parasites are widespread in the water column, aquatic insects are able to use some sort of ability to detect parasitoids and adopt spatial avoidance strategies to evade them. Detection of parasitoid signals may trigger responses such as evasive or aggressive behaviors in hosts (Gross 1993). Volatile cues, along with visual, tactile cues such as nonvolatile chemicals and waterborne cues such as vibrations and chemicals, could be detected by potential hosts that may change their behavior to evade parasitoidism (Hirayama and Kasuya 2015). The detection of chemical compounds, called necromones, produced by dead or infected animals, may trigger behavioral responses to avoid affected habitats or food sources and infected mates, conspecifics, and prey (Behringer et al. 2018). Females of *Ae. aegypti* avoid oviposition in water containing or that previously contained larvae parasitized by *Plagiorchis elegans* (Rudolphi) (Trematoda: Plagiorchiidae), which suggests a response to chemical signals produced by the parasitic infection (Lowenberger and Rau 1994). In high parasitoid wasp densities, which represents a higher parasitism risk, female water striders *Aquarius paludum insularis* (Motschulsky) (Heteroptera: Gerridae) tend to increase oviposition depth (Hirayama and Kasuya 2009).

Ultimately, adult and larval forms of aquatic insects must have well-adapted immunological responses to parasites. However, the response to infection can be dependent on resource allocation and adopted behavioral strategies, which may be sex dependent. There is a delicate balance between energetic investment in defense against pathogens and reproductive traits, causing a sex-biased difference favoring females (Stoehr and Kokko 2006; Klein and Flanagan 2016). For example, male odonates tend to allocate more resources to mating success than to defense against parasites, which cause higher infection levels and lower survival rates than females (Córdoba-Aguilar and Munguía-Steyer 2013).

### 5.1.3 Competition

Competition occurs when individuals must share resources. In freshwater environments, such resources are often related to microhabitat occupancy, oviposition sites, and food exploration. These interactions are crucial for the maintenance of biodiversity, since sympatric species must develop ways to live in equilibrium or exclude each other. Competition may also occur between individuals of the same species;

however, this topic will be better addressed in another chapter of this book. Here, we exemplify the most striking relationship between sympatric species; for instance, mosquitoes that live in phytotelmata are good models for studies on interspecific competition because several species are sympatric, at least in the larval stage, and directly compete for resources. Yee and Skiff (2014) carried out a study to determine potential ecological effects of *Culex coronator* (Dyar and Knab) (Diptera: Culicidae) on other container-dwelling mosquitoes. In this study, larvae of *Ae. albopictus* (Skuse) and *C. quinquefasciatus* (Say) (Diptera: Culicidae) were placed with *C. coronator* in three different resource scenarios. Survival rates, male and female body mass, and development time were impacted by interspecific competition. Female mass reduction may directly affect egg production, and hence mating and oviposition.

Competition can be so intensive that results in cannibalism (Fig. 5.5a). Eating conspecifics is a behavior that usually is expressed in stressful situations, such as in high population densities and reduced resources. The cannibal benefits not just from nutritional input, but also by attenuating competition in the future (Fincke 1994). Consequently, cannibalism aids populations to endure stressful situations (Parajulee and Phillips 1995). In well-delimited habitats, such as phytotelmata and ponds, the effects of cannibalism can be well studied. The larval *Megaloprepus coerulatus* Drury (Odonata: Pseudostigmatinae) live in water-filled tree holes where, independently of prey availability, larger larvae regularly prey on smaller conspecifics. In this case, cannibalism is the major cause of mortality in the first life stages (Fincke 1994).

Intraguild predation (IGP) occurs when individuals of one species prey on other species that use similar resources, reducing potential competition (Fig. 5.5b). IGP also has an important role in poor nutrient habitats, as seen in the caddisfly *Asynarchus nigriculus* (Banks). These trichopterans live in subalpine wetlands, in temporary ponds, and present intraguild predation and cannibalism independent of conspecific or heterospecific density, taking advantage of a head start in development that propitiates larger body sizes (Wissinger et al. 1996). In temporary ponds, ephemeral habitats, individuals have to emerge fast, hence, animals are much more aggressive than heterospecifics that live in other habitats. Caddisflies

**Fig. 5.5** Cannibalism in *Homeoura* Kennedy (Odonata: Coenagrionidae) in Peru (**a**) and a case of intraguild predation by *Gomphus pulchellus* Selys (Odonata: Gomphidae) devouring a male *Ceriagrion tenellum* (De Villiers) (Odonata: Coenagrionidae) (**b**). Photographs by Adolfo Cordero Rivera

from permanent wetlands were much less aggressive than caddisflies from temporary ponds, suggesting that IGP can play an important role in complementing the detritivores diet. Predating other insects of the same guild leads to earlier emergence, something important for an animal that lives in unpredictable habitats (Wissinger et al. 2004).

### *5.1.4 Harmonic Interactions*

Aquatic insects can take a hike on other organisms and can be used as phoretic transports by others. Chironomids are known to adhere to other aquatic insects such as Megaloptera, dragonflies, black flies, stoneflies, damselflies, mayflies, belostomatids, and naucorids, and also hydrozoans, fish, and gastropods (Roque et al. 2004; Boonsoong 2016). Some are even known to associate with freshwater sponges (Fusari et al. 2014). Such associations can result in mutualistic interactions, for instance, *Cricotopus nostocicola* (Wirth) (Diptera: Chironomidae) and *Nostoc parmelioides* Kützing (Cyanobacteria) are involved in a mutualistic relationship. The chironomid benefits from eating the mutualist cells, and *Nostoc* colonies that contain mosquito larvae are larger and present higher photosynthetic rates (Ward et al. 1985).

Ecological facilitation may occur when shredders live in sympatry (Bronstein 2009). This type of interaction benefits at least one species and cause harm to neither. When sympatry, shredder caddisflies facilitate food acquisition towards other species. As a result, there may be positive effects on emergence rate, wing length, and biomass, resulting in higher survival and fecundity of one caddisfly (Westveer et al. 2018). Rarely, other mutualistic relationships, such as pollination, are recorded for aquatic insects. For example, some stoneflies visit flowers and are known to act as facultative pollinators (Wong Sato and Kato 2017).

Some aquatic insects rely on the association with plants for growth and survival. Several species use macrophytes as oviposition sites and shelter for larvae. Water scorpions (Hemiptera: Nepidae), naucorids, odonate larvae and diving beetles actively occupy and seek shelter in macrophyte leaves and roots and use them as perch to stalk prey (Fig. 5.6). Other aquatic insects depend on plants for camouflage, or to defend a territory when adults. Evidence suggests, for example, that some damselfly species may exhibit preference and site fidelity in relation to the species of plant on the environment (Guillermo-Ferreira and Del-Claro 2011). Caddisfly larvae use leaves and plant material to build their cases, and some aquatic Lepidoptera have a close relationship with macroalgae that are cultivated in their silk cocoons.

**Fig. 5.6** Predatory aquatic insects usually seek shelter in macrophytes and use the perch to stalk prey, for example, water scorpions *Ranatra linearis* (Hemiptera: Nepidae) (**a**), creeping water bugs (Hemiptera: Naucoridae) (**b**), diving beetles (Coleoptera: Dytiscidae) (**c**), and dragonfly larvae (Odonata: Aeshnidae). Photographs (**a**) by Adolfo Cordero Rivera and (**b–d**) by Frederico Salles

### *5.1.5 Bacteria*

Bacterial activities are essential to nutrient cycles on ecosystems and for the main tenance of food web dynamics (Baines and Pace 1991; Elser et al. 2000; Godwin et al. 2017). Some of the most commonly found species in lakes and rivers are Proteobacteria, Cyanobacteria, Actinobacteria, Verrucomicrobia, and Cytophaga-Flavobacterium-Bacteroides groups (Zwart et al. 2002). These organisms degrade the organic matter dissolved in the water and directly affect the entire trophic networks (Logue et al. 2016). For example, most aquatic bacteria constitute an important element of freshwater biofilms, consisting in a matrix that may be formed by communities of prokaryotic and eukaryotic organisms that grow on rock surfaces (Zancarini et al. 2017) and plants (Marsollier et al. 2004). Some mayflies are scrapers and feed on the biofilm on rock surfaces. Chironomid larvae feed directly upon bacteria that degrade methane (Kiyashko et al. 2004), and some species of leaf shredders depend on bacterial enzymes acquired through ingestion of detritus or produced by endobiont bacteria that live in their guts (Sinsabaugh et al. 1985).

Aside from decomposition agents, some bacteria are in fact pathogenic and widespread among freshwater environments. The proteobacteria *Wolbachia* Hertig and Wolbach, an obligate intracellular parasite whose incidence in aquatic insects was subject to recent surveys, is found in 52% of the species. Ephemeroptera,

Plecoptera, and Trichoptera have the lowest incidence of infection, while infection prevalence is higher in Diptera, Hemiptera, Odonata, and Coleoptera (Sazama et al. 2017). Some strains of *Wolbachia* are vital in mutualistic associations, where the growth and nutrition of the host are dependent on B vitamins provided by these endosymbionts, but most strains are facultative or parasitic associates (Nikoh et al. 2014). Certain strains of this bacteria, for example, have strong deleterious effects or additive effects on egg production and viability, survival, adult body size, and fecundity, and seem to be transmitted to offspring whenever the female is infected (e.g., gut bacteria—Coon et al. 2016; Axford et al. 2016).

### 5.1.6 Fungi

There are also interactions between sympatric aquatic insects and fungi. For instance, fungi greatly contribute to the growth of caddisfly larvae (Chung and Suberkropp 2009). Symbiotic fungi (Trichomycetes) are found in the guts of aquatic insect larvae, such as shredder caddisflies that need this association to break leaf material (Ferrington Jr et al. 2005). The interaction between shredders, fungi, and bacteria involved in the breakdown of leaf litter is widely recognized, especially involving aquatic hyphomycetes colonizing leaf material (for reviews, see Cummins and Klug 1979; Bärlocher 1985; Anderson and Cargill 1987). Some studies have shown that fungal colonization on leaves can affect the food choice and growth of shredders by changing the palatability and the nutritional quality of leaves (Suberkropp 1992). Fungi can even play a greater role than bacteria in the food choice of shredders. For example, *Pycnopsyche* Banks caddisflies exhibit a greater preference for fungal-colonized than for bacterial-colonized leaf material (Mackay and Kalff 1973). Leaf material colonized by fungi is a high-quality food resource and contributes to the growth of *Pycnopsyche* caddisflies (Chung and Suberkropp 2009). Limnephilid caddisfly larvae may even exhibit preference for leaves colonized by two specific fungi, when leaves that were individually colonized by ten different fungal species were offered to larvae, showing that shredders can even discriminate among different fungal species (Arsuffi and Suberkropp 1988).

Nevertheless, fungi can also act as natural enemies. *Lagenidium giganteum* Couch is commercially produced as a control agent of mosquitoes (Scholte et al. 2004). *Ophiocordyceps odonatae* (Kobayasi) and *Hymenostilbe odonatae* Kobayasi are known as entomopathogenic fungi that affect dragonflies (Ke and Ju 2015). Therefore, freshwater communities could face large-scale effects caused by fungal infections, which are major regulators of population dynamics (Kohler and Hoiland 2001). For instance, the dominant caddisfly *Glossosoma nigrior* Banks outcompetes other grazers and filterers, except when microsporidian outbreaks affect *G. nigrior* populations (Kohler and Wiley 1997).

### 5.1.7 Herbivory

Herbivory is the "ingestion of living plant tissue from algae or vascular plants" (Cummins 1973). Shredder herbivores feed on living tissues of rooted plants, producing holes in leaves (Cummins 2016). Examples of shredders are beetles (Coleoptera), aquatic moths (Lepidoptera), and some representatives of Diptera and Trichoptera (Merritt et al. 1996). Aquatic insects that feed on algae and associated material on surfaces where the periphyton accumulates are called scrapers. This is the case of Leptophlebiidae and Heptageniidae mayfly nymphs (Cummins 2016). To feed on surfaces, these insects need to maintain their position in the stream, and have adaptations for this purpose, such structures to attach to the substrate (Cummins 2016).

Aquatic insects may also form galls and induce overgrowth of plant tissue by chemical compounds found in salivary and oviposition fluids (Bartlett and Connor 2014). Among the few records of aquatic insects forming galls in aquatic plants, there are chironomids in Podostemaceae macrophytes (Jäger-Zürn et al. 2013), and gall midges (Cecidomyiidae: Diptera) were found inducing galls in rhizomes of Erichhornia azurea (Sw.) Kunth (Ponteridaceae) (Peláez-Rodriguez et al. 2003; Urso-Guimarães et al. 2014). Larvae of some species of chironomids burrow their way inside the host plant while feeding, mining the tissue and using cavities as shelter to complete development (McGaha 1952).

### 5.1.8 Concluding Remarks

Freshwater ecosystems harbor an extraordinarily rich and sensitive biota that depend on ecological interactions to thrive. Here in this chapter, we sought to show part of this biodiversity and to give an insight into how ecological interactions may shape aquatic insect assemblages. The great majority of the 126,000 species of freshwater animals are insects (60.4%) (Balian et al. 2008); however, the study of biodiversity lacks even the most basic data on this group (see Cardoso et al. 2011 for an overview). Acquiring knowledge on different species or phylogenetic lineages is a fundamental step in documenting biodiversity, mainly by understanding the behavioral, ecological, and evolutionary processes associated with diversity patterns and consequently support conservation strategies. Finally, we suggest that future studies should address the ecological and evolutionary patterns of interaction networks involving tropical aquatic insects to unravel the driving forces of their evolution and distribution.

## References


Allan JD (1995) Predation and its consequences, Stream Ecology. Springer, Dordrecht, pp 163–185

Akira Shimizu, (1992) Nesting behavior of the semi-aquatic spider wasp,Anoplius eous, which transports its prey on the surface film of water (Hymenoptera, Pompilidae). Journal of Ethology 10 (2):85-102

Amelia K. Ward, Clifford N. Dahm, Kenneth W. Cummins, NOSTOC (CYANOPHYTA) PRODUCTIVITY IN OREGON STREAM ECOSYSTEMS: INVERTEBRATE INFLUENCES AND DIFFERENCES BETWEEN MORPHOLOGICAL TYPES1. Journal of Phycology 21 (2):223-227

Anderson NH, Cargill AS (1987) Nutritional ecology of aquatic detritivorous insects. Nutritional Ecology of Insects, Mites, Spiders, and Related Invertebrates, 903–925

Anderson RA, Koellaf JC, Hurd H (1999) The effect of *Plasmodium yoelii* nigeriensis infection on the feeding persistence of Anopheles stephensi Liston throughout the sporogonic cycle. Proc R Soc Lond B Biol Sci 266(1430):1729–1733

Andrade MR, Albeny-Simões DANIEL, Breaux JA, Juliano SA, Lima E (2017) Are behavioural responses to predation cues linked across life cycle stages? Ecol Entomol 42(1):77–85

Arsuffi TL, Suberkropp K (1988) Effects of fungal mycelia and enzymatically degraded leaves on feeding and performance of caddisfly (Trichoptera) larvae. J N Am Benthol Soc 7(3):205–211

Auld JR, Relyea RA (2011) Adaptive plasticity in predator-induced defenses in a common freshwater snail: altered selection and mode of predation due to prey phenotype. Evol Ecol 25(1):189–202

Axford JK, Ross PA, Yeap HL, Callahan AG, Hoffman AA (2016) Fitness of wAlbB Wolbachia Infection in *Aedes aegypti*: parameter estimates in an outcrossed background and potential for population invasion. Am J Trop Med Hyg 94(3):507–516

Babbitt K, Jordan F (1996) Predation on *Bufo terrestris* tadpoles: effects of cover and predator identity. Copeia 1996:485–488

Bartlett, L., & Connor, E. F. (2014). Exogenous phytohormones and the induction of plant galls by insects. Arthropod-Plant Interactions, 8(4), 339-348.

Baines SB, Pace ML (1991) The production of dissolved organic matter by phytoplankton and its importance to bacteria: patterns across marine and freshwater systems. Limnol Oceanogr 36(6):1078–1090

Balian EV, Segers H, Leveque C, Martens K (2008) An introduction to the freshwater animal diversity assessment (FADA) project. Hydrobiologia 595:3–8

Bärlocher F (1985) The role of fungi in the nutrition of stream invertebrates. Bot J Linn Soc 91(1–2):83–94

Behringer DC, Karvonen A, Bojko J (2018) Parasite avoidance behaviours in aquatic environments. Philos Trans R Soc B Biol Sci 373(1751)

Benard MF (2004) Predator-induced phenotypic plasticity in organisms with complex life histories. Annu Rev Ecol Evol Syst 35:651–673

Benfield EF (1974) Autohemorrhage in two stoneflies (Plecoptera) and its effectiveness as a defense mechanism. Ann Entomol Soc Am 67(5):739–742

Bennett, A. M. (2007). Global diversity of hymenopterans (Hymenoptera; Insecta) in freshwater. In Freshwater Animal Diversity Assessment (pp. 529-534). Springer, Dordrecht.

Boonsoong B (2016) Phoretic associations between *Nanocladius asiaticus* (Diptera, Chironomidae) and its hosts *Gestroiella* (Heteroptera, Naucoridae) and *Euphaea masoni* (Odonata, Euphaeidae) in streams in Western Thailand. Ann Limnol Int J Limnol 52:163–169. EDP Sciences

Böttger K (1972) Biological and ecological studies on the life cycle of freshwater-mites II. The life cycle of *Limnesia maculata* and *Unionicola crassipes*. Int Rev Ges Hydrobiol Hydrogr 57(2):263–319

Böttger K (1976) Types of parasitism by larvae of water mites (Acari: Hydrachnellae). Freshw Biol 6(6):497–500

Bronstein JL (2009) The evolution of facilitation and mutualism. J Ecol 97(6):1160–1170

Cardoso P, Erwin TL, Borges PAV, New TR (2011) The seven impediments in invertebrate conservation and how to overcome them. Biol Conserv 144:2647–2655
Chae J-s, Pusterla N, Johnson E, DeRock E, Lawler SP, Madigan JE (2000) Infection of aquatic insects with trematode metacercariae carrying Ehrlichia Risticii, the cause of potomac horse fever. J Med Entomol 37(4):619–625
Chase J (1999) Food web effects of prey size refuge: variable interactions and alternative stable equilibria. Am Nat 154:559–570
Chung N, Suberkropp K (2009) Contribution of fungal biomass to the growth of the shredder, *Pycnopsyche gentilis* (Trichoptera: Limnephilidae). Freshw Biol 54(11):2212–2224
Coccia C, Boyero L, Green AJ (2014) Can differential predation of native and alien corixids explain the success of *Trichocorixa verticalis verticalis* (Hemiptera, Corixidae) in the Iberian Peninsula? Hydrobiologia 734(1):115–123
Coon KL, Brown MR, Strand MR (2016) Gut bacteria differentially affect egg production in the anautogenous mosquito Aedes aegypti and facultatively autogenous mosquito *Aedes atropalpus* (Diptera: Culicidae). Parasit Vectors 9(1):375
Corbet PS (1999) Dragonflies: behavior and ecology of Odonata. Harley Books, Colchester, p 829
Córdoba-Aguilar A, Munguía-Steyer R (2013) The sicker sex: understanding male biases in parasitic infection, resource allocation and fitness. PLoS One 8(10):e76246. https://doi.org/10.1371/journal.pone.0076246
Cummins KW, Klug MJ (1979) Feeding ecology of stream invertebrates. Annu Rev Ecol Syst 10(1):147–172
Dahl J, Peckarsky BL (2002) Induced morphological defenses in the wild: predator effects on a mayfly, *Drunella coloradensis*. Ecology 83(6):1620–1634
Del-Claro K, Torezan-Silingardi HM (2012) Ecologia das interações plantas-animais: Uma abordagem ecológico-evolutiva. Technical Books Editora
Del-Claro K, Rico-Gray V, Torezan-Silingardi HM, Alves-Silva E, Fagundes R, Lange D, Dáttilo W, Vilela AA, Aguirre A, Rodriguez-Morales D (2016) Loss and gains in ant–plant interactions mediated by extrafloral nectar: fidelity, cheats, and lies. Insect Soc 63(2):207–221
DeWitt TJ (1998) Costs and limits of phenotypic plasticity: tests with predator-induced morphology and life history in a freshwater snail. J Evol Biol 11(4):465–480
Duffield RM, Blum MS, Wallace JB, Lloyd HA, Regnier FE (1977) Chemistry of the defensive secretion of the caddisfly *Pycnopsyche scabripennis* (Trichoptera: Limnephilidae). J Chem Ecol 3(6):649–656
Elliott JM (1982) The life cycle and spatial distribution of the aquatic parasitoid Agriotypus armatus (Hymenoptera: Agriotypidae) and its caddis host *Silo pallipes* (Trichoptera: Goeridae). J Anim Ecol 51:923–941
Elser JJ, Fagan WF, Denno RF, Dobberfuhl DR, Folarin A, Huberty A, Interlandi S et al (2000) Nutritional constraints in terrestrial and freshwater food webs. Nature 408(6812):578–580
Esch GW, Barger MA, Fellis KJ (2002) The transmission of digenetic trematodes: style, elegance, complexity. Integr Comp Biol 42(2):304–312
Eveland LL, Bohenek JR, Silberbush A, Resetarits WJ Jr. (2016) Detection of fish and newt kairomones by ovipositing mosquitoes. Chemical Signals in Vertebrates 13 (ed. by B. A.)
Feitosa MCB, Querino RB, Hamada N (2016) Association of *Anagrus amazonensis* Triapitsyn, Querino & Feitosa (Hymenoptera, Mymaridae) with aquatic insects in upland streams and floodplain lakes in central Amazonia, Brazil. Rev Brasil Entomol 60:267–269
Ferrington LC Jr, Lichtwardt RW, Hayford B, Williams MC (2005) Symbiotic Harpellales (Trichomycetes) in Tasmanian aquatic insects. Mycologia 97(1):254–262
Fincke OM (1984) Giant damselflies in a tropical forest: reproductive biologyof Megaloprepus coerulatus with notes on Mecistogaster (Zygoptera: Pseudostigmatidae). Adv Odonatol 2(1):13–27
Fincke OM (1994) Population regulation of a tropical damselfly in the larval stage by food limitation, cannibalism, intraguild predation and habitat drying. Oecologia 100(1–2):118–127

Fusari, L. M., Roque, F. D. O., & Hamada, N. (2014). Systematics of Oukuriella Epler, 1986, including a revision of the species associated with freshwater sponges. Insect Systematics & Evolution, 45(2), 117-157.
E. S. Gabitzsch, C. D. Blair, B. J. Beaty, (2006) Effect of La Crosse Virus Infection on Insemination Rates in Female Aedes triseriatus (Diptera: Culicidae). Journal of Medical Entomology 43 (5):850-852
Godwin CM, Whitaker EA, Cotner JB (2017) Growth rate and resource imbalance interactively control biomass stoichiometry and elemental quotas of aquatic bacteria. Ecology 98(3):820–829
González-Tokman D, Córdoba-Aguilar A, González-Santoyo I, Lanz-Mendoza H (2011) Infection effects on feeding and territorial behaviour in a predatory insect in the wild. Anim Behav 81(6):1185–1194
Grabner DS (2017) Hidden diversity: parasites of stream arthropods. Freshw Biol 62:52–64
Gross P (1993) Insect behavioral and morphological defenses against parasitoids. Annu Rev Entomol 38(1):251–273
Guillermo-Ferreira R, Del-Claro K (2011) Oviposition site selection in *Oxyagrion microstigma* Selys, 1876 (Odonata: Coenagrionidae) is related to aquatic vegetation structure. Int J Odonatol 14(3):275–279
Guillermo-Ferreira R, Vilela DS (2013) New records of *Forcipomyia* (Pterobosca) *incubans* (Diptera: Ceratopogonidae) parasitizing wings of Odonata in Brazil. Biota Neotrop 13(1):360–362
Hayashi M, Ohba S-Y (2018) Mouth morphology of the diving beetle *Hyphydrus japonicus* (Dytiscidae: Hydroporinae) is specialized for predation on seed shrimps. Biol J Linn Soc 125(2):315–320
Henrikson B-I (1988) The absence of antipredator behavior in the larvae of Leucorrhinia dubia (Odonata) and the consequences for their distribution. Oikos:179–183
Hering D, Plachter H (1997) Riparian ground beetles (Coleoptera, Carabidae) preying on aquatic invertebrates: a feeding strategy in alpine floodplains. Oecologia 111(2):261–270
Hirayama H, Kasuya E (2009) Oviposition depth in response to egg parasitism in the water strider: high-risk experience promotes deeper oviposition. Anim Behav 78(4):935–941
Hirayama H, Kasuya E (2015) Parasitoid avoidance behavior is not triggered by airborne cues in a semi-aquatic bug. Hydrobiologia 745:195–200
Hirayama T, Yoshida T, Nagasaki O (2014) The life history and host-searching behaviour of the aquatic parasitoid wasp Apsilops japonicus (Hymenoptera: Ichneumonidae), a parasitoid of the aquatic moth Neoshoenobia testacealis (Lepidoptera: Crambidae). J Nat Hist 48(15–16):959–967
Honkavaara J, Rantala MJ, Suhonen J (2009) Mating status, immune defence, and multi-parasite burden in the damselfly Coenagrion armatum. Entomol Exp Appl 132(2):165–171
Hopkins GR, Gall BG, Brodie ED (2011) Ontogenetic shift in efficacy of antipredator mechanisms in a top aquatic predator, *Anax junius* (Odonata: Aeshnidae). Ethology 117(12):1093–1100
Hoverman, J. T., Auld, J. R., & Relyea, R. A. (2005). Putting prey back together again: integrating predator-induced behavior, morphology, and life history. Oecologia, 144(3), 481-491.
Ingley SJ, Bybee SM, Tennessen KJ, Whiting MF, Branham MA (2012) Life on the fly: phylogenetics and evolution of the helicopter damselflies (Odonata, Pseudostigmatidae). Zool Scr 41(6):637–650
Jackson DJ (1966) Observations on the Biology of Caraphractus Cinctus Walker (Hymenoptera: Mymaridae), a Parasitoid of the Eggs of Dytiscidae (Coleoptera): III. The Adult Life and Sex Ratio. Trans R Entomol Soc Lond 118(2):23–49
Jackson BT, Brewster CC, Paulson SL (2012) La Crosse virus infection alters blood feeding behavior in Aedes triseriatus and *Aedes albopictus* (Diptera: Culicidae). J Med Entomol 49(6):1424–1429
Jäger-Zürn I, Spies M, Philbrick CT Bove CP, Mora-Olivo A (2013) Plant galls (cecidia) in the neotropical water plant family Podostemaceae induced by larvae of Chironomidae. Spixiana 36(1):97–112

Johansson F (2002) Reaction norms and production costs of predator-induced morphological defences in a larval dragonfly (Leucorrhinia dubia: Odonata). Can J Zool 80(5):944–950
Johansson F, Samuelsson L (1994) Fish-induced variation in abdominal spine length of Leucorrhinia dubia (Odonata) larvae? Oecologia 100(1–2):74–79
Ke YH, Ju YM (2015) Two rare ophiocordycipitaceous fungi newly recorded in Taiwan. Bot Stud 56(1):30
Kenneth W. Cummins, (1973) Trophic Relations of Aquatic Insects. Annual Review of Entomology 18 (1):183-206
Kenneth W. Cummins, (2016) Combining taxonomy and function in the study of stream macroinvertebrates. Journal of Limnology 75 (s1)
Kehr A, Schnack J (1991) Predator prey relationship between Giant water bugs (Belostoma oxyurum) and larval anurans (*Bufo arenarum*). Alytes 9:61–69
Kiyashko SI, Imbs AB, Narita T, Svetashev VI, Wada E (2004) Fatty acid composition of aquatic insect larvae *Stictochironomus Pictulus* (Diptera: Chironomidae): evidence of feeding upon methanotrophic bacteria. Comp Biochem Physiol B Biochem Mol Biol 139(4):705–711
Klein SL, Flanagan KL (2016) Sex differences in immune responses. Nat Rev Immunol 16(10):626
Knight TM, McCoy MW, Chase JM, McCoy KA, Holt RD (2005) Trophic cascades across ecosystems. Nature 437(7060):880
Koella JC et al (2002) Stage-specific manipulation of a mosquito's host-seeking behavior by the malaria parasite Plasmodium gallinaceum. Behav Ecol 13:816–820
Kohler SL (2008) The ecology of host–parasite interactions in aquatic insects. In: Lancaster J, Briers R, Macadam C (eds) Aquatic insects: challenges to populations. CAB International, Wallingford, pp 55–80
Kohler SL, Hoiland WK (2001) Population regulation in an aquatic insect: the role of disease. Ecology 82(8):2294–2305
Kohler SL, Wiley MJ (1997) Pathogen outbreaks reveal large-scale effects of competition in stream communities. Ecology 78(7):2164–2176
Kula RR, Gates MG, Buffington ML, Harms NE (2017) Parasitoid Wasps (Hymenoptera: Apocrita) Associated with *Sagittaria latifolia* Willd. and *Sagittaria platyphylla* (Engelm.) J. G. sm. (Alismatales: Alismataceae) in the Nearctic Region. Proc Entomol Soc Wash 19(2):215–217
Lima, S. L., & Dill, L. M. (1990). Behavioral decisions made under the risk of predation: a review and prospectus. Canadian journal of zoology, 68(4), 619-640.
Lima-Camara TN, Bruno RV, Luz PM, Castro MG, Lourenço-de-Oliveira R, Sorgine MHF (2011) Dengue infection increases the locomotor activity of *Aedes aegypti* females. PLoS One 6(3):e17690
Logue JB, Stedmon CA, Kellerman AM, Nielsen NJ, Andersson AF, Laudon H, Lindström ES, Kritzberg ES (2016) Experimental insights into the importance of aquatic bacterial community composition to the degradation of dissolved organic matter. ISME J 10(3):533–545
Lowenberger CA, Rau ME (1994) *Plagiorchis elegans*: emergence, longevity and infectivity of cercariae, and host behavioural modifications during cercarial emergence. Parasitology 109(1):65–72
Luttbeg B, Kerby JL (2005) Are scared prey as good as dead? Trends Ecol Evol 20(8):416–418
Mackay RJ, Kalff J (1973) Ecology of two related species of caddis fly larvae in the organic substrates of a woodland stream. Ecology 54(3):499–511
Majdi N, Traunspurger W, Richardson JS, Lecerf A (2015) Small stonefly predators affect microbenthic and meiobenthic communities in stream leaf packs. Freshw Biol 60(9):1930–1943
Marden JH (1989) Bodybuilding dragonflies: costs and benefits of maximizing flight muscle. Physiol Zool 62(2):505–521
Marsollier L, Stinear T, Aubry J, Saint André JP, Robert R, Legras P, Manceau AL, Audrain C, Bourdon S, Kouakou H, Carbonnelle B (2004) Aquatic plants stimulate the growth of and biofilm formation by *Mycobacterium ulcerans* in axenic culture and harbor these bacteria in the environment. Appl Environ Microbiol 70(2):1097–1103

Martens A, Grabow K (2011) Early stadium damselfly larvae (Odonata: Coenagrionidae) as prey of an aquatic plant, Utricularia australis. Int J Odonatol 14(1):101–104
Y. J. McGaha, (1952) The Limnological Relations of Insects to Certain Aquatic Flowering Plants. Transactions of the American Microscopical Society 71 (4):355
Merritt, R. W., Cummins, K. W., & Berg, M. B. (1996). Aquatic insects of North America. Kendall Hunt, Dubuque.
Menke A (1979) Family Belostomatidae - Giant water bugs. In: Menke A (ed) The semiaquatic and aquatic Hemiptera of California (Heteroptera: Hemiptera). University of California Press, Berkeley, CA, pp 76–86
Moore KA, Williams DD (1990) Novel strategies in the complex defense repertoire of a stonefly (*Pteronarcys dorsata*) nymph. Oikos:49–56
Nair GA, Morse JC, Marshall SA (2015) Aquatic insects and their societal benefits and risks. J Entomol Zool Stud 3(3)
Nakano S, Murakami M (2001) Reciprocal subsidies: dynamic interdependence between terrestrial and aquatic food webs. Proc Natl Acad Sci 98(1):166–170
Nikoh N, Hosokawa T, Moriyama M, Oshima K, Hattori M, Fukatsu T (2014) Evolutionary origin of insect–Wolbachia nutritional mutualism. Proc Natl Acad Sci 111(28):10257–10262
Nislow KH, Molles MC Jr (1993) The influence of larval case design on vulnerability of *Limnephilus frijole* (Trichoptera) to predation. Freshw Biol 29(3):411–417
Paetzold A, Schubert CJ, Tockner K (2005) Aquatic terrestrial linkages along a braided-river: riparian arthropods feeding on aquatic insects. Ecosystems 8(7):748–759
Parajulee, M. N., & Phillips, T. W. (1995). Survivorship and cannibalism in Lyctocoris campestris (Hemiptera: Anthocoridae): effects of density, prey availability, and temperature. Journal of Entomological Science, 30(1), 1-8.
Peckarsky BL (1980) Predator-prey interactions between stoneflies and mayflies: behavioral observations. Ecology 61(4):932–943
Peckarsky BL (1987) Mayfly cerci as defense against stonefly predation: deflection and detection. Oikos:161–170
Peckarsky BL, Dodson SI (1980) Do stonefly predators influence benthic distributions in streams? Ecology 61(6):1275–1282
Peláez-Rodríguez, M., Trivinho-Strixino, S., & Urso-Guimarães, M. V. (2003). Galls in rhizome of an aquatic macrophyte, Eichhornia azurea (Swartz) kunth (Pontederiaceae), in Jataí Ecological Station, Luiz Antônio, SP, Brazil. Brazilian Journal of Biology, 63(4), 723-726.
Poulin R (1995) "Adaptive" changes in the behaviour of parasitized animals: a critical review. Int J Parasitol 25(12):1371–1383
Poulin R, Morand S (2000) The diversity of parasites. Q Rev Biol 75(3):277–293
Proctor H, Pritchard G (1989) Neglected predators: water mites (Acari: Parasitengona: Hydrachnellae) in freshwater communities. J N Am Benthol Soc 8(1):100–111
Querino, R. B., & Hamada, N. (2009). An aquatic microhymenopterous egg-parasitoid of Argia insipida Hagen in Selys (Odonata: Coenagrionidae) and biological observations in the Central Amazon, Brazil. Neotropical entomology, 38(3), 346-351.
Ready PD (2008) Leishmania manipulates sandfly feeding to enhance its transmission. Trends Parasitol 24(4):151–153
Resetarits Jr WJ, Binckley CA (2009) Spatial contagion of predation risk affects colonization dynamics in experimental aquatic landscapes. Ecology, 90(4), 869-876.
Roberts D (2017) Mosquito larvae can detect water vibration patterns from a nearby predator. Bull Entomol Res 107(4):499–505
Rogers ME (2012) The role of Leishmania proteophosphoglycans in sand fly transmission and infection of the mammalian host. Front Microbiol 3:223
Rogers ME, Bates PA (2007) Leishmania manipulation of sand fly feeding behavior results in enhanced transmission. PLoS Pathog 3(6):e91
Roque FDO, Trivinho-Strixino S, Jancso M, Fragoso EN (2004) Records of Chironomidae larvae living on other aquatic animals in Brazil. Biota Neotrop 4(2):1–9

Rossignol PA, Ribeiro JMC, Spielman ANDA (1986) Increased biting rate and reduced fertility in sporozoite-infected mosquitoes. Am J Trop Med Hyg 35(2):277–279

Sabo JL, Bastow JL, Power ME (2002) Length–mass relationships for adult aquatic and terrestrial invertebrates in a California watershed. J N Am Benthol Soc 21(2):336–343

Salinas AS, Costa RN, Orrico VG, Solé M (2018) Tadpoles of the bromeliad-dwelling frog *Phyllodytes luteolus* are able to prey on mosquito larvae. Ethol Ecol Evol 30(6):485–496

Santolamazza S, Baquero E, Cordero-Rivera A (2011) Incidence of *Anagrus obscurus* (Hymenoptera: Mymaridae) egg parasitism on *Calopteryx haemorrhoidalis* and *Platycnemis pennipes* (Odonata: Calopterygidae: Platycnemididae) in Italy. Entomol Sci 14(3):366–369

Sanzone DM et al (2003) Carbon and nitrogen transfer from a desert stream to riparian predators. Oecologia 134(2):238–250

Sazama EJ, Bosch MJ, Shouldis CS, Ouellette SP, Wesner JS (2017) Incidence of Wolbachia in aquatic insects. Ecol Evol 7:1165–1169

Schilder RJ, Marden JH (2007) Metabolic syndrome in insects triggered by gut microbes. J Diabetes Sci Technol 1(5):794–796

Scholte EJ, Knols BG, Samson RA, Takken W (2004) Entomopathogenic fungi for mosquito control: a review. J Insect Sci 4(1)

Schwartz, A., & Koella, J. C. (2001). Trade-offs, conflicts of interest and manipulation in Plasmodium–mosquito interactions. Trends in parasitology, 17(4), 189-194.

Sih A (1982) Foraging strategies and the avoidance of predation by an aquatic insect, *Notonecta hoffmanni*. Ecology 63(3):786–796

Silberbush A, Resetarits WJ Jr (2017) Mosquito female response to the presence of larvivorous fish does not match threat to larvae. Ecol Entomol 42(5):595–600

Sinsabaugh RL, Linkins AE, Benfield EF (1985) Cellulose digestion and assimilation by three leaf-shredding aquatic insects. Ecology 66(5):1464–1471

Smallegange, R. C., van Gemert, G. J., van de Vegte-Bolmer, M., Gezan, S., Takken, W., Sauerwein, R. W., & Logan, J. G. (2013). Malaria infected mosquitoes express enhanced attraction to human odor. PloS one, 8(5), e63602.

Smith B (1988) Host-parasite interaction and impact of larval water mites on insects. Annu Rev Entomol 33(1):487–507

Soluk DA, Clifford HF (1985) Microhabitat shifts and substrate selection by the psammophilous predator *Pseudiron centralis* McDunnough (Ephemeroptera: Heptageniidae). Can J Zool 63(7):1539–1543

Stechmann DH (1978) Eiablage, Parasitismus und postparasitische Entwicklung von *Arrenurus*-Arten (Hydrachnellae, Acari). Z Parasitenkd 57(2):169–188

Stoehr AM, Kokko H (2006) Sexual dimorphism in immunocompetence: what does life-history theory predict? Behav Ecol 17(5):751–756

Stoks R, Córdoba-Aguilar A (2012) Evolutionary ecology of Odonata: a complex life cycle perspective. Annu Rev Entomol 57:249–265

Suberkropp K (1992) Interactions with invertebrates. In: Bärlocher F (ed) The ecology of aquatic hyphomycetes, Ecological Studies, vol 94. Springer-Verlag, Berlin, pp 118–134

Swart C, Felgenhauer B (2003) Structure and function of the mouthparts and salivary gland complex of the Giant Waterbug, *Belostoma lutarium* (Stal) (Hemiptera: Belostomatidae). Ann Entomol Soc Am 96:870–882

Turlings, T. C., & Erb, M. (2018). Tritrophic interactions mediated by herbivore-induced plant volatiles: mechanisms, ecological relevance, and application potential. Annual review of entomology, 63, 433-452.

Urso-Guimarães, M. V. (2014). New species of Lopesia (Cecidomyiidae, Diptera) associated with Eichhornia azurea (Sw.) Kunth (Pontederiaceae) from Brazil. Iheringia Série Zoologia, 104(4).

Vance SA, Peckarsky BL (1997) The effect of mermithid parasitism on predation of nymphal *Baetis bicaudatus* (Ephemeroptera) by invertebrates. Oecologia 110(1):147–152

Ward AK, Dahm CN, Cummins KW (1985) Nostoc (Cyanophyta) productivity in oregon stream ecosystems: invertebrate influences and differences between morphological types 1. Journal of Phycology 21(2):223–227

Westveer JJ, Verdonschot PF, Verdonschot RC (2018) Biotic interactions enhance survival and fitness in the caddisfly *Micropterna sequax* (Trichoptera: Limnephilidae). Hydrobiologia 818(1):31–41

White DA (1969) The infection of immature aquatic insects by larval *Paragordius* (Nematomorpha). Great Basin Nat 29(1):10

Wildermuth H, Martens A (2007) The feeding action of Forcipomyia paludis (Diptera: Ceratopogonidae), a parasite of Odonata imagines. Int J Odonatol 10(2):249–255

Wiles CM, Bolek MG (2015) Damselflies (Zygoptera) as paratenic hosts for *Serpinema trispinosum* and its report from turtle hosts from Oklahoma, USA. Folia Parasit 62:019

Williams D, Williams S (2017) Aquatic insects and their potential to contribute to the diet of the globally expanding human population. Insects 8(3):72

Wissinger SA, Sparks GB, Rouse GL, Brown WS, Steltzer H (1996) Intraguild predation and cannibalism among larvae of detritivorous caddisflies in subalpine wetlands. Ecology 77(8):2421–2430

Wissinger S, Steinmetz J, Alexander JS, Brown W (2004) Larval cannibalism, time constraints, and adult fitness in caddisflies that inhabit temporary wetlands. Oecologia 138(1):39–47

Wissinger SA et al (2006) Predator defense along a permanence gradient: roles of case structure, behavior, and developmental phenology in caddisflies. Oecologia 147(4):667–678

Wong Sato AA, Kato M (2017) Pollination system of Corylopsis gotoana (Hamamelidaceae) and its stonefly (Plecoptera) co-pollinator. Plant Species Biol 32(4):440–447

Yang E et al (2016) Water striders adjust leg movement speed to optimize takeoff velocity for their morphology. Nat Commun 7:13698

Yee DA, Skiff JF (2014) Interspecific competition of a new invasive mosquito, Culex coronator, and two container mosquitoes, Aedes albopictus and Cx. quinquefasciatus (Diptera: Culicidae), across different detritus environments. J Med Entomol 51(1):89–96

Zancarini A, Echenique-Subiabre I, Debroas D, Taïb N, Quiblier C, Humbert JF (2017) Deciphering Biodiversity and Interactions between Bacteria and Microeukaryotes within Epilithic Biofilms from the Loue River, France. Sci Rep 7(1):1–13002E

Zwart G, Crump BC, Kamst-van Agterveld MP, Hagen F, Han SK (2002) Typical freshwater bacteria: an analysis of available 16S rRNA gene sequences from plankton of lakes and rivers. Aquatic microbial ecology 28(2):141–155

# Chapter 6
# Drumming for Love: Mating Behavior in Stoneflies

**J. Manuel Tierno de Figueroa, Julio M. Luzón-Ortega, and Manuel Jesús López-Rodríguez**

**Abstract** This chapter describes the current knowledge on the mating behavior of stoneflies (Insecta, Plecoptera). Arctoperlarian stoneflies use intersexual vibrational communication as a mechanism for encountering mates and pair forming, acting as a reproductive isolation mechanism. These vibrational signals can be produced by drumming, rubbing, tremulation, or a combination of these methods and are species specific; therefore, they can be used as a method to confirm/infer phylogenies or define species. Of particular interest is our understanding of their evolution from ancestral to derived models. Currently, the vibrational signals of approximately 200 species, mainly from North America, Europe, and, of less concern, Asia, have been described. For those species that do not use vibrational communication (principally Antarctoperlaria) or that occur in particular habitats where it is not effective, other mate-encountering mechanisms are discussed. In relation to non-vibrational mate-encountering mechanisms, some studies seem to show that certain substrates could act as encounter sites by aggregating sexes, both for Arctoperlaria and Antarctoperlaria. Finally, other reproductive behaviors, such as mating (sperm transmission, female stimulation, etc.), formation of mating balls, or mate-guarding strategies (both contact and noncontact guarding mechanisms), are also addressed.

**Keywords** Plecoptera · Vibrational communication · Encounter sites · Reproductive strategies · Mate guarding

In stoneflies (order Plecoptera), most biological studies have focused on nymphs because this life stage is much longer than the adult stage. Nevertheless, although the adult biology of stoneflies, and particularly their reproductive behavior, has been scarcely studied (except perhaps their employment of intersexual vibrational

J. M. Tierno de Figueroa (✉) · J. M. Luzón-Ortega
Facultad de Ciencias, Departamento de Zoología, Universidad de Granada, Granada, Spain
e-mail: jmtdef@ugr.es

M. J. López-Rodríguez
Facultad de Ciencias, Departamento de Ecología, Universidad de Granada, Granada, Spain
e-mail: manujlr@ugr.es



K. Del-Claro, R. Guillermo (eds.), *Aquatic Insects*,
https://doi.org/10.1007/978-3-030-16327-3_6

communication, which has received greater attention in the last decades), we now know that they exhibit complex strategies to facilitate mate encounters, mate evaluation, copulation, mate guarding, etc., making its study a bewitching topic of research. In this chapter, we try to elaborate on a brief compilation of the extant knowledge.

## 6.1 Vibrational Communication

Vibrational communication seems to be a very ancient system in animals; it is probably the primary channel of communication in some groups and is ubiquitous in arthropods and vertebrates (Hill 2008; Cocroft et al. 2014). Although sound communication has been well studied in many animals, vibrational communication using low-frequency, substrate-borne vibrations has been considerably less explored, despite its importance (Stewart 1997; Cocroft et al. 2014). In fact, many insect groups communicate by means of the production and reception of vibrational signals through solid substrates, and this is particularly the case in some aquatic insects, such as Plecoptera, Trichoptera, Megaloptera, and Neuroptera Sisyridae (e.g., Rupprecht 1968, 1975; Stewart 1997). However, among insects, it is in stoneflies (Plecoptera), particularly Arctoperlaria, where this communication system has evolved into its most complex form as a mechanism for mate encounter and pair forming (Stewart 1997, 2001). Nonetheless, among stoneflies, vibrational communication is unknown in Antarctoperlaria and has not been reported in Arctoperlaria that inhabit the Southern Hemisphere (Notonemuridae and part of Perlidae), in either Scopuridae or Styloperlidae families (Stewart 2001; Boumans and Johnsen 2015; DeWalt et al. 2015).

In Plecoptera, this communication system was first quantified by Rupprecht (1965, 1968), but as compiled by Rupprecht (1968) it was known in the nineteenth century (Newport 1851) and had been reported in the second half of the nineteenth century and first half of the twentieth century by authors such as Hagen (1877), MacNamara (1926), and Gaunitz (1935). Since then, the vibrational signals of approximately 200 species have been described, mainly in North America, Europe, and, to a lesser extent, Asia (Boumans and Johnsen 2015, DeWalt et al. 2015, updated data).

Stoneflies produce vibrational signals by drumming, rubbing, tremulation, or a combination of these methods (Stewart 1997). Drumming, or percussion, consisting of tapping (beating) on substrates with the ventral portion of the abdomen or some kind of specialized structure on it (such as ventral lobes, hammers, knobs, or vesicles, Fig. 6.1), is the most widely employed signaling method among stoneflies (Stewart 1997). It was Rupprecht (1976) who first demonstrated, by means of observational and experimental studies, that hammers and vesicles were used for drumming, acting in the production, internal monitoring, and control of the signals (in Abbott and Stewart 1997). A complete review of the evolutionary relationships between these structures and the vibrational communication behaviors can be found in Stewart and Maketon (1991). Rubbing consists of scraping the specialized

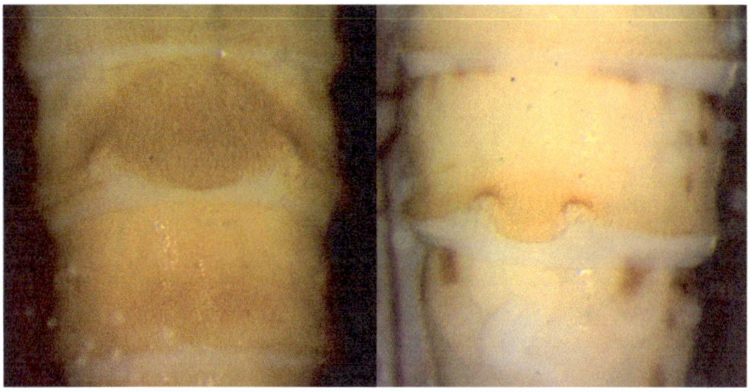

**Fig. 6.1** Ventral view of the male abdomen of *Hemimelaena flaviventris* (left) and *Isoperla morenica* (right) showing the lobe

abdominal structure on a substrate, producing a more continuous vibration. According to Stewart (1997), rubbing is an "abdominal-substrate stridulation". Finally, tremulation involves rocking (jerking) the abdomen or entire body and transmitting the resulting vibrations into a solid substrate through the legs. Examples of species that produce signals by drumming can be found in all Plecoptera families in which vibrational signals have been described (Stewart 1997), e.g., males and females of *Perla marginata* (Rupprecht 1965), *Isoperla phalerata* (Szczytko and Stewart 1979), and *Protonemura meyeri* (Tierno de Figueroa et al. 2014). Rubbing signals, reported in Perlidae, Perlodidae, and Peltoperlidae (DeWalt et al. 2015), are produced by some species, such as the males of *Peltoperla tarteri* (Maketon and Stewart 1988), *Flavoperla hagiensis* (formerly *Gibosia hagiensis*) (Hanada and Maruyama 2001), and *Isoperla bipartita* (Tierno de Figueroa and Luzón-Ortega 2002). Some species produce their signals through a combination of rubbing and tapping (drumming), such as the male of *Peltoperla arcuata* (Maketon and Stewart 1988) or the female of *Isoperla bipartita* (Tierno de Figueroa and Luzón-Ortega 2002). Tremulation in stoneflies was described by Rupprecht (1981) and, until now, had only been described in a few Chloroperlidae species, such as *Siphonoperla torrentium, Siphonoperla montana*, and *Suwalia pallidula* (Rupprecht 1981; Alexander and Stewart 1997).

Adult stoneflies (as some other arthropods, e.g., Bradbury and Vehrencamp 2011) can communicate by introducing waves into plant tissues (or other substrates able to transmit waves), and receivers that are some distance away from the sender can detect these waves as they pass under their feet when they are stationary. In fact, Rupprecht (1968) proved, through experiments in which he eliminated or stimulated certain parts of the body, that the subgenual organs [specialized chordotonal or scolopophorous organs in the proximal part of the tibiae (Chapman 1998) and one of the possible sense organs for vibrational reception] are the perceiving sense organs in stoneflies. Detailed explanations on the mechanistic basis of signal detection and transmission and other aspects of vibrational communication in insects can be found in Chapman (1998), Bradbury and Vehrencamp (2011), and, particularly, Hill (2008) and Cocroft et al. (2014).

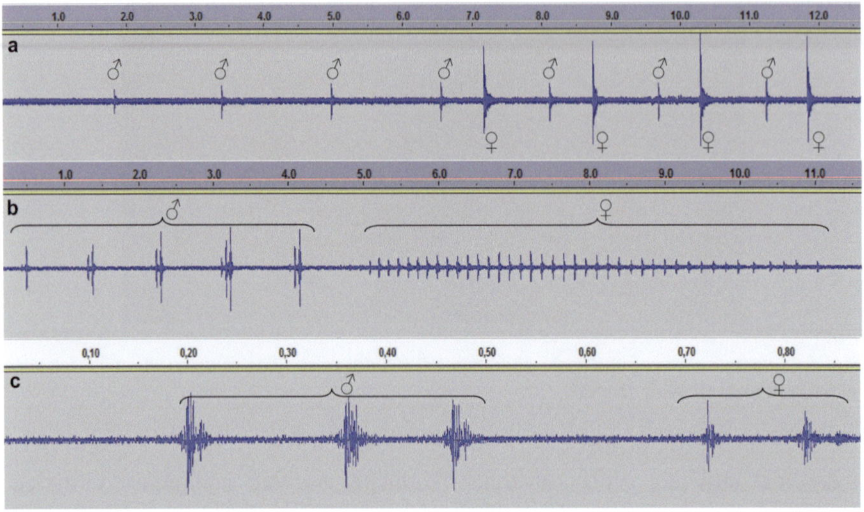


**Fig. 6.2** Oscillograms showing duets of (**a**) *Capnioneura mitis* (fam. Capniidae), (**b**) *Protonemura meyeri* (fam. Nemouridae), and (**c**) *Hemimelaena flaviventris* (fam. Perlodidae). Time (horizontal axis) in seconds. Images obtained using Audacity v1.2.6 software (Free Software Foundation, Boston, MA, USA)

The signal is usually initiated by the male ("call") and followed by the female ("answer"), but in some species a three-way signal (male-female-male) is produced (the last signal of the male is called the "reply"). It is accepted that only virgin females produce vibrational signals (Rupprecht 1968, 1969). Duets (Fig. 6.2) are typically repeated sequences of two-way or three-way conversations, or symphonies of grouped calls interspersed by grouped answers (Stewart and Sandberg 2006). Exceptionally, females can produce signals when males are not present (Maketon and Stewart 1988; Tierno de Figueroa et al. 2014), and, therefore, the female signal can also be considered a call and not only an answer, as usually pointed out in stoneflies. Female signals are usually short [even composed by only one beat, as in *Hansonoperla appalachia* and *Perlinella drymo* (Zeigler and Stewart 1977) or *Guadalgenus franzi* (Tierno de Figueroa et al. 2002)] and less complex than male signals, but the female signals of some Perlidae (Stewart and Sandberg 2006) and Nemouridae (Luzón-Ortega and Tierno de Figueroa 2003; Tierno de Figueroa et al. 2014) are composed of a high number of beats, and cases like that of *Isoperla bipartita* female, which has a signal composed of a beat and a rub (Tierno de Figueroa and Luzón-Ortega 2002) and that of *Suwallia pallidula*, in which the female answers the male's tremulation by tremulating (Stewart and Sandberg 2006), among others, are exceptional. The generally higher complexity of male signals could be a consequence of intrasexual competition and intersexual conflict, the latter when part of male signal mimics the female answer (Boumans and Johnsen 2015).

Vibrational signals produced by stoneflies can be as simple as a two-beat signal, such as the one described in the male of *Guadalgenus franzi* by Tierno de Figueroa

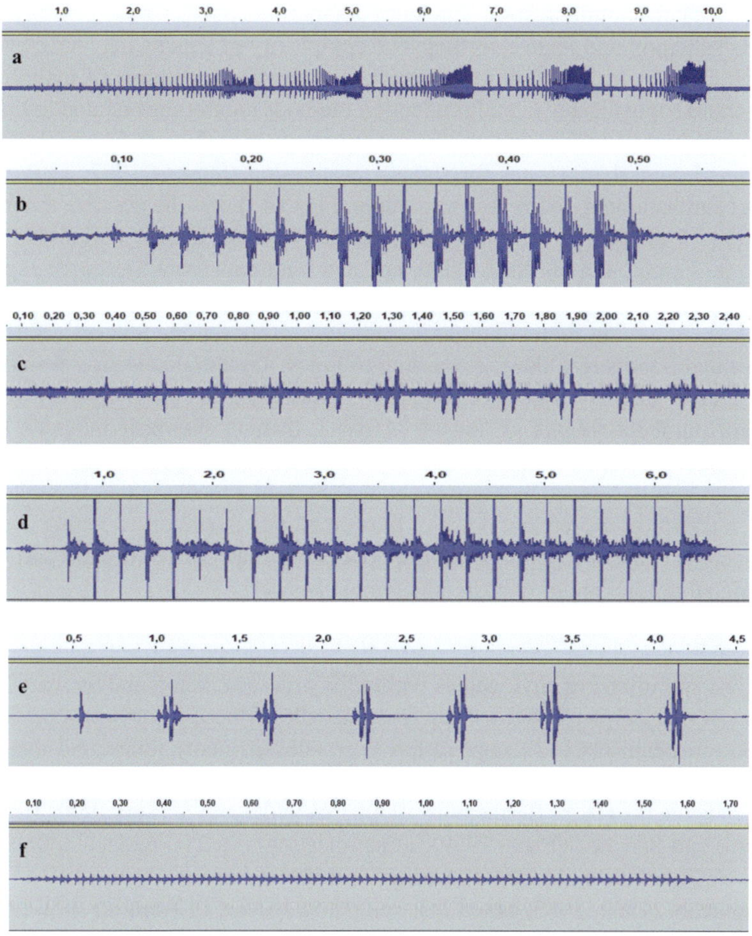


**Fig. 6.3** Oscillograms of the male vibrational signals of (**a**) *Isoperla nevada* (fam. Perlodidae), (**b**) *Eoperla ochracea* (fam. Perlidae), (**c**) *Tyrrhenoleuctra minuta* (fam. Leuctridae), (**d**) *Capnioneura libera* (fam. Capniidae), (**e**) *Protonemura meyeri* (fam. Nemouridae), and (**f**) *Rhabdiopteryx thienemanni* (fam. Taeniopterygidae). Time (horizontal axis) in seconds. Images obtained using Audacity v1.2.6 software (Free Software Foundation, Boston, MA, USA)

et al. (2002), or as complex as the one composed of diphasic sequences with its first sequence being different from the following ones (e.g., the male of *Isoperla nevada,* Tierno de Figueroa and Sánchez-Ortega 1999) or the one that involves different call production methods (e.g., *Peltoperla arcuata,* Maketon and Stewart 1988, as reported above) (Fig. 6.3). Understanding the evolution of these signals within Plecoptera from ancestral to complex derived models is of particular interest. According to Maketon and Stewart (1988), the ancestral method of signal production was percussion, and those ancestral signals were monophasic repetitions of evenly spaced beats. Increasingly complex signals were favored by natural selection, allowing for greater communication efficiency for pair encounters and, possibly, male fitness evaluation and mate selection. The complexity of signals may have evolved through changes in the rhythmic patterning of signals or patterns of male-female duets, more sophisticated signal production methods (sometimes related to the specialized coevolution of ventral abdominal structures in males), and possibly the use of selected natural substrates for signal transmission (Stewart 1997, 2001, 2009; Stewart and Sandberg 2006). As an example, Fig. 6.4 shows the possible theoretical evolution of vibrational call complexity in the Southern Iberian Perlodidae (from Tierno de Figueroa et al. 2002, 2009, updated with data from Tierno de Figueroa et al. 2011a, 2013, and unpublished data on a Spanish population of *Perlodes microcephalus*), following the evolutionary paradigm proposed by Stewart (2001) and Stewart and Sandberg (2006). According to Fig. 6.4, natural selection would have acted in favor of derivation from an ancestral, monophasic call to new signals with different numbers (higher or lower) of beats, distinct interbeat intervals (often phased and/or grouped to form sequences), or different methods of call production (such as rubbing) (Tierno de Figueroa et al. 2002). In a recent paper, Boumans and Johnsen (2015) proposed that much of the complexity in stonefly duetting could be explained by male competition and pre-copulatory mate guarding with part of the male signal mimicking or masking the female answer.

The stonefly vibrational signals are genetically programmed and species specific, which make them potentially useful lines of evidence for resolving systematic questions and inferring phylogenies within the order (Maketon and Stewart 1984a, b; Stewart and Zeigler 1984b). Therefore, the role of these signals as reproductive isolation mechanisms in Plecoptera has been stressed many times, and they have thus been repeatedly used to solve systematic questions and clarify the taxonomic status of some taxa (e.g., Berthélemy 1979b; Szczytko and Stewart 1979; Tierno de Figueroa and Sánchez-Ortega 1999), sometimes supported by molecular data (e.g., Tierno de Figueroa et al. 2011a; Murányi et al. 2014).

Rupprecht (1969, 1983) noted that sympatric species of *Isoperla* showed more different call models than the allopatric species. The great diversity of calls sometimes detected among closely related species, which mostly coexist in common areas and have very similar flight periods, may be, in part, the result of selective pressures that favor such diversification and avoid erroneous mating (acting, therefore, as a mechanism of prezygotic reproductive isolation) (e.g., Tierno de Figueroa et al. 2002, 2009 for Plecoptera, Matthews and Matthews 2010 for insects in general). This affirmation was pointed out in a recent study, in which Tierno de Figueroa et al. (2011b) analyzed the drumming signals of the four species of *Capnioneura* present in Southern Spain. The call patterns were homogeneous among the four species, only differing in mean interbeat intervals [in fact, in stoneflies, the ability of an unmated female to recognize a signal from a male of the same species appears to be mainly based on the interbeat intervals (Stewart and Zeigler 1984a, Zeigler and Stewart 1985b) and/or, of lesser concern, intersequence intervals, as seems to occur in at least some *Protonemura* species (Tierno de Figueroa et al. 2014), or other parameters (Stewart and Maketon 1990, see below)]. Although the four species have overlapping flight periods, coexistence in the same area has only been reported in *C. mitis* and

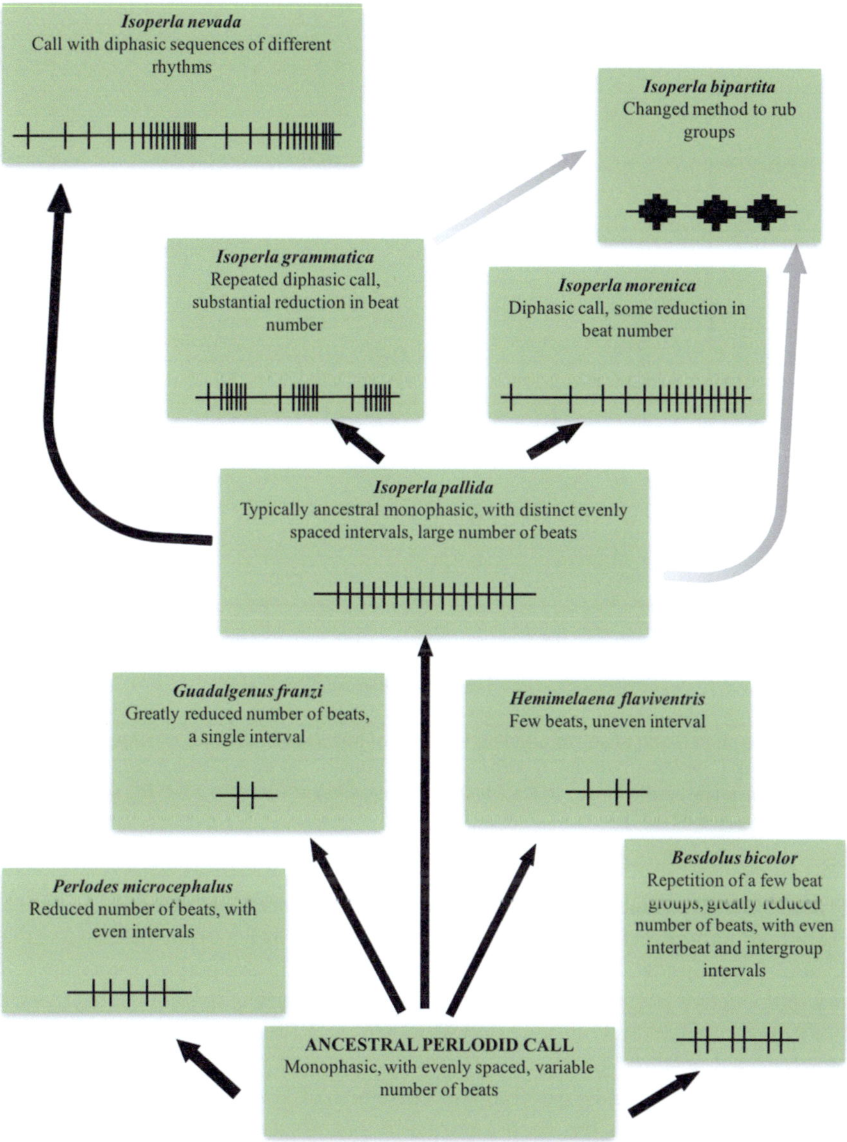


**Fig. 6.4** Possible evolution of vibrational call complexity in Southern Iberian Perlodidae (modified from Tierno de Figueroa et al. (2002, 2009) and updated with data from Tierno de Figueroa et al. (2011a, 2013) and unpublished data on the drumming call of a Spanish population of *Perlodes microcephalus*). As the drumming call of the Andalusian population of *Perlodes microcephalus* has not been yet recorded, unpublished data on a *Perlodes microcephalus* population from Central Spain has been used. Gray arrows indicate alternative possibilities

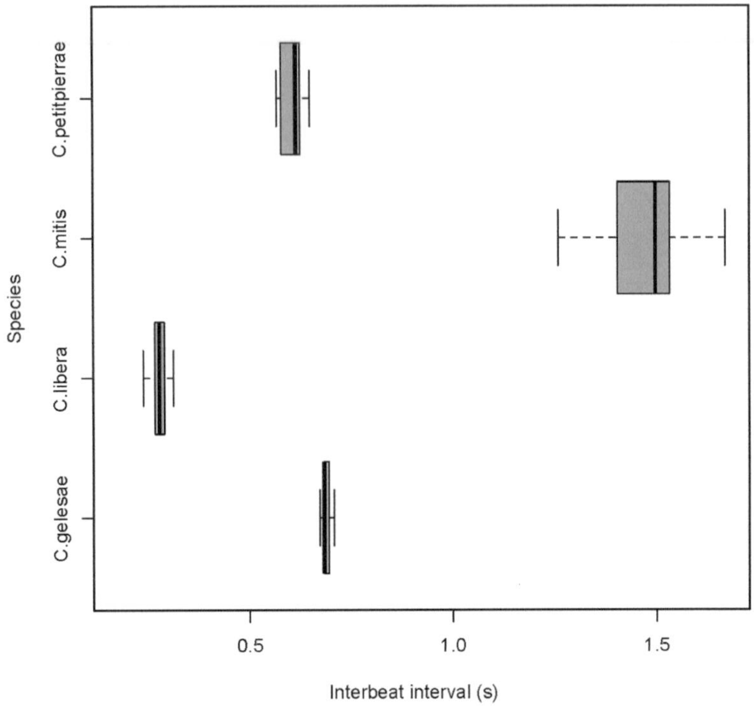


**Fig. 6.5** Graphical comparison of interbeat intervals of four Iberian *Capnioneura* species: *C. gelesae*, *C. libera*, *C. petitpierreae* and *C. mitis* (modified from Tierno de Figueroa et al. 2011b). Boxes represent the interquartile range (IQR), whiskers represent data within the 1.5·IQR, and the middle line is the median of the dataset

*C. petitpierreae*, on the one hand, and *C. mitis* and *C. libera*, on the other. As seen in Fig. 6.5, drumming calls were more different in species that coexist than in those that do not, supporting the hypothesis that stonefly drumming can play an important role as a reproductive isolating mechanism. Nonetheless, cases of hybridization (Hanson 1960) and mitochondrial introgression (Boumans and Tierno de Figueroa 2016) have been detected in stoneflies.

Despite being species specific, the existence of dialects among populations inside the same species has been reported in stoneflies, such as *Diura bicaudata*, which was the first known case of dialects in vibrational signals among insects (Rupprecht 1972). It is also known that temperature can affect the beat rate (and consequently the duration of interbeat intervals) of drumming signals in stoneflies (e.g., Zeigler and Stewart 1977), but females respond to male calls registered at temperatures similar to those where the female is (in Zwick 1980). This coincides with what has been pointed out in a spider species, in which each female appeared to prefer vibrational signals from males that were produced at her current ambient temperature (Shimizu and Barth 1996, in Hill 2008). Experimental studies (with modified computer-simulated male calls) have proven that the female answer depends on (1) a minimum threshold of number of beats or rubs, (2) a discriminative range of interbeat intervals, and (3) other specific parameters, such as general patterns of signals (with sequences, phases, etc.) (Stewart and Maketon 1990). Finally, and despite the short life span of adult stoneflies, age-related effects on drumming have also been described (Maketon and Stewart 1984a; Zeigler and Stewart 1985a), with decreasing numbers of calls per hour in older individuals and no effect on the signal structure. As reported by Zeigler and Stewart (1985a), those differences could be related to age and/or differences in nutritional state.

It was Rupprecht (1968) who first reported that mate encounter is favored by the establishment of a duet between a male and female. Therefore, when a female answers the male's call, he can become oriented by the intensity of the signal. The importance of drumming in mate locating and male searching behavior has been experimentally studied on artificial and natural substrates (Abbott and Stewart 1993; Alexander and Stewart 1996b), and these studies have demonstrated that males could integrate the information received from the answer of the female to correct wrong turns and reinforce correct turns toward her. The ability of certain stoneflies, particularly *Isoperla*, to effectively transmit their drumming signals in the presence of streamside extraneous noises, including rushing water, has been shown as well (Szczytko and Stewart 1979), and experiments with *Perlinella drymo* have demonstrated that signals can be transmitted and received even at a distance of 8 m through a wooden substrate (Stewart and Zeigler 1984b). It has also been reported that the female of *Pteronarcella badia* can discern the fitness of particular males that duet with her by measuring the time required for them to find her, and, if the time requirement is not met, she leaves her position (Abbott and Stewart 1993, in Stewart 1997).

## 6.2 Encounter Sites

It has been repeatedly pointed out that adults of some species of stoneflies show clear preferences for particular substrates (Zwick 1980) where they aggregate (Fig. 6.6). As vibrational communication is only effective on resonant substrates, and males and females need to be relatively close for the correct signal transmission, the aggregation of individuals at particular sites is an important prerequisite for duet formation (Stewart 1994, 1997). In fact, encounter sites acting as aggregation mechanisms for mating have been described as important events during the imaginal life of Plecoptera (Stewart 1994), as well as of other insects (Parker 1978). These aggregation sites in stoneflies (encounter sites sensu Stewart 1994) can be a consequence of the selection of particular places for emergence by nymphs or the active search for places by adults, mainly through visual patterns, either for reproductive purposes or for places to obtain food or protection (Stewart 1994; Tierno de Figueroa et al. 2000b). Nonetheless, as pointed out by Stewart (1994), some encounter sites, such as bridges, may be such because they are structures that intercept stoneflies when flying or walking.

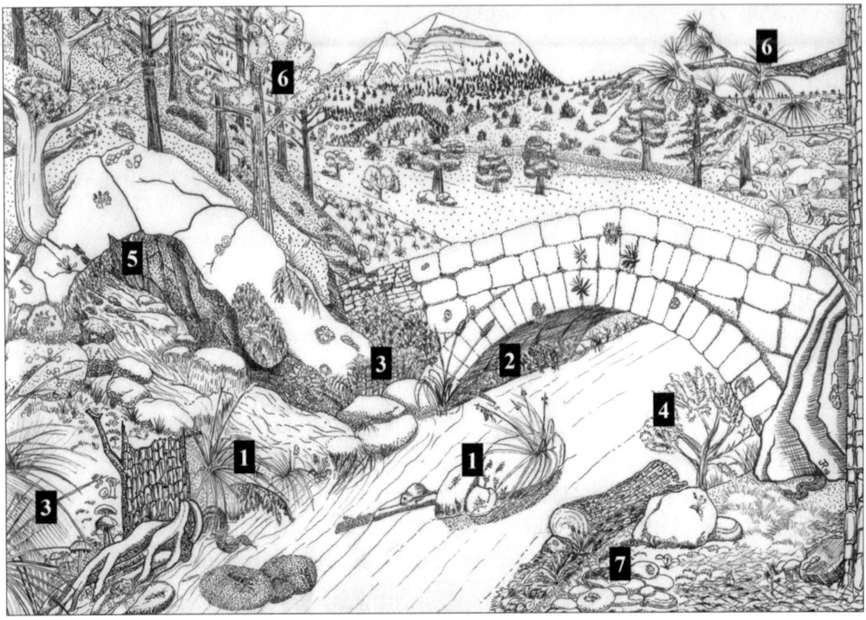


**Fig. 6.6** Drawing illustrating a representation of the main aggregation sites of some stonefly species from the Southern Iberian Peninsula based on the authors' observations (this is a hypothetical river putting together species from different areas). *1 Dinocras cephalotes*, *2 Protonemura alcazaba*, *3 Isoperla nevada*, *4 Guadalgenus franzi*, *5 Protonemura gevi*, *6 Capnioneura mitis*, *7 Tyrrhenoleuctra minuta*

Some studies seem to show that emergence substrates could act as encounter sites to aggregate sexes. Alexander and Stewart (1996a), Hanada et al. (1997), and Jáimez-Cuéllar and Tierno de Figueroa (2005) studied three stonefly species, *Claassenia sabulosa*, *Microperla brevicauda*, and *Dinocras cephalotes*, respectively, and concluded that emergence sites seemed to function as encounter sites for mating in these species, in which both sexes were able to mate relatively soon after emergence (additionally, *C. sabulosa* and *D. cephalotes* males are brachypterous and, consequently, have a low dispersal ability). In *D. cephalotes*, Jáimez-Cuéllar and Tierno de Figueroa (2005) found that in the different selected emergence substrates, alternative strategies could be employed to maximize the mating encounter effectiveness, including emerging in stream stones, which favors random mate encounters because of the small surface but renders it difficult due to the ineffectiveness of drumming transmission (and enemy exposure is high), and riparian vegetation, where drumming calls can be transmitted without difficulty, but its greater surface area makes random encounters difficult.

In many cases, stoneflies must actively move to encounter sites, directly or previously aggregating at feeding sites after a postemergence maturation toward mating conditions (Stewart 1994). It is known that many stoneflies need a period of egg maturation (Hynes 1976; Zwick 1980), and adult feeding is also very important for most species (Hynes 1976; Rupprecht 1990; Zwick 1990), except for those with the largest sizes (Tierno de Figueroa and Fochetti 2001; Fenoglio and Tierno de Figueroa 2003).

The existence of encounter sites where individuals of both sexes can aggregate for mating is particularly important in stoneflies from the Southern Hemisphere, which do not produce vibrational signals, and for other species that inhabit particular habitats. Some of these encounter sites could be places where individuals go initially with another aim, such as feeding (Stewart 1994). On the other hand, Stewart (2001) hypothesized that those stoneflies could exhibit particular aggregation behaviors using very specific encounter sites, as well as employ other intersexual communication channels (visual, olfactory, etc.) that are useful for encountering a mate (in Stewart and Sandberg 2006). In this sense, Andersson (1986) experimentally tested the role of pheromones in mate location in the Arctoperlarian *Isoperla grammatica* and concluded that they did not play a significant function in this behavior [although this species employs drumming signals (Rupprecht 1969)]. Nonetheless, Rebora et al. (2016), based on an ultrastructural investigation on the Arctoperlarian *Dinocras cephalotes*, suggested that stonefly antennae were rich in sensilla of different kinds, opening new perspectives into the study of the behavior of these insects during their adult stages, and Abbott and Stewart (1993) pointed out that in *Pteronarcella badia*, when close, males find females by tactile contact. A further electrophysiological study of Rebora et al. (2017) in males and females of *D. cephalotes* demonstrated that they could perceive olfactory cues through their antennae and hypothesized an involvement of the sense of smell in finding the encounter sites.

Although adult behavior data of Southern Hemisphere stoneflies are scarce, Hynes (1974) pointed out that species in different families of Australian Antarctoperlaria fly or climb to tall trees, bushes, or tree ferns. These places probably act as aggregation and encounter sites, and perhaps as feeding places. In relation to the latter, Smith and Collier (2000) pointed out that some adult gripopterygid stoneflies from New Zealand moved toward riparian vegetation to feed after emergence.

As reported by López-Rodríguez and Tierno de Figueroa (2012), the exclusively cavernicolous species *Protonemura gevi* (Fig. 6.7) seems to rely mainly on aggregation sites to maximize pair encounters, considering that, opposite to other epigean species of the genus (Tierno de Figueroa et al. 2014), this species does not seem to produce drumming signals and, even if it could produce them, the substrate inside the cave would not transmit the vibrations. Moreover, the total absence of light is also an important problem for mate searching; therefore, cues other than visual must be implicated in aggregation at particular places. In relation to this, the study of the antennal sensilla of this and two other *Protonemura* species (Piersanti et al. 2017) showed that no differences existed in the types and morphologies of antennal sensilla among the species, but antennae had a wide variety and quantity of sensilla that could help in mate encounter, wherever the habitat.

Another particular species is the wingless stonefly, *Capnia lacustra,* which only inhabits the deep waters in Lake Tahoe associated with *Chara* sp. (Caires et al. 2016).

**Fig. 6.7** Male of *Protonemura gevi* in its natural habitat

This is the only known species of this order of insects that does not emerge from the water for mating and laying eggs (Jewett 1963), but unfortunately the knowledge on its mate-encountering behavior is scarce.

## 6.3 Mating and Mate Guarding

While intersexual vibrational communication has been widely studied in stoneflies, other important aspects of their reproductive biology, such as other strategies for mate finding, copulation, displacement attempts during copulation, and mate guarding, have received scarce attention (e.g., Brinck 1956; Stewart 1994; Alexander and Stewart 1996a; Tierno de Figueroa 2003).

It is known that in some species individuals can mate soon after emergence, while others need a period of a few to several days before mating (Khoo 1964; Hynes 1976; Zwick 1980). Although other factors can also be involved, the degree of gamete development at the moment of emergence could influence how much time is needed. Thereby, some species of stoneflies emerge with mature eggs and can oviposit hours after emergence, while others need to complete egg maturation during the adult stage (Zwick 1980). Similarly, Fausto et al. (2002) observed that in the testes of adults of two species of Perloidea with a large size and long life cycles and that do not feed during the imaginal stage, the spermatogenetic process is complete, and follicular cysts contain only spermatozoa ready to enter into the ducts (as previously reported by other authors, such as Matthey and Aubert 1946), while in some smaller Perloidea species with univoltine life cycles and in which adults actively feed during the adult stage different stages of spermatogenesis can be observed in their follicular cysts.

Although the emergence period (and consequently the flight period) of males and females is almost identical, a protandry (i.e., males emerging earlier than females) has been detected many times in stoneflies with seasonal life cycles that inhabit temperate and cold areas (e.g., Hynes 1976; Zwick 1990; Tierno de Figueroa et al. 2003). A possible advantage of this protandry is that males could search and occupy potential encounter sites while waiting for female emergence (Alexander and Stewart 1996a), enhancing their reproductive success in relation to the fact that females usually mate only once and then become non-receptive, while males can mate many times (in Tierno de Figueroa et al. 2003). Nevertheless, the fact that females copulate several times may be more common than previously reported in the literature (e.g., Brinck 1949; Stewart et al. 1969; Luzón-Ortega and Tierno de Figueroa 2003; Tierno de Figueroa 2003), as discussed below. In a study on *Agnetina capitata*, Moreira and Peckarsky (1994) pointed out that, while females are mature soon after they become adults, the maximum activity and quality of male drumming calls occur some days after emergence, coinciding with the delay observed between male and female emergence. Tierno de Figueroa et al. (2003) studied and quantified the degree of protandry of 11 species with different flight period models. Although all species exhibited a positive protandry value, it was less pronounced in species with very extended flight periods, on the one hand, and particularly brief flight periods or highly concentrated flight maximums, on the other. This is because, in the first case, protandry does not bring benefits, considering that there would be a continuous availability of young females, and in the second displaced flight periods could hinder encounters between the sexes. These authors also obtained that the degree of protandry did not correlate with the degree of sexual dimorphism in size, contradicting the theoretical prediction in other insect orders that dimorphism is positively correlated with protandry and showing that more pronounced protandry is not simply a physiological consequence of differential growth between sexes.

The mating of stoneflies occurs on the substrate (stones, branches, etc.) when males and females meet, and while vibrational signals are very useful for mate encounters they are not an indispensable prerequisite for mating, even in species that employ this communication system (Berthélemy 1979a; Alexander and Stewart 1996a; Tierno de Figueroa et al. 1998, 2000a). Brinck (1956) and Zwick (1973), among a few others, carried out comprehensive studies in stoneflies that related the morphology and anatomy of the reproductive system to copulation. Despite the great extant diversity of stonefly genitalia, particularly the structures involved in sperm transmission (Brinck 1956), the approach and initiation of copulation are very similar in all studied species (e.g., Wu 1923; Frison 1929, 1935; Stewart et al. 1969; Benedetto 1970; Burkantis and Peckarsky 1985; Tierno de Figueroa et al. 2000a; Yoshimura et al. 2003). When a male and a female meet (under conditions of light or darkness, depending on the pattern of diel activity of each species), the male tries to mount the female's back. Some authors (e.g., Khoo 1964; Rupprecht 1968; Zeigler 1990; Tierno de Figueroa 2003) have observed that non-receptive females escape or raise their abdomen or wings to avoid male copulation attempts. If the female is receptive to copulation, the male mounts her and bends his abdomen in an S-shape under the female's abdomen, allowing the contact of his copulatory organ

**Fig. 6.8** *Perla* sp. mating

with the female genital opening (Fig. 6.8). Sperm transference in stoneflies can occur through spermatophores (in Antarctoperlaria and some Arctoperlaria such as *Pteronarcys reticulata*, *P. sachalina*, and several Taeniopterygidae) or free sperm (direct semen transfer) (in most Arctoperlaria) (Zwick 2000; Madsen and Aagaard 2016). Stewart and Stark (1977) described a particular external sperm transference in *Hydroperla crosbyi,* consisting of an active aspiration of sperm by the female (by means of telescoping contractions of the abdomen's apical segments) from an external pocket (a cuplike depression formed by the male's epiproct around the female's genital opening) where the sperm was previously deposited and after a cercal brushing or tapping by the male on the female. During the copulation, the stonefly male mounts the female, adopting different positions ranging from parallel (e.g., *Leuctra hippopus* or *Isoperla nevada*) to X-shaped (e.g., *Capnioneura mitis*), passing for V-shaped, more or less oblique (e.g., *Isoperla morenica*, formerly included in *I. curtata*) (Brinck 1956; Tierno de Figueroa 1998; Tierno de Figueroa et al. 2000a).

Some authors (Brinck 1949; Stewart et al. 1969; Benedetto 1970; Tierno de Figueroa 1998; Tierno de Figueroa et al. 1998, 2000a), studying species from Europe, North America, and South America, pointed out the existence of tactile contacts between sexes during copulation, consisting of the male rubbing its mouthparts on the female pronotum or both partners touching each other with the antennae. These contacts have been interpreted as possible tactile stimulation, but they could also be a way of obtaining information about the partner. As noted by Rebora et al. (2016), the presence of olfactory sensilla, gustatory sensilla, and mechanoreceptors in the antennae could play an important role in mate recognition and evaluation. Moreover, it has been reported in some stonefly species that males begin to actively drum just after mating (Brinck 1949). Particularly, in *Isoperla morenica*, formerly included in *Isoperla curtata*, Tierno de Figueroa et al. (2000a) detected this action in all cases in which mating occurred, and they proposed that this behavior could be useful for the female's evaluation of males in terms of cryptic female choice (Eberhard 1997), especially considering that females of this species usually re-mate with other males. Conversely, Zeigler (1990), based on the scarce receptivity of the females of the species that he studied to re-mate or answer male drumming calls, along with their short adult life span and the fact that they probably require mating only once for fertilization of most or all their eggs, suggested that females are probably poorly selective. More studies in different stonefly species are necessary to evaluate the importance of the selectivity of the females in this order of insects.

In the studied stonefly species, males were not selective and, despite having some inactive periods, tried to mate with as many females as possible. In fact, males trying to mate with females of a different species, other males, dead females, or even mature nymphs near emergence have been repeatedly observed in Plecoptera, both in nature and in the laboratory (e.g., Brinck 1949; Khoo 1964; Zeigler 1990; Tierno de Figueroa 1998), as well as in other insects (Thornhill and Alcock 1983). According to Zeigler (1990), this lack of selectivity could be related to the low male investment in sperm transmission (only 1.12% of the body weight of the species he studied, *Pteronarcella badia*). In this sense, it would be expected that males that produce spermatophores were more selective than those that do not, but this hypothesis should be tested in stoneflies.

The time that the couple stays in the mating position is very variable among species (as well as within some species), ranging from just 1 min to several hours or even days (e.g., Brinck 1949; Khoo 1964; Benedetto 1970; Berthélemy 1979a; Alexander and Stewart 1996a; Tierno de Figueroa et al. 2000a; Luzón-Ortega and Tierno de Figueroa 2003; Tierno de Figueroa 2003), and, as reported by most of these authors, long periods in the mating position could correspond with a post-copulatory mate-guarding mechanism (Parker 1970; Thornhill and Alcock 1983). In fact, when a male leaves after mating, the female is susceptible to being fertilized by a new male that could displace the sperm of the first (Zeigler 1991). In relation to this, cases of males attempting to displace other males that are in the mating position are frequently reported, but such attempts are usually unsuccessful (Zeigler 1991; Alexander and Stewart 1996a; Tierno de Figueroa et al. 1998, 2000a; Tierno de Figueroa 2003). Furthermore, Zeigler (1991) observed that couples of *Pteronarcella badia* remained in mating positions for considerably more time when other individuals were present and pointed out that, while his observations were incomplete, they suggested mate guarding by the male during copulation. Afterwards, Tierno de Figueroa (2003) statistically demonstrated that in *Leuctra fusca*, the duration of the male in the mating position is longer when other males interfere, trying to displace him and, from research on this and other species (*Leuctra andalusiaca*, *Capnioneura mitis*, and *Isoperla nevada*), obtained results that seem to support the idea that natural selection has favored a contact mate-guarding mechanism (consisting of long periods in the mating position or the possibility to prolong mating when displacement attempts occur) in males whose females are occasionally polyandric but not in those whose females copulate only once (Tierno de Figueroa 2003; Tierno de Figueroa et al. 2009). In addition to the previously

**Fig. 6.9** Mating balls of *Dinocras cephalotes* (left) and *Hemimelaena flaviventris* (right)

referred contact-guarding type, post-copulatory, noncontact guarding mechanisms have also been reported in stoneflies, such as the existence of sperm plugs to prevent subsequent inseminations (Berthélemy 1979a; Zwick 1990).

Finally, another mating strategy displayed by stoneflies is the formation of mating balls, i.e., when some males simultaneously attempt copulation with one female, and they pile onto and around her forming a ball (Tierno de Figueroa et al. 2006). Mating balls have been described in the stoneflies, *Dinocras cephalotes* and *Hemimelaena flaviventris* (Fig. 6.9), and they were previously known in other animal groups, particularly in colubrids and anurans, which share some characteristics, such as the nonexistence of male-male combat, high population density in a relatively concentrated reproductive period, tendency to aggregate, and females that are larger than males (Tierno de Figueroa et al. 2006), suggesting that the same strategy has evolved through convergence in phylogenetically distant groups (Tierno de Figueroa et al. 2009).

## 6.4 Conclusions

Stoneflies exhibit a wide array of complex behaviors of intersexual communication, mate encountering and copulation, as well as mate guarding, which we have just begun to understand. In fact, as can be concluded after reading this chapter, most of the extant information (except that regarding vibrational communication) comes from the study of one or a few species. Therefore, we only know the tip of the iceberg. We hope that future research focuses on this interesting topic and that our knowledge on these insects continues to increase.

**Acknowledgments** The authors thank Drs. Kleber Del Claro and Rhainer Guillermo, editors of this book, for inviting us to collaborate in this exciting project and Mr. J. Ramón Fernández Cardenete for the realization of the drawing in Fig. 6.5 and permission to use the photo in Fig. 6.6. The authors would like to dedicate this chapter to those classic researchers, and particularly to the deceased Drs. Stewart, Szczytko and Rupprecht, that spent decades of their lives studying, observing, and enjoying the adult biology of stoneflies, whose valuable contributions are spread in wonderful books and papers that younger generations should consult and appreciate in order to continue to increase our knowledge on this stimulating topic.

## References

Abbott JC, Stewart KW (1993) Male search behavior of the stonefly, *Pteronarcella badia* (Hagen) (Plecoptera: Pteronarcyidae), in relation to drumming. J Insect Behav 6:467–481

Abbott JC, Stewart KW (1997) Drumming of three *Mesocapnia* species (Capniidae) and *Soliperla thyra* (Peltoperlidae) from California, USA. In: Landolt P, Satori M (eds) Ephemeroptera & Plecoptera biology ecology systematics. MTL, Fribourg, pp 88–92

Alexander KD, Stewart KW (1996a) Description and theoretical considerations of mate finding and other adult behaviors in a Colorado population of *Claasenia sabulosa* (Plecoptera: Perlidae). Ann Entomol Soc Am 89:290–296

Alexander KD, Stewart KW (1996b) The mate searching behavior of *Perlinella drymo* (Newman) (Plecoptera: Perlidae) in relation to drumming on a branched system. Mitt schweiz ent Ges 69:121–126

Alexander KD, Stewart KW (1997) Further considerations of mate searching behavior and communication in adult stoneflies (Plecoptera); first report of tremulation in *Suwallia* (Chloroperlidae). In: Landolt P, Satori M (eds) Ephemeroptera & Plecoptera biology ecology systematics. MTL, Fribourg, pp 107–112

Andersson KG (1986) Bäcksländornas kärleksliv en beteendeekologisk studie. Fauna Flora 81:207–214

Benedetto L (1970) Notes on the biology of *Jewettoperla munoai* Benedetto. Limnologica 7:383–389

Berthélemy C (1979a) Accouplement, période d'incubation et premiers stades larvaires de *Brachyptera braueri* et de *Perlodes microcephalus* (Plecoptera). Ann Limnol 15:317–335

Berthélemy C (1979b) Mating call and taxonomy in Pyrenean *Isoperla*. Gewäss Abwass 64:71–72

Boumans L, Johnsen A (2015) Stonefly duets: vibrational sexual mimicry can explain complex patterns. J Ethol 33:87–107

Boumans L, Tierno de Figueroa JM (2016) Introgression and species demarcation in western European *Leuctra fusca* and *L. digitata* (Plecoptera, Leuctridae). Aquat Insects 37:115–126

Bradbury JW, Vehrencamp SL (2011) Principles of animal communication, 2nd edn. Sinauer Associates, Inc., Sunderland, MA

Brinck P (1949) Studies on Swedish stoneflies. Opusc Entomol Suppl 11:1–250

Brinck P (1956) Reproductive system and mating in Plecoptera. Opusc Entomol 21:57–127

Burkantis RT, Peckarsky BL (1985) Observations on the emergence and habits of adults of *Agnetina capitata* and *Acroneuria carolinensis*. Am Midl Nat 114:200–204

Caires AM, Chandra S, Nelson CR (2016) Unique reproductive characteristics of Lake Tahoe's *Capnia lacustra* (Plecoptera: Capniidae), a stonefly in decline. Freshwater Sci 35:1291–1299

Chapman RF (1998) The insects. Structure and function, 4th edn. Cambridge University Press, Cambridge

Cocroft RB, Gogala M, Hill PS, Wessel A (eds) (2014) Studying vibrational communication. Springer, Berlin/Heidelberg

DeWalt RE, Kondratieff BC, Sandberg JB (2015) Order Plecoptera. In: Thorp J, Rogers DC (eds) Ecology and general biology: Thorp and Covich's freshwater invertebrates. Academic Press, London, pp 933–949

Eberhard WG (1997) Sexual selection by cryptic female choice in insects and arachnids. In: Choe JC, Crespi BJ (eds) The evolution of mating systems in insects and arachnids. Cambridge University Press, Cambridge, pp 32–57
Fausto AM, Belardinelli M, Fochetti R, Tierno de Figueroa JM, Mazzini M (2002) Comparative spermatology in Plecoptera (Insecta). II. An ultrastructural investigation on four species of Systellognatha. Arthropod Struct Dev 31:147–156
Fenoglio S, Tierno de Figueroa JM (2003) Observations on the adult feeding of some *Neoperla* and *Anacroneuria* species (Plecoptera, Perlidae). Afr Entomol 11:138–139
Frison HT (1929) Fall and winter stoneflies, or Plecoptera, of Illinois. Ill Nat Hist Surv Bull 18:345–409
Frison HT (1935) The stoneflies or Plecoptera of Illinois. Nat Hist Surv Bull 20:277–471
Gaunitz D (1935) *Nephelopteryx nebulosa* L. Ent Tidskr 56:41–50
Hagen HA (1877) Beiträge Zur Kenntniss Von *Pteronarcys*. Stet Entomol Z 38:477–489
Hanada S, Maruyama H (2001) Morphology of the hammer and drumming signals in two stonefly species, *Gibosia hagiensis* (Okamoto) and *Kiotina pictetii* (Klapálek) (Perlidae). In: Domínguez E (ed) Trends in research in Ephemeroptera and Plecoptera. Kluwer Academic/Plenum Publishers, New York, pp 453–457
Hanada S, Isobe Y, Oishi T (1997) Emergence sites as mate-encounter locations in *Microperla brevicauda* Kawai (Plecoptera, Peltoperlidae). In: Landolt P, Satori M (eds) Ephemeroptera & Plecoptera biology ecology systematics. MTL, Fribourg, pp 99–106
Hanson JF (1960) A case of hybridization in Plecoptera. Bull Brookl Entomol Soc 55:25–34
Hill PSM (2008) Vibrational communication in animals. Harvard University Press, Cambridge
Hynes HBN (1974) Observations on the adult and eggs of Australian Plecoptera. Aust J Zool 29:37–52
Hynes HBN (1976) Biology of Plecoptera. Annu Rev Entomol 21:135–153
Jáimez-Cuéllar P, Tierno de Figueroa JM (2005) Local emergence distribution patterns in *Dinocras cephalotes* (Plecoptera, Perlidae). Biologia 60:539–543
Jewett SG (1963) A stonefly aquatic in the adult stage. Science 139:484–485
Khoo SG (1964) Studies on the biology of the stoneflies. Ph.D. thesis, University of Liverpool
López-Rodríguez MJ, Tierno de Figueroa JM (2012) Life in the dark: on the biology of the cavernicolous stonefly *Protonemura gevi* (Insecta, Plecoptera). Am Nat 180:684–691
Luzón-Ortega JM, Tierno de Figueroa JM (2003) Some aspects of behaviour of *Nemoura fulviceps* (Plecoptera: Nemouridae). In: Gaino E (ed) Research update on Ephemeroptera & Plecoptera. University of Perugia, Perugia, pp 47–53
Macnamara C (1926) The drumming of stoneflies (Plecoptera). Can Entomol 58:53–54
Madsen BL, Aagaard P (2016) Behaviour and dispersal in adult *Taeniopteryx nebulosa* (Linnaeus, 1758) and *Perlodes microcephalus* (Pictet, 1833) (Plecoptera) in Denmark. Aquat Insects 37:37–57
Maketon M, Stewart KW (1984a) Drumming behavior in four North American Perlodidae (Plecoptera) species. Ann Entomol Soc Am 77:621–626
Maketon M, Stewart KW (1984b) Further studies of the drumming behavior of North American Perlidae (Plecoptera). Ann Entomol Soc Am 77:770–778
Maketon M, Stewart KW (1988) Pattern and evolution of drumming behavior in the stonefly families Perlidae and Peltoperlidae. Aquat Insects 10:77–98
Matthews RW, Matthews JR (2010) Insect behavior. Springer, Dordrecht
Matthey R, Aubert J (1946) Les chromosomes des Plécoptères. Bull Biol Fr Belg 81:1–246
Moreira GRP, Peckarsky BL (1994) Multiple developmental pathways of *Agnetina capitata* (Plecoptera: Perlidae) in a temperate forest stream. J N Am Benthol Soc 13:19–29
Murányi D, Gamboa M, Orci JM (2014) *Zwicknia* gen. n., a new genus for the *Capnia bifrons* species group, with descriptions of three new species based on morphology, drumming signals and molecular genetics, and a synopsis of the West Palaearctic and Nearctic genera of Capniidae (Plecoptera). Zootaxa 3812:001–082
Newport G (1851) On the anatomy and affinities of *Pteronarcys regalis*, Newm.: with a postscript containing descriptions of some American Perlidae, together with notes on their habits. Trans Linn Soc 20:425–451

Parker GA (1970) Sperm competition and its evolutionary consequences in the insects. Biol Rev Camb Philos Soc 45:525–567
Parker GA (1978) Evolution of competitive mate searching. Annu Rev Entomol 23:173–196
Piersanti S, Rebora M, López-Rodríguez MJ, Tierno de Figueroa JM (2017) A comparison between the antennal sensilla of the cavernicolous stonefly *Protonemura gevi* and other epigean *Protonemura* species (Plecoptera: Nemouridae) in a biological context. Ann Soc Ent Fr 53:47–54
Rebora M, Tierno de Figueroa JM, Piersanti S (2016) Antennal sensilla of the stonefly *Dinocras cephalotes* (Plecoptera: Perlidae). Arthropod Struct Dev 45:552–561
Rebora M, Piersanti S, Frati F, Salerno G (2017) Antennal responses to volatile organic compounds in a stonefly. J Insect Physiol 98:231–237
Rupprecht R (1965) Trommeln als verständigungsmittel bei steinfliegen (Plecoptera). Z Naturforschg B 20:1258–1260
Rupprecht R (1968) Das trommeln der Plecopteren. Z vergl Physiol 59:38–71
Rupprecht R (1969) Zur Artspezifität der Trommelsignale der Plecopteren (Insecta). Oikos 20:26–33
Rupprecht R (1972) Dialektbildung bei den Trommelsignalen von *Diura* (Plecoptera). Oikos 22:410–412
Rupprecht R (1975) Die kommunikation von *Sialis* (Megaloptera) durch vibrationssignale. J Insect Physiol 21:305–320
Rupprecht R (1976) Struktur und funktion der bauchblase und des hammers von plecopteren. Zool Jb Anat Bd 95:9–80
Rupprecht R (1981) A new system of communication within Plecoptera and a signal with a new significance. Biol Inl Wat 2:30–35
Rupprecht R (1983) Kommunicationssignale von Arten der Gattung *Isoperla* (Plecoptera) und deren Einung zur Abgrenzung von Arten. Verh Dtsch Zool Ges 1983:198
Rupprecht R (1990) Can adult stoneflies utilize what they eat? In: Campbell IC (ed) Mayflies and stoneflies: life histories and biology. Kluwer Academic Publishers, Dordrecht, pp 119–123
Shimizu I, Barth FG (1996) The effect of temperature on the temporal structure of the vibratory courtship signals of a spider (*Cupiennius salei* Keys.). J Comp Physiol A 179:363–370
Smith BJ, Collier KJ (2000) Interactions of adult stoneflies (Plecoptera) with riparian zones II. Diet. Aquat Insects 22:285–296
Stewart KW (1994) Theoretical considerations of mate finding and other adult behaviors of Plecoptera. Aquat Insects 16:95–104
Stewart KW (1997) Vibrational communication in insects. Epitome in the language of stoneflies? Am Entomol 43:81–91
Stewart KW (2001) Vibrational communication (drumming) and mate-searching behavior of stoneflies (Plecoptera); evolutionary considerations. In: Domínguez E (ed) Trends in research in Ephemeroptera and Plecoptera. Kluwer Academic/Plenum Publishers, New York, pp 217–225
Stewart KW (2009) Plecoptera. In: Resh VH, Cardé RT (eds) Encyclopedia of insects, 2nd edn. Academic Press, Burlington, pp 810–813
Stewart KW, Maketon M (1990) Intraspecific variation and information content of drumming in three Plecoptera species. In: Campbell IC (ed) Mayflies and stoneflies biology. Kluwer Academic Publishers, Dordrecht, pp 259–268
Stewart KW, Maketon M (1991) Structures used by Nearctic stoneflies (Plecoptera) for drumming, and their relationship to behavioral pattern diversity. Aquat Insects 13:33–53
Stewart KW, Sandberg JB (2006) Vibrational communication and mate searching behavior in stoneflies. In: Drosopoulos S, Claridge MF (eds) Insect sounds and communication. Physiology, behaviour, ecology and evolution. Taylor & Francis, Boca Raton, pp 179–186
Stewart KW, Stark BP (1977) Reproductive system and mating of *Hydroperla crosbyi*: a newly discovered method of sperm transfer in Insecta. Oikos 28:84–89
Stewart KW, Zeigler DD (1984a) Drumming behavior of twelve North American stonefly (Plecoptera) species: first description in Peltoperlidae, Taeniopterygidae and Chloroperlidae. Aquat Insects 6:49–61

Stewart KW, Zeigler DD (1984b) The use of larva morphology and drumming in Plecoptera systematics, and further studies of drumming behavior. Ann Limnol 20:105–114

Stewart KW, Atmar GL, Solon BM (1969) Reproductive morphology and mating behavior of *Perlesta placida*. Ann Entomol Soc Am 62:1433–1438

Szczytko SW, Stewart KW (1979) Drumming behavior of four Western Nearctic *Isoperla* (Plecoptera) species. Ann Entomol Soc Am 72:781–786

Thornhill R, Alcock J (1983) The evolution of insect mating systems. Harvard University Press, Cambridge

Tierno de Figueroa JM (1998) Biología imaginal de los plecópteros (Insecta, Plecoptera) de Sierra Nevada (Granada, España). Ph.D. thesis, University of Granada

Tierno de Figueroa JM (2003) Mate guarding and displacement attempts in stoneflies (Insecta, Plecoptera). Biologia 58:925–928

Tierno de Figueroa JM, Fochetti R (2001) On the adult feeding of several European stoneflies (Insecta, Plecoptera). Entomol News 112:130–134

Tierno de Figueroa JM, Luzón-Ortega JM (2002) The drumming call of *Isoperla bipartita* Aubert, 1962 (Plecoptera, Perlodidae). Aquat Insects 24:87–90

Tierno de Figueroa JM, Sánchez-Ortega A (1999) The male drumming call of *Isoperla nevada* Aubert, 1952 (Plecoptera, Perlodidae). Aquat Insects 21:33–38

Tierno de Figueroa JM, Luzón-Ortega JM, Sánchez-Ortega A (1998) Imaginal biology of *Hemimelaena flaviventris* (Pictet, 1841) (Plecoptera, Perlodidae). Ann Zool Fenn 35:225–230

Tierno de Figueroa JM, Luzón-Ortega JM, Sánchez-Ortega A (2000a) Male calling, mating and oviposition in *Isoperla curtata* (Plecoptera, Perlodidae). Eur J Entomol 97:171–175

Tierno de Figueroa JM, Palomino Morales JA, Luzón-Ortega JM (2000b) Spatial distribution in river banks of *Isoperla nevada* (Plecoptera, Perlodidae), *Chloroperla nevada* (Plecoptera, Chloroperlidae) and *Sericostoma cf. vittatum* (Trichoptera, Sericostomatidae). Ital J Zool 67:355–358

Tierno de Figueroa JM, Luzón-Ortega JM, Stewart KW (2002) The drumming of *Isoperla pallida* Aubert and *Guadalgenus franzi* (Aubert) (Plecoptera, Perlodidae), and review and evolutionary considerations of Southern Iberian Peninsula Perlodidae. Zool Sci 19:871–875

Tierno de Figueroa JM, Luzón-Ortega JM, Sánchez-Ortega A (2003) Protandry and its relationship with adult size in some Spanish stoneflies species (Plecoptera). Ann Entomol Soc Am 96:560–562

Tierno de Figueroa JM, Luzón-Ortega JM, López-Rodríguez MJ (2006) Mating balls in stoneflies (Insecta, Plecoptera). Zoologica Baetica 17:93–96

Tierno de Figueroa JM, Luzón-Ortega JM, López-Rodríguez MJ (2009) Estudios sobre el comportamiento reproductivo y la alimentación adulta en plecópteros (Insecta). In: Dopazo H, Navarro A (eds) Evolución y adaptación 150 años después del Origen de las Especies. Sociedad Española de Biología Evolutiva (SESBE). Obrapropia S.L., Valencia, pp 133–141

Tierno de Figueroa JM, Gaetani B, Luzón-Ortega JM, López-Rodríguez MJ, Fochetti R (2011a) On the identity of *Isoperla curtata* (Plecoptera: Perlodidae): behavioural and molecular approaches show the existence of two separate species. Zootaxa 3000:49–58

Tierno de Figueroa JM, Luzón-Ortega JM, López-Rodríguez MJ (2011b) Intraspecific and interspecific variation in drumming signals in the genus *Capnioneura* (Plecoptera: Capniidae). Aquat Insects 33:335–341

Tierno de Figueroa JM, Luzón-Ortega JM, López-Rodríguez MJ (2013) Description of the male drumming signals of *Besdolus bicolor* (Navás, 1909) (Plecoptera, Perlodidae). Ann Soc Ent Fr 49:242–243

Tierno de Figueroa JM, Luzón-Ortega JM, López-Rodríguez MJ (2014) First record of the drumming signals of stoneflies *Capnopsis* Morton, 1896 and *Protonemura* Kempny, 1898 genera (Plecoptera, Capniidae and Nemouridae). Entomol Sci 17:302–308

Wu CF (1923) Morphology, anatomy and ethology of *Nemoura*. Bull Lloyd Libr Ent Ser 3:1–46

Yoshimura M, Isobe Y, Oishi T (2003) The relation among emergence date, days until first effective mating, fecundity and adult longevity in *Isoperla aizuana* and *Sweltsa* sp. (Plecoptera). Zool Sci 20:471–479

Zeigler DD (1990) Observations pertinent to the role of sexual selection in the stonefly *Pteronarcella badia* (Plecoptera: Pteronarcyidae). Entomol News 101:283–287

Zeigler DD (1991) Passive choice and possible mate guarding in the stonefly *Pteronarcella badia* (Plecoptera: Pteronarcyidae). Fla Entomol 74:335–340

Zeigler DD, Stewart KW (1977) Drumming behavior of eleven Nearctic stonefly (Plecoptera) species. Ann Entomol Soc Am 70:495–505

Zeigler DD, Stewart KW (1985a) Age effects on drumming behavior of *Pteronarcella badia* (Plecoptera) males. Entomol News 96:157–160

Zeigler DD, Stewart KW (1985b) Drumming behavior of five stonefly (Plecoptera) species from Central and Western North America. Ann Entomol Soc Am 78:717–722

Zwick P (1973) Insecta: Plecoptera Phylogenetisches System und Katalog. Das Tierreich, Berlin

Zwick P (1980) Plecoptera (Steinfliegen). In: Beier M (ed) Handbuch der Zoologie 4(2), 2/7. Walter de Gruyter, Berlin, pp 1–115

Zwick P (1990) Emergence, maturation and upstream oviposition flights of Plecoptera from the Breitenbach, with notes on the adult phase as a possible control of stream insect populations. Hydrobiologia 194:207–223

Zwick P (2000) Phylogenetic system and zoogeography of the Plecoptera. Annu Rev Entomol 45:709–746

# Chapter 7
# Aquatic Insect Sensilla: Morphology and Function

**Manuela Rebora, Gianandrea Salerno, and Silvana Piersanti**

**Abstract** Insect extraordinary evolutionary success is due to different reasons among which their ability to receive and respond to a great variety of sensory cues thanks to their developed sense organs encompassing a high number of diversified sensilla, mainly located on their antennae. The successful invasion of lotic and lentic freshwaters by terrestrial insect required physiological constraints also regarding sensory systems. This chapter reviews the present knowledge about antennal sensory equipment in adult and aquatic stages of Ephemeroptera, Odonata, Plecoptera and Trichoptera. These insect orders have aquatic larval stages well adapted to this environment, while they move to the terrestrial habitat as adults. To be able to receive sensory cues in two very different biotopes during the young and the adult stage is a fundamental prerequisite for these insects. The data reported are mainly based on morphological investigations under scanning and transmission electron microscope (SEM, TEM), and behavioural and electrophysiological investigations (the latter available only for Odonata and Plecoptera). The chapter considers separately the main sensory capacities located on the antennae in the above-reported aquatic insect orders, in particular mechanoreception, chemoreception, thermo-hygroreception and their modifications from the aquatic to the adult stage.

**Keywords** Antennae · Mayflies · Dragonflies · Stoneflies · Caddisflies

M. Rebora · S. Piersanti (✉)
Dipartimento di Chimica, Biologia e Biotecnologie, University of Perugia, Perugia, Italy
e-mail: silvana.piersanti@unipg.it

G. Salerno
Dipartimento di Scienze Agrarie, Alimentari e Ambientali, University of Perugia, Perugia, Italy



K. Del-Claro, R. Guillermo (eds.), *Aquatic Insects*,
https://doi.org/10.1007/978-3-030-16327-3_7

## 7.1 Introduction

Different factors contributed to make insects the most evolutionarily successful group of organisms on Earth. One of them is probably the ability to respond quickly to external and internal cues, thanks to a great variety of sensory systems able to evolve over relatively short time (Gullan and Cranston 2006).

Insects rely mainly on compound eyes for photoreception while the other senses are located inside special organs defined sensilla constituted of a cuticular and a cellular component. The latter is composed of modified epidermal cells, called auxiliary cells, and bipolar sensory neurons enveloped by a dendrite sheath. Thanks to a specialised sensory cuticle, sensilla can capture specific types of signals, such as mechanical and chemical cues, and convert them into changes in action potentials, which are then carried by sensory neurons to the brain where the information is integrated (primary and secondary pathways). We can distinguish mechanoreceptors, chemoreceptors and thermo-hygroreceptors on the basis of their internal and external morphological and molecular organisation reflecting a different sensory role. Sensilla with pores (multiporous, uniporous) generally serve as chemosensilla, whereas sensilla associated with touch-, thermo- and hygroreception are predominantly aporous (Fig. 7.1). Sensilla can be named after the structure of their cuticular parts (Schneider 1964; Zacharuk 1985), and we can distinguish trichoid sensilla with a long thin hair; chaetic sensilla with a long thick hair; basiconic sensilla with a usually shorter hair and a thin cuticle; styloconic sensilla showing a wide stylus bearing an apical sensory cone; coeloconic sensilla with a short peg-like structure usually situated in a pit; and placoid sensilla situated in a plate-like cuticular structure (Fig. 7.2) (Zacharuk 1985; Hallberg and Hansson 1999).

Nearly 100,000 species from 12 insect orders spend one or more life stages in freshwater (Dijkstra et al. 2014). In consideration that all aquatic insect groups are the result of the invasion of freshwaters by terrestrial groups, adaptation to life in running (lotic) and standing (lentic) water bodies requires interesting physiological constraints, not only about thermo- and osmoregulation, respiration, feeding and locomotion but also regarding sensory systems.

In consideration that antennae are the main insect sensory organs for intra- and interspecific communication and for the detection of environmental cues, the aim of this chapter is to review the present knowledge about antennal sensory equipment in adult and aquatic stages of Ephemeroptera, Odonata, Plecoptera and Trichoptera. These orders (together with Megaloptera, a very small insect order here disregarded) are considered purely aquatic insects because they are almost exclusively restricted to freshwater by an aquatic larval stage showing a closed tracheal system with gills, in agreement with a well-developed adaptation to this environment, while they move to the terrestrial habitat as adults. To be able to receive sensory cues in two biotopes during the young and the adult stage is a fundamental prerequisite for these insects. During metamorphosis (incomplete in Ephemeroptera, Odonata and Plecoptera, complete in Trichoptera), the insect nervous system must change to accommodate the sensory requirements of the new environment.

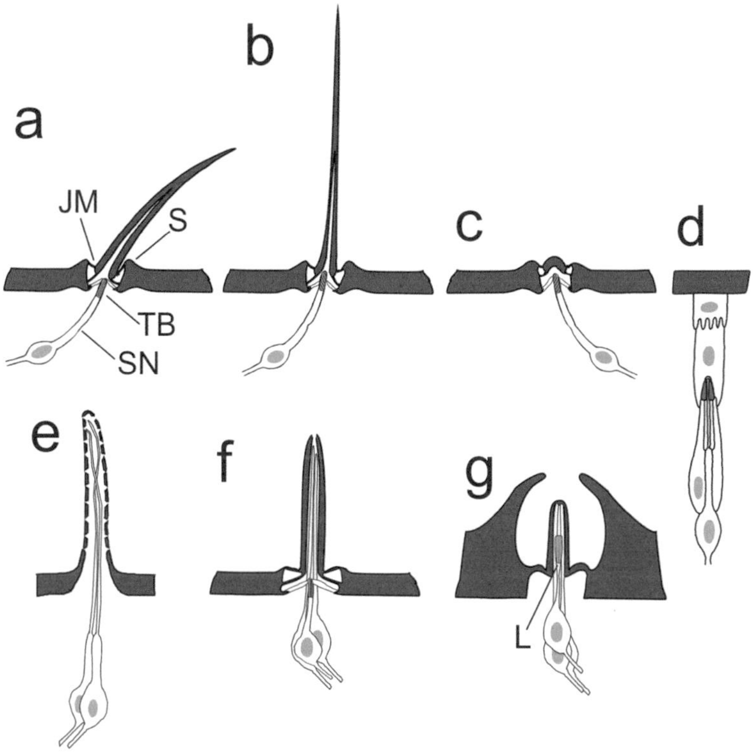


**Fig. 7.1** External and internal morphology of mechanoreceptors (**a–d**), chemoreceptors (**e**, **f**) and thermo-hygroreceptors (**g**). Chemoreceptors can be multiporous olfactory sensilla (**e**) or uniporous gustatory sensilla (**f**) whereas mechanoreceptors (**a–d**) and thermo- and hygroreceptors (**g**) are predominantly aporous. A bipolar sensory neuron/s (SN) innervates each sensillum. Cuticular mechanoreceptors can be hairs (**a**, **b**), campaniform sensilla (**c**) or chordotonal organs (**d**). Hairs can be functionally divided into bristles (**a**) and filiform hairs (**b**). Note that in mechanoreceptors the sensory cuticle (typically a hair) is connected via an elastic "joint membrane" (JM) to a circular or oval socket (S) in the body cuticle. The sensory neuron ends at the base of the sensory cuticle and contains tightly packed microtubules constituting a distal structure called the tubular body (TB). Gustatory sensilla typically have a mechanosensory neuron ending at the base of the sensillum while the remaining sensory cells reach the terminal pore. In olfactory sensilla the sensory neurons enter the sensory cuticle reaching the apical region of the hair. Thermo-hygroreceptors are typically located inside cuticle invaginations, and have inflexible socket and a dendritic sheath connected with the sensory cuticle; thermoreceptive unit typically differentiates lamellae (L) in the outer dendritic segment

Although the morphology of antennae varies greatly among insects, they are composed of three basic segments including scape, pedicel and a sub-segmented flagellum. Traditionally, Ephemeroptera and Odonata have been classified as Paleoptera (old wings), while Plecoptera and Trichoptera belong to Neoptera (new wings). In particular, Plecoptera are basal Neoptera, while Trichoptera have a strongly supported sister taxon relationship with Lepidoptera (moths and butterflies)

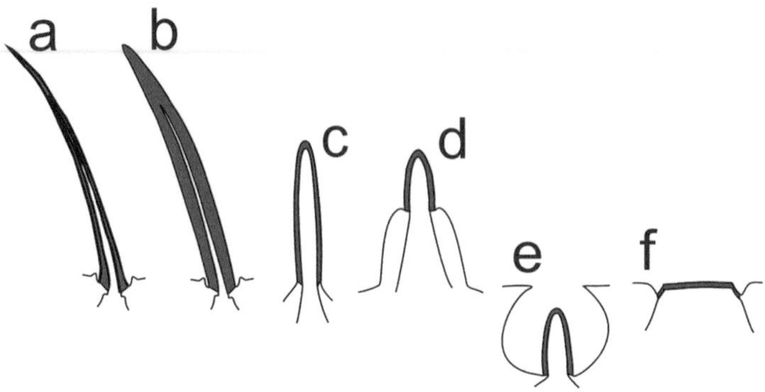


**Fig. 7.2** Sensilla categorisation on the basis of their cuticular parts; we can distinguish trichoid sensilla with a long thin hair (**a**); sensilla chaetica with a long thick hair (**b**); basiconic sensilla with a usually shorter hair and a thin cuticle (**c**); styloconic sensilla showing a wide stylus bearing an apical sensory cone (**d**); coeloconic sensilla with a short peg-like structure usually situated in a pit (**e**); placoid sensilla situated in a plate-like cuticular structure (**f**)

(Zhou et al. 2016). Together with their inability to fold the wings over the abdomen, adult Odonata and Ephemeroptera share other common features such as big eyes and short, reduced antennae compared with those present in their larval stage. A different situation, with young and adult stages showing well-developed antennae, is typical of Plecoptera, while Trichoptera show fairly long and threadlike antennae at the adult stage and short antennae in case-bearing species.

In reviewing the knowledge on antennal sensilla in Ephemeroptera, Odonata, Plecoptera and Trichoptera it is worth to remember that Odonata are the best-researched group in this field, while less data are available for Ephemeroptera and Plecoptera, especially at the larval stage; some data are available for Trichoptera at the adult stage while almost nothing is known for the larval stages. The data here reported are mainly based on morphological investigations under scanning and transmission electron microscope (SEM, TEM), and behavioural and electrophysiological investigations (the latter available only for Odonata and Plecoptera).

The chapter considers separately the main sensory capacities located on the antennae in the above-reported aquatic insect orders, in particular mechanoreception, chemoreception, thermo-hygroreception and their modifications from the aquatic to the adult stage.

## 7.2 Mechanoreception

Mechanoreception is the sense that allows insects to detect movements of objects in the environment and sounds or provides proprioceptive cues (sensory input about the position or orientation of the body and its appendages) (for reviews see Keil 1997; Tuthill and Wilson 2016). The sensory cuticle (typically a hair) is connected via an elastic “joint membrane” to a circular or oval socket in the body cuticle (Fig. 7.1). The joint membrane, due to its resilin-dominated material composition (Michels et al. 2016), is soft and flexible and allows movement and bending of the hair shaft resulting in a stimulation of the receptor. A bipolar sensory neuron is closely apposed to the base of the sensory cuticle, and its dendrite contains tightly packed microtubules constituting a distal structure called the tubular body (Fig. 7.1). The hair acts as a lever arm that exerts forces on the tip of the sensory neuron dendrites, where mechanotransduction channels open to produce electrical currents (Thurm 1965). Cuticular mechanoreceptors can be subdivided into three major groups, hairs, campaniform sensilla, or chordotonal organs. Hairs can be in the form of long, thin, or shorter hairs (trichoid sensilla) or pegs and are functionally divided into filiform hairs and bristles (Fig. 7.1). The bristle seems to be the most common type of mechanoreceptor, usually responding to direct touch, while filiform hairs or trichobothria are sensilla deflected by faint air currents and low-frequency sound, or vibrations of the surrounding environment (Keil 1997). Campaniform (dome-shaped) sensilla are frequently found on the outer surface, particularly near the joints, where they detect stress in the cuticle. Chordotonal receptors are generally found further beneath the integument and can be connected to the cuticle by attachment structures. They serve several functions, including hearing and joint movement detection.

### 7.2.1 Aquatic Stage

The density of water allows vibrations to be carried much farther than in atmosphere, with the consequence that sounds and movements are more readily detectable than in air. This is the reason why fishes, and many aquatic organisms such as insects, rely greatly on the ability to detect vibrations in water through mechanoreception. Moreover, mechanoreception may be the most important source of sensory information for aquatic insects in consideration that vision underwater can be very challenging in turbidity conditions. The importance of mechanoreceptors in the larval stages of aquatic insects has been highlighted in different behavioural investigations in relation to prey or predator detection (e.g. Peckarsky and Wilcox 1989; Scrimgeour et al. 1994). In Odonata larvae mechanical stimuli have a predominant role in predation (Caillère 1964, 1965, 1968; Corbet 1999; Rebora et al. 2004), as shown in the larva of *L. depressa* which relies on a non-contact mechanical sense for the detection of the prey (Rebora et al. 2004). The ability to respond to faint movements of the surrounding medium is allowed by long filiform hairs characterised by a joint membrane reduced to a very thin cuticular lamella and a complex socket septum with two projections in close connection with the tubular body (Keil 1997; Gaino and Rebora 2001) (Fig. 7.3b). These sensilla have been extensively studied in terrestrial arthropods (Tautz 1977; Shimozawa and Kanou 1984; Devarakonda et al. 1996) and Decapod Crustacea (Laverack 1962; Wiese 1976). In aquatic insects this kind of mechanoreceptors has been reported in all the

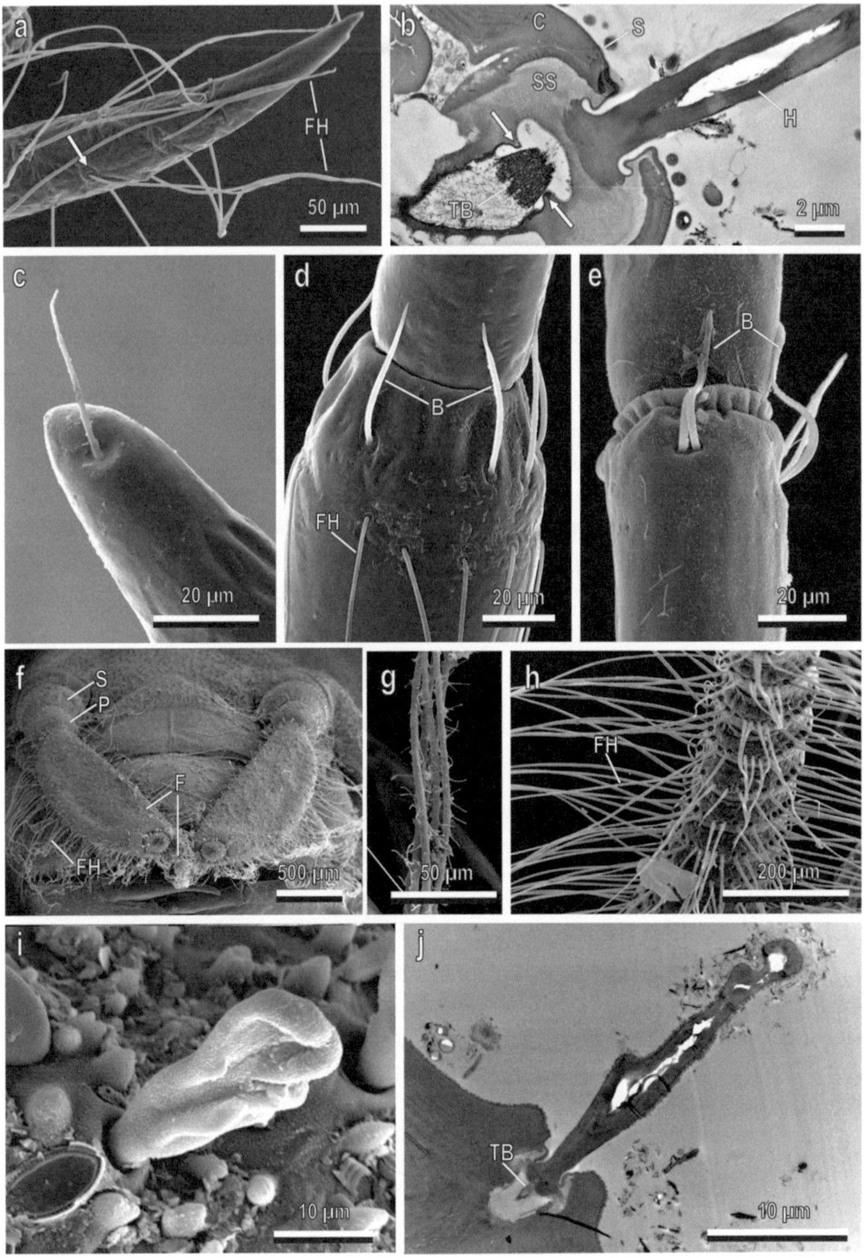


**Fig. 7.3** Mechanoreceptors in the aquatic stage of different species belonging to Odonata (**a**–**d**, **f**, **g**, **i**, **j**) and Ephemeroptera (**h**) under SEM (**a**, **c**–**i**) and under TEM (**b**, **j**). (**a**) Filiform hairs (FH) on the antennal flagellum of the dragonfly *Libellula depressa*. Arrow points out the socket; (**b**) longitudinal section of a filiform hairs of *L. depressa*. Note the complex socket (S) and three-socket septum (SS) with two projections (arrows) in close connection with the tubular body (TB). C, cuticle; H, hair; (**c**) apical portion of the antennal flagellum of the dragonfly *Aeshna cyanea*

dragonflies and damselflies larval antennae observed so far (Piersanti and Rebora 2018), belonging to different families such as Libellulidae (Gaino and Rebora 2001), Gomphidae (Faucheux and Meurgey 2009; Rebora et al. 2015), Aeshnidae (Faucheux and Meurgey 2007), Petaluridae (Faucheux and Meurgey 2007), Coenagrionidae (Meurgey and Faucheux 2006a), Lestidae (Meurgey and Faucheux 2006b) and also in the relic dragonfly *E. superstes* (Faucheux 2007) belonging to the family Epiophlebiidae (suborder Anisozygoptera) (Fig. 7.3a, b, d, f, g). Filiform hairs are particularly developed in Odonata species living in lentic environments, while they seem shorter in those living in lotic water (Piersanti and Rebora 2018), probably in consideration that faint movements are easier to be detected in lentic environments. In this regard, in insect orders specialised for living in lotic waters such as Ephemeroptera and Plecoptera, filiform hairs have been described only in a mayfly burrowing species (see below) and in the stonefly *Paragnetina media* (only under SEM) (Kapoor 1985).

As far as bristles are concerned, in Odonata a well-developed mechanosensory seta is located at the apex of the flagellum in *Aeshna cyanea*, *Calopteryx haemorrhoidalis* and *Ischnura elegans* (Fig. 7.3c and 7.5a, b) (Piersanti and Rebora 2018). In different Odonata species bristles are frequently visible along the distal portion of each flagellar segment (Fig. 7.3d) suggesting a function as mechanoreceptors sensing the movements of the different flagellomeres, one in relation to the other (Piersanti and Rebora 2018). In particular, they have been observed in different species of coenagrionids (Meurgey and Faucheux 2006a; Piersanti and Rebora 2018), in *C. haemorrhoidalis* (Piersanti and Rebora 2018), in a lestid (Meurgey and Faucheux 2006b) and in *Aeshna cyanea* (Faucheux and Meurgey 2007). Mechanosensory neurons inside hairs located at the end of each flagellar segment have been observed also in the nymphs of different species belonging to Ephemeroptera (Gaino and Rebora 1997, 1998; Rebora and Gaino 2008) (Fig. 7.3e). Sensilla trichodea located along the distal portion of each flagellar segment are present also in Plecoptera nymphs (Kapoor 1985). Some of them could be mechanoreceptors.

Interesting adaptations are visible in the antennal sensilla of aquatic burrowing species living in running waters under gravel or sandy sediments to avoid the effects of current and predators (Suhling et al. 2015). In this regard, one of the most peculiar antennae in Odonata is that of gomphids, which is composed of four short and flat segments (Fig. 7.3f). The antenna of the gomphid *Onychogomphus forcipatus*

**Fig. 7.3** (continued) showing an apical bristle; (**d**) bristles (B) located along the distal portion of each flagellar segment of the antenna of the damselfly *Ischnura elegans*; FH, filiform hairs; (**e**) bristles (B) located along the distal portion of each flagellar segment of the antenna of the mayfly *Ephemera danica* on the ventral side; (**f**) dorsal view of the antennae of the burrowing dragonfly *Onychogomphus forcipatus*, which is composed of four short and flat segments, scape (S), pedicel (P) and flagellum (F) with two segments. FH, filiform hairs; (**g**) detail of the thin thorny trichoid sensilla of *O. forcipatus*; (**h**) dorsal side of the antenna of the burrowing mayfly *Ephemera danica*. Note the developed filiform hairs (FH) located mainly on the lateral portion of the flagellum; (**i**) club-shaped mechanoreceptor with the typical structure of bristle on the antennal flagellum of *O. forcipatus*; (**j**) longitudinal section of the club-shaped sensillum. Note the tubular body (TB) has been observed in detail revealing a great diversity of mechanoreceptors (Rebora et al. 2015). In particular club-shaped sensilla (Fig. 7.3i, j), sensilla chaetica and treelike sensilla show the typical structure of bristles, usually responding to direct touch, while numerous long thin thorny trichoid sensilla of different length (some of them reaching 800 μm) show a morphology recalling the structure of filiform hairs (Fig. 7.3f, g). Since hair length is one of the several parameters determining a hair's mechanical sensitivity (Humphrey et al. 1993), the different length of the numerous filiform hairs of *O. forcipatus* (and the presence of thorns on some of them (Fig. 7.3g)) could serve different ranges of sensitivity, helping in the detection of variation and intensity of water flow. In this regard it is interesting to remember the ability of gomphid larvae to reach safe shelters by burrowing deeper in the sediment during flood periods (Suhling and Müller 1996). Very developed filiform hairs (reaching 800 μm) are present also on the antenna of the burrowing mayfly *Ephemera danica* (Rebora and Gaino 2008) (Fig. 7.3h). These nymphs live in "U"-shaped burrows in sandy sediments and maintain a sediment-free space beneath the head and the body through which water flows taking feeding particles. The filiform hairs located mainly on the lateral portion of the flagellum can be fundamental to inform the insect on the intensity of the water flow inside the burrow.

### 7.2.2 Adult Stage

The reduced antennae of Paleoptera (Odonata and Ephemeroptera) do not show particular mechanoreceptors on their flagellum except for the presence of occasional campaniform sensilla (Rebora et al. 2009a, b; Piersanti et al. 2010), proprioceptors sensing the bending of the antenna.

On the contrary, a detailed investigation on the antennal sensilla of the stonefly *Dinocras cephalotes* revealed that the most represented sensilla on the developed antenna of the adult are mechanoreceptors (shorter or longer) with a grooved surface with the typical structure of bristles (Rebora et al. 2016) (Fig. 7.6g). These mechanoreceptors could help the insect to explore the surrounding environment (e.g. to find a suitable surface to perform drumming). Indeed, both sexes continuously touch the surface with the antennae during walking on the substrate (Rebora et al. 2016). They could be important also to get information about the partner because it has been reported for some species that during the mating process the male continuously touches the female with its antennae (Tierno de Figueroa et al. 1998).

Trichoptera show a well-developed antenna only at the adult stage. A comparative analysis (based on SEM investigations) on the antennal sensilla type in representatives of various families revealed that straight trichoid sensilla are located on the lateral surface of the antenna on each flagellomere and most likely are mechanoreceptors (Melnitsky and Ivanov 2011).

## 7.3 Chemoreception

Chemoreception allows organisms to respond to chemical cues in their environments and depends on the senses of taste and smell. Taste is typically used for detection of aqueous chemicals, and smell, for airborne chemical cues—but the distinction is vague. Alternative terms are contact (taste, gustatory) and distant (smell, olfactory) chemoreception. In insects living in water, as in other aquatic animals, all chemicals sensed are in aqueous solution, and strictly all chemoreception should be termed "taste." However, the distinction is still used on the basis of the structure, response or particular location of the sensilla or the animal's behavioural response (Zacharuk 1980).

In insects, chemoreception is mediated by transmembrane receptors (olfactory receptors, ORs; gustatory receptors, GRs; and ionotropic receptors, IRs) located on olfactory and gustatory sensory neurons hosted inside olfactory and gustatory sensilla (see reviews in Vosshall et al. 2000; Galizia and Rössler 2010; Hansson and Stensmyr 2011).

Gustatory sensilla are typically uniporous sensilla responding to taste and tactile stimuli (touch and taste function). One neuron is mechanosensory and ends at the base of the sensillum while the remaining sensory cells reach the terminal pore, and function as contact chemoreceptors (the dendrite sheath stops at the base of the sensory hair) (Fig. 7.1). Olfactory sensilla are typically multiporous and their sensory neurons enter the sensory cuticle not enveloped by the dendrite sheath (Fig. 7.1); on the basis of their morphology they can be distinguished in single-walled sensilla and double-walled sensilla (Altner et al. 1977). Single-walled sensilla can be shaped as hairs, pegs or plates; their pore system has been described in detail by Steinbrecht (1997), and consists of cuticular canals through the hair wall, which on the inner side is connected to pore tubules (Fig. 7.4). Double-walled sensilla show typically a short hair, and are apically provided with slit-like pores that penetrate the cuticle while pore tubules are lacking (Hunger and Steinbrecht 1998) (Fig. 7.4).

### *7.3.1 Aquatic Stage*

Chemical stimuli propagating in water, as when released in air, form a plume that can be preserved at great distances from the source (Murlis et al. 2003). Moreover, freshwater environmental conditions, such as turbidity of water and reduced light transmission, prompt aquatic insects to use chemical cues for intra- and interspecific relationships such as foraging or predator avoidance (Brönmark and Hansson 2000; Wisenden 2000). The ability to perceive chemical stimuli in water has been demonstrated in behavioural observations in larvae of many aquatic insects (see review in Crespo 2011) belonging to Odonata (Chivers et al. 1996; Wisenden et al. 1997; Stoks 2001; Crumrine 2006; Mortensen and Richardson 2008; Siepielski et al. 2016), Ephemeroptera (McIntosh and Peckarsky 1996, 2004; McIntosh et al. 1999; Alvarez et al. 2014), Plecoptera (Williams 1987) and Trichoptera (Gall and Brodie Jr 2009; Okano et al. 2017).

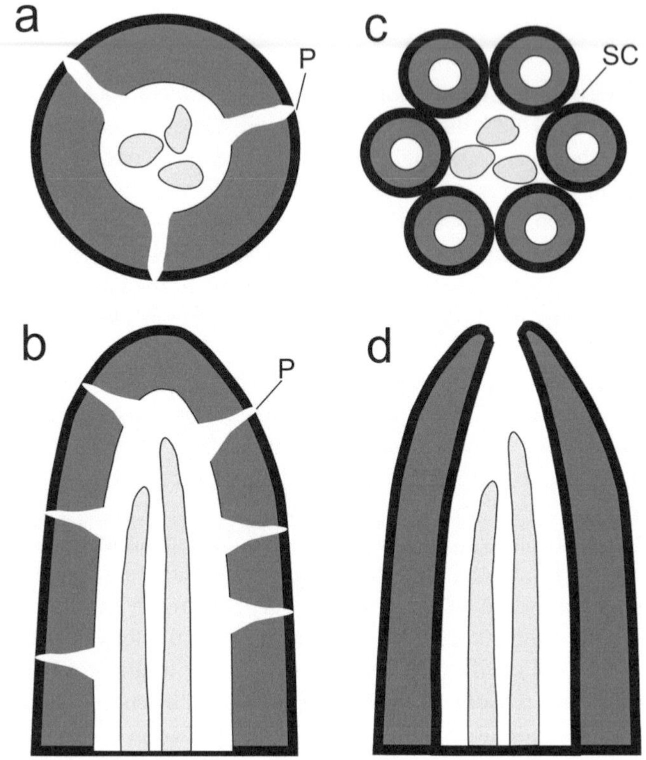


**Fig. 7.4** Schematic view of single-walled sensilla (**a**, **b**) and double-walled sensilla (**c**, **d**) in transversal (**a**, **c**) and longitudinal (**b**, **d**) sections. Note in the single-walled sensilla the cuticular canals (pores, P) through the hair wall, which on the inner side is connected to pore tubules. Double-walled sensilla are apically provided with slit-like pores (spoke channels, SC) that penetrate the cuticle while pore tubules are lacking (modified from Steinbrecht 1997)

In Odonata, a composed coeloconic sensillum deriving from the fusion of two individual receptors, each innervated by a group of three neurons entering the peg without dendrite sheath, has been described in the apical portion of the antennal flagellum of all the dragonfly and damselfly species so far investigated and belonging to different families, such as *L. depressa* (Gaino and Rebora 2001), *O. forcipatus* (Rebora et al. 2015), *Epiophlebia superstes* (Anisozygoptera) (Faucheux 2007), *C. haemorrhoidalis*, *I. elegans* (Piersanti and Rebora 2018) and *A. cyanea* (Fig. 7.5a–e). In all the species observed under TEM, pores or pore tubules are not clearly visible in this sensillum but the sensory cuticle of the peg shows an irregular coat with wide channels, underlain by a more or less compact electron-dense layer surrounding the dendrites (Fig. 7.5e). Among the Odonata antennal sensilla, this apical sensillum is the best candidate to represent a chemoreceptor (Rebora et al. 2015; Piersanti and Rebora 2018). The absence of pores on the sensory cuticle could not preclude a chemosensory function; in fact, the aesthetascs, the olfactory organs in Decapod crustaceans, have a thin cuticle without any obvious pores (Tuchina et al. 2014; Derby et al. 2016). The question if and how airborne odour molecules

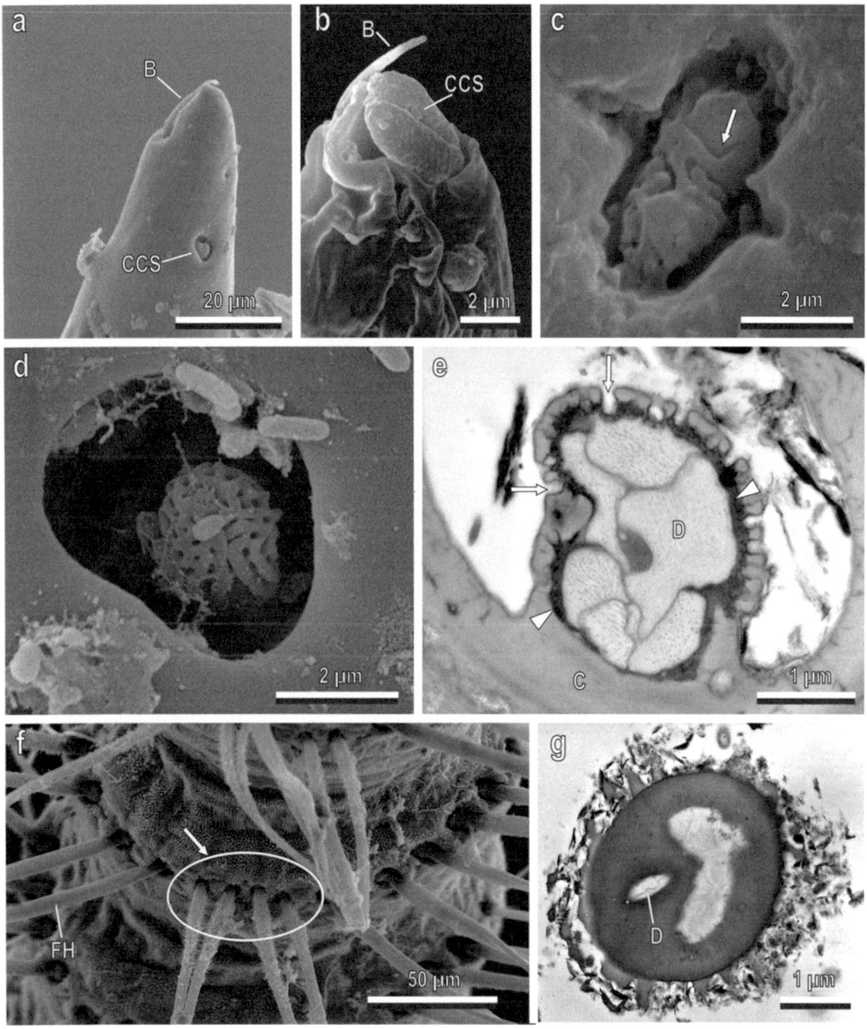


**Fig. 7.5** Chemoreceptors in the aquatic stage of different species belonging to Odonata (**a–e**) and Ephemeroptera (**f**, **g**) under SEM (**a–d**, **f**) and under TEM (**e**, **g**). (**a**) Apical portion of the antennal flagellum of the dragonfly *Aeshna cyanea*. Note the subapical composed coeloconic sensillum (CCS). B, apical bristle; (**b**) apical portion of the antennal flagellum of the damselfly *Calopteryx haemorrhoidalis*. Note the apical composed coeloconic sensillum (CCS). B, apical bristle; (**c**) detail of the composed coeloconic sensillum on the flagellum of the dragonfly *Onychogomphus forcipatus*. Note the sensory cuticle of the peg showing an irregular coat with wide channels (arrow); (**d**) same sensillum in the subapical portion of the antennal flagellum of the dragonfly *Libellula depressa*; (**e**) longitudinal section of the composed coeloconic sensillum of *L. depressa*. Note the two groups of three dendrites (D) entering the peg without dendrite sheath. The sensory cuticle of the peg shows an irregular coat with wide channels (arrows), underlain by a more or less compact electron-dense layer (arrowhead) surrounding the dendrites. C, cuticle; (**f**) dorsal side of the antenna of the burrowing mayfly *Ephemera danica*. Note the group of four chemo-mechanoreceptors (arrow). FH, filiform hairs; (**g**) cross section of a chemo-mechanoreceptor of *E. danica*. Note the two dendrites extending along the shaft

penetrate the aesthetasc cuticle in crustaceans is not answered yet, even if in the past it was hypothesised that in water chemicals could simply diffuse through the cuticle (Zacharuk et al. 1971). This hypothesis on chemoreception in an aquatic environment is surely interesting and the composed coeloconic sensillum of Odonata larvae represents a good model to perform electrophysiological investigations to test it (Piersanti and Rebora 2018).

As far as Ephemeroptera are concerned, chemo-mechanoreceptors with an apical pore, a tubular body and dendrites extending along the shaft have been described on the antennae of many species belonging to different families such as Baetidae, Heptageniidae, Leptophlebiidae and Ephemeridae (Gaino and Rebora 1996, 1997, 1998, 1999) (Fig. 7.5f, g). In some families such as Heptageniidae and Baetidae they show a spatulate apical portion and they have been defined flat tipped. The interaction between chemical and tactile cues in mayfly detection of predators has been highlighted in behavioural investigations (Ode and Wissinger 2006).

As far as Plecoptera nymphs are concerned, *Paragnetina media* shows at the distal end of each flagellar segments tufts of multiporous hairs; each hair is innervated by two bipolar neurons (Kapoor 1987). In addition, uniporous basiconic and coeloconic pegs have been described on the antennal flagellum of the nymph of the same species (Kapoor 1985) but no information on their internal structure is available.

Concerning Trichoptera, there is only one paper (Akent'eva 2012) reporting the presence of a sensory area with a porous cuticle at the apex of the antenna of *Limnephilus centralis* (Limnephilidae), but further investigations are necessary to clarify the sensory function of the reduced antenna.

### 7.3.2 Adult Stage

As above reported, adult Paleoptera have short reduced antennae with a low number of sensilla and a very reduced role in mechanoreception (Rebora et al. 2010). Otherwise, the chemosensory role of these antennae has been supposed for the first time many years ago, because of the presence of porous coeloconic sensilla (Slifer 1977; Slifer and Sekhon 1972), and has been clearly demonstrated by our ultrastructural, electrophysiological and behavioural investigations in the last 10 years.

Damselflies and dragonflies possess multiporous coeloconic sensilla on the latero-ventral side of the antennal flagellum, whose internal structure strongly resembles that of single-walled olfactory sensilla (Rebora et al. 2008; Rebora et al. 2009a; Piersanti et al. 2010) (Fig. 7.6a–c). In detail, these sensilla are innervated by

**Fig. 7.6** (continued) its surface, representing an olfactory sensillum; (**f**) longitudinal section of a sensillum coeloconicum of the mayfly *R. semicolorata*. Note the dendrites (D) inside the peg and the pores (P); (**g**) antennal flagellum of the stonefly *Dinocras cephalotes* revealing a high amount of trichoid sensilla (arrows) representing mechanoreceptors, uniporous sensilla chetica (SC) representing gustatory sensilla, digitated pegs (DP) and porous pegs (PP), both representing olfactory sensilla; (**h**) detail of **g**, showing a digitated peg; (**i**) detail of **g** showing a porous peg. P, pores

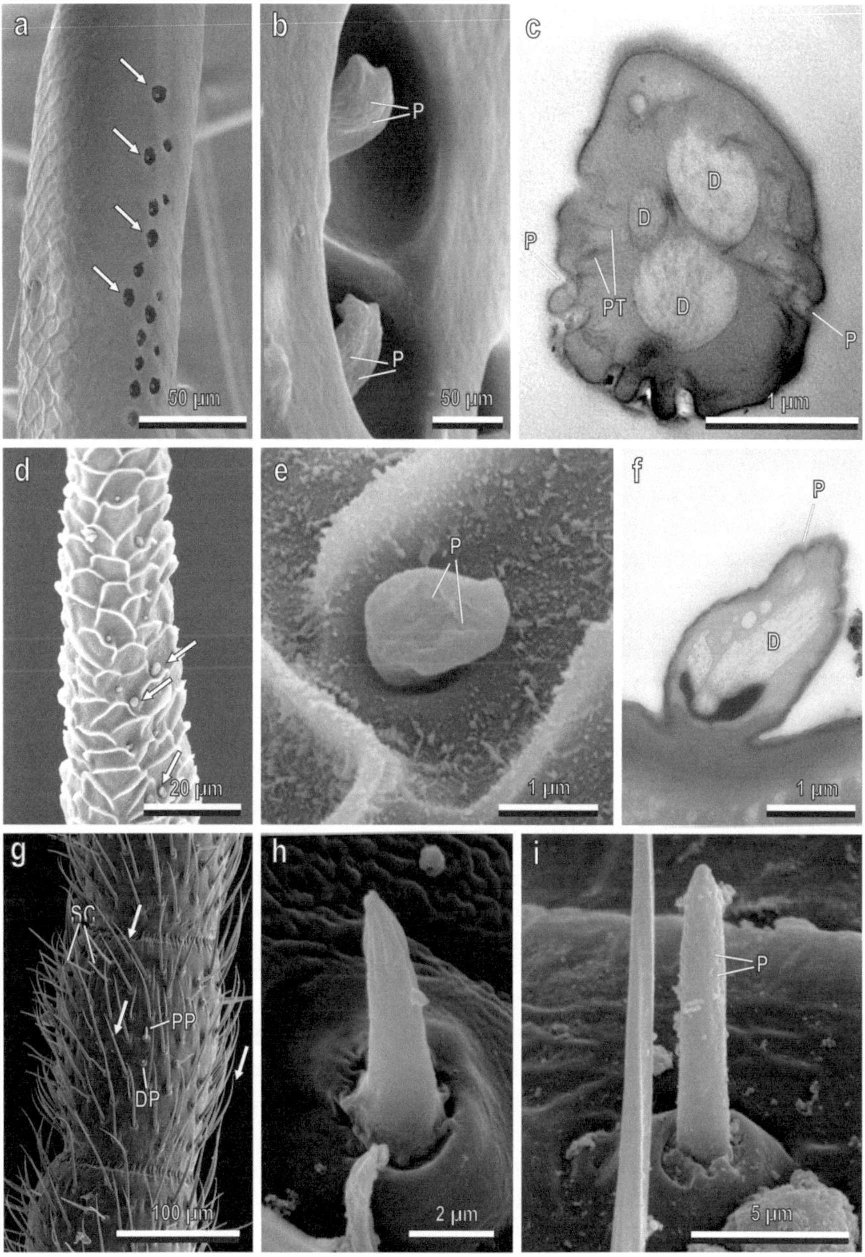


**Fig. 7.6** Chemoreceptors in the adult stage of different species belonging to Odonata (**a**–**c**) and Ephemeroptera (**d**, **f**) and Plecoptera (**g**–**i**) under SEM (**a**, **b**, **d**, **e**, **g**–**i**) and under TEM (**c**, **f**). (**a**) Latero-ventral side of the antennal flagellum of the dragonfly *Libellula depressa* revealing sensilla located in pits (arrows); (**b**) detail of **a**, showing the multiporous coeloconic sensilla with pores (P) representing olfactory sensilla; (**c**) cross section of a coeloconic sensillum at the level of the peg. Note the three unbranched dendrites (D) that enter the peg, the pores (P) and the pore tubules (PT); (**d**) antennal flagellum of the mayfly *Rhithrogena semicolorata* revealing sensilla coeloconica (arrows); (**e**) detail of a sensillum coeloconicum of the mayfly *R. semicolorata* with pores (P) on

three unbranched dendrites that enter the peg in direct touch with the sensillum lymph, and contact the environment through actual pores with pore tubules in the sensory cuticle (Rebora et al. 2008, 2009a) (Fig. 7.6c). Single-walled olfactory sensilla in the form of porous pegs innervated by three neurons have been described also in the two-winged stages of mayflies, the sexually mature *imago* and the transitory immature *subimago* (Rebora et al. 2009a, 2010) (Fig. 7.6d–f).

Paleoptera (dragonflies, damselflies and mayflies) have been traditionally considered visual-dependent insects not able to perceive odours, also in consideration that their brain lacks glomerular antennal lobes and mushroom body calices, which in Neoptera are crucial neuropils involved in odour sensing (Strausfeld et al. 1998, 2009; Farris 2005). Our investigations demonstrated that dragonfly and damselfly antennae are able to perceive several odorants, encompassing mostly amines, carboxylic acids and aldehydes (Piersanti et al. 2014a; Rebora et al. 2012), and that these olfactory inputs reach the brain in an aglomerular antennal lobe provided with spherical knots that probably represent the functional connections between the afferent olfactory sensory neurons and the antennal lobe interneurons (Rebora et al. 2013). Secondary pathway of olfaction in Odonata is still unknown but we also demonstrated that olfactory cues are able to drive relevant behaviour in laboratory, such as attraction towards potential prey (Piersanti et al. 2014b) (Fig. 7.7a), recognition of suitable oviposition sites (Frati et al. 2016) and male attraction towards females (Frati et al. 2015). To investigate the role of dragonflies and damselflies olfaction in the field is much more challenging, probably also in view of the dominant role that sight plays in these insects (Corbet 1999) and of the typically noisier natural conditions. Indeed, olfactory cues from females are able to attract males in one-way olfactometer (Frati et al. 2015) but are not enough, for free-flying males, to detect and recognise potential mates along the shoreline of a natural pond (Rebora et al. 2018). The female odour acting as a cue able to elicit electrophysiological responses in male damselflies could be produced by a gland or more probably a cuticular hydrocarbon (Frati et al. 2015), but further chemical investigations are needed to clarify this aspect, totally disregarded so far. Considering the short lifespan of adult Ephemeroptera, in which the main task is reproduction, and some differences in the number and distribution of olfactory sensilla in males and females, it is plausible to hypothesise that mayfly olfaction could be involved in mating, but no behavioural assays or electrophysiological recordings have ever been performed to investigate this topic, and until now the use of chemical signals for sex attraction in Ephemeroptera is only speculative (Crespo 2011).

Despite the re-evaluation of the antennae as olfactory organs in Paleoptera, dragonfly, damselfly and mayfly smell is surely poor in comparison with the most part of Neoptera, and it is probably quite similar to what it was in the common ancestor. This is in consideration of the dominant role of sight in both orders (Corbet 1999; Kriska et al. 2007), of the simple organisation of the antennal lobe (Strausfeld et al. 1998, 2009; Farris 2005; Rebora et al. 2013), of the small number of the flagellar olfactory sensilla (Rebora et al. 2009b, 2010) and of the similarity in the response profile of damselflies and dragonflies (Piersanti et al. 2014a). In addition, Odonata antennae have been demonstrated to be sensitive to several compounds active on

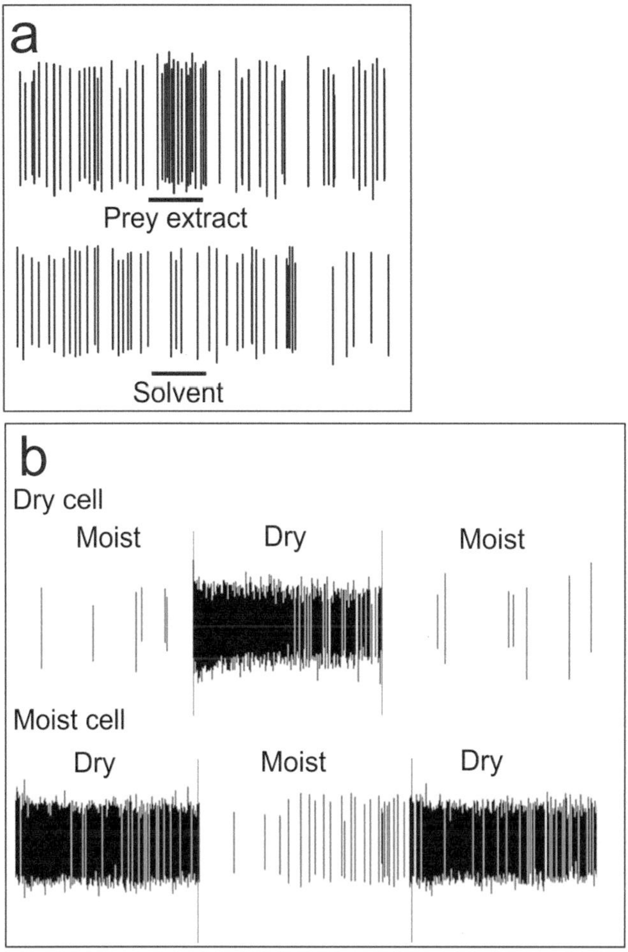


**Fig. 7.7** Example of single-cell recording of sensory neurons in adult (**a**) and nymph (**b**) of *Libellula depressa*. (**a**) Response of an olfactory neuron to prey extract in an acetone and hexane blend or to the solvent blend; (**b**) response of hygro-sensory neurons to rapid changes in air relative humidity (RH)

ionotropic receptors (IRs) located in coeloconic sensilla and/or grooved peg of *Drosophila melanogaster* and other insects (Yao et al. 2005), and the odour-response properties observed in dragonflies and damselflies are very similar to those of IR-expressing neurons and very different from the OR-expressing ones (Piersanti et al. 2014a). Considering that only IRs are employed in Archaeognatha olfaction (Missbach et al. 2014) while ORs are typically present in all Neoptera investigated, it could be very interesting to perform molecular investigations also on Paleoptera.

As far as olfaction in Plecoptera is regarded, their long antennae bear porous and digitated pegs (Fig. 7.6g–i) whose internal structure is coherent with that described for single-walled and double-walled olfactory receptors, respectively (Rebora et al. 2016, 2017). In detail, the porous pegs are innervated by two neurons whose outer dendritic segments are branched and fill the lumen of the peg, the dendrite sheath stops at the base of the peg and actual pores characterise the sensory cuticle (Rebora et al. 2016; Piersanti et al. 2017). The digitated pegs can be considered olfactory sensilla in consideration that they are unsocketed and that the outer dendritic segments of the three dendrites fill the peg without dendrite sheath around; moreover, spoke channels between the fingers connect the lumen of the sensillum, containing the dendrites, with the external environment (Rebora et al. 2016; Piersanti et al. 2017). In agreement with this morpho-functional hypothesis, antennae of the stonefly *Dinocras cephalotes*, stimulated with various chemicals in an electroantennographic research, demonstrated to be able to perceive olfactory cues from some acids and ammines, and responses were very similar in males and females (Rebora et al. 2017).

In addition to olfactory receptors, the antennae of Plecoptera bear numerous sensilla chaetica with an apical pore whose internal structure reveals a typical "touch and taste" function (Rebora et al. 2016; Piersanti et al. 2017) (Fig. 7.6g). In detail, they are innervated by five or six neurons and in both cases one of them stops at the base of the shaft forming a tubular body (mechanosensory unit), while the others fill the shaft reaching the apical pore (gustatory units) (Rebora et al. 2016). Considering that the main task of adult stoneflies is reproduction and that the sexes need to encounter in stone rivers, that typically are very heterogeneous and dark habitats, it is possible to hypothesise that chemoreception is involved in this moment of their life cycle. In particular, males and females could use chemical cues to aggregate in sites where they can effectively perform their typical vibrational mating communication, called drumming, and later, after the encounter, to recognise and evaluate each other before mating (Rebora et al. 2016, 2017; Piersanti et al. 2017). At the moment no behavioural data are available on this topic and, as far as we know, no detailed neuroanatomical investigation has been performed on Plecoptera brain. These researches, together with parallel investigations in Paleoptera, could be very interesting, also considering that the adaptation to a highly derived style of life could have produced a secondary loss of the glomerular antennal lobe in terrestrial Neoptera, such as in cicadas (Strausfeld et al. 2009).

The very conspicuous antennae of adult Trichoptera vary in length according to species and sex and are involved in the major olfactory communication channel in the order, mediated by volatiles secreted by the sternal glands (Crespo 2011). The behaviour related with pheromones, the structure of the exocrine glands and the pheromone production has been extensively investigated in the last 30 years, while the antennal structure, and especially the olfactory sensilla situated on the antennal flagellum, has been disregarded so far (Crespo 2011; Melnitsky et al. 2018). Exocrine glands can be present in abdominal sternite of both males and females, and probably produce compounds that have different sexual roles in males and females (e.g. Ansteeg and Dettner 1991), together with common additional roles such as defence against predators (Duffield et al. 1977; Duffield 1981). Morphological investigations of the external structure of antennal sensilla in Limnephilidae (Slifer and Sekhon 1971), Leptoceridae (Tozer 1982), different species of Philopotamidae

(Melnitsky et al. 2018) and other families (Melnitsky and Ivanov 2011) report the presence of thick-walled chemoreceptor pegs, thin-walled chemoreceptor pegs and a special type of thin-walled chemoreceptor called plate organs. These structures are potentially involved in chemical perception, and can be uniformly or differently distributed between sexes depending on the species, but no more detailed data are available on their internal organisation and innervation. Electrophysiological experiments confirmed that identified compounds from the female exocrine glands are perceived by male's antenna (e.g. Löfstedt et al. 1994, 2008; Jewett et al. 1996; Bergmann et al. 2001), and some of these experiments also proved that active compounds produced by males can be perceived by male's antennae and similarly for females (e.g. Jewett et al. 1996; Bergmann et al. 2001). These data suggest that, notwithstanding the close phylogenetic relationship between Lepidoptera and Trichoptera (Zhou et al. 2016), the pheromone system of the latters is much less specific compared with that of the formers. On this account a recent paper (Yuvaraj et al. 2018) suggests a shift in major types of olfactory sensilla, from a high proportion of sensilla placodea or auricillica in Trichoptera and the most basal moth lineages, respectively, to sensilla trichodea in the more derived Lepidoptera, which parallels the change in the types of sex pheromones used. Although moths are typical models for insect neuroanatomy and neurophysiology, there are very few data on Trichoptera brain (Ehnbom 1948; Dacks et al. 2006), reporting that olfactory neurons from the antennae reach the brain in a glomerular antennal lobe, except the genus *Mystacides* that show long antennae and an aglomerular antennal lobe. Molecular investigations on olfaction in these aquatic Neoptera, also in comparison with other aquatic insects and model Lepidoptera, could be particularly interesting to reconstruct the evolutionary pathways of insect olfaction.

## 7.4 Thermo-Hygroreception

Thermoreception and hygroreception are very important for small terrestrial animals such as insects, which need to cope with a high rate of water loss, due to the large ratio of surface area to body size, have a small storage capacity for water, and depend on temperature of environment for their activity being poikilothermal. Despite its relevance, thermo- and hygro-sensitivity in insects has been poorly investigated in comparison with other sensory abilities such as olfaction (Meritt 2007). In addition to a general role in controlling water balance, to sense humidity and temperature variation can also drive specific behaviours, such as host searching for haematophagous insects (Enjin 2017). Similarly, aquatic insects that lay eggs in water or humid places must detect humidity gradient, and consequently temperature gradient, in air to search for adequate oviposition sites.

Typically, hygro- and thermoreceptors in arthropods are not numerous, located on the antennae and housing together, in a single sensillum, a physiologic triad of receptor neurons: moist receptor, dry receptor and hygrocool receptor (defined according to the input able to increase the firing rate of the neuron) (Altner and Loftus 1985; Yokohari 1999; Enjin 2017). On the other hand, different assortments of sensory neurons have been described in different insects, such as *Periplaneta americana*, which has a hygrocool receptor associated with olfactory receptors in a double-walled sensillum (Nishikawa et al. 1992); *Periplaneta, Locusta* and *Leucophaea*, which have an additional hygrocool cell (Loftus 1976, 1978; Altner et al. 1977; Schaller 1982; Nishikawa et al. 1992); or *Aedes*, which possess both warm and cold receptors in the same sensillum (Davis and Sokolove 1975).

Morphologically, the first report of modality-specific structures in a hygro- and thermoreceptive sensillum, which was unambiguously identified electrophysiologically, is the description of the sensillum capitulum on the antenna of *Periplaneta* (Yokohari et al. 1975). This is a non-perforated cone with a mushroom-shaped inner apparatus. Later, thermo-hygroreceptors have been identified with variable external morphologies in different insects, such as coeloconic in some Orthoptera or styloconic in some Lepidoptera, but some peculiar characters, both in external and internal structure (see review in Altner and Loftus 1985): they typically are protected from wind by cuticle invaginations, and they have inflexible socket, not perforated cuticle, clogged molten pore, clefts and/or cavities in the sensory cuticle and electron-dense material and or lipid-like deposits in the outer lymph cavity, and a dendritic sheath fuse with the sensory cuticle; thermoreceptive units typically differentiate lamellae and/or cylindrical branches in the outer dendritic segments (reviewed in Tichy and Loftus 1996; Steinbrecht 1998; Yokohari 1999; Enjin 2017) (Fig. 7.1).

These features are in agreement with the most accredited transduction models for thermo-hygroreceptors (Yokohari 1999), such as the mechanosensor model that hypothesises a sensory cuticle working as a hygroscopic material, but the transduction mechanism of these sense organs is still unknown (Enjin 2017; Tichy et al. 2017).

Recent molecular and neuroanatomical investigations on hygro-thermoreceptors on the antennae of *D. melanogaster* (Enjin et al. 2016; Enjin 2017) demonstrated that in fruit flies hygro- and thermoreceptor neurons express ionotropic receptors (IR) and the information is processed in a glomerular region of the antennal lobe called posterior antennal lobe. Dry, moist and hygrocool glomeruli are intermingled with hot and cool glomeruli, innervated by pure thermosensory neurons (Gallio et al. 2011), and probably information is combined in the brain to guide fly navigation through the environment. In adult Odonata, the aglomerular antennal lobe is divided in an anteroventral lobe and a larger posterodorsal lobe, probably receiving, separately, olfactory and thermo-hygroreceptor neurons (Rebora et al. 2013). This is the only neuroanatomical study on the organisation of the antennal lobe in aquatic insects and further researches are needed to test this hypothesis and investigate more deeply the topic.

### *7.4.1 Aquatic Stage*

Larval stages of aquatic insects typically live in water, so it is quite obvious to consider as redundant hygroreception in such environment. Otherwise, many aquatic larvae, specially belonging to dragonflies, live in small shallow ponds, sometimes artificial irrigation pools, in temperate climates, and these environments can easily dry out, or reduce to a mosaic of small pools and mud, during very hot summer or because of human activities. On this account, the last larval stages of *L. depressa* (Odonata, Libellulidae) gave evidence of water deprivation tolerance and hygro-positive responses out of water in experimental conditions (Rebora et al. 2007a), and in field experiments they showed a great ability to seek for residual water in response to the drying of their pond (Piersanti et al. 2007). In agreement with these behaviours, electrophysiological recordings (Rebora et al. 2007b) on the apical antennae of these larvae demonstrated the presence of moist and dry cells responding antagonistically to humidity changes (Fig. 7.7b), while thermal stimuli were not tested. These responses are related with the presence of two pegs in pits showing typical features of thermo-hygroreceptors at the antennal apex (Gaino and Rebora 2001; Rebora et al. 2007b). They have a sensory cuticle without pores but irregularly folded, a dendrite stopping at the base of the peg where it splits into several branches, and two dendrites entering the peg wrapped by an electron-dense dendrite sheath in close contact with the sensory cuticle around (Gaino and Rebora 2001; Rebora et al. 2007b). Similar coeloconic sensilla, putative hygro-thermoreceptors, have been described in other dragonfly and damselfly larvae (Rebora et al. 2015, Piersanti et al. 2018; personal observations in *Aeshna cyanea*). On this account, it could be very interesting to evaluate how much these sensory structures are widespread in Odonata, also considering the ability of some species to breed in small inconstant aquatic environments, and the need of all species to leave the water, partly or wholly and intermittently or irrevocably, during the days before the emergence (Corbet 1999). On this account it is interesting to remember that dragonfly larvae are able to breathe in air through their gills protected in the anal basket, while damselfly larvae are not, and that odonatan mesothoracic spiracles start to open at the last development stage (Gaino et al. 2007). The timing and duration of emergence also involve responses to temperature in many aquatic insects and precise temperature thresholds for ecdysis have been reported for some aquatic insects (Ward and Stanford 1982). Migration of mature larvae to shorelines where temperatures are more sensitive to terrestrial conditions has been reported for lentic and lotic species (Ward and Stanford 1982).

In Ephemeroptera, coeloconic non-porous sensilla have been reported on the antennae of Baetidae (Gaino and Rebora 1999), but their function cannot be hypothesised because the internal structure has not been described.

As far as Plecoptera is concerned ultrastructural investigations on the antennae of the stonefly nymph *Paragnetina media* describe non-porous sensilla coeloconica, located in shallow pits and innervated by a single neuron entering the peg, that strongly resembles the typical morphology of thermo-hygroreceptors (Kapoor 1991).

No references are available on thermo-hygroreceptors in the greatly reduced antennae of Trichoptera larvae.

### 7.4.2 Adult Stage

It is well known that many aquatic insects can "see" the presence of water, also when it is covered by bushes or is very far away, on the basis of the horizontally polarised light reflected from the water surface and perceived by the ventral region of the eye (Bernáth et al. 2002; Schwind 1991). Despite this, the ability to perceive humidity and temperature as environmental cues can obviously contribute in adult habitat selection, and can be very important for hygro-thermoregulation in "sunny insects", as dragonflies, that are very active at high temperature and spend many hours perching in sunny places. Hygro-thermoreception could also be useful in taxis behaviour with respect to water; indeed many teneral dragonflies and damselflies show a negative taxis and roost in the landscape until sexual maturation, while they come back to water to reproduce (Hardersen 2007). For fragile insects as Ephemeroptera and Trichoptera, which can dry out quickly during moulting and swarming (Kriska et al. 1998), to be able to orient in suitable hygro-thermoenvironmental conditions is fundamental.

On this account, dry, moist and cold receptor neurons have been identified on the antennal flagellum of the adult dragonfly *L. depressa* (Piersanti et al. 2011). Unfortunately, it was not possible to perform single sensillum recordings on these antennae, because the three types of sensilla present in the flagellum are hosted in pits and are not visible on the cuticular surface (Piersanti et al. 2011) (Fig. 7.8a–c). The authors suggest that the thermo- and hygroreceptive cells identified could belong to deeply sunken sensilla styloconica, described in *L. depressa* (Rebora et al. 2008) and other dragonflies (Rebora et al. 2009a) and damselflies (Piersanti et al. 2010) and "sharing common features typical of thermo-hygroreceptors" (Fig. 7.8a–d). In particular, type 2 deeply sunken sensillum is the most probable thermo-hygroreceptive unit, because it has a rugged cuticular surface with a clogged moulting pore and is innervated by three unbranched neurons, one stopping at the base of the cone and the others filling the stylus, closely adherent to each other and enveloped by a tightly adherent dendrite sheath, in close contact with the sensory cuticle characterised by internal clefts (Rebora et al. 2008, 2009a) (Fig. 7.8b–d).

In mayflies (Rebora et al. 2009b, 2010) and stoneflies (Rebora et al. 2016, 2017; Piersanti et al. 2017) antennal flagella are provided with small pegs whose structure resembles that of thermo-hygroreceptors. In Ephemeroptera they are visible on the antennal surface and innervated by two or four neurons (Fig. 7.8e), while in Plecoptera they are innervated by three neurons and protected in small domes (Fig. 7.8f). No behavioural data are available about the ability of these insects to orient in humidity and temperature gradients, but the most probable function for these sensilla is related with habitat selection and oviposition (Rebora et al. 2009b, 2010, 2016; Piersanti et al. 2017).

Antennal sensory structures in adult Trichoptera, as above reported, show a significant diversity and relevant differences in number, type and distribution between families (Ivanov and Melnitsky 2011, 2016; Melnitsky and Ivanov 2011, 2016; Melnitsky et al. 2018). Then there are no reasons to suppose that caddisfly

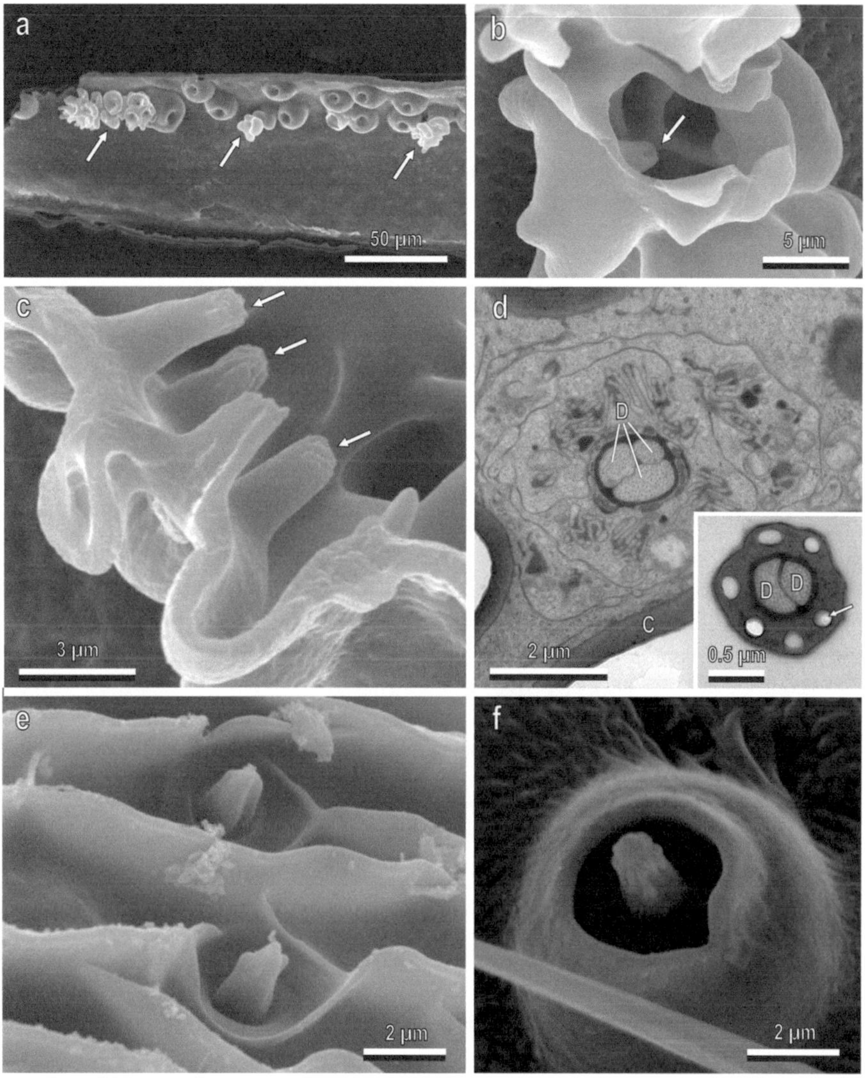


**Fig. 7.8** Thermo-hygroreceptors in the adult stage of different species belonging to Odonata (**a**–**d**), Ephemeroptera (**e**) and Plecoptera (**f**) under SEM (**a**–**c**, **e**, **f**) and under TEM (**d**). (**a**) Internal cuticular surface of the antennal flagellum of the dragonfly *Libellula depressa* (treated with KOH to remove the internal tissues (for procedure, see Rebora et al. 2008). Note the convolute deep cavities (arrows) housing the deeply sunken sensilla styloconica probably representing thermo-hygroreceptors; (**b**) detail of a broken cavity showing the presence of a sensillum styloconicum (arrow) inside; (**c**) sensilla styloconica (arrows) inside the cavity. Note the rugged cuticular surface; (**d**) cross section of a sensillum styloconicum at the base of the peg. Note the three dendrites (D) innervating it. C, cuticle. Inset shows a cross section of the same sensillum at a more distal level. Note that one dendrite stops at the base of the peg while two dendrites (D) extend along the peg in close connection with the sensory cuticle characterised by internal clefts (arrow); (**e**) detail of small aporous pegs on the antennal flagellum of the mayfly *Rhithrogena semicolorata* whose internal structure resembles that of thermo-hygroreceptors; (**f**) aporous peg located in a small dome on the antennal flagellum of the stonefly *Dinocras cephalotes* whose internal structure resembles that of thermo-hygroreceptors

antennae cannot perceive relevant environmental cues such as humidity and temperature. Small sensilla of the "lower level," as defined by Melnitsky et al. (2018), probably encompass thermo-hygroreceptors, but no data are available on the internal structure of these sensilla and these sensory abilities have never been investigated in the order.

## 7.5 Conclusions

The knowledge on the sensory systems underlying the neuroethology of aquatic insect behaviours and the evolutionary pressures that may have shaped them is still poorly known. Further investigations in this field could be particularly relevant for many reasons. A better knowledge of aquatic insect sensilla could play a significant role in unravelling insect perception and adaptability of species to different environmental conditions. To investigate the sensory systems of hemimetabolous insects in particular could shed light on the morphological and synaptic reorganisation of the nervous systems allowing to perform new functions in the transition from the aquatic to the terrestrial environment. Particularly interesting problems such as the reduction of olfaction in adult Paleoptera would deserve attention in relation with the evolution of insect olfaction. Moreover, in consideration that insects have evolved a set of extremely effective sensory systems which are structurally simple, functionally versatile, powerful, highly distributed and noise and fault tolerant (Ma and Krings 2009), to study aquatic insect sensilla can be highly inspirational for the engineering design and biosensor development in the aquatic environment.

## References

Akent'eva NA (2012) Multimodal sensory organs in larvae of some insect species. Zool Zh 91:38–48

Altner H, Loftus R (1985) Ultrastructure and function of insect thermo- and hygroreceptors. Annu Rev Entomol 30:273–295

Altner H, Sass H, Altner I (1977) Relationship between structure and function of antenna chemo-, hygro-, and thermoreceptive sensilla in *Periplaneta americana*. Cell Tissue Res 176:389–405

Alvarez M, Landeira-Dabarca A, Peckarsky B (2014) Origin and specificity of predatory fish cues detected by *Baetis larvae* (Ephemeroptera; Insecta). Anim Behav 96:141–149

Ansteeg O, Dettner K (1991) Chemistry and possible biological significance of secretions from a gland discharging at the 5th abdominal sternite of adult caddisflies (Trichoptera). Entomol Gen 15:303–312

Bergmann J, Löfstedt C, Ivanov VD et al (2001) Identification and assignment of the absolute configuration of biologically active methyl-branched ketones from limnephilid caddisflies. Eur J Org Chem 2001:3175–3179

Bernáth B, Szedenics G, Wildermuth H et al (2002) How can dragonflies discern bright and dark waters from a distance? The degree of polarization of reflected light as a possible cue for dragonfly habitat selection. Freshw Biol 47:1707–1719

Brönmark C, Hansson L-A (2000) Chemical communication in aquatic systems: an introduction. Oikos 88:103–109
Caillère L (1964) Contribution au comportement de capture des larves d'*Agrion splendens* Harris 1782 (Odonates, Zygopteres): role des antennes dans le declenchement du reflexe de capture. 89° Congr Nat Soc Sav Lyon 435–442
Caillère L (1965) Description du reflexe de capture chez la larve d'*Agrion splendens* Harris 1782 (Insectes, Odonates, Zygopteres). Bull Mens Soc Linn Lyon 34:424–434
Caillère L (1968) Role des organes des sens dans le comportement de capture chez la larve d'*Agrion splendens* Harris 1782 (Insectes, Odonates, Zygopteres). Bull Mens Soc Linn Lyon 37:25–34
Chivers DP, Wisenden BD, Smith RJF (1996) Damselfly larvae learn to recognize predators from chemical cues in the predator's diet. Anim Behav 52:315–320
Corbet PS (1999) Dragonflies behaviour and ecology of Odonata. Harley Books, Colchester
Crespo JG (2011) A review of chemosensation and related behavior in aquatic insects. J Insect Sci 11:1–39
Crumrine PW (2006) Age specific behavioral responses of odonate larvae to chemical and visual cues from predators. J Fresh Ecol 21:9–16
Dacks AM, Christensen TA, Hildebrand JG (2006) Phylogeny of a serotonin immunoreactive neuron in the primary olfactory center of the insect brain. J Comp Neurol 498:727–746
Davis EE, Sokolove PG (1975) Temperature responses of antennal receptors of the mosquito *Aedes aegypti*. J Comp Physiol 96:223–236
Derby CD, Kozma MT, Senatore A et al (2016) Molecular mechanisms of reception and perireception in crustacean chemoreception: a comparative review. Chem Sens 41:381–398
Devarakonda R, Barth FG, Humphrey JAC (1996) Dynamics of arthropod filiform hairs. IV. Hair motion in air and water. Philos Trans R Soc Lond Ser B Biol Sci 351:933–946
Dijkstra KD, Monaghan MT, Pauls SU (2014) Freshwater biodiversity and aquatic insect diversification. Annu Rev Entomol 59:143–163
Duffield RM (1981) 2-Nonanol in the exocrine secretion of the Nearctic caddisfly, *Rhyacophila fuscula* (Walker) (Rhyacophilidae: Trichoptera). Proc Entomol Soc Wash 83:60–63
Duffield RM, Blum MS, Wallace JB et al (1977) Chemistry of the defensive secretion of the caddisfly *Pycnopsyche scabripennis* (Trichoptera: Limnephilidae). J Chem Ecol 3:649–656
Ehnbom K (1948) Studies on the central and sympathetic nervous system and some sense organs in the head of neuroptid insects. Opuscula Entomol Suppl 8:1–80
Enjin A (2017) Humidity sensing in insects-from ecology to natural processing. Curr Opin Insect Sci 24:1–6
Enjin A, Zaharieva EE, Frank DD et al (2016) Humidity sensing in Drosophila. Curr Biol 26:1352–1358
Farris SM (2005) Development organization of the mushroom bodies of *Thermobia domestica* (Zygentoma, Lepismatidae): insights into mushroom body evolution from a basal insect. Evol Dev 7:150–159
Faucheux MJ (2007) Multiporous and aporous sensilla on the larval antennae of the relict dragonfly *Epiophlebia superstes* (Selys, 1889) (Odonata: Anisozygoptera: Epiophlebiidae). Bull Inst R Sci Nat Belg Entomol 77:121–128
Faucheux MJ, Meurgey F (2007) Sensilles chètiformes et filiformes sur les antennes larvaires d'*Uropelata chiltoni* Tillyard, 1930 (Odonata, Anisoptera, Petaluridae). Martinia 23:127–132
Faucheux MJ, Meurgey F (2009) Les sensilles antennaires d'une larve fouisseuse, *Ophiogomphus cecilia* (Geoffroy in Fourcroy, 1785) (Odonata, Anisoptera, Gomphidae). Martinia 25:85–92
Frati F, Piersanti S, Rebora M et al (2015) Scent of a Dragonfly: sex recognition in a polymorphic Coenagrionid. PLoS One 10:e0136697. https://doi.org/10.1371/journal.pone.0136697
Frati F, Piersanti S, Rebora M et al (2016) Volatile cues can drive oviposition behaviour in Odonata. J Insect Physiol 91-92:34–38
Gaino E, Rebora M (1996) Fine structure of flat-tipped antennal sensilla in three species of mayflies (Ephemeroptera). Invertebr Biol 115:145–149

Gaino E, Rebora M (1997) Antennal cuticular sensilla in some mayflies (Ephemeroptera). In: Landolt P, Sartori M (eds) Ephemeroptera & Plecoptera: Biology-Ecology-Systematics. MTL, Fribourg, pp 317–325

Gaino E, Rebora M (1998) Ultrastructure of the antennal sensilla of the mayfly *Baetis rhodani* (Pictet) (Ephemeroptera: Baetidae). Int J Insect Morphol Embryol 27:143–149

Gaino E, Rebora M (1999) Larval antennal sensilla in water-living insects. Microsc Res Tech 47:440–457

Gaino E, Rebora M (2001) Apical antennal sensilla in nymphs of *Libellula depressa* (Odonata: Libellulidae). Invertebr Biol 120:162–169

Gaino E, Piersanti S, Rebora M (2007) Ultrastructural organization of the larval spiracles in *Libellula depressa* L. (Anisoptera: Libellulidae). Odonatologica 36:373–379

Galizia CG, Rössler W (2010) Parallel olfactory systems in insects: anatomy and function. Annu Rev Entomol 55:399–420

Gall BG, Brodie Jr ED (2009) Behavioral avoidance of injured conspecific and predatory chemical stimuli by larvae of the aquatic caddisfly *Hesperophylax occidentalis*. Can J Zool 87:1009–1015

Gallio M, Ofstad TA, Macpherson LJ et al (2011) The coding of temperature in the Drosophila brain. Cell 144:614–624

Gullan PJ, Cranston PS (2006) The insects: an outline of entomology. Chapman & Hall, London

Hallberg E, Hansson BS (1999) Arthropod sensilla: morphology and phylogenetic considerations. Microsc Res Tech 47:428–439

Hansson B, Stensmyr MC (2011) Evolution of insect olfaction. Neuron 72:698–711

Hardersen S (2007) Telemetry of Anisoptera after emergence- first results (Odonata). Int J Odonatol 10:189–202

Humphrey JAC, Devarakonda R, Iglesias J et al (1993) Dynamics of arthropod filiform hairs. I. Mathematical modelling of the hair and air motions. Philos Trans R Soc B 340:423–444

Hunger T, Steinbrecht RA (1998) Functional morphology of a double-walled multiporous olfactory sensillum: the sensillum coeloconicum of *Bombyx mori* (Insecta. Lepidoptera). Tissue Cell 30:14–29

Ivanov VD, Melnitsky SI (2011) Structure and morphological types of the antennal olfactory sensilla in Phryganeidae and Limnephilidae (Insecta: Trichoptera). Zoosymposia 5:210–234

Ivanov VD, Melnitsky SI (2016) Diversity of the olfactory sensilla in caddis-flies (Trichoptera). Zoosymposia 10:224–233

Jewett DK, Brigham DL, Bjostad LB (1996) *Hesperophylax occidentalis* (Trichoptera: Limnephilidae): electroantennogram structure-activity study of sex pheromone component 6-methylnonan-3-one. J Chem Ecol 22:123–137

Kapoor NN (1985) External morphology and distribution of the antennal sensilla of the stonefly, *Paragnetina media* (Walker) (Plecoptera: Perlidae). Int J Insect Morphol Embryol 14:273–280

Kapoor NN (1987) Fine structure of the coniform sensillar complex on the antennal flagellum of the stonefly nymph *Paragnetina media* (Plecoptera: Perlidae). Can J Zool 65:1827–1832

Kapoor NN (1991) Antennal campaniform and coeloconic sensilla of the stonefly nymph, *Paragnetina media* (Walker) (Plecoptera: Perlidae). In: Alba-Tercedor J, Sanchez-Ortega A (eds) Overview and strategies of Ephemeroptera and Plecoptera. Sandhill Crane Press, Inc., Gainesville, FL, pp 39–46

Keil TA (1997) Functional morphology of insect mechanoreceptors. Microsc Res Tech 39:506–531

Kriska G, Horváth G, Andrikovics S (1998) Why do mayflies lay their eggs en masse on dry asphalt roads? Water-imitating polarized light reflected from asphalt attracts Ephemeroptera. J Exp Biol 201:2273–2286

Kriska G, Bernáth B, Horváth G (2007) Positive polarotaxis in a mayfly that never leaves the water surface: Polarotactic water detection in *Palingenia longicauda* (Ephemeroptera). Naturwissenschaften 94:148–154

Laverack MS (1962) Responses of cuticular sense organs of the lobster, *Homarus vulgaris*. II. Hair-fan organs as pressure receptors. Comp Biochem Physiol 6:137–145

Löfstedt C, Hansson BS, Petersson E et al (1994) Pheromonal secretions from glands on the 5th abdominal sternite of hydropsychid and rhyacophilid caddisflies (Trichoptera). J Chem Ecol 20:153–170

Löfstedt C, Bergmann J, Francke W et al (2008) Identification of a sex pheromone produced by sternal glands in females of the (Trichoptera) caddisfly *Molanna angustata* Curtis (Trichoptera, Molannidae). J Chem Ecol 34:220–228

Loftus R (1976) Temperature-dependent dry receptor on antenna of *Periplaneta*. J Comp Physiol 111:153–170

Loftus R (1978) Peripheral thermal receptors. In: Ali MA (ed) Sensory ecology: reviews and perspectives, NATO Advanced Study Institutes Series, vol 18. Plenum, New York/London, pp 439–466

Ma ZS, Krings AW (2009) Insect sensory systems inspired computing and communications. Ad Hoc Netw 7:742–755

McIntosh AR, Peckarsky BL (1996) Differential behavioural responses of mayflies from streams with and without fish to trout odour. Freshw Biol 35:141–148

McIntosh AR, Peckarsky BL (2004) Are mayfly anti-predator responses to fish odour proportional to risk? Arch Hydrobiol 160:145–151

McIntosh AR, Peckarsky BL, Taylor BW (1999) Rapid size-specific changes in the drift of *Baetis bicaudatus* (Ephemeroptera) caused by alterations in fish odour concentration. Oecologia 118:256–264

Melnitsky SI, Ivanov VD (2011) Structure and localization of sensilla on antennaeof caddisflies (Insecta: Trichoptera). J Evol Biochem Physiol 47:593–602

Melnitsky SI, Ivanov VD (2016) Structure of the antennal olfactory sensilla in the genus Molanna (Trichoptera: Molannidae). Zoosymposia 10:292–300

Melnitsky SI, Vladimir D, Ivanov VD et al (2018) Comparison of sensory structures on the antenna of different species of Philopotamidae (Insecta: Trichoptera). Arthropod Struct Dev 47:45–55

Meritt DJ (2007) The organule concept of insect sense organs: sensory transduction and organule evolution. Adv Insect Physiol 33:192–242

Meurgey F, Faucheux MJ (2006a) Vibroreceptors and proprioceptors on the larval antennae of *Erythromma lindenii* (Sélys) (Zygoptera: Coenagrionidae). Odonatologica 35:255–264

Meurgey F, Faucheux MJ (2006b) Organes sensoriels des antennes de la larve de *Chalcolestes viridis* (Van der Linden, 1825) (Odonata, Zygoptera, Lestidae). Martinia 22:167–171

Michels J, Appel E, Gorb SN (2016) Functional diversity of resilin in Arthropoda. Beilstein J Nanotech 7:1241–1259

Missbach C, Dweck HKM, Vogel H et al (2014) Evolution of insect olfactory receptors. elife 3:1–22

Mortensen L, Richardson JML (2008) Effects of chemical cues on foraging in damselfly larvae, *Enallagma antennatum*. J Insect Behav 21:285–295

Murlis J, Elkinton JS, Cardé RT (2003) Odor plumes and how insects use them. Annu Rev Entomol 37:505–532

Nishikawa M, Yokohari F, Ishibashi T (1992) Response characteristics of two types of cold receptors on the antennae of the cockroach, *Periplaneta Americana* L. J Comp Physiol A 171:299–307

Ode PR, Wissinger S (2006) Interaction between chemical and tactile cues in mayfly detection of stoneflies. Freshw Biol 30:351–357

Okano JI, Tayasu I, Nakano SI et al (2017) Differential responses of two ecologically similar case-bearing caddisfly species to a fish chemical cue: implications for a coexistence mechanism. Zool Sci 34:461–467

Peckarsky BL, Wilcox RS (1989) Stonefly nymphs use hydrodynamic cues to discriminate between prey. Oecologia 79:265–270

Piersanti S, Rebora M (2018) The antennae of damselfly larvae. Arthropod Struct Dev 47:36–44

Piersanti S, Rebora M, Salerno G et al (2007) Behavioural strategies of the larval dragonfly *Libellula depressa* (Odonata: Libellulidae) in drying pools. Ethol Ecol Evol 19:127–136

Piersanti S, Rebora M, Gaino E (2010) A scanning electron microscope study of the antennal sensilla in adult Zygoptera. Odonatologica 39:235–241
Piersanti S, Rebora M, Almaas TJ et al (2011) Electrophysiological identification of thermo- and hygro-sensitive receptor neurons on the antennae of the dragonfly *Libellula depressa*. J Insect Physiol 57:1391–1398
Piersanti S, Frati F, Conti E et al (2014a) The sense of smell in Odonata: an electrophysiological screening. J Insect Physiol 70:49–58
Piersanti S, Frati F, Conti E et al (2014b) First evidence of the use of olfaction in Odonata behaviour. J Insect Physiol 62:26–31
Piersanti S, Rebora M, Lopez Rodriguez MJ et al (2017) A comparison between the adult antennal sensilla of the cavernicolous stonefly *Protonemoura gevi* and other epigean *Protonemura* species (Plecoptera: Nemouridae) in a biological context. Ann Soc Entomol Fr 53:47–54
Rebora M, Gaino E (2008) The antennal sensilla of the nymph of *Ephemera danica*. In: Hauer FR, Stanford JA, Newell RL (eds) International advances in the ecology, zoogeography and systematics of mayflies and stoneflies. University of California Press, Berkeley, CA, pp 307–312
Rebora M, Piersanti S, Gaino E (2004) Visual and mechanical cues in prey detection by the larva of *Libellula depressa* (Odonata: Libellulidae). Ethol Ecol Evol 16:133–144
Rebora M, Piersanti S, Salerno G, Conti E, Gaino E (2007a) Water deprivation tolerance and humidity response in a larval dragonfly: a possible adaptation for survival in drying ponds. Physiol Entomol 32:121–126
Rebora M, Piersanti S, Almaas TJ et al (2007b) Hygroreceptors in the larva of *Libellula depressa* (Odonata: Libellulidae). J Insect Physiol 53:550–558
Rebora M, Piersanti S, Gaino E (2008) The antennal sensilla of the adult of *Libellula depressa* (Odonata: Libellulidae). Arthropod Struct Dev 37:504–510
Rebora M, Piersanti S, Gaino E (2009a) A comparative investigation of the antennal sensilla of adult Anisoptera. Odonatologica 38:329–340
Rebora M, Piersanti S, Gaino E (2009b) The antennal sensilla of adult mayflies: *Rhithrogena semicolorata* as a case study. Micron 40:571–576
Rebora M, Piersanti S, Gaino E (2010) The antennal sensory function in the oldest pterygote insects: an ultrastructural overview. In: Méndez Vilas A, Díaz J (eds) Microscopy: science, technology, applications and education. Formatex Research Center, Badajoz, pp 137–145
Rebora M, Salerno G, Piersanti S, Dell'Otto A, Gaino E (2012) Olfaction in dragonflies: electrophysiological evidence. J Insect Physiol 58:270–277
Rebora M, Dell'Otto A, Rybak J et al (2013) The antennal lobe of *Libellula depressa* (Odonata, Libellulidae). Zoology 116:205–214
Rebora M, Piersanti S, Salerno G, Gorb S (2015) The Antenna of a burrowing dragonfly larva, *Onychogomphus forcipatus* (Anisoptera,Gomphidae). Arthropod Struct Dev 44:595–603
Rebora M, Tierno de Figueroa JM, Piersanti S (2016) Antennal sensilla of the stonefly *Dinocras cephalotes* (Plecoptera: Perlidae). Arthropod Struct Dev 45:552–561
Rebora M, Piersanti S, Frati F et al (2017) Antennal responses to volatile organic compounds in a stonefly. J Insect Physiol 98:231–2376
Rebora M, Frati F, Piersanti S et al (2018) Field tests of multiple sensory cues in sex recognition and harassment of a colour polymorphic damselfly. Anim Behav 136:127–136
Schaller D (1982) Structural and functional classification of antennal sensilla of the cockroach, *Leucophea maderae*. Cell Tissue Res 225:129–142
Schneider D (1964) Insect antennae. Annu Rev Entomol 9:103–122
Schwind R (1991) Polarization vision in water insects and insects living on a moist substrate. J Comp Physiol A 169:531–540
Scrimgeour GJ, Culp JM, Cash KJ (1994) Anti-predator responses of mayfly larvae to conspecific and predator stimuli. J N Am Benthol Soc 13:299–309
Shimozawa T, Kanou M (1984) The aerodynamics and sensory physiology of range fractionation in the cercal filiform sensilla of the cricket *Gryllus bimaculatus*. J Comp Physiol A 155:495–505
Siepielski AM, Fallon E, Boersma K (2016) Predator olfactory cues generate a foraging–predation trade-off through prey apprehension. R Soc Open Sci 3:150537

Slifer EH (1977) Sense organs on the antennal flagellum of Mayflies (Ephemeroptera). J Morphol 153:355–362
Slifer EH, Sekhon SS (1971) Structures on the antennal flagellum of a caddisfly, *Frenesia missa* (Trichoptera, Limnephilidae). J Morphol 135:373–388
Slifer EH, Sekhon SS (1972) Sense organs on the antennal flagella of damselflies and dragonflies (Odonata). Int J Insect Morphol Embryol 1:289–300
Steinbrecht RA (1997) Pore structures in insect olfactory sensilla: a review of data and concepts. Int J Insect Morphol Embryol 26:229–245
Steinbrecht RA (1998) Bimodal thermo-and hygrosensitive sensilla. In: Harrison FW, Locke M (eds) Microscopic Anatomy of Invertebrates 11B. Wiley-Liss, New York, pp 405–422
Stoks R (2001) Food stress and predator induced stress shape developmental performance in a damselfly. Oecologia 127:222–229
Strausfeld NJ, Hansen L, Li Y et al (1998) Evolution, discovery and interpretations of arthropod mushroom bodies. Learn Mem 5:11–37
Strausfeld NJ, Sinakevitch I, Brown SM et al (2009) Ground plan of the insect mushroom body: functional and evolutionary implications. J Comp Neurol 513:265–291
Suhling F, Müller O (1996) Die Flußjungfern Europas. Magdeburg: Westarp-Wiss. Spektrum Akad. Verlag, Heidelberg
Suhling F, Sahlen G, Gorb S et al (2015) Order Odonata. In: Thorp J, Rogers DC (eds) Ecology and general biology: Thorp and Covich's Freshwater Invertebrates. Academic Press, New York, pp 893–932
Tautz J (1977) Reception of medium vibration by thoracal hairs of caterpillars of *Barathra brassicae* L. (Lepidoptera, Noctuidae). I Mechanical properties of the receptors hairs. J Comp Physiol 118:13–31
Thurm U (1965) An insect mechanoreceptor. I. Fine structure and adequate stimulus. Cold Spring Harb Symp Quant Biol 30:75–82
Tichy H, Loftus R (1996) Hygroreceptors in insects and a spider: humidity transduction models. Naturwissenschaften 83:255–263
Tichy H, Hellwig M, Kallina W (2017) Revisiting theories of humidity transduction: a focus on electrophysiological data. Front Physiol 8:1–5
Tierno de Figueroa JM, Luzon-Ortega JM, Sànchez-Ortega A (1998) Imaginal biology of *Hemimelaena flaviventris* (Pictet, 1841) (Plecoptera, Perlodidae). Ann Zool Fenn 35:225–230
Tozer W (1982) External antennal morphology of the adult and larva of *Nectopsyche albida* (Walker) (Trichoptera: Leptoceridae). Freshw Invertebr Biol 1:35–43
Tuchina O, Groh KG, Talarico G et al (2014) Morphology and histochemistry of the aesthetasc associated epidermal glands in terrestrial hermit crabs of the genus *Coenobita* (Decapoda: Paguroidea). PLoS One 9(5):e96430
Tuthill JC, Wilson RI (2016) Mechanosensation and adaptive motor control in insects. Curr Biol 26(20):1022–1038
Vosshall LB, Wong AM, Axel R (2000) An olfactory sensory map in the fly brain. Cell 102:147–159
Ward JV, Stanford JA (1982) Thermal responses in the evolutionary ecology of aquatic insects. Annu Rev Entomol 27:97–117
Wiese K (1976) Mechanoreceptors for near-field water displacements in crayfish. J Neurophysiol 39:816–833
Williams DD (1987) A laboratory study of predator-prey interactions of stoneflies and mayflies. Freshw Biol 17:471–490
Wisenden BD (2000) Olfactory assessment of predation risk in the aquatic environment. Philos Trans R Soc Lond Ser B Biol Sci 355(1401):1205–1208
Wisenden BD, Chivers DP, Smith RJF (1997) Learned recognition of predation risk by *Enallagma* damselfly larvae (Odonata, Zygoptera) on the basis of chemical cues. J Chem Ecol 23:137–151
Yao CA, Ignell R, Carlson JR (2005) Chemosensory Coding by Neurons in the Coeloconic Sensilla of the *Drosophila* Antenna. J Neurosci 25:8359–8367
Yokohari F (1999) Hygro- and thermoreceptors. In: Eguchi E, Tominaga Y (eds) Atlas of arthropod sensory receptors: dynamic morphology in relation to function. Springer, Berlin, pp 191–210

Yokohari F, Tominaga Y, Ando M et al (1975) An antennal hygroreceptive sensillum of the cockroach. J Electron Microsc 24:291–293
Yuvaraj JK, Andersson MN, Anderbrant O et al (2018) Diversity of olfactory structures: a comparative study of antennal sensilla in Trichoptera and Lepidoptera. Micron 111:9–18
Zacharuk R (1980) Ultrastructure and function of insect chemosensilla. Annu Rev Entomol 25:27–48
Zacharuk R (1985) Antennae and sensilla. In: Gilbert LI, Kerkut GA (eds) Comprehensive insect physiology, biochemistry and pharmacology, vol 6. Pergamon Press, Oxford, pp 1–70
Zacharuk R, Ru-Siu Yin L, Blue SG (1971) Fine structure of the antenna and its sensory cone in larvae of *Aedes aegypti* (L.). J Morphol 135:273–298
Zhou X, Frandsen PB, Holzenthal RW et al (2016) The Trichoptera barcode initiative: a strategy for generating a species-level Tree of Life. Phil Trans R Soc B 371:20160025

# Chapter 8
# Territoriality in Aquatic Insects

**Gregory F. Grether**

**Abstract** Research on territoriality has barely progressed beyond the descriptive stage in most aquatic insects, but some territorial species have been studied intensively and served as model organisms for testing certain aspects of evolutionary theory. After providing a brief introduction to territoriality in general, I review the taxonomic distribution and types of territoriality reported in aquatic insects, before delving into more theoretical topics. Larval feeding territories have been described in aquatic insects of several orders (Diptera, Ephemeroptera, Odonata, Plecoptera, Trichoptera), while adult feeding and mating territories have only been reported in Odonata (damselflies and dragonflies) and Hemiptera (water striders and ripple bugs). With mostly Odonata examples, I review the following subjects more critically: (1) determinants of territorial status and the outcome of territorial contests, (2) territorial signals and cues, (3) persistence of interspecific territoriality, and (4) agonistic character displacement. Important advances have been made in each of these areas using aquatic insects, but persistent methodological issues have also impeded progress. I offer some general advice for studying territoriality and conclude by identifying areas where more research is needed.

**Keywords** Territorial behavior · Contests · Character displacement · Agonistic interactions

> For a large number of problems there will be some animal of choice or a few such animals on which it can be most conveniently studied—(Krogh 1929)

G. F. Grether (✉)
Department of Ecology and Evolutionary Biology, University of California, Los Angeles, Los Angeles, CA, USA
e-mail: ggrether@ucla.edu



K. Del-Claro, R. Guillermo (eds.), *Aquatic Insects*,
https://doi.org/10.1007/978-3-030-16327-3_8

## 8.1 Introduction

Territoriality is difficult to define in a completely satisfactory way. Simple definitions, such as "defense of space," mean different things to different people, but attempts to define it more precisely often leave out phenomena that should be included (Kaufmann 1983). Of course, biology is messy and there are bound to be cases that are difficult to categorize, but there are also important distinctions to make between territoriality and related phenomena. I will attempt to clarify what does and does not qualify as territoriality while avoiding the pitfall of being overly narrow.

Territoriality is a form of social dominance in which the resident individual (or group) has priority of access to resources (e.g., food, mates, refuges) at a particular location. Dominant-subordinate interactions that are independent of location are not considered territoriality. Not all aggressive interactions are territorial, and not all territorial interactions involve aggression. Most territorial animals have some means of advertising their presence and warding off intruders without direct confrontation (e.g., song, scent marks, visual signals). In some species, territorial aggression is seldom observed, but site fidelity alone should not be equated to territoriality.

Territories range widely in size, function, and exclusivity. Some territorial animals defend multipurpose territories that supply all their daily needs and only occasionally venture outside their territory boundaries. In other species, territories are used only for specific purposes (e.g., mating, feeding, roosting) and cover a small portion of the home range. The distinction between territory and home range is that the territory is the portion of the home range, if any, which is defended against intrusion or resource utilization by some class of individuals. The excluded class typically includes conspecifics of the same age category and sex as the resident (e.g., sexually mature males), but can be broader or narrower. In some species, residents tolerate same-sex conspecifics if they behave in particular ways (e.g., Davies and Houston 1981; Alcock 1982).

Territoriality also varies temporally, such that residents are only active in defense at certain times of day or under particular weather conditions, and it is common for territory defense to be abandoned at very low or high resource levels or population densities (e.g., Gill and Wolf 1975; Wilson et al. 1978; Davies and Houston 1981; Wilcox and Ruckdeschel 1982). The difficulty of observing territorial interactions (in some taxa), combined with spatial, temporal, and situational variation in defense, helps account for conflicting reports about whether particular species are territorial. Territoriality can also vary geographically, however, such that some populations of a species are territorial while others are not (Brown 1964).

Much has been written about the functions, and fitness costs and benefits, of territoriality (Hinde 1956; Brown 1964; Baker 1983; Kaufmann 1983; Stamps 1994). Priority of access to resources is usually at the top of the list of potential benefits, followed by spacing and site fidelity benefits. Spacing benefits can be couched in group selection or individual selection terms. The idea that territoriality reduces population density and thereby decreases the likelihood that a species will exhaust its resources is a group selection hypothesis. But there are several possible benefits of spacing at the individual level, such as avoidance of predators or parasites (Stamps 1994). Site fidelity benefits include being familiar with the location of resources and habitat features, such as efficient travel routes or refuges to escape from predators or inclement weather. While animals can be site faithful without being territorial, non-territorial animals might need to range over larger areas than territorial animals, because spacing reduces exploitative resource competition. Thus, the benefits of spacing, site fidelity, and priority of access to resources often come together as a package in territorial species.

Costs and benefits of territoriality are traditionally considered from the perspective of an undefeatable territory holder (Brown 1964; Gill and Wolf 1975; Adams 2001; Tanaka et al. 2011). This simplifies the problem to whether the benefit of having priority of access to resources in a space offsets the energetic and other costs of defending it. But in reality, territory holders are not undefeatable, and challenging intruders puts residents at risk of losing access to the space. A thorough consideration of the evolution of territoriality requires comparing the expected fitness payoff of territory defense to that of non-territorial tactics, such as scramble competition (i.e., nonaggressive resource exploitation). The payoffs are likely to be frequency dependent, meaning that the best tactic depends on what other individuals in the population are doing. This is not a trivial problem, theoretically or empirically (Grafen 1987; Kokko et al. 2006; Mesterton-Gibbons and Sherratt 2014; Sherratt and Mesterton-Gibbons 2015; Kemp 2018).

In many species, territorial behavior coexists with one or more non-territorial tactics. Much of the theoretical literature on this topic has focused on exploring scenarios in which territorial and non-territorial phenotypes represent different genotypes, and there are some well-documented examples of such genetic polymorphisms (reviewed in Taborsky and Brockmann 2010). However, in the vast majority of cases, the tactics are conditional; individuals develop into territorial or non-territorial morphs depending on their condition (e.g., body size, growth rate) at some critical stage or switch between tactics as their condition or social environment changes (Taborsky and Brockmann 2010). Usually, the territorial tactic has the highest fitness payoff for individuals with higher than average fighting ability (also known as resource holding power [RHP]) while the non-territorial tactic is the best available option for individuals with low RHP, but there are also known cases in which territoriality is the lower fitness tactic (Gross 1996).

## 8.2 Territoriality in Insects

In a forward-looking review, Baker (1983) asserted that every facet of territoriality in vertebrates is also exhibited by insects and that many questions about territoriality could probably be answered more easily with insects than with vertebrates. Baker (1983) did not explain the latter assertion, probably because it seemed self-evident, but at the risk of stating the obvious most insects are more abundant, shorter lived, easier to catch, easier to manipulate experimentally, and also easier to get permission to study than most vertebrates. Some insects are also extremely easy to observe in the wild without disrupting their natural behavior. Thirty-five years later, the literature on territoriality is still vertebrate biased, but insects have featured prominently in testing certain aspects of theory relating to animal contests (Suhonen et al. 2008; Vieira and Peixoto 2013; Kemp 2018) and interspecific aggression (Grether et al. 2013; Tinghitella et al. 2018).

To the best of my knowledge, the only general review of territoriality in aquatic insects is that of Hart (1987). For most groups of aquatic insects, the relevant literature consists mainly of descriptions of the behavior of particular species. Research on the ecological and evolutionary causes and consequences of territorial behavior is rare in most groups of aquatic insects, the chief exceptions being adult damselflies and dragonflies (Zygoptera and Anisoptera: Odonata; Suhonen et al. 2008) and semiaquatic bugs (Heteroptera: Hemiptera; Arnqvist 1997).

In the first section below, I review what is known about territoriality in aquatic insect larvae before turning to what is known, with greater certainty, about territoriality in aquatic insects at the adult stage.

## 8.3 Territoriality in Aquatic Insect Larvae

Our knowledge of territoriality in aquatic insect larvae does not appear to have advanced much since Hart's (1987) review. Hart's (1987) Table 1 lists cases of larval feeding territories in damselflies and dragonflies (Odonata), caddisflies (Trichoptera), stoneflies (Plecoptera), mayflies (Ephemeroptera), moths (Lepidoptera) and black flies (Simuliidae; Diptera). According to Hart (1987) the only orders of aquatic insects in which larval feeding territories have not been reported are dobsonflies and alderflies (Megaloptera), spongeflies (Neuroptera), and beetles (Coleoptera). Hart (1987) concluded that feeding territories are seen primarily in species with small foraging areas and high resource renewal rates, and argued that this pattern matches theoretical predictions based on economic defendability (Brown 1964).

Hart (1987) was admittedly rather liberal in classifying species as territorial, and some of the evidence for larval feeding territories has since been disputed. Behavioral studies on aquatic insect larvae are often carried out in the laboratory under conditions of questionable relevance to the natural environment, and it is often unclear whether the aggressive interactions are site dependent (territorial) or site independent (non-territorial). Testing for site dependence requires tracking individuals, and this is rarely done in studies of aquatic insect larvae. In the absence of data on individuals, it is still possible to make reasonably robust inferences about whether aggression results in spacing, however.

Corkum (1978) observed aggression in one of the two species of larval mayflies that graze on periphyton (attached algae) under still-water conditions in the laboratory. She described larvae of the aggressive species (*Paraleptophlebia mollis*) striking each other laterally with their abdomens or facing each other and touching each other with forelegs and antennae until one individual withdrew by a few millimeters. In some instances, one larva chased the other briefly. Based on these observations, Corkum (1978) suggested that aggression and drifting serve as alternative spacing mechanisms in mobile grazing mayflies. Williams (1987) also reported that aggressive interactions appeared to increase spacing among mobile grazing mayfly larvae under low-flow conditions in the laboratory.

Kohler (1992) drew a distinction between sessile grazers and mobile grazers, and questioned whether aggression among mobile grazers occurs under lotic (flowing water) conditions in the wild. Based on field and laboratory experiments, he concluded that spacing among mobile grazing mayflies (*Baetis tricaudatus*) and caddisflies (*Glossosoma nigrior*) in a stream in Michigan results from exploitative competition (i.e., periphyton depletion) alone, and that interference between mobile grazers and filter-feeding blackflies (*Simulium* spp.) and chironomids (*Rheotanytarsus* spp.) results from non-aggressive physical contact (as grazers move around, they bump into and displace filter feeders). By Kohler's (1992) account, only sessile grazers have been shown to exhibit territoriality under natural (i.e., lotic) conditions. For a particularly well-documented example of territoriality in sessile, grazing caddisflies, see Hart (1985).

Aggression appears to increase spacing among filter-feeding black fly larvae (Simuliidae: Diptera). Hart (1986) reported that *Simulium piperi* larvae were most aggressive toward upstream neighbors; filter-feeding rates increased after upstream neighbors were displaced; and aggression decreased when food availability was experimentally increased. Some filter-feeding caddisfly larvae have also been reported to be territorial. Matczak and Mackay (1990) varied food concentration and current velocity in the laboratory and found that the larvae were less aggressive and more closely spaced at higher food concentrations and flow rates. They suggested that spacing is maintained by a combination of aggression (bites and lunges) and signaling (stridulation). Some fights ended with one larva killing and partially consuming the other, but usually one larva was displaced (set adrift) (Matczak and Mackay 1990). Body-size differences and residency (possession of a retreat) both appeared to play a role in the outcome of interactions, and some larvae attempted to take over the retreats of other larvae.

Based on laboratory experiments, Sjöström (1985) concluded that predatory larval stoneflies (Plecoptera) are territorial. The experiments involved placing two or three size-matched *Dinocras cephalotes* larvae together in aquaria and observing them compete over shelters. The evidence for territoriality is that the larvae responded aggressively to each other and the first larva introduced to an aquarium usually dominated newcomers. There are other reports of aggression between larval stoneflies, but whether the behaviors observed qualify as territoriality is not clear. In a laboratory experiment with unlimited food, Lieske and Zwick (2008) inferred that intraspecific interference competition influences the growth and development of the biofilm-consuming stonefly *Nemurella pictetii* and observed aggressive interactions between the larvae, but did not specifically document territorial behavior.

Larval odonates have a well-documented repertoire of agonistic behaviors. The larvae of some species have been described as territorial (Rowe 1980; Harvey and Corbet 1986; Corbet 1999), while others have been described as non-territorial (Baker 1981, 1983; Johnson 1991; Fincke 1996). In the clearest case of territoriality that I found in the literature, Rowe (1980) observed individually color-marked *Xanthocnemis zealandica* damselfly larvae in aquaria and reported that they used the same hunting perches for several days and repelled intruders with a combination of displays (abdomen swinging) and physical attacks (labium strikes). Most such encounters ended with the original occupant keeping the perch and the intruder retreating. Fincke (1996) questioned this and other reports of territoriality in larval odonates, mostly on the grounds that the experimental designs did not allow territorial behavior to be distinguished from non-territorial dominance interactions. Fincke (1996) studied *Megaloprepus caerulatus* damselfly larvae in containers and varied the availability of food and cover. She concluded that spacing within the containers resulted from a kind of size-dependent, site-independent dominance—not territoriality. The key evidence against territoriality is that individual larvae were not site faithful (most moved at least once per day within the containers) and size differences, not prior occupancy, determined the outcome of encounters (larger larvae dominated smaller larvae). Rowe (2006) accepted Fincke's (1996) argument against territoriality in *M. caerulatus* but argued that *X. zealandica* larvae are territorial. More research is needed to establish whether territoriality is common in larval odonates.

## 8.4 Taxonomic Distribution of Territoriality in Adult Aquatic Insects

Most aquatic insects are not territorial at the adult stage. Adult male mating territories appear to only have been documented in damselflies (Zygoptera: Odonata), dragonflies (Anisoptera: Odonata), and semiaquatic bugs (Heteroptera: Hemiptera). In other groups of aquatic insects, aggression between adults appears to be non-territorial. For example, male-male aggression has been reported in captive Mexican dobsonflies (*Corydalus bidenticulatus*; Megaloptera), but males only responded aggressively to each other in the immediate vicinity of females (Álvarez et al. 2017). If territoriality has been documented in other aquatic insects at the adult stage, I hope this review encourages the researchers to publish their findings.

## 8.5 Types of Territoriality Reported in Adult Aquatic Insects

(a) Odonata (damselflies and dragonflies): The types of territories that have been reported in Odonata can be divided into feeding (hunting) territories and mating territories. While mating territories have only been documented convincingly in males, both sexes can defend feeding territories (Corbet 1999). In species that hunt from perches, aggressive defense of preferred perches, which are typically higher than surrounding perches (Gorb 1994), is common, and females often dominate males in this context (Corbet 1999).

Mating territories are usually found at water bodies where females come to oviposit, but in some species males defend territories at other sexual rendezvous sites (e.g., forest gaps), and mating pairs travel in tandem (the male clasping the female) to the aquatic habitat where oviposition occurs (e.g., Conrad and Pritchard 1988). The length of time males spend on their territory varies widely, as does the degree of site fidelity exhibited across days. In some species, a male will spend a large proportion of his reproductive life defending the same territory, while in other species males establish new territories daily (reviewed in Corbet 1999).

Territorial odonates can be divided roughly into perchers and patrollers (also known as fliers). In perching species, males adopt a sit-and-wait strategy and rarely leave their perches except to evict male intruders or pursue females. In patrolling species, males regularly fly in a circuit around their territory, searching for females and intruding males. Many species exhibit both modes of territoriality, to varying extents (Corbet 1999). Species that spend most of their time patrolling generally have larger territories than species that spend most of their time perching (Corbet 1999), perhaps because males can only defend what they see. In typical perching species, territories are only a few square meters in area, while the territories of some patrolling species are over 100 square meters (Corbet 1999).

Territorial defense by patrolling is an energetically costly behavior that can limit the length of time males spend on their territories to a small portion of the day. In *Plathemis lydia* (Libellulidae), seven males in succession have been observed patrolling the same territory (portion of a pond perimeter) on a given day (Koenig 1990). Individual males tend to arrive around the same time and defend similar sites on subsequent days, resulting in a kind of temporal partitioning or "time sharing" arrangement (Koenig 1990). By contrast, in some perching species, such as *Hetaerina americana* (Calopterygidae), individual males can reliably be found on their territories throughout the day, except when guarding an ovipositing female (Johnson 1962; Weichsel 1987; Grether 1996b).

(b) Heteroptera (water striders, ripple bugs): Feeding and male mating territories have also been documented in water striders and other semiaquatic bugs (Kaitala and Dingle 1993; Arnqvist 1997). The feeding territories consist of an area on the surface of a water body where the resident has priority of access to food (live or dead arthropods). Feeding territories are seen in both sexes, but females are generally larger and dominant over males and juveniles (nymphs). Experimental manipulations have shown that defense of feeding territories is dynamic and responsive to environmental and social conditions. For example, by manipulating food availability in the field and lab, Wilcox and Ruckdeschel (1982) showed that adult *Gerris remigis* of both sexes defend feeding territories when food availability is between two thresholds. Territoriality is abandoned when food availability is above the upper threshold or below the lower threshold. By manipulating the age and sex composition of pools, Wilson et al. (1978) showed that both sexes and age classes of *Rhagovelia scabra* will attempt to defend territories on the inflow side of the pools, where food availability is highest, but females outcompete males for these positions; at high adult female densities, juveniles and males abandon feeding and cluster together.

Male mating territories are most commonly seen in water strider species with "Type II" mating behavior (Arnqvist 1997). Typically, males anchor themselves to oviposition sites and attract females by sending out "calling" signals by rippling the water surface (Jablonski and Wilcox 1996). When females approach, males switch to "courtship" signals and the female inspects the oviposition site. The female is free to leave, but if she stays the pair copulates and the male usually guards the female from a short distance while she oviposits. Approaching males elicit aggressive "repel" signals and, if necessary, residents will fight physically to defend oviposition sites. By contrast, in species with "Type I" mating behavior, males usually exhibit non-territorial mate searching and attempt to mount and copulate with females without courtship. However, the correlation between territoriality and mating behavior is not perfect. In some territorial species, males attempt to mount females forcibly (Nummelin 1988) and in some non-territorial species males court females (Olosutean 2018). Alternative male mating tactics are found in some species, and males flexibly switch between tactics as conditions change (Arnqvist 1997).

## 8.6 Determinants of Territorial Status and the Outcome of Territorial Contests

A large body of research on territorial insects has been done in the service of testing animal contest theory (Suhonen et al. 2008; Vieira and Peixoto 2013; Kemp 2018). Much of this research can be characterized as attempting to identify the morphological and physiological determinants of RHP (Vieira and Peixoto 2013). This subject was recently reviewed for odonates by Suhonen et al. (2008), for insects in general by Kemp (2018), and for arthropods in general by Vieira and Peixoto (2013). One overall conclusion is that winners and losers of territorial contests usually differ from each other phenotypically, but the traits that correlate with winning are quite variable among species and cannot be predicted from the mode of fighting (Vieira and Peixoto 2013; Kemp 2018). In a meta-analysis of studies on the determinants of fight outcomes in arthropods, persistence-related traits, such as energy reserves, were not consistently better at predicting the outcome of contests than strength-related traits, such as body mass, or vice versa, regardless of whether the contests involve physical contact or instead appear to be pure endurance contests (Vieira and Peixoto 2013), which is a rather puzzling result.

Damselflies (Zygoptera) have featured prominently in this sector of the literature, but I am not convinced that we have truly learned much about how territorial contests are resolved in damselflies. I will illustrate the problem with a gem of the animal contest literature, a paper that has been cited over 300 times (Marden and Waage 1990). Taking advantage of Waage's (1988) discovery that neighboring *Calopteryx maculata* territory holders can be induced to engage in escalated fights by gradually merging their oviposition sites (clumps of bur weed, *Sparganium* spp.), Marden and Waage (1990) recorded the outcome of such fights, as well as some naturally occurring fights, and then recaptured the males, made mass and size measurements, and carried out whole-body lipid extractions. Winners usually had more remaining fat than losers but did not differ from losers in body size, wing loading, or flight muscle ratio. Since contestants mirrored each other in flight and presumably expended similar amounts of energy, winners must also have had more fat prior to fighting. Based on this, Marden and Waage (1990) inferred that "escalated damselfly territorial contests are energetic wars of attrition." Marden and Rollins (1994) subsequently replicated the study on another *C. maculata* population, with similar results. Most subsequent discussions have focused on questions such as whether male damselflies can assess each other's fat reserves and how they might do so (Marden and Rollins 1994; Mesterton-Gibbons et al. 1996; Payne 1998; Fitzstephens and Getty 2000). However, it is important to realize that while Marden et al. controlled the circumstances of staged fights, they had no control over attributes of the paired contestants. Contest outcomes might therefore have depended on factors correlated with fat reserves, instead of fat reserves per se. Young mature male *Calopteryx* usually contain more fat than older males and are also more likely to win escalated fights (Forsyth and Montgomerie 1987; Marden and Waage 1990; Marden and Rollins 1994; Plaistow and Siva-Jothy 1996). Do young males win because they have more fat, or do fat males win because they are younger? Out of 38 contests for which age estimates were available, 32 (84%) were won by the fatter male and 30 (79%) were either won by the younger male or there was no "discernable" age difference between the males (Marden and Rollins 1994). Taken at face value, these numbers suggest that fat is more important than age, but fat was measured much more precisely than age (Marden and Waage 1990; Marden and Rollins 1994). Furthermore, a male's chronological age, even if known precisely, might not correlate perfectly with age-related factors that affect endurance. In *Homo sapiens*, for example, young mature males can, on the average, outlast old males in endurance contests, but not all males senesce at the same rate (pers. obs.). Body fat reserves may well be one of the critical factors that determine the outcome of escalated contests in damselflies, but experiments disassociating fat content from other age-related factors would be required to test this rigorously.

Several other morphological and physiological variables have also been shown to correlate with territorial status and contest outcomes in damselflies (reviewed in Suhonen et al. 2008; Vieira and Peixoto 2013; Córdoba-Aguilar and González-Tokman 2014; Kemp 2018). The physiological correlates of territoriality have been particularly intensively studied in *Hetaerina* spp. damselflies (reviewed in Córdoba-Aguilar and González-Tokman 2014). Males classified as "territorial" and "non-territorial" have been shown to differ in muscle mass, energy reserves, and immune response, and in all respects territorial males appear to be in better physiological condition than non-territorial males (reviewed in Córdoba-Aguilar and González-Tokman 2014). As in *Calopteryx*, a male's ability to hold a territory first increases and then decreases with age postemergence (Guillermo-Ferreira and Del-Claro 2011; Fig. 8.1), but the relationship between physiological condition and age appears not to have been studied yet in *Hetaerina.*

I am not aware of any published studies in which morphological or physiological variables have been manipulated experimentally to evaluate their influence on the outcome of territorial contests in aquatic insects. Manipulative experiments have

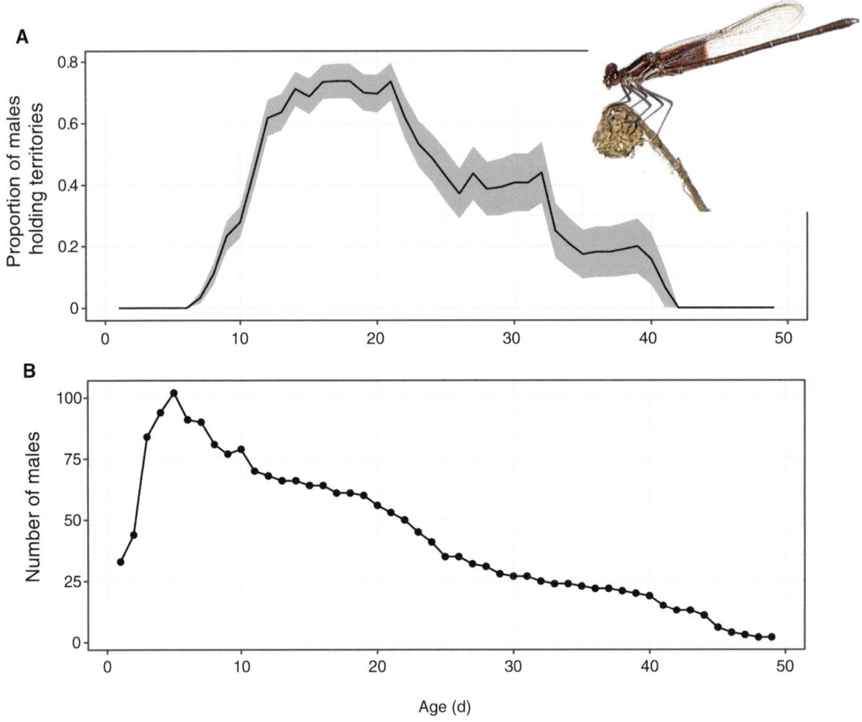


**Fig. 8.1** The rise and fall in territorial fighting ability in the damselfly *Hetaerina americana.* (**a**) Variation in the probability of territory ownership with age (0 is the day of adult emergence). The shaded area shows the standard error of the proportions. Most males first acquired territories between 10 and 15 days of age. Territory-holding ability declined sharply after age 21, but some males managed to acquire a second or third territory, which probably accounts for the stepped shape of the declining function. (**b**) Sample sizes on which the proportions and standard errors in (**a**) are based. These data were collected from August 14 to October 22, 1991, on a cohort of $N = 140$ males on a stream in Northern California (for further details, see Grether 1996b, **a**). Age was either known precisely to 1 day or estimated using a multiple regression equation based on the known-age males and phenotypic traits that change with age (Grether 1996a). Sample sizes increase between ages 1 and 5 due to variation in the age at which males were marked and included in the study. Males that were sighted outside the study area, or were missing for one or more days after age 9 until their final disappearance or documented death, were excluded from the cohort. Inset photo of a male *H. americana* courtesy of Neil Losin

been carried out on other territorial insects to test for effects of specific traits on contest outcomes. For example, Kemp et al. (2006) reared butterflies (*Pararge aegeria*) under different thermal and photoperiod treatments to increase variation in morphological traits related to flight performance and staged contests between individuals with different flight morphologies. Interestingly, the only significant predictor of contest outcome was age (Kemp et al. 2006).

## 8.7 Territorial Signals and Cues

In the animal signaling literature, a cue is a trait that has an effect on the behavior of receivers and a signal is a trait that evolved to have a particular effect on the behavior of receivers (Laidre and Johnstone 2013). Cues and signals affect the behavior of other individuals because they convey information about the signaler, or to be more specific because they reduce uncertainty about the signaler's attributes or future actions (Maynard Smith and Harper 1995). In many if not most cases, signals are likely to have evolved from cues. For example, the disturbance of the water surface caused by a water strider's legs might have originally served merely as a cue that a conspecific is nearby, but over time become modified by selection to serve different functions, such as calling females, courting females, and repelling males (Arnqvist 1997). In this case, we can recognize that there are three distinct signals each with a different function. In other cases, the same trait might serve as a signal in multiple behavioral contexts. For example, color patches on the wings of *Hetaerina* damselflies are used for sex recognition (i.e., distinguishing between the sexes; Grether et al. 2015), male mate recognition (i.e., distinguishing between conspecific and heterospecific females; Drury et al. 2015), competitor recognition within species (i.e., distinguishing between immature and mature males; Grether et al. 2015), and competitor recognition between species (i.e., distinguishing between conspecific and heterospecific males; Anderson and Grether 2010a), and might also serve as indicators of territorial fighting ability (Grether 1996b; Contreras-Garduño et al. 2006; Contreras-Garduño et al. 2007; Guillermo-Ferreira and Del-Claro 2011; Álvarez et al. 2013; González-Santoyo et al. 2014; but see Raihani et al. 2008; González-Santoyo et al. 2010). It is not surprising that one trait (or the homologous trait in males and females) can serve multiple functions, but this is different from showing that it evolved in multiple contexts. For example, did pink wing spots evolve in immature male *Hetaerina* to reduce aggression from mature males, or is pink just a necessary transition in the development of red wing spots? The burden of proof for showing that a trait is a signal is considerably higher than that for showing that it is a cue.

Showing that a trait is a territorial cue (if not also a signal) requires showing that it affects the behavior of other individuals in a territorial context, and the most direct way to do that is to manipulate the trait experimentally. For example, after finding that male *Mnesarete pudica* (Zygoptera, Calopterygidae) with larger red wing spots are more likely to win territorial contests, Guillermo-Ferreira et al. (2015b) presented territory holders with male intruders with and without experimentally augmented red wing spots. They found that territory holders with relatively large wing spots responded more aggressively to experimental males than to control males, whereas territory holders with relatively small wing spots did not differ in their responses to the two treatment groups. Why males with larger red spots have an advantage in territorial fights remains to be determined, but this study leaves little doubt that the red spots are involved somehow. Numerous studies have provided correlational or experimental evidence that coloration affects the outcome of territorial contests in Odonata (Grether 1996b; Tsubaki et al. 1997; Hooper et al. 1999; Plaistow and Tsubaki 2000; Tynkkynen et al. 2004, 2005, 2006; Cordoba-Aguilar et al. 2007; Contreras-Garduno et al. 2008; Guillermo-Ferreira and Del-Claro 2011; Guillermo-Ferreira et al. 2014, 2015a, b; Iyengar et al. 2014; Vilela et al. 2017; Pestana et al. 2018).

There are several possible ways that territorial signals could evolve (Box 8.1). A systematic review of the evidence for, and against, the various models in aquatic insects (or more generally) would be valuable, but is beyond the scope of this chapter. However, there is good evidence that the wing coloration of male odonates directly affects their ability to hold territories (see previous paragraph), is costly for survival (Grether and Grey 1996; Grether 1997), and correlates positively with various measures of phenotypic condition (reviewed in Suhonen et al. 2008; Vieira and Peixoto 2013; Córdoba-Aguilar and González-Tokman 2014; Kemp 2018), which seems most consistent with the handicap model (Box 8.1).

**Box 8.1: Classic Models for the Evolution of Territory Signals**

**Priority model**: Territorial signals evolve from cues about the territory owner's presence and increase in conspicuousness to a level that balances the benefits of being detected by conspecifics with the costs of producing the trait and being conspicuous to predators and prey. Territory owners and receivers benefit from avoiding unnecessary interactions, but all the signal conveys to receivers is that the owner is present on the territory (Hansen and Rohwer 1986; Butcher and Rohwer 1989).

**Handicap model**: Traits with no effect on fighting ability evolve to become honest indicators of fighting ability because larger values of the trait are more costly and the cost of a given value of the trait is lower for individuals with higher fighting ability (Zahavi 1977; Grafen 1990).

**Uncorrelated asymmetry model**: The size of the trait signals aggressiveness, not fighting ability. For this type of signal to be evolutionarily stable, aggressive individuals must incur a cost independent of any particular contest, the value of the contested resource must be low relative to the costs of fighting, and individuals with deceptively large values of the trait must be punished when they encounter aggressive conspecifics (Maynard Smith and Harper 1988; Johnstone and Norris 1993).

(continued)

**Box 8.1: (continued)**

**Arbitrary identity badge model**: Territory signals evolve from cues that initially gave good fighters an advantage by making them more recognizable, but then spread through the population by mimicry; however, once such a trait is fixed in the population, it is no longer useful as a signal of fighting ability (Rohwer and Røskaft 1989).

## 8.8 Interspecific Territoriality

Territorial aggression between species is quite common in Odonata. In some cases, aggressive responses to other species can be explained as recognition errors (Grether 2011). For example, territorial male *Perithemis tenera* (Libellulidae) chase horseflies and butterflies that are similar in size, color, and flight height to the dragonflies (Schultz and Switzer 2001). Schultz and Switzer (2001) considered multiple alternative hypotheses and concluded that the mistaken identity hypothesis (also known as the misdirected intraspecific aggression hypothesis; Murray 1981) is the most plausible explanation. While male *P. tenera* do not seem to benefit from chasing horseflies and butterflies, the encounters are infrequent, and the costs of failing to respond to conspecific intruders might prevent improvements in discrimination ability from evolving. Similar explanations for the persistence of interspecific aggression have been offered for *Leucorrhinia* spp. dragonflies (Singer 1989, 1990) and *Calopteryx* spp. damselflies (Tynkkynen et al. 2004, 2005, 2006).

A different explanation has been proposed for interspecific territoriality in *Hetaerina* spp. damselflies. Drury et al. (2015) found that levels of territorial aggression between sympatric species in this genus are positively correlated with levels of reproductive interference (i.e., interspecific mating attempts; Fig. 8.2). Pre-clasping courtship is absent in *Hetaerina*, and males initiate mating by clasping females in flight. Females invariably reject heterospecific males, but in the time that takes the female can be transported outside the territory where she was clasped. Thus, from the standpoint of a territory holder, tolerating males of another species could be disadvantageous if they fail to discriminate between females of the two species. Evolutionary simulations based on the damselfly system showed that when males of two species compete locally for females, interspecific territoriality can evolve and be maintained indefinitely (Drury et al. 2015).

## 8.9 Agonistic Character Displacement

When costly territorial aggression occurs frequently between species, and there are no countervailing benefits, selection should cause the species to diverge in ways that reduce the frequency and costs of the interaction, an evolutionary process known as agonistic character displacement (ACD; Grether et al. 2009, 2013, 2017). Territorial damselflies have provided some of the best evidence for ACD.

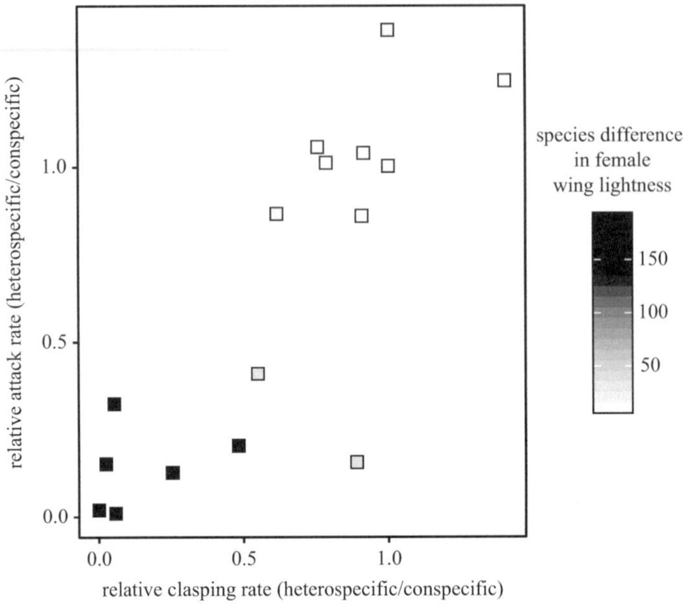


**Fig. 8.2** Correlation between reproductive interference and interspecific aggression in *Hetaerina* damselflies. Relative attack rate, a measure of interspecific aggression, was obtained by presenting live, tethered males to territory holders and dividing the number of attacks elicited by heterospecific male intruders by the number of attacks elicited by conspecific male intruders. Relative clasping rate, a measure of reproductive interference, was obtained by presenting live, tethered females to territory holders and dividing the proportion of females that elicited sexual responses in trials with heterospecific males by the proportion of tethered females that elicited sexual responses in trials with conspecific males. Grayscale: Species differences in female wing lightness, as measured by reflectance spectrometry. Each point represents a population at a sympatric site. From Drury et al. (2015) (authors are allowed to reuse their own figures without permission from the publisher)

ACD has been reported in two pairs of *Calopteryx* species, one in Europe and the other in North America (Tynkkynen et al. 2004; Iyengar et al. 2014). Males of one species in each pair have fully black wings (*C. virgo* and *C. maculata*) and males of the other species have partially black wings (*C. splendens* and *C. aequabilis*). In both cases, the species with partially black wings shows a geographic pattern of character displacement in male wing coloration (Waage 1979; Tynkkynen et al. 2004; Mullen and Andres 2007; Honkavaara et al. 2011; but see Hassall 2014; Suhonen et al. 2018). Simulated territory intrusion experiments revealed that males of the species with fully black wings respond more aggressively to heterospecific males with larger amounts of black wing pigmentation (Tynkkynen et al. 2004; Iyengar et al. 2014). As further evidence that geographic wing color shifts in the European species pair were caused by ACD, Tynkkynen et al. (2005) found that survival selection on *C. splendens* wing spots switched from positive to negative as the relative abundance of *C. virgo* increased. Experiments in which *C. virgo* was removed provided further evidence that interspecific aggression is responsible for the character displacement pattern in *C. splendens* (Tynkkynen et al. 2005, 2006). Nevertheless, reproductive character displacement (RCD) is a viable alternative hypothesis for the same geographic patterns because female mate recognition is also based on male wing coloration (Honkavaara et al. 2011; Kuitunen et al. 2011), and evolutionary simulations suggest that RCD would dominate ACD in this situation (Okamoto and Grether 2013).

Two pairs of *Hetaerina* species show geographic patterns indicative of divergent ACD in both wing coloration and competitor recognition (Anderson and Grether 2010a, b). The alternative hypothesis that these patterns were caused by RCD was rejected based on evidence that female mate recognition is not based on male wing coloration (Grether 1996b; Drury and Grether 2014). The key evidence was obtained in a field experiment in which male wing coloration in one species, *H. americana*, was manipulated to resemble that of a sympatric congener, *H. titia*. Compared to controls, male *H. americana* with *H. titia* wing coloration were involved in more interspecific fights and had reduced territory tenure and survival, but experienced no reduction in any component of mating success potentially under female control (Drury and Grether 2014).

Another potential case of ACD has been reported in *Mnais* damselflies (Fig. 8.3). In allopatry, *M. costalis* and *M. pruinosa* both exhibit a male color polymorphism

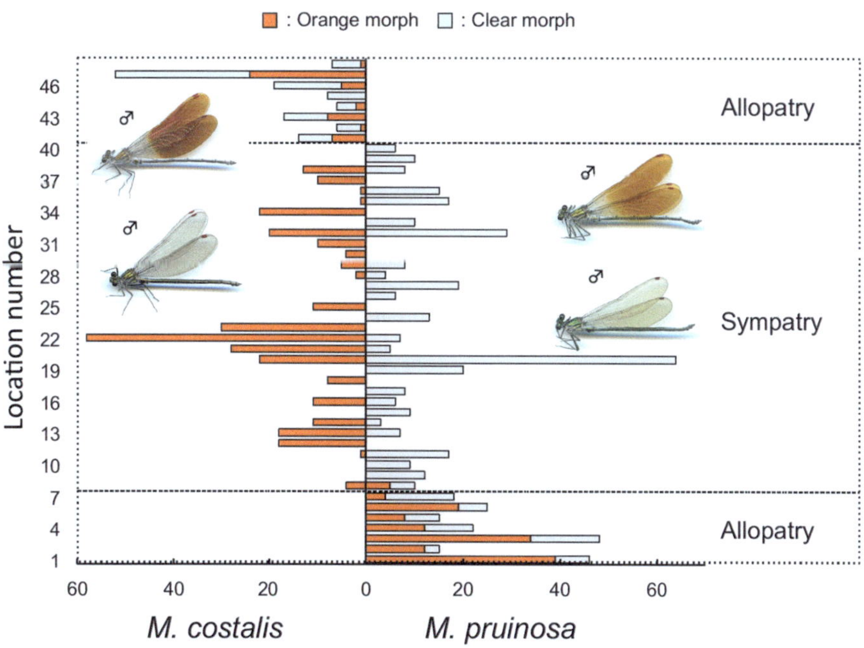


**Fig. 8.3** Character displacement pattern in wing color morph ratios of *Mnais* damselflies in Japan. Bars show the number of males of the clear and orange morphs collected at each sampling location. Inset photos show orange and clear male morphs of the two species. From Tsubaki and Okuyama 2016, with permission from Springer Nature

linked to territorial behavior: males with orange wings are territorial while males with clear wings usually pursue a non-territorial sneaking tactic but can switch to territorial behavior if orange males are absent (reviewed in Tsubaki and Okuyama 2016). In sympatry, however, *M. costalis* exhibits only the orange morph and *M. pruinosa* usually exhibits only the clear morph (Fig. 8.3). Apparently, the selection trade-off that maintains the color polymorphisms in allopatry (Tsubaki et al. 1997; Plaistow and Tsubaki 2000; Tsubaki 2003) is disrupted in sympatry (Tsubaki and Okuyama 2016). Whether the character displacement pattern has resulted from aggressive or reproductive interference between the species, both kinds of interference, or some other unknown factor, remain to be determined (Tsubaki and Okuyama 2016; Tsubaki and Samejima 2016).

## 8.10 General Advice for Studying Territorial Behavior

Studying territorial behavior does not usually require highly specialized techniques or instruments, but it does require attention to certain details that are often overlooked, or disregarded, in studies of aquatic insects. I do not wish to criticize specific papers, but counterexamples to each of the common-sense suggestions below are easy to find in the literature. This complicates making comparisons between species, and even between studies of the same species by different researchers. While different methods might be warranted in studies with different goals, the vast majority of studies of territoriality would benefit from the following practices.

First and foremost, it is essential to have a robust method for distinguishing between individuals. Assuming that animals perching, skating, or flying in the same area at different times are the same individual is totally invalid, as demonstrated, for example, by Koenig's (1990) discovery that seven or more dragonflies can occupy the same territory in succession, on a single day. With rare exceptions, reliably distinguishing between individuals requires marking them with unique codes. It is important to choose a marking technique that allows individuals to be identified without recapturing them or otherwise affecting their behavior, and also to verify that the marks themselves do not affect territorial behavior, mating success or survival (Anderson et al. 2011; Álvarez-Covelli et al. 2015).

Second, it is important to have clearly defined, empirically based criteria for classifying individuals as territory holders and non-territory holders. Ideally, this would involve multiple observations of territory defense, but at a minimum should include establishing that an individual is site faithful on an appropriate timescale (which varies by species). Verifying territory defense does not necessarily require waiting for naturally occurring encounters; it is often possible to test for aggressive responses by simulating territory intrusion (e.g., by presenting conspecifics, models, or other stimuli that evoke territorial responses). It is also important to recognize that being a territory holder is a temporary state, not a characteristic of the individual. In some species, all individuals of a particular age and sex class are territorial in the sense that they would occupy and defend space if they had the opportunity and ability to do so.

Third, while it might be convenient to assume that individuals that have disappeared from the study area have died, this is a very tenuous assumption for flying insects, unless the study area encompasses the dispersal range of the species. Equating disappearance to death is particularly problematic in studies correlating territorial behavior and associated phenotypic traits with components of fitness, such as survival and mating success (i.e., selection studies), because individuals that fail to secure a territory in the study area might succeed elsewhere. Survival analysis is a statistical method that allows "censored" observations to be included without bias, and is often a good choice for analyzing these kinds of data.

Finally, as is always the case in science, it is important not to confuse a statistical correlation between two variables with evidence that one variable has a causal influence on the other. In a true experiment, the researcher has control over one or more independent variables. Just because a manipulation of some sort has been carried out does not necessarily mean that the hypothesis of interest has been subjected to an experimental test. This issue has surfaced repeatedly in studies attempting to identify the determinants of territorial status or the outcome of territorial contests. Unless the predictor variable has been manipulated experimentally, it might be confounded with other variables that also affect territorial contests (Grether 1996a; Suhonen et al. 2008).

Overcoming the methodological issues described above is relatively easy with damselflies and water striders, and some species of dragonflies are highly tractable too. In addition to being easy to catch and mark individually, these insects habituate to humans and consequently can be observed directly at very close range. If more behavioral ecologists followed Krogh's (1929) principle, I think territorial aquatic insects would be even more popular for ecological and evolutionary studies than they already are.

## 8.11 Future Directions

Most behavioral ecologists would probably rank territoriality among the best understood and least controversial subjects in their field. But it is rather easy to identify important questions about territoriality that have received very little rigorous scientific attention. Why are some species territorial while others are not? Why do some species defend multipurpose territories while others only defend mating or feeding territories? What effects does territoriality have on dispersal and population connectivity? Does territoriality regulate population size? What role does territoriality play in speciation? Under what circumstances is interspecific territoriality adaptive? How do territorial interactions between species affect their coexistence and evolution? How does interspecific territoriality affect the structure of ecological communities? Does territoriality affect whether introduced species become invasive? Should territoriality be taken into account when projecting effects of climate change on species ranges? None of these questions is specific to aquatic insects, but as Baker (1983) suggested they could probably be answered more easily with insects than vertebrates.

**Acknowledgments** I thank Rhainer Guillermo-Ferreira and Kleber Del-Claro for inviting me to write this chapter and an anonymous reviewer for helpful comments on a previous draft. My research on *Hetaerina* damselflies has been funded by multiple grants from the National Science Foundation and the University of California.

## References

Adams ES (2001) Approaches to the study of territory size and shape. Annu Rev Ecol Syst 32:277–303

Alcock J (1982) Post-copulatory mate guarding by males of the damselfly *Hetaerina vulnerata*-Selys (Odonata, Calopterygidae). Anim Behav 30:99–107

Álvarez HA, Serrano-Meneses MA, Reyes-Márquez I et al (2013) Allometry of a sexual trait in relation to diet experience and alternative mating tactics in two rubyspot damselflies (Calopterygidae: *Hetaerina*). Biol J Linn Soc 108:521–533. https://doi.org/10.1111/j.1095-8312.2012.02031.x

Álvarez HA, García-Chávez JH, Cebada-Ruiz JA (2017) Do social environments affect the use of exaggerated traits in the dobsonfly *Corydalus bidenticulatus*? Entomol Sci 20:168–176. https://doi.org/10.1111/ens.12249

Álvarez-Covelli C, Álvarez-Covelli MA, Palacino-Rodríguez F (2015) Abdomen or wings? Comparing two body places for marking in *Mesamphiagrion laterale* (Odonata: Coenagrionidae). Odonatologica 44:343–348

Anderson CN, Grether GF (2010a) Interspecific aggression and character displacement of competitor recognition in *Hetaerina* damselflies. Proc R Soc B 277:549–555. https://doi.org/10.1098/rspb.2009.1371

Anderson CN, Grether GF (2010b) Character displacement in the fighting colours of *Hetaerina* damselflies. Proc R Soc B 277:3669–3675. https://doi.org/10.1098/rspb.2010.0935

Anderson CN, Cordoba-Aguilar A, Drury JP, Grether GF (2011) An assessment of marking techniques for odonates in the family Calopterygidae. Entomol Exp Appl 141:258–261. https://doi.org/10.1111/j.1570-7458.2011.01185.x

Arnqvist G (1997) The evolution of water strider mating systems: causes and consequences of sexual conflicts. In: Choe JC, Crespi BJ (eds) The evolution of mating systems in insects and arachnids. Cambridge Univ. Press, Cambridge, pp 146–163

Baker RL (1981) Behavioural interactions and use of feeding areas by nymphs of *Coenagrion resolutum* (Coenagrionidae: Odonata). Oecologia 49:353–358. https://doi.org/10.1007/BF00347597

Baker RR (1983) Insect Territoriality. Annu Rev Entomol 28:65–89

Brown JL (1964) The evolution of diversity in avian territorial systems. Wilson Bull 76:160–169. https://doi.org/10.1017/CBO9781107415324.004

Butcher GS, Rohwer S (1989) The evolution of conspicuous and distinctive coloration for communication in birds. Curr Ornithol 6:51–108

Conrad KF, Pritchard G (1988) The reproductive behavior of *Argia vivida* Hagen: an example of a female-control mating system (Zygoptera: Coenagrionidae). Odonatologica 17:179–185

Contreras-Garduño J, Canales-Lazcano J, Córdoba-Aguilar A (2006) Wing pigmentation, immune ability, fat reserves and territorial status in males of the rubyspot damselfly, *Hetaerina americana*. J Ethol 24:165–173. https://doi.org/10.1007/s10164-005-0177-z

Contreras-Garduño J, Lanz-Mendoza H, Córdoba-Aguilar A (2007) The expression of a sexually selected trait correlates with different immune defense components and survival in

males of the American rubyspot. J Insect Physiol 53:612–621. https://doi.org/10.1016/j.jinsphys.2007.03.003
Contreras-Garduno J, Buzatto BA, Serrano-Meneses MA et al (2008) The size of the red wing spot of the American rubyspot as a heightened condition-dependent ornament. Behav Ecol 19:724–732. https://doi.org/10.1093/beheco/arn026
Corbet PS (1999) Dragonflies: behavior and ecology of Odonata. Cornell University Press, Ithaca, NY
Córdoba-Aguilar A, González-Tokman DM (2014) The behavioral and physiological ecology of adult rubyspot damselflies (*Hetaerina*, Calopterygidae, Odonata). Adv Study Behav 46:311–341
Cordoba-Aguilar A, Lesher-Trevino AC, Anderson CN (2007) Sexual selection in *Hetaerina titia* males: a possible key species to understand the evolution of pigmentation in calopterygid damselflies (Odonata: Zygoptera). Behaviour 144:931–952
Corkum LD (1978) The influence of density and behavioural type on the active entry of two mayfly species (Ephemeroptera) into the water column. Can J Zool 56:1201 1206. https://doi.org/10.1139/z78-164
Davies NB, Houston AI (1981) Owners and satellites: the economics of territory defence in the pied wagtail, *Motacilla alba*. J Anim Ecol 50:157–180. https://doi.org/10.2307/4038
Drury JP, Grether GF (2014) Interspecific aggression, not interspecific mating, drives character displacement in the wing coloration of male rubyspot damselflies (*Hetaerina*). Proc R Soc B 281:20141737. https://doi.org/10.1098/rspb.2014.1737
Drury JP, Okamoto KW, Anderson CN, Grether GF (2015) Reproductive interference explains persistence of aggression between species. Proc R Soc B 282:20142256
Fincke OM (1996) Larval behaviour of a giant damselfly: territoriality or size-dependent dominance? Anim Behav 51:77–87
Fitzstephens D, Getty T (2000) Colour, fat and social status in male damselflies, *Calopteryx maculata*. Anim Behav 60:851–855. https://doi.org/10.1006/anbe.2000.1548
Forsyth A, Montgomerie RD (1987) Alternative reproductive tactics in the territorial damselfly *Calopteryx maculata*: sneaking by older males. Behav Ecol Sociobiol 21:73–81
Gill FB, Wolf LL (1975) Economics of feeding territoriality in the golden-winged sunbird. Ecology 56:333–345. https://doi.org/10.2307/1934964
González-Santoyo I, Córdoba-Aguilar A, González-Tokman DM, Lanz-Mendoza H (2010) Phenoloxidase activity and melanization do not always covary with sexual trait expression in *Hetaerina* damselflies (Insecta: Calopterygidae). Behaviour 147:1285–1307. https://doi.org/10.1163/000579510X516777
González-Santoyo I, González-Tokman DM, Munguía-Steyer RE, Córdoba-Aguilar A (2014) A mismatch between the perceived fighting signal and fighting ability reveals survival and physiological costs for bearers. PLoS One 9:e84571. https://doi.org/10.1371/journal.pone.0084571
Gorb SN (1994) Female perching behaviour in *Sympetrum sanguineum* (Mueller) at feeding places (Anisoptera: Libellulidae). Odonatologica 23:341–353
Grafen A (1987) The logic of divisively asymmetric contests: respect for ownership and the desperado effect. Anim Behav 35:462–467
Grafen A (1990) Biological signals as handicaps. J Theor Biol 144:517–546
Grether GF (1996a) Sexual selection and survival selection on wing coloration and body size in the rubyspot damselfly *Hetaerina americana*. Evolution 50:1939–1948
Grether GF (1996b) Intrasexual competition alone favors a sexually dimorphic ornament in the rubyspot damselfly *Hetaerina americana*. Evolution 50:1949–1957
Grether GF (1997) Survival cost of an intrasexually selected ornament in a damselfly. Proc R Soc B 264:207–210. https://doi.org/10.1098/rspb.1997.0029
Grether G (2011) The neuroecology of competitor recognition. Integr Comp Biol 51:807–818
Grether GF, Grey RM (1996) Novel cost of a sexually selected trait in the rubyspot damselfly *Hetaerina americana*: conspicuousness to prey. Behav Ecol 7:465–473. https://doi.org/10.1093/beheco/7.4.465

Grether GF, Losin N, Anderson CN, Okamoto K (2009) The role of interspecific interference competition in character displacement and the evolution of competitor recognition. Biol Rev 84:617–635. https://doi.org/10.1111/j.1469-185X.2009.00089.x

Grether GF, Anderson CN, Drury JP et al (2013) The evolutionary consequences of interspecific aggression. Ann N Y Acad Sci 1289:48–68. https://doi.org/10.1111/nyas.12082

Grether GF, Drury JP, Berlin E, Anderson CN (2015) The role of wing coloration in sex recognition and competitor recognition in rubyspot damselflies (*Hetaerina* spp.). Ethology 121:674–685. https://doi.org/10.1111/eth.12382

Grether GF, Peiman KS, Tobias JA, Robinson BW (2017) Causes and consequences of behavioral interference between species. Trends Ecol Evol. https://doi.org/10.1016/j.tree.2017.07.004

Gross MR (1996) Alternative reproductive strategies and tactics: diversity within sexes. Trends Ecol Evol 11:92–98

Guillermo-Ferreira R, Del-Claro K (2011) Resource defense polygyny by *Hetaerina rosea* Selys (Odonata: Calopterygidae): influence of age and wing pigmentation. Neotrop Entomol 40:78–84

Guillermo-Ferreira R, Therézio EM, Gehlen MH et al (2014) The role of wing pigmentation, UV and fluorescence as signals in a neotropical damselfly. J Insect Behav 27:67–80. https://doi.org/10.1007/s10905-013-9406-4

Guillermo-Ferreira R, Bispo PC, Appel E et al (2015a) Mechanism of the wing colouration in the dragonfly *Zenithoptera lanei* (Odonata: Libellulidae) and its role in intraspecific communication. J Insect Physiol 81:129–136. https://doi.org/10.1016/j.jinsphys.2015.07.010

Guillermo-Ferreira R, Gorb SN, Appel E et al (2015b) Variable assessment of wing colouration in aerial contests of the red-winged damselfly *Mnesarete pudica* (Zygoptera, Calopterygidae). Sci Nat 102. https://doi.org/10.1007/s00114-015-1261-z

Hansen AJ, Rohwer S (1986) Coverable badges and resource defence in birds. Anim Behav 34:69–72

Hart DD (1985) Causes and consequences of territoriality in a grazing stream insect. Ecology 66:404–414

Hart DD (1986) The adaptive significance of territoriality in filter-feeding larval black flies (Diptera Simuliidae). Oikos 46:88–92

Hart DD (1987) Feeding territoriality in aquatic insects: cost-benefit models and experimental tests. Integr Comp Biol 27:371–386. https://doi.org/10.1093/icb/27.2.371

Harvey IF, Corbet PS (1986) Territorial interactions between larvae of the dragonfly *Pyrrhosoma nymphula*: outcome of encounters. Anim Behav 34:1550–1561. https://doi.org/10.1016/S0003-3472(86)80224-X

Hassall C (2014) Continental variation in wing pigmentation in *Calopteryx* damselflies is related to the presence of heterospecifics. PeerJ 2:e438. https://doi.org/10.7717/peerj.438

Hinde A (1956) The biological significance of the territories of birds. Ibis (Lond 1859) 98:340–369. https://doi.org/10.1111/j.1474-919X.1956.tb01419.x

Honkavaara J, Dunn DW, Ilvonen S, Suhonen J (2011) Sympatric shift in a male sexual ornament in the damselfly Calopteryx splendens. J Evol Biol 24:139–145. https://doi.org/10.1111/j.1420-9101.2010.02146.x

Hooper RE, Tsubaki Y, Siva-Jothy MT (1999) Expression of a costly, plastic secondary sexual trait is correlated with age and condition in a damselfly with two male morphs. Physiol Entomol 24:364–369

Iyengar VK, Castle T, Mullen SP (2014) Sympatric sexual signal divergence among North American *Calopteryx* damselflies is correlated with increased intra- and interspecific male–male aggression. Behav Ecol Sociobiol 68:275–282. https://doi.org/10.1007/s00265-013-1642-2

Jablonski PG, Wilcox RS (1996) Signalling asymmetry in the communication of the water strider *Aquarius remigis* in the context of dominance and spacing in the non-mating season. Ethology 102:353–359

Johnson C (1962) A description of territorial behavior and a quantitative study of its function in males of *Hetaerina americana* (Fabricius) (Odonata: Agriidae). Can Entomol 94:178–191
Johnson DM (1991) Behavioral ecology of larval dragonflies and damselflies. Trends Ecol Evol 6:8–13. https://doi.org/10.1016/0169-5347(91)90140-S
Johnstone RA, Norris K (1993) Badges of status and the cost of aggression. Behav Ecol Sociobiol 32:127–134
Kaitala A, Dingle H (1993) Wing dimorphism, territoriality and mating frequency of the water strider *Aquarius remigis* (Say). Ann Zool Fenn 30:163–168
Kaufmann JH (1983) On the definitions and functions of dominance and territoriality. Biol Rev 58:1–20. https://doi.org/10.1111/j.1469-185X.1983.tb00379.x
Kemp DJ (2018) Habitat selection and territoriality. In: Córdoba-Aguilar A, González-Tokman D, Isaac González S (eds) Insect behavior: from mechanisms to ecological and evolutionary consequences. Oxford University Press, Oxford, pp 80–97
Kemp DJ, Wiklund C, Gotthard K (2006) Life history effects upon contest behaviour: age as a predictor of territorial contest dynamics in two populations of the speckled wood butterfly, *Pararge aegeria* L. Ethology 112:471–477
Koenig WD (1990) Territory size and duration in the white-tailed skimmer *Plathemis lydia* (Odonata: Libellulidae). J Anim Ecol 59:317–333
Kohler SL (1992) Competition and the structure of a benthic stream community. Ecol Monogr 62:165–188
Kokko H, Lopez-Sepulcre A, Morrell LJ (2006) From hawks and doves to self-consistent games of territorial behavior. Am Nat 167:901–912. https://doi.org/10.1086/504604
Krogh A (1929) The progress of physiology. Am J Phys 90:243–251
Kuitunen K, Kotiaho JS, Luojumaki M, Suhonen J (2011) Selection on size and secondary sexual characters of the damselfly Calopteryx splendens when sympatric with the congener Calopteryx virgo. Can J Zool 89:1–9. https://doi.org/10.1139/z10-090
Laidre ME, Johnstone RA (2013) Animal signals. Curr Biol 23:R829–R833. https://doi.org/10.1016/j.cub.2013.07.070
Lieske R, Zwick P (2008) Effects of intraspecific competition on the life cycle of the stonefly, *Nemurella pictetii* (Plecoptera: Nemouridae). BMC Ecol 8:1–8. https://doi.org/10.1186/1472-6785-8-5
Marden JH, Rollins RA (1994) Assessment of energy reserves by damselflies engaged in aerial contests for mating territories. Anim Behav 48:1023–1030
Marden JH, Waage JK (1990) Escalated damselfly territorial contests are energetic wars of attrition. Anim Behav 39:954–959
Matczak TZ, Mackay RJ (1990) Territoriality in filter-feeding caddisfly larvae: laboratory experiments. J N Am Benthol Soc 9:26–34. https://doi.org/10.2307/1467931
Maynard Smith J, Harper DGC (1988) The evolution of aggression: can selection generate variability? Phil Trans R Soc Lond B 319:557–570
Maynard Smith J, Harper GGC (1995) Animal signals: models and terminology. J Theor Biol 177:305–311
Mesterton-Gibbons M, Sherratt TN (2014) Bourgeois versus anti-Bourgeois: a model of infinite regress. Anim Behav 89:171–183. https://doi.org/10.1016/j.anbehav.2014.01.002
Mesterton-Gibbons M, Marden J, Dugatkin LA (1996) On wars of attrition without assessment. J Theor Biol 181:65–83
Mullen SP, Andres JA (2007) Rapid evolution of sexual signals in sympatric *Calopteryx damselflies*: reinforcement or "noisy-neighbour" ecological character displacement? J Evol Biol 20:1637–1648
Murray BG (1981) The origins of adaptive interspecific territorialism. Biol Rev Camb Philos Soc 56:1–22
Nummelin M (1988) The territorial behavior of four Ugandan water strider species (Heteroptera: Gerridae). Ann Entomol Fenn 54:121–134

Okamoto KW, Grether GF (2013) The evolution of species recognition in competitive and mating contexts: the relative efficacy of alternative mechanisms of character displacement. Ecol Lett 16:670–678. https://doi.org/10.1111/ele.12100

Olosutean H (2018) Does population density determine the mating strategy in males of the water strider *Aquarius paludum*? Entomol News 127:349–360. https://doi.org/10.3157/021.127.0407

Payne RJH (1998) Gradually escalating fights and displays: the cumulative assessment model. Anim Behav 56:651–662

Pestana GC, Caromano TG, Guillermo-Ferreira R (2018) Sexual ornamentation triggers rival aggressiveness in the Neotropical damselfly *Hetaerina longipes* (Odonata: Calopterygidae). Odonatologica 47:121–132. https://doi.org/10.5281/zenodo.1239951

Plaistow S, Siva-Jothy MT (1996) Energetic constraints and male mate-securing tactics in the damselfly *Calopteryx splendens xanthostoma* (Charpentier). Proc Biol Sci 263:1233–1239

Plaistow SJ, Tsubaki Y (2000) A selective trade-off for territoriality and non-territoriality in the polymorphic damselfly *Mnais costalis*. Proc R Soc B Biol Sci 267:969–975. https://doi.org/10.1098/rspb.2000.1098

Raihani G, Serrano-Meneses MA, Córdoba-Aguilar A (2008) Male mating tactics in the American rubyspot damselfly: territoriality, nonterritoriality and switching behaviour. Anim Behav 75:1851–1860. https://doi.org/10.1016/j.anbehav.2007.11.002

Rohwer S, Røskaft E (1989) Results of dyeing male yellow-headed blackbirds solid black: implications for the arbitrary identity badge hypothesis. Behav Ecol Sociobiol 25:39–48

Rowe RJ (1980) Territorial behaviour of a larval dragonfly *Xanthocnemis zealandica* (McLachlan) (Zygoptera: Coenagrionidae). Odonatologica 9:285–292

Rowe RJ (2006) Patterns and processes in freshwater systems: the social dimension. N Z Nat Sci 31:59–70

Schultz JK, Switzer PV (2001) Pursuit of heterospecific targets by territorial amberwing dragonflies (*Perithemis tenera* say): a case of mistaken identity. J Insect Behav 14:607–620

Sherratt TN, Mesterton-Gibbons M (2015) The evolution of respect for property. J Evol Biol 28:1185–1202. https://doi.org/10.1111/jeb.12648

Singer F (1989) Interspecific aggression in *Leucorrhinia* dragonflies - a frequency-dependent discrimination threshold hypothesis. Behav Ecol Sociobiol 25:421–427

Singer F (1990) Reproductive costs arising from incomplete habitat segregation among three species of *Leucorrhinia* dragonflies. Behaviour 115:188–202

Sjöström P (1985) Territoriality in nymphs of *Dinocras cephalotes* (Plecoptera). Oikos 45:353–357

Stamps JA (1994) Territorial behavior: testing the assumptions. Adv Study Behav 23:173–232

Suhonen J, Rantala MJ, Honkavaara J (2008) Territoriality in odonates. In: Córdoba-Aguilar A (ed) Dragonflies and damselflies: model organisms for ecological and evolutionary research. Oxford Univ Press, Oxford, pp 203–216

Suhonen J, Ilvonen S, Dunn DW et al (2018) Parasitism affects variation in a male damselfly sexual ornament. Ethol Ecol Evol 30:256–266. https://doi.org/10.1080/03949370.2017.1354920

Taborsky M, Brockmann H (2010) Alternative reproductive tactics and life history phenotypes. In: Kappeler P (ed) Animal behaviour: evolution and mechanisms. Springer, Berlin/Heidelberg, pp 537–586

Tanaka Y, Iguchi K, Yoshimura J et al (2011) Historical effect in the territoriality of ayu fish. J Theor Biol 268:98–104. https://doi.org/10.1016/j.jtbi.2010.09.040

Tinghitella RM, Lackey ACR, Martin M et al (2018) On the role of male competition in speciation: a review and research agenda. Behav Ecol:1–15. https://doi.org/10.1093/beheco/arx151

Tsubaki Y (2003) The genetic polymorphism linked to mate-securing strategies in the male damselfly *Mnais costalis* Selys Odonata: Calopterygidae. Popul Ecol 45:263–266. https://doi.org/10.1007/s10144-003-0162-8

Tsubaki Y, Okuyama H (2016) Adaptive loss of color polymorphism and character displacements in sympatric *Mnais* damselflies. Evol Ecol 30:811–824. https://doi.org/10.1007/s10682-015-9778-3

Tsubaki Y, Samejima Y (2016) Hot males live fast and die young: habitat segregation, reproductive output, and lifespan of sympatric *Mnais* damselflies. Behav Ecol Sociobiol 70:725–732. https://doi.org/10.1007/s00265-016-2095-1

Tsubaki Y, Hooper RE, SivaJothy MT (1997) Differences in adult and reproductive lifespan in the two male forms of *Mnais pruinosa costalis* Selys (Odonata: Calopterygidae). Res Popul Ecol 39:149–155

Tynkkynen K, Rantala MJ, Suhonen J (2004) Interspecific aggression and character displacement in the damselfly *Calopteryx splendens*. J Evol Biol 17:759–767

Tynkkynen K, Kotiaho JS, Luojumaki M, Suhonen J (2005) Interspecific aggression causes negative selection on sexual characters. Evolution 59:1838–1843

Tynkkynen K, Kotiaho JS, Luojumaki M, Suhonen J (2006) Interspecific territoriality in *Calopteryx* damselflies: the role of secondary sexual characters. Anim Behav 71:299–306

Vieira MC, Peixoto PEC (2013) Winners and losers: a meta-analysis of functional determinants of fighting ability in arthropod contests. Funct Ecol 27:305–313. https://doi.org/10.1111/1365-2435.12051

Vilela DS, Tosta TAA, Rodrigues RR et al (2017) Colours of war: visual signals may influence the outcome of territorial contests in the tiger damselfly, *Tigriagrion aurantinigrum*. Biol J Linn Soc 121:786–795. https://doi.org/10.1093/biolinnean/blx024

Waage JK (1979) Reproductive character displacement in *Calopteryx* (Odonata, Calopterygidae). Evolution 33:104–116

Waage JK (1988) Confusion over residency and the escalation of damselfly territorial disputes. Anim Behav 36:586–595

Weichsel JI (1987) The life history and behavior of *Hetaerina americana* (Fabricus) (Odonata: Calopterygidae). University of Michigan

Wilcox RS, Ruckdeschel T (1982) Food threshold territoriality in a water strider (*Gerris remigis*). Behav Ecol Sociobiol 11:85–90. https://doi.org/10.1007/BF00300096

Williams DD (1987) A laboratory study of predator-prey interactions of stoneflies and mayflies. Freshw Biol 17:471–490

Wilson DS, Leighton M, Leighton DR (1978) Interference competition in a tropical ripple bug (Hemiptera: Veliidae). Biotropica 10:302–306

Zahavi A (1977) Reliability in communication and the evolution of altruism. In: Stonehouse B, Perrins CM (eds) Evolutionary ecology. Macmillan, London, pp 253–259

# Chapter 9
# Defenses of Water Insects

**Konrad Dettner**

**Abstract** This chapter compiles active and passive defensive mechanisms of aquatic and semiaquatic developmental stages of all insect orders against various predators. Mainly escape reactions, mechanical defense, defensive stridulation, and especially chemical defenses are described, illustrated, and tabulated. Apart from the large aquatic groups of ephemeropteran, Odonata or Trichoptera larvae especially aquatic bugs and water beetles are considered by even including small groups from Collembola up to Mecoptera.

Differences between defensive mechanisms and strategies in aquatic and terrestrial insects are described. Aquatic insects especially rely on escape, mechanical defenses, defensive stridulation, and chemical defenses. Exocrine glands are mainly restricted to large taxa with both terrestrial and aquatic representatives (adephagan beetles, Heteroptera) and not invented in aquatic groups. Chemically aquatic insects especially evolved biosynthesis of aromatic and few aliphatic compounds against microorganisms. In contrast mainly steroids are targeted against cold-blooded vertebrates such as fishes and amphibians. As compared with terrestrial insects, aquatic representatives lack many mechanisms of defense such as reflex bleeding, incorporation of toxic compounds from plants, freshwater animals, or microorganisms. Exocrine secretions of water insects are usually externalized by secretion grooming in order to receive a clean body surface, to achieve an optimal breathing, and to modify the wettability of the body surface. Generally there exists a considerable lack of knowledge concerning bionomy and especially defenses of aquatic insects.

**Keywords** Mechanical · Chemical defense · Compilation aquatic insect orders

K. Dettner (✉)
Department of Animal Ecology II, University of Bayreuth, Bayreuth, Germany
e-mail: k.dettner@uni-bayreuth.de



K. Del-Claro, R. Guillermo (eds.), *Aquatic Insects*,
https://doi.org/10.1007/978-3-030-16327-3_9

## 9.1 Introduction

Adults and developmental stages of aquatic insects are eaten by many predators (e.g., carnivorous fishes, predatory insects, various invertebrates), suffer from parasitoids, and have problems with insect pathogenic microorganisms (Protozoa, Bacteria) and fungi. In extreme cases water insects may suffer from snake attacks (Peddle and Larson 1999) and reversally prey on vertebrates (McCormick and Polis 1982). Therefore water insects which are found in various sometimes unique habitats (Heckman 2018) have to defend against these natural enemies and evolved various strategies of defense, sometimes even multiple defenses against these target organisms.

There exist many important general books dealing with defense mechanisms of animals or insects: Curio (1976), Edmunds (1974), Evans and Schmidt (1990), and Ruxton et al. (2004). In 1990 Witz studied antipredator mechanisms of hundreds of arthropods from terrestrial and aquatic habitats and showed that active or secondary defenses are most important. These defenses are energetically expensive mechanisms, which increase the probability of surviving attacks of predators. They include for example chemical defenses (46%), fighting (11%), escape reactions (8%), postural defenses (4%), armors (3%), or defensive stridulation (2%). In contrast Witz (1990) showed that passive or primary defenses such as cryptic appearance (9%) or group size (4%) are of minor importance. These defenses seem relatively inexpensive and reduce the probability of an encounter between predator and prey.

The main aim of this review is to compile such active and passive defenses in all aquatic insect orders and aquatic or semiaquatic developmental stages (especially larvae) of water insects. Also aquatic representatives of usually terrestrial orders are considered, because now many data from these and other taxa are available due to publications on biodiversity and many valuable books, for example "Thorp and Covich's Freshwater Invertebrates" (Thorp and Rogers 2015, 2016; Hamada et al. 2018). With regard to insect species numbers in different biogeographical regions (Stork 2018) especially the incredible species diversity in the neotropics (1,620,000 insect species) as compared with the Palaearctic (524,165 insect species; albeit terrestrial taxa) were helpful for this compilation, because the various strategies of defense increase towards tropical areas.

Since theoretical data on predator-prey interactions are abundant (e.g., Resh and Rosenberg 1984; Kerfoot and Sih 1987; Williams and Feltmate 1992; Lampert and Sommer 2007; Lancaster and Downes 2013) these interesting aspects were not considered. The same applies to chemical signals such as pheromones or kairomones which were compiled by Chivers and Smith (1998), Burks and Lodge (2002), Sotka et al. (2009), Ferrari et al. (2010), and Brönmark and Hansson (2012), although our knowledge on chemistry of these interesting signal molecules is very poor. For this review it is highly interesting to consider and to compare with valuable papers dealing with defense of water insects such as from Peckarsky (1984), Scrimshaw and Kerfoot (1987), Crespo (2011), or Kicklighter (2012). Also chemical defenses of lower freshwater animals (apart from insects) may be of interest (Dettner 2010) because these animals might be fed by aquatic insects (Bay 1974).

Unfortunately data on bionomy of eggs, egg deposition (Hinton 1981; Hilker and Meiners 2002), and especially pupation in most holometabolous taxa are scarce or unknown especially with respect to aquatic and semiaquatic taxa. Therefore these data are nearly unavailable for this compilation. Moreover it seems interesting for forthcoming evaluations to consider the significance of attachment devices (Gorb 2007) and the role of adhesives (Betz and Kölsch 2004; Betz 2010) as defense mechanisms of aquatic insects. Apart from the dytiscid beetles (Dettner 2014) the same lack of knowledge concerns interactions between symbiontic and parasitic internal but also external microorganisms (bacteria, fungi), protozoa (peritrichic Ciliata, Laboulbeniales), and in/external parasites (mites, trematodes) and their water insect hosts.

Depending on the number of aquatic taxa within each insect order, the data on defenses of aquatic insects are presented on the level of the order or alternatively on the family level. The percentual value of aquatic or semiaquatic species as compared with the total species number per taxon is indicated after each insect order. Bold numbers in the text and legends indicate numbers of compounds (see Tables 9.1, 9.2, and 9.3).

I hope that further studies of biology and bionomy of water insects may help to complete our fragmentary knowledge concerning defense mechanisms of these interesting animals, especially if tropical taxa are additionally taken into consideration.

A lot of phylogenetic trees indicate that the elder arrangement of entognathous taxa as insects (e.g., Dettner and Peters 2010) has to be corrected. Rather now ectognathous taxa represent Insecta whereas Entognatha + Ectognatha are named Hexapoda (e.g., Beutel et al. 2014). Nevertheless semiaquatic Collembola as representatives of Entognatha are treated in this chapter.

**COLLEMBOLA (springtails; about 45% hydrophilous)**: With worldwide approximately 9000 species (Stork 2018: 8140; according to Cipola et al. 2018 probably 50,000 species) this is the largest group of apterygote and endognathous insects or hexapods. Due to their jumping ability with furca and retinaculum (springtails) and detachable hairs and scales this is a well-defended taxon. Unusually for insects springtails continue to molt (up to 52 stages) after reaching sexual maturity. However there is a gap of knowledge with respect to collembolan biology which might be correlated to their small body size and the difficulty to determine all species. With respect to their habitats about 45% of all Collembola species are hydrophilous, represent riparian species, are associated with wet habitats as shores of ponds and lakes, or are frequent in marine littoral zones (Thorp and O'Neill 2015). Cipola et al. (2018) divided Collembola into primary or water-dependent and secondary (epigean hydrophilous) aquatic-associated species. However no species has developed a permanent underwater lifestyle, wherefore no special adaptations for swimming or diving have evolved. Instead most species have hydrophobic hairs, a type of wax and/or a specific surface geometry like microtubercles. Even under high pressure, with the addition of surfactants or other organic fluids Collembola are not

**Table 9.1** Constituents **1–23** of pygidial glands in genera (number of species) of predacious diving beetles (Dytiscidae; for species data see Dettner 2014)

| | |
|---|---|
| **Dytiscidae, Hydroporinae** | |
| *Hyphydrus* (2 sp.). | (2), (3), 11, 12, 13, 16 |
| *Hydrovatus* (1 sp.) | (3), (11), 12, 13, 16, 17, (18), 21, 22 |
| *Hydroglyphus* (1 sp.) | 11, 13, 16 |
| *Geodessus* (1 sp.) | 11 |
| *Hygrotus* (2 sp.) | (6), 7, 11, 12, 13, (16), 17, 22 |
| *Deronectes* (4 sp.) | (1), (3), (5), 7, 11, 16 |
| *Graptodytes* (1 sp.) | (3), 11, 12, 13, 16 |
| *Suphrodytes* (1 sp.) | (3), 11, 12, (16) |
| *Hydroporus* (13 sp.) | (2), (3), (5), (6), 11, 12, 13, 16, 19 |
| *Oreodytes* (1 sp.) | (3), (6), 11, 13 |
| *Nebrioporus* (2 sp.) | (2), (3), 11, 12, (13), 16 |
| *Scarodytes* (1 sp.) | (3), (5), 11, 12, 13, 16 |
| *Stictonectes* (1 sp.) | (5), (6), 11, (16) |
| *Stictotarsus* (1 sp.) | 11, 12, 13, 16 |
| **Dytiscidae: Colymbetinae** | |
| *Liopterus* (2 sp.) | (1), 2, 3, 5, 6, 11, 13 |
| *Agabus* (18 sp.) | 1, 2, 3, 5, 6, 7, (11), (13), 15 |
| *Ilybiosoma* (1 sp.) | 5, 6 |
| *Platambus* (1 sp.) | (1), 2, 3, 5, 6, (7), (11), 18, 19 |
| *Colymbetes* (2 sp.) | (1), 2, 3, 5, 6, 7, (11) |
| *Ilybius* (10 sp.) | 1, 2, 3, 4, 5, 6, 7, (10), (11), 23 |
| *Meladema* (2 sp.) | 1, 2, 3, 4, 5, 6, (7), (11), (13) |
| *Rhantus* (4 sp.) | 1, 2, 3, 4, 5, 6, 7, (11), (13) |
| **Dytiscidae, Dytiscinae** | |
| *Eretes* (1 sp.) | (2), 3, 5, 6, (7), (11) |
| *Hydaticus* (2 sp.) | 2, 3, 5, 6, 10, (11) |
| *Acilius* (5 sp.) | 1, 2, 3, 5, 6 |
| *Graphoderus* (2 sp.) | 1, (2), 3, 5, 6, 7 |
| *Dytiscus* (4 sp.) | 1, 2, 3, 5, 6, 7, (11), (13), 15 |
| *Cybister* (3 sp.) | 3, 5, 6, 7, 8, 10, 11 |
| **Dytiscidae, Laccophilinae** | |
| *Laccophilus* (2 sp.) | (2), (3), (5), (6), 14, 17, 19, 20, 22 |

**1:** 4-Hydroxybenzoic acid, **2:** hydroquinone, **3:** benzoic acid, **4:** benzoic acid ethyl ester, **5:** 4-hydroxybenzaldehyde, **6:** 4-hydroxybenzoic acid methyl ester, **7:** 3,4-dihydroxybenzoic acid methyl ester, **8:** 3,4-dihydroxybenzoic acid ethyl ester, **9:** 2,5-dihydroxyphenylacetic acid methyl ester, **10:** phenylpropionic acid, **11:** phenylacetic acid, **12:** 4-hydroxyphenylacetic acid, **13:** phenylpyruvic acid, **14:** 3,4-dihydroxyphenylacetic acid methyl ester, **15:** marginalin (= 4′5-dihydrox y-benzalisocumaranone), **16:** 3-indoleacetic acid, **17:** 3-hydroxyoctanoic acid, **18:** octanoic acid, **19:** *Z*-3-octenoic acid, **20:** 3-hydroxynonanoic acid, **21:** nonanoic acid, **22:** 3-hydroxydecanoic acid, **23:** tiglic acid

(): minor component; without brackets: major component (see Dettner 2014)

**Table 9.2** Constituents **24–108** of prothoracic glands in genera (number of species) of predacious diving beetles (Dytiscidae; for species data see Dettner 2014)

| | |
|---|---|
| **Dytiscidae, Colymbetinae** | |
| *Ilybiosoma* (1 sp.) | 43 |
| *Agabus* (11 sp.) | 25, 29, 34, 38, 41, 42, 43, 44, 45, 46, 47, 48, 50, 52, 55, 59, 61, 62, 64, 66, 68, 70, 73, 74, 76, 77, 78, 80, 81, 82, 83, 84, 85, 86, 102, 103, 104, 105 |
| *Platambus* (1 sp.) | 54, 80, 101 |
| *Colymbetes* (1 sp.) | Colymbetin |
| *Ilybius* (6 sp.) | 24,25,29,26,27,28,29,30,31,32,33,35,44,50,51,53,54,57,63,65,69,71,72,87,88,89,90,91,92,93,94,95,96,97,98,99 |
| **Dytiscidae, Dytiscinae** | |
| *Acilius* (4 sp.) | 43, 44, 49, 50, 51, 56, 57, 63, unknown steroid |
| *Graphoderus* (2 sp.) | 43, 67, 75, 78 |
| *Dytiscus* (2 sp.) | 36, 37, 43, 50, 52, 53, 56, 57, 68 |
| *Cybister* (6 sp.) | 3, 39, 40, 43, 50, 51, 57, 58, 63, 64, pentadecanoic acid, octadecanoic acid |
| *Thermonectes* (1 sp.) | 57, 60 |
| **Dytiscidae, Laccophilinae** | |
| *Laccophilus* (1 sp.) | 73, 77 |
| **Dytiscidae, Hydroporinae** | |
| *Hyphydrus* (1sp.) | 106, 107, 108 |

**24:** 3-Hydroxyestra-1,3,5(10)-trien-17-one; estrone, **25:** 3,17ß-dihydroxyestra-1,3,5(10)-triene; 17ß-estradiol, **26:** 17ß-hydroandrost-1,4-dien-3-one; boldenon, **27:** 2α,17ß-dihydroxy-androst-4-ene-3-one; 2α-hydroxytestosterone, **28:** 2ß,17ß-dihydroxyandrost-4-ene-3-one; 2ß-hydroxytestosterone **29:** 17ß-hydroxyandrost-4-ene-3-one; testosterone, **30:** 17ß,19-di-hydroxyandrost-4-ene-3-one; 19-hydroxytestosterone, **31:** 3ß,17ß, dihydroxyandrost-5-ene, **32:** 3α,17ß, dihydroxyandrost-5-ene, **33:** 3ß,17α, dihydroxyandrost-5-ene, **34:** androst-4-ene-3,17-dione, **35:** 3ß,16ß,18-trihydroxyandrost-5-ene-17-one, **36:** pregn-4-ene-3-one, **37:** pregn-4-ene-3,20-dione; progesterone, **38:** 5ß-hydroxy-pregn-4-ene-3,20-dione; 6ß-hydroxy-progesterone, **39:** 12ß-hydroxypregne-4-ene-3,20-dione; 12ß-hydroxyprogesterone, **40:** 12ß-(3′pentenoyl-)pregn-4-ene-3,20-dione, **41:** 17α-hydroxypregn-4-ene-3,20-dione; 17α-hydroxy-progesterone, **42:** 11,15-dihydroxypregn-4-ene-3,20-dione; 11,15-di-hydroxy-progesterone, **43:** 21-hydroxypregn-4-ene-3,20-dione; 21-hydroxyprogesterone; cortexone; deoxycorticosterone, **44:** 11α,21-dihydroxypregn-4-ene-3,20-dione; 11α,21-dihydroxy-progesterone; **45:** 11ß,21-dihydroxypregn-4-ene-3,20-dione; 11ß,21-dihydroxyprogesterone; corticosterone, **46:** 18,21-dihydroxypregn-4-ene-3,20-dione; 18,21-dihydroxyprogesterone, **47:** 15α-hydroxy-

(continued)

**Table 9.2** (continued)

progesterone-7α-isobutyrate, **48:** 15α-hydroxyprogesterone-7α-hydroxy-isobutyrate, **49:** pregn-4-ene-3,19,20-trione; 19-oxoprogesterone, **50:** 20α-hydroxypregn-4-ene-3-one, **51:** 20ß-hydroxy-pregn-4-ene-3-one, **52:** 20 α,21-dihydroxypregn-4-ene-3-one, **53:** 20ß,21-dihydroxypregn-4-ene-3-one, **54:** 15α,20ß-dihydroxypregn-4-ene-3-one, **55:** 16ß, 20ß-dihydroxypregn-4-ene-3-one, **56:** pregna-4,6-diene-3,20-dione;6-dehydro-progesterone, **57:** 20α-hydroxypregna-4,6-diene-3-one; cybisterone, **58:** 12ß-hydroxypregna-4,6-diene-3,20-dione; cybisterol, **59:** 15α-hydroxypregna-4,6-diene-3,20-one, **60:** 18-hydroxypregna-4,6-diene-3-one-18,20-hemiketal; mirasorvone, **61:** pregna-4,6-diene-3,20-dione-15α-isobutyrate, **62:** 15α-hydroxypregna-4,6-diene-3-on-20-isobutyrate, **63:** 21-hydroxypregna-4,6-diene-3,20-dione; 6,7-dehydrocortexone, **64:** 12ß,20α-dihydroxypregna-4,6-diene-3-one, **65:** 20ß,21-dihydroxypregna-4,6-diene-3-one, **66:** 15a,20ß-dihydroxypregna-4,6-diene-3-one-20-isobutyrate, **67:** 3ß-hydroxypregn-5-ene-20-one; pregnenolone, **68:** 3ß,20α-dihydroxypregn-5-ene, **69:** 3ß,20α, 21-trihydroxypregn-5-ene, **70:** 3ß,21-dihydroxypregn-5-ene-20-one; 21-hydroxypregnenolone, **71:** 3ß,16α,21-trihydroxypregn-5-ene-20-one; 16α,21-dihydroxypregnenolone, **72**: 3ß-hydroxypregn-5-ene-7,20-dione; 7-oxopregnenolone, **73:** 3α-hydroxy-5ß-pregnane-20-one, **74:** 3α,20α -dihydroxy-5α-pregnane, **75:** 11α-hydroxy-5ß-pregnane-3,20-dione, **76:**21-hydroxy-5ß-pregnane-3,20-dione, **77:** 3α,12 α -dihydroxy-5ß-pregnane-20-one, **78:** 3α,11α-dihydroxy-5ß-pregnane-20-one, **79:** 3α,11ß-dihydroxy-5ß-pregnane-20-one, **80:** 3α,21-dihydroxy-5α-pregnane-20-one, **81:** 3α,21 -dihydroxy-5ß-pregnane-20-one, **82:** 3ß,11ß,15α-trihydroxy-5α-pregnane-20-one, **83**: 3α,11ß,15α-tri-hydroxy-5α-pregnane-20-one, **84:** 3α,11ß,15ß,20ß-tetrahydroxy-5α-pregnane, **85:** 3ß-hydroxy-5α-cholestane, **86:** 3ß-hydroxy-5ß-cholestane, **87**: methylisobutanoate, **88:** α-copaene, **89:** α-muurolene, **90:** γ-muurolene, **91:** ε-muurolene, **92:** γ-cadinene, **93:** δ-cadinene, **94:** α-cubebene, **95:** β-cubebene, **96:** torreyol, **97:** trans-β-farnesene, **98:** γ-gurjunene, **99:** α-cedrene, **100:** methyl-8-hydroxy-quinolinecarboxylate, **101:** 7α,5ß,10α-endesin-4(14)-ene-12,6α-diol; platambin, **102:** 2-(*cis*-5,8,11,14-eicosatetraenoyl)glycerol (2-ara-gl), **103:** 2-(*cis*-5,8,11,14,17-eicosapentaenoyl)glycerol (2-epa-gl), **104:** 1-(*cis*-5,8,11,14-eicosatetraenoyl)glycerol (1-ara-gl), **105:** 1-(*cis*-5,8,11,14,17-eicosapentaenoyl)glycerol (1-epa-gl), **106:** 1-hexadecanoate-glycerol, **107:** 2-octadecanoate-glycerol, **108:** 1-octa-decanoate-glycerol

A selection of gland constituents is presented in Fig. 9.8

**Table 9.3** Pygidial gland constituents of whirligig beetles (Gyrinidae) and crawling water beetles (Haliplidae)

| | |
|---|---|
| **Gyrinidae, Gyrinini** | |
| *Gyrinus frosti* | 110, 111 (Newhart and Mumma 1979) |
| *Gyrinus minutus* | 109, 110, 111, 114, 115, 118, 120 (Dettner and Luthardt unpubl.) |
| *Gyrinus natator* | 110, 114, 115 (Blum 1981) |
| *Gyrinus substriatus* | 109–118, 120 (Dettner and Luthardt unpubl.; Ivarsson et al. 1996) |
| *Gyrinus ventralis* | 110 (Dettner 1985) |
| *Gyrinus aeratus* | 114, 115 (Ivarsson et al. 1996) |
| *Gyrinus marinus* | 109–111, 114, 115, 118, 120 (Dettner and Luthardt unpubl.) |
| *Gyrinus paykulli* | 109–111, 114, 115 (Dettner and Luthardt unpubl.) |
| *Gyrinus urinator* | 109–111, 114, 115, 118, 120 (Dettner and Luthardt unpubl.) |
| **Gyrinidae, Enhydrini** | |
| *Dineutus assimilis* | 110–113 (Miller et al. 1975; Newhart and Mumma 1979) |
| *Dineutus discolor* | 113, 119 (Wheeler et al. 1972; Blum 1981) |
| *Dineutus hornii* | 110 (Eisner and Aneshansley 2000) |
| *Dineutus nigrior* | 11, 12, 110–113 (Miller et al. 1975) |
| *Dineutus serrulatus* | 110 |
| *Macrogyrus oblongus* | Various unknown terpenes, not 110–113 (Dettner and Luthardt unpubl.) |
| **Gyrinidae, Orectochilini** | |
| *Orectochilus villosus* | Odor like *Viola,* no known products apart from octadecan, heptadecan, n-heneicosan, n-docosan (Dettner and Luthardt unpubl.) |
| **Haliplidae** | |
| *Peltodytes caesus* | 11, 12, 17, 22 (Dettner and Böhner 2009) |
| *Peltodytes rotundatus* | 11, 12, 22, 122 (Dettner and Böhner 2009) |
| *Brychius elevates* | 3, 11, 17, 22 (Dettner and Böhner 2009) |
| *Haliplus (Neohaliplus) lineatocollis* | 11, 12, 17, 22 (Dettner and Böhner 2009) |
| *Haliplus (Haliplus) ruficollis* | 11, 12, 17, 22, 121 (122) (Dettner and Böhner 2009) |
| *Haliplus (Liaphlus) laminatus* | 11, 17, 22, 121 (Dettner and Böhner 2009) |
| *Haliplus (Liaphlus) flavicollis* | 11, 17, 22, 121 (Dettner and Böhner 2009) |
| *Haliplus (Liaphlus) fulvus* | 11, 17, 22 (Dettner and Böhner 2009) |
| *Haliplus (Haliplus) heydeni* | 3, 11, 17, 22, 121 (?) (Dettner and Böhner 2009) |

**109:** phenylacetaldehyde, **110:** gyrinidal, **111:** isogyrinidal, **112:** gyrinidione, **113:** gyrinidone **114:** 3-methyl-1-butanal, **115:** 3-methyl-1-butanol, **116:** 2-methyl-1-propanol, **117:** 6-methyl-5-hept-2-on, **118:** 3-methyl-2-butanol, **119:** octanal, **120:** 3-methylbutyric acid, **121:** phenyl- lactid acid, **122:** 3-hydroxy phenylacetic acid
For numbers below **109** see Table 9.1
For numbers of compounds see Table 9.1 (**1–23**), Table 9.2 (**24–108**), and Table 9.3 (**109–122**)

immersed due to their unwettable cuticle. Only representatives of genera *Spinactaletes*, *Arlesminthurus*, and *Pseudobourletiella* are capable of being submerged (Cipola et al. 2018). Interestingly young springtails after molting lose their nonwetting properties for a short time. If they stand on the water surface this is due to their nonwetting claws. If forced to submerge larvae will sink, have cutaneous respiration, and can also be invaded by microorganisms (Chang 1966; Thorp and O'Neill 2015). These underwater larvae will not molt. If gravid females are disturbed expelled eggs float on the water surface and later sink down. The underwater development continued and most eggs hatched (Chang 1966). An interesting movement in order to escape predators was observed in *Anurida maritima* (Fig. 9.1/4) on coastal pools and *Podura aquatica* (Fig. 9.1/3) at the surface of ponds. They lay on the water surface and pull up with a central wetting tube or ventral tube (collophore) at the same time as nonwetting head and abdominal tip are pushed down. When they release the wetting tube from the surface film, the animal is propelled into the air (Fig. 9.1/4, Bush et al. 2008). This posture by forming a meniscus also enables them to attract neighbors over a small distance.

Some semiaquatic species live and feed on water surfaces. Especially *Podura aquatica* (Fig. 9.1/3, Poduridae), *Isotomurus palustris* (Isotomidae), *Sminthurides aquaticus* (Sminthuridae) and *S. malmgreni* live on open water surfaces. Other more specialized species such as representatives of *Arrhopalites* (Sminthuridae) live on surface films of cave waters or other species prefer surfaces of aquatic plants such as *Lemna* or *Nasturtium*). Interestingly Collembola which are mainly found on water surfaces are characterized by distinctly paddle-like broadened mucrones, the claws of the abdominal springing organ (*Podura aquatica*; *Proisotoma crassicauda*, *P. borealis* Isotomidae; *Archisotoma* Isotomidae; *Sminthurides*; Schulze 1924). These mucrones prevent the breakthrough of the surface tension during a jump (Hopkin 1997).

Chemical defense is obvious in very different Collembola taxa and not only significant for those species which are eyeless or have lost their jumping ability (Dettner 2015). Defensive compounds were especially found in representatives of Onychiuridae (Fig. 9.1/1 *Onychiurus* spec. depletes secretion from a pseudocellus), Hypogastruridae, Neanuridae, or Tullbergiidae. These allomones are usually externalized by integumental pores, so-called pseudocells, (Fig. 9.1/2 pseudocellus from *Tetrodontophora* spec.), but there exist also springtails exhibiting warning colorations or exposing club-like defensive glands. Among the semiaquatic families there exist chemical data from Poduridae and Isotomidae. From *Podura aquatica* body surfaces a new hydrocarbon named poduran, with an unusual tricyclo(6.2.0.0) decane system was identified (Fig. 9.1/5; Schulz et al. 1997). Finally representatives of Isotomidae such as *Folsomia fimetaria* contain unknown hemolymph toxins which negatively influence the reproduction of the main predatory species, the spider *Erigone atra* (Marcussen et al. 1999).

**EPHEMEROPTERA (mayflies; 100% aquatic)**: Mayflies represent the oldest order of winged insects. About 3500 species worldwide (Salles et al. 2018; Stork 2018: 3240) are characterized by aquatic eggs, phytophagous larvae (nymphs), and 2 terrestrial winged stages (subimago, imago) (Sartori and Brittain 2015a, b). The

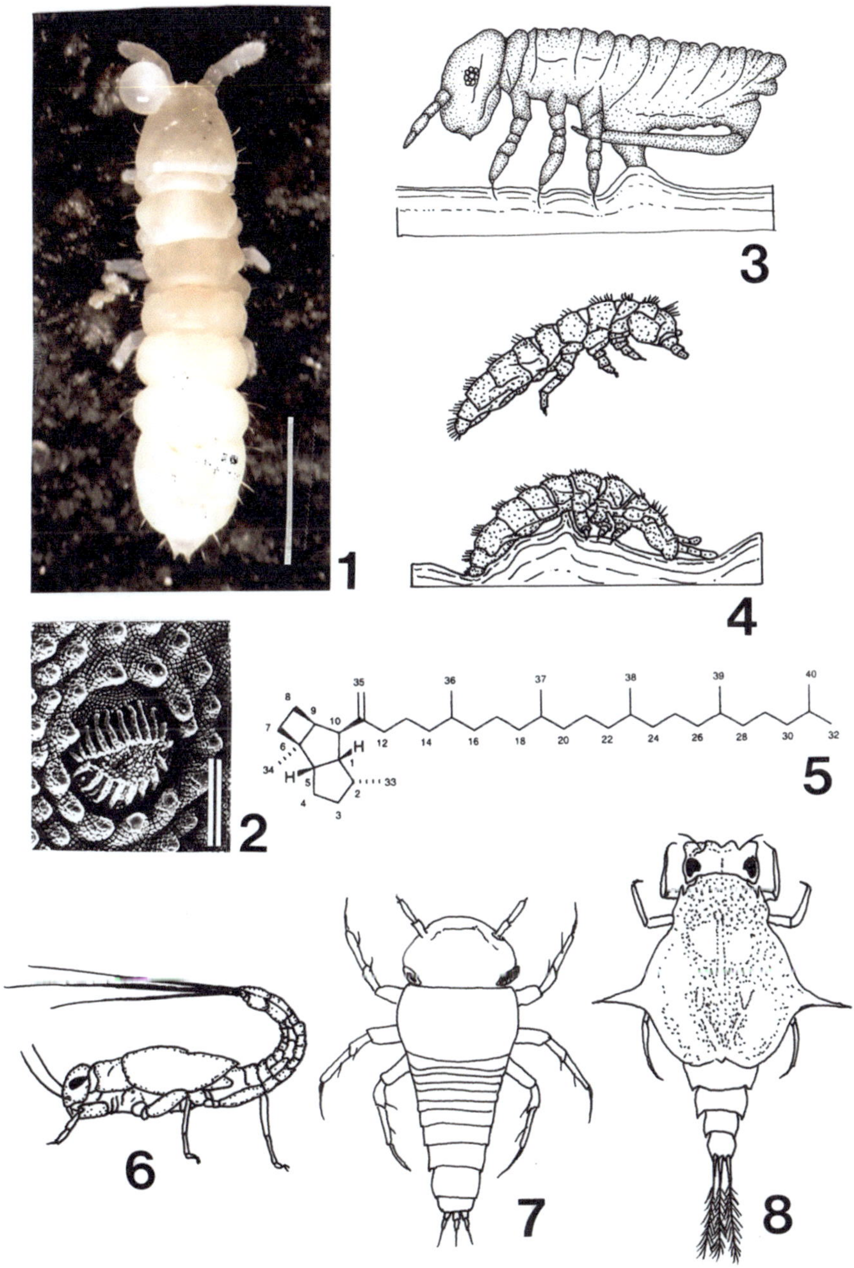


**Fig. 9.1** Defenses of Collembola (1–5), Ephemeroptera (6–8). 1. *Onychiurus* spec. depletes defensive droplet from pseudocell (scale: 0.5 mm). 2. Pseudocell of *Tetrodontophora bielanensis* (scale: 10 um). 3. *Podura aquatica* on water surface. 4. Springtail *Anurida maritima* escapes predators by using its ventral tube. 5. Unusual hydrophobous hydrocarbon poduran from *P. aquatica*. 6. High-intensity scorpion posture of *Ephemerella subvaria*. 7. First larval instar of *Baetisca rogersi.* 8. Last larval instar of *Baetisca* spec. Sources: 1, 2, 3: Dettner and Peters 2010. 4. Modified after Lancaster and Downes 2013. 5. After Schulz et al. 1997. 6. Modified after Peckarsky 1987. 7. Modified after Pescador and Peters (1974). 8. Modified after Pescador et al. (2009)

unique subimago is covered by hydrofugous hairs which allows this stage to escape the water surface tension (Sartori and Brittain 2015a, b). There are 9–45 larval stages which live 3 weeks to 2 years, whereas adults live from 1 h to few days (Morse 2017). Mayfly larvae and subimagines which may be very diverse with respect to morphology represent a major part of the macroinvertebrate biomass and are therefore eaten by a wide range of aquatic invertebrates (Plecoptera, Trichoptera, Sialidae, Odonata, aquatic bugs and water beetles, leeches, triclads, crayfish) and vertebrates (fishes, specialized birds; Grant 2001). 42 fish families analyzed fed on aquatic insects and in Nearctic plecopteran genera 50% of the gut contents consisted of mayfly larvae (Bauernfeind and Soldan 2012). As compared with larvae of stoneflies which often show a greater degree of chitinization, larvae of Ephemeroptera are more soft and often have large lamellar gills (Brittain 1990). It is astonishing that there exist no chemical or acoustical defense mechanisms of mayfly developmental stages (neither stridulation, Aiken 1985, nor chemical defense, Dettner 2015) against predators, parasitoids, or pathogenic organisms. Instead depending on the presence of predators mayflies have evolved behavioral defensive mechanisms such as drifting, swimming, crawling away, hiding, scorpion posturings, or timing of activities (mass emergence, mating swarms) in order to reduce the chance of predation (Peckarsky 1996). As shown by Crespo (2011) for mayfly larvae chemical signals are highly important. Synchronous emergence could saturate potential predators whereas dispersed emergence eventually lowers the possibility of predator-prey encounters (Sartori and Brittain 2015a, b). After contact with predacious stonefly larvae *Ephemerella infrequens* (Ephemerellidae) exhibited a scorpion posture (Fig. 9.1/6). This reaction was usually evoked by touch of body parts; sometimes the reaction was shown when predators approached from upstream (Peckarsky 1987). In contrast *Baetis bicaudatus* (Baetidae) showed a "tail curl" posture in flexing cerci (which represent sensory structures, Crespo 2011) and posterior abdominal segments against predators. Peckarsky (1980, 1987) reports that *Ephemerella* deflects stoneflies because the posture may increase the apparent size of the mayflies which are then rejected by tactile predators. On the other hand *Baetis* detects predator's wave disturbances and enables the prey to avoid predator encounters. In other species defence reactions such as crawling evasions and tactile (chemotactile) stimuli were observed as being responsible for these behaviors (Peckarsky 1980). Moreover chemical interactions between predator and mayfly-larvae-prey were supposed, because larvae of *Baetis* actively left the vicinity of stonefly larvae (see Bauernfeind and Soldan 2012). Elder larvae of Baetiscidae (Fig. 9.1/8) as compared with freshly hatched stages (Fig. 9.1/7) have a unique and bizarre appearance due to the thoracic notum and the presence of large spines (two lateral, two dorsal projections) and various lobes and humps ("armored mayflies"; Pescador et al. 2009). In addition these larvae exhibit cuticles which are harder than most mayfly larvae. The thoracic notum of Baetiscidae is fused, covers various abdominal segments (5), and encloses the gills underneath. According to Pescador and Peters (1974) the mesonotal shield helps to protect the gills and the lateral spines act as balancers and maintain the dorsoventral position of the larvae. Here I interpret the spines and the hardened cuticle as a typical mechanical defense against larger predators. Moreover it is noteworthy that *Baetisca* larvae exhibited thanatosis after mechanical molestation and show a coloration closely resembling their habitat (Pescador and Peters 1974). Both cryptic colorations (other larval genera such as *Baetis* and *Rhithrogena*) and aposematic color patterns (*Ecdyonurus, Heptagenia, Ephemerella*) are observed (Bauernfeind and Soldan 2012). However the presence of toxic natural compounds in mayflies has not been observed at all. Apart from thanatosis or scorpion postures common other defensive reactions observed in mayfly larvae are escape into the substrate (Bauernfeind and Soldan 2012). There are no data available on how mayflies defend against parasitic or commensalic organisms. However mayflies parasitized by mermithid nematodes showed increased defensive activities against predaceous stoneflies (Benton & Pritchard 1990).

## 9.2 ODONATA (damselflies, dragonflies; 99–100% aquatic)

Worldwide there exist about 6000 species (Stork 2018: 5899; another 1000 to 1500 species are expected to exist per Pessaq et al. 2018) of Odonata which comprise the three suborders Zygoptera (damselflies), Anisoptera (dragonflies), and Anisozygoptera (damsel dragons) (Suhling et al. 2015; Morse 2017; St. Quentin and Beier 1968). Eggs and larvae are aquatic and are found in various types of freshwater habitats including water-filled bromeliads. Few species are semiaquatic, whereas adults are terrestrial flying near water. Depending on species larvae need few weeks to 5 years to complete their development comprising 10–15 instars. Aquatic larvae and aerial adults of Odonata are predators with extrudable mouthparts of larvae and developed sight and flight in adults. Odonata represent one of the best known insect orders concerning taxonomy, zoogeography, or studies on sexual selection and conflicts by sperm displacement (Cordoba-Aguilar 2008; Dijkstra et al. 2014).

In addition predators of Odonata larvae, their prey, and their various defense mechanisms against opponents are well known in spite of the fact that there was not realized any chemical defense mechanism in larvae or adults of this insect order (Corbet 1999; Suhling et al. 2015). The risk of fish predation is reduced by various behavioral and morphological features of the odonate larvae. As was demonstrated by Wohlfahrt et al. (2005) these traits may be fixed and occur both in the presence and absence of predators. On the other hand on response to predators these traits may be flexible. Main predators of Odonata larvae are fishes which are significant in shaping odonate community structures as was demonstrated in North American *Enallagma* species (Johnson 1991). Larvae of certain species reduce their movements and their foraging activity when fishes are present, whereas other species escape predators by swimming away (Johnson 1991). As was shown by Crespo (2011), larvae of Odonata can sense infochemicals from various predators and even learn to associate them with predator presence. Invertebrate predators include other Odonata larvae including conspecifics (cannibalism), dytiscid beetles (larvae and adults), crayfish, belostomatid bugs, and several water birds (see Suhling et al.

2015). Other larvae exhibit nocturnal circadian rhythms in order to reduce predation pressure (Pierce 1988). It was also reported from zygopteran larvae that avoidance responses after contacts with predators may change during ontogeny (Corbet 1999). Many Odonata larvae show cryptic colorations which correspond to the microhabitat background, being greenish, yellowish, darkened, or with dark spots on pale ground (mixture or mixtures of sediments). In young instars of *Anax* transverse dark bands are observed followed by pale stripes. As in adults also in larvae there are abilities to change coloration. Larvae of other species may bear many setae on their body surface; adhering to fine detritus may result in an excellent camouflage. Many Odonata larvae hide in dense vegetation (Thompson 1987) or burrow in suitable substrates (Suhling and Müller 1996) in order to reduce predation pressure.

Larvae of *Epiophlebia* (Anisozygoptera) which represent living fossils are unusual in producing sound and sometimes exhibiting thanatosis when molested. The production of sounds after molestation was also reported in *Uropetala carvei* (Petaluridae; see Corbet 1999). Larvae with 14 instars need maximally 8 years (longest record for Odonata) for their development. Elder larvae are terrestrial for several months and breathe via thoracic stigma. In laboratory experiments it could be shown that larval stridulation was produced in response to mechanical disturbance (Aiken 1985; Asahina 1950). In order to stridulate lateral stridulatory files on abdominal segments 3–7 (Fig. 9.2/1) are pressed against metafemur serrations. Sounds are produced when the abdomen is twisted and telescoped. Stridulatory files are first evident in the sixth instar before eclosion (Aiken 1985). *Epiophlebia* larvae which appear like larvae from the Anisoptera are unable to use the anisopteran jet propulsion mode of escape but walk. In elder anisopteran larvae a kind of jet propulsion as a rapid escape mechanism was described (Corbet 1999) which is elicited by stimulation of paraproct nerves. These larvae may suddenly deplete their water-filled hindgut. In elder larvae this is the only way of locomotion, because leg muscles degenerate before larvae leave the water. When zygopteran larvae are irritated they may lose their caudal appendages (Stoks 1999), which possess breaking joints (Corbet 1999). Thus influx of water or loss of hemolymph is prevented. Autotomized body parts regenerate during succeeding moltings. These abdominal appendages are sometimes broadened in order to increase swimming speed (McPeek 2000). In other genera such as *Aeshna*, *Enallagma*, or *Ischnura* also whole legs are autotomized (Corbet 1999). In many cases larvae of Odonata are immobilized and show thanatosis (Fig. 9.2/6, 7) when they are molested (Wildermuth 2000). In *Somatochlora* larvae either extended their legs laterally, held them obliquely upwards (Fig. 9.2/6), or folded them close to their bodies. Among 18 investigated anisopteran species larvae of *Brachytron pratense* and three *Somatochlora* species revealed an obligatory reaction to tactile stimulation (Wildermuth 2000). Corbet (1999) reports reflex immobilization from 11 families of Odonata (Aeshnidae, Coenagrionidae, Cordulegastridae, Corduliidae, Epiophlebiidae, Gomphidae, Hemiphlebiidae, Libellulidae, Neopetaliidae, Petaluridae, and Pseudolestidae). Within ten species duration of immobilization ranged from more than 20 s to about 889 s; the maximum recorded time was more than 30 min. Many zygopteran larvae show distinct "unnatural" postures mainly of their caudal appendages (*Xanthocnemis*, Corbet 1999). Various anisoptera larvae have extremely flexible abdomina. Due to

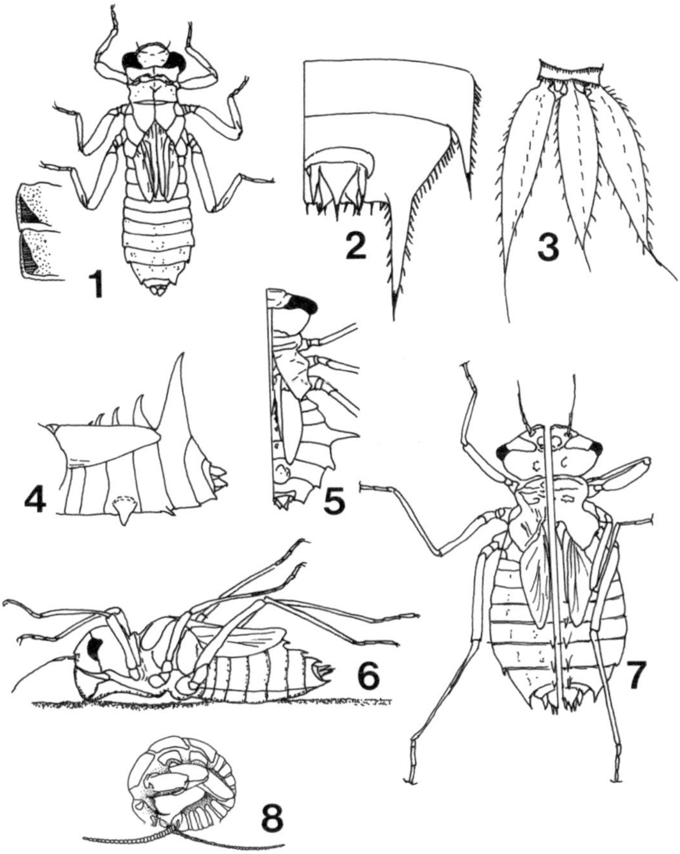


**Fig. 9.2** Defenses of Odonata (1–7), Plecoptera (8). 1. Last-stage larva of *Epiophlebia superstes* and stridulatory files on tergites of abdominal segments 5/6 (left). 2. Apex half of abdomen larva of *Aeschnosoma* sp. (dorsal view). 3. Apex of abdomen with caudal appendages of female larva of *Heteragrion consors* (Megapodagrionidae, dorsal view). 4. Left lateral view of abdomen of last-stage larva of *Orionothemis felixorioni*. 5. Half last-stage larva (dorsal view) of *Orionothemis felixorioni*. 6. Thanatosis of *Somatochlora flavomaculata* after molestation (side view). 7. *Somatochlora flavomaculata* last instar larva (dorsal view), normal sitting position (left), thanatosis after molestation (right). 8. Plecopteran larva of family Taeniopterygidae after molestation in curling defence position. Sources: 1. Modified after Aiken 1985. 2. Modified after Neiss et al. 2018. 3. Modified after Anjos-Santos et al. 2018. 4, 5. Modified after Fleck et al. 2009. 6, 7. Modified after Wildermuth 2000. 8. Modified after Zwick 1980

spines which are arranged laterally and dorsally various predators and especially fishes are warded off (Johansson and Mikolajewski 2008). Spines of anisopteran larvae occur especially on the last third of the abdomina and form a spiky pyramid in order to "sting" potential predators. Remarkable examples for abdominal "weapons" are known from representatives of the anisopteran genera *Aeshnosoma* (Fig. 9.2/2), *Melanocacus*, *Orionothemis* (Fig. 9.2/4, 5), or *Paracordulia* and even from zygopteran species such as *Heteragrion consors* (Fig. 9.2/3; Pessacq et al.

2018). Sometimes large anisopteran larvae as *Tanypteryx pryeri* or *Anax junius* use their mandibles in order to bite aggressors (Corbet 1999).

Recently there were published several results in order to test hypotheses about parasite (water mites, gregarines) mediated selection with odonate hosts (Forbes and Robb 2008; Honkavaara et al. 2009). Probably mite infections may compromise a male's resistance against further infections by parasites which is achieved by hemolymph encapsulation in *Coenagrion armatum*. In addition the pseudopeptide antibiotic amicoumacin was isolated from *Bacillus* species which were isolated from *Coenagrion* guts. However it is not known if the bacteria produce this compounds in the Zygoptera larva (see Dettner 2015).

**PLECOPTERA (stoneflies; 100% aquatic)**: As basal order of the Neoptera stoneflies have worldwide about 4000 species (Stork 2018: 3743) and are characterized by terrestrial adults and aquatic eggs and larvae (Avelino-Capistrano et al. 2018). Plecoptera are usually associated with cool and cold waters and possess 10–22 larval instars (Morse 2017). Stoneflies and especially their larvae have a lot of enemies and parasites ranging from salmonid fishes, birds, and predatory arthropods to gregarines, insect-pathogenic fungi, and water mites (Zwick 1980). Communication between males and females is often achieved by tapping on substrate ("drumming"; DeWalt et al. 2015). Crespo (2011) demonstrated in stonefly larvae both mechanosensory and chemical cues, which means nonvisual modalities were involved in detection of prey. In various especially young adult species (*Pteronarcys proteus*, *Peltoperla maria*, Benfield 1974; *Perla burmeisteriana*, *Acroneuria arenosa*, Zwick 1980) and one larval species (*P. dorsata*; Moore and Williams 1990) reflex bleeding behavior was recorded after molestations. Beaty (2015) reported that preserved larvae of Pteronarcyidae showed white exudates originating from thoracic nota and coxae. Also in various Plecoptera adults reflex bleeding was observed; however no active compound could be isolated (Stocks 2004). Zwick (1980) suggested that plecopteran hemolymph acts as an adhesive. Zwick (1980) reported that hemolymph of *Pteronarcys* can be sprayed away up to 25 cm by reflex bleeding, a behavior which is accompanied by an audible popping sound. Apparently hemolymph of stoneflies is not malodorous and represents no deterrent (Zwick 1980). The basis for this autohemorrhage was presumed to be increased hemolymph pressure (Stocks 2004). Moore and Williams (1990) were the first who demonstrated that the larvae of *Pteronarcys dorsata* show different responses which depends on the predators. Reflex bleeding and retreat were observed after contact with benthic crayfish predators. In contrast freezing and thanatosis were observed after contacts with benthic and pelagic fish predators. Beaty (2015) reported that *Pteronarcys* larvae exhibit thanatosis and turn into a ball. Further predator-prey studies were performed by Peckarsky (1980). Finally it is well known that larvae of Taeniopterygidae (e.g., *Brachyptera*) exhibit a characteristic of curled thanatosis position (Fig. 9.2/8; Zwick 1980). In various plecopteran taxa coiling movements were observed in order to escape from carnivorous insect larvae such as Odonata or Dytiscidae (Zwick 1980). Few larvae are also characterized by spines which might be located on thorax (*Kempnyia*, *Gripopteryx*) or abdomen

(*Neopentura*) and one species may be covered by detritus (*Pelurgoperla personata*) (Avelino-Capistrano et al. 2018).

**ORTHOPTERA (bush crickets, crickets, grasshoppers, locusts; 0.3% aquatic)**: Among the over 28,000 orthopteran species (Stork 2018: 23,855) about 80 species with adaptations to aquatic life are aquatic whereas additional 110 species are water dependant (associated with aquatic habitats; Cover and Bogan 2015; Heckman 2018). Adaptations of aquatic Orthoptera are paddle-shaped hind tibiae (adapted for swimming), use of a plastron when submerged (Acrididae: Leptysminae), and diving in order to feed on submerged plants and to escape predators (Gutjahr and Braga 2018). Following families include Nearctic aquatic species: grasshoppers (Acrididae), pygmy grasshoppers (Tetrigidae), pygmy mole grasshoppers (Tridactylidae), katydids (Tettigoniidae), crickets (Gryllidae), and mole crickets (Gryllotalpidae). Also from Asia aquatic representatives are registered from the subfamilies Oxyinae (Acrididae), Tropidopolinae, and Hemiacridinae. Interestingly one aquatic and South American acridid species *Cornops aquaticum* feeds on water hyacinths and is being bred for possible release as a way to control this weed (Cover and Bogan 2015). Other South American species from the genera *Paulinia* and *Marellia* exclusively consume aquatic plants and are used in biocontrol of aquatic weeds in Africa and Australia (Cover and Bogan 2015). Unfortunately biology and ecology of these aquatic species are mainly unknown. However the semi-aquatic grasshopper *Paulinia acuminata* and its food plant *Salvinia auricularia* exhibit mimicry and ultrastructural analogy in order to achieve water-repellent surfaces (Barthlott et al. 1994). Generally among Orthoptera a lot of primary and secondary defensive mechanisms such as crypsis, aposematism, jumping, biting, flight, stridulation, or autotomy are realized (Dettner 2015). Employed chemical defenses are regurgitation, defecation, glandular defensive secretions, and internal toxins (Dettner 2015).

Within Tetrigoidea (about 1000 species), members of Scelimini are fully aquatic and can swim effectively (Rentz and Su 2003). Usually Tetrigidae are common near water, to which they readily take to escape from predators. They swim below the surface, and cling to plants until the danger has passed. Honma et al. (2006) studied the death-feigning posture in *Criotettix japonicus* against the frog *Rana nigromaculata*, which represents a sit-and-wait, gape-limited predator. The characteristic posture of this pygmy grasshopper enlarges its body size because pronotum, hind legs, and lateral spines project in different directions. Therefore the predator is unable to swallow the prey. Therefore it was shown that thanatosis does not generally means to mimic the posture of dead animals.

Concerning aquatic Orthoptera only few defensive mechanisms have been described. Members of Tridactylidae prefer damp habitats, where they construct nests out of mud and debris. Many species are active swimmers. Larvae and both sexes of adult Tridactylidae possess paired sternal glands with openings and opening muscles associated with the second abdominal sternum. When disturbed *Tridactylus japonicus* discharges a characteristic scent. Chemically the defensive secretions are unknown but they effectively repel *Tetramorium* ants and *Pardosa* spiders (Moriya and Ichinose 1988; Moriya 1989). In addition within mesotibiae of

*Tridactylus variegatus* further exocrine glands were identified (Messner 1969). Probably the mucus secretion serves to strengthen the wall of the nest tunnels of the pygmy mole grasshopper. Stridulation in which the tegmina are pulled against the hind wings occurs in some species. Members of Pauliniidae (Acridoidea) represent small- to medium-sized aquatic grasshoppers from South America. They can skate on water surface, they dive, and they can swim. Egg deposition takes place below water on water plants. Another species *Paulinia acuminata* was introduced into Africa for control of *Salvinia* (Rentz and Su 2003). There were reported further defensive mechanisms of selected orthopterans (Heckman 2018). When attacked by insectivorous fishes various *Cornops* species (Acrididae) fly away. Other species such as Southamerican cricket *Argizala* spec. (Gryllidae) or *Marilia* spec. dive beneath water surface and conceal among roots of water plants. Few species such as *Paraneonotus* spec. (Stenopelmatidae) dive as long as 12 m and remain motionless until the danger has departed.

**BLATTODEA (roaches, termites; 0.8% aquatic)**: Among the more than 6400 described cockroach species (Stork 2018: 7314) including termites, most representatives of the aquatic and amphibious Blattodea belong to subfamily Epilamprinae (Blaberidae; Cover and Bogan 2015). Many species inhabit small ecosystems as phytotelmata in neotropics and Indomalaya and about 60 cockroach species were reported from leaf bases of bromeliads. Especially representatives of the genera *Epilampra, Phlebonotus, Poeciloderrhis, Opisthoplatia, Rhabdoblatta*, and *Rhicnoda* have been reported as aquatic/amphibious in spite of the fact that little is known on both the biology of these phytotelmata species and those taxa which are found along streams and rivers. When disturbed the rapidly swimming species *Epilampra maya* from Central America enters the aquatic habitat and remain submerged up to 15 min (Crowell 1946). Females of *Phlebonotus pallens* from India protect their larvae when they are crawling along the stream bottom (Bell et al. 2007). Larvae and apterous females of the Indian species *Rhicnoda natatrix* are active swimmers and hide under submerged stones (Nesemann et al. 2010).

Reports on defensive mechanisms especially in aquatic and amphibious cockroaches are not available. Generally Blattodea show a diverse array of defensive mechanisms. Sometimes they are camouflaged and resemble dead or green leafs; otherwise they mimic toxic insects (Dettner 2015). In addition they show evasive behavior; they stridulate; they fight, fly, or run away; and they dive into and quickly swim in water. Finally they cause allergic symptoms in humans and many species have exocrine glands where a lot of allomones are produced (Brossut 1993; Schal et al. 1982; Dettner 2015).

**HETEROPTERA (true bugs; 13% aquatic)**: Heteroptera represent an insect order (about 40,500 sp.; Stork (2018): Hemiptera: 103,590) within Hemiptera. Numbers of special heteropteran taxa were from Lytle (2015) and Moreira et al. (2018); careful morphological characterizations of complex exocrine glands are compiled by Slater (1982); chemical data concerning complex glands were given by Aldrich (1988), Blum (1981), Dazzini-Valcurone and Pavan (1978), Weatherston and Percy (1978), and Millar (2005). Data on maxillary glands are based on Puchkova (1965). Moreira et al. (2018) report that representatives of Gerromorpha,

Nepomorpha, and Leptopodomorpha (the three infraorders have more than 4800 species) are primarily associated with freshwater and sometimes with saltwater. Biological data on Heteroptera are especially based on Wesenberg-Lund (1943), Andersen and Weir (2004), Scholtz and Holm (1986), Schuh and Slater (1995), and Scholtz and Holm (1986).

**Gerromorpha (semiaquatic bugs)**: According to this compilation there exist 2846 Gerromorpha species (Moreira et al. (2018): 2100 sp.). Members of this specialized group use the surface water film and glide on it rapidly in search for prey. Several data on Gerromorpha were compiled by Andersen (1982). In accordance with Moreira et al. (2018) only eight families (not superfamilies, and Madeoveliidae) are considered. Antipredator adaptations in Gerromorpha include cryptic coloration, countershading, and defensive behavior (Andersen 1996). The metasternal secretions of Gerromorpha are chemically unknown but may have repellent functions against predators (Andersen 1996). Nearly nothing is known on the composition of salivary glands (Walker et al. 2016). Schmidt (1982) reports that representatives of Gerromorpha (e.g., Veliidae, Gerridae) possess lethal and painful venoms. Aldrich (1988) speculates that the widespread occurrence of communication via waves may explain the frequent loss of exocrine glands especially in these semiaquatic bugs.

**Mesoveliidae** (water treaders; 49 sp.): Metasternal scent glands are present as a single median reservoir which is at least partially lined with glandular cells. Larvae have a somewhat reduced dorsal scent gland opening between terga III and IV.

**Hydrometridae** (water measurers; 129 sp.): Representatives of *Hydrometra* exhibit thanatosis and have an extreme slender body. Combined with rhythmical, vertical movements probably this may have the function of obscuring the outline of the insect (Andersen 1996). Hydrometridae possess a rapidly lethal venom within their salivary glands (Schmidt 1982). The metasternal scent glands have glandular cells opening into the collecting ducts and there is a reservoir with a dorsal accessory gland. Larvae lack dorsal abdominal scent glands.

**Hebridae** (velvet shore bugs; 233 sp.): The metasternal scent gland is a single simple median pouch with a narrow slit opening and without differentiated evaporative area. Larvae have a somewhat reduced dorsal scent gland opening between abdominal terga III and IV.

**Paraphrynoveliidae** (two sp.): The abdominal scent gland is situated between terga III and IV. No data are available with respect to the metasternal glands.

**Macroveliidae** (Macroveliid shore bugs; 3 sp.): The metasternal scent glands open into collecting ducts and contain a reservoir with a dorsal accessory gland and an opening of the median elevation on the posterior part. There are no fully developed lateral scent gland channels. There is a dorsal abdominal scent gland opening on the tubercle on tergum IV.

**Veliidae** (riffle bugs; 903 sp.): Representatives of this family can hop or jump or remain motionless (Scrimshaw and Kerfoot 1987). The escape movements called expansion skating might be the most elaborate defense mechanism of Heteroptera at all. It is observed in the genera *Velia* and *Microvelia* (Andersen 1996) and is comparable to the staphylinid beetles of Steninae (Betz et al. 2018). The detergent saliva which is ejected through the rostrum lowers the surface tension of the water. The metasternal scent glands on the sternum open into collecting ducts and have a dorsal accessory gland. The ostiolar channels on the metapleuron are usually elongated and prolonged on metapleura, ending in elongated evaporative areas with hair tufts in front of the hind coxae. Henrikson and Stenson (1993) concluded that aggregations of the dark, highly contrasting neustonic *Velia caprai* in exposed areas advertise their bad taste, which represents a case of aposematism. Species of *Velia caprai* were rejected by *Salmo trutta* and *Lepomis macrochirus* (Scrimshaw and Kerfoot 1987).

**Madeoveliidae** (one sp.): These predacious bugs live among wet rocks. Lateral metathoracic scent gland channels are present.

**Gerridae** (water striders; 713 sp.): Within genus *Gerris* death feigning and thanatosis are observed (Andersen 1996). Water striders can also hop and jump and often have countershading to reduce shadows (Scrimshaw and Kerfoot 1987). In order to defend optimally against bottom-striking predators it was found that long midlegs were especially important. The genetic basis of this interaction between a Hox protein Ubx and a new target gene (*gilt*) was studied by Armisen et al. (2015). The metasternal scent gland apparatus consists of two lateral glands and a median reservoir discharging through the median ventral aperture. Obviously the secretions evaporate due to many setae (Scrimshaw and Kerfoot 1987). Lateral secretory channels are "retained" in *Brachymetra* and *Cylindrostethus*. Usually there are no dorsal abdominal scent glands present in adults or larvae. Bioassays showed that some adult gerrids were rejected by bluegills (*Lepomis macrochirus*; Scrimshaw and Kerfoot 1987). Interestingly diapausing representatives of *Gerris* develop a diapause secretion that means there is a coat of densely packed microparticles on their body surfaces (Fig. 9.3/8) before leaving the water surface (Hauser 1985). It was speculated that these particles either represent fungicides or prevent freezing.

**Hermatobatidae** (coral treaters, 12 sp.): In this family the abdomen is extremely short and has a single scent gland opening on the fourth mediotergite.

**Leptopodomorpha (shore bugs)**: There exist 382 sp. worldwide. Chemical data from metathoracic glands are not available. Maxillary glands are present in representatives of Leptopodomorpha (Puchkova 1965). According to Moreira et al. (2018) Saldidae and Aepophilidae represent sister groups to Omaniidae and Leptopodidae.

**Fig. 9.3** (continued) emission of secretion is indicated by arrow. 5. Secretion grooming of *Sigara arguta* on the water surface prior to flying away. 6. Glands of *Ranatra linearis*: Paired maxillary glands (max) and bipartite paired labial glands (only one gland system figured) consisting of an accessory gland (ag) and a main gland (mg). Reservoirs of maxillary glands open at the base of proboscis (paired arrows); emission of secretion of labial glands occurs at the apex of proboscis (arrow). 7. Foreleg of male *Anisops stali* with stridulation organ (arrow). 8. Diapause secretion (stippled) according to REM figure of *Gerris costae* 5 weeks after imaginal molt. 9. Adult larva of *Nevrorthus fallax*; side view (left), after molestation, side view (right). Gland constituents: **1.** (E)-4-oxo-hex-2-enal; **2.** (E)2-hexenal, **3.** (E)-2-octenal, **4.** hydrogen peroxide, **5.** deoxycorticosterone, **6.** 4-hydroxy benzaldehyde. Sources: 1–3. Modified according to Staddon and Thorne (1973). 4. Modified according to Kovac et al. 1991. 5. Modified according to Kovac and Maschwitz (1991). 6. Modified according to Pawlowsky (Pawlowsky 1927) and A. Böttcher (Böttcher and Dettner unpubl.). 7–9. Modified according to Andersen and Weir (7. Andersen and Weir 2004, 8. Hauser 1985, and 9. Zwick 1967).

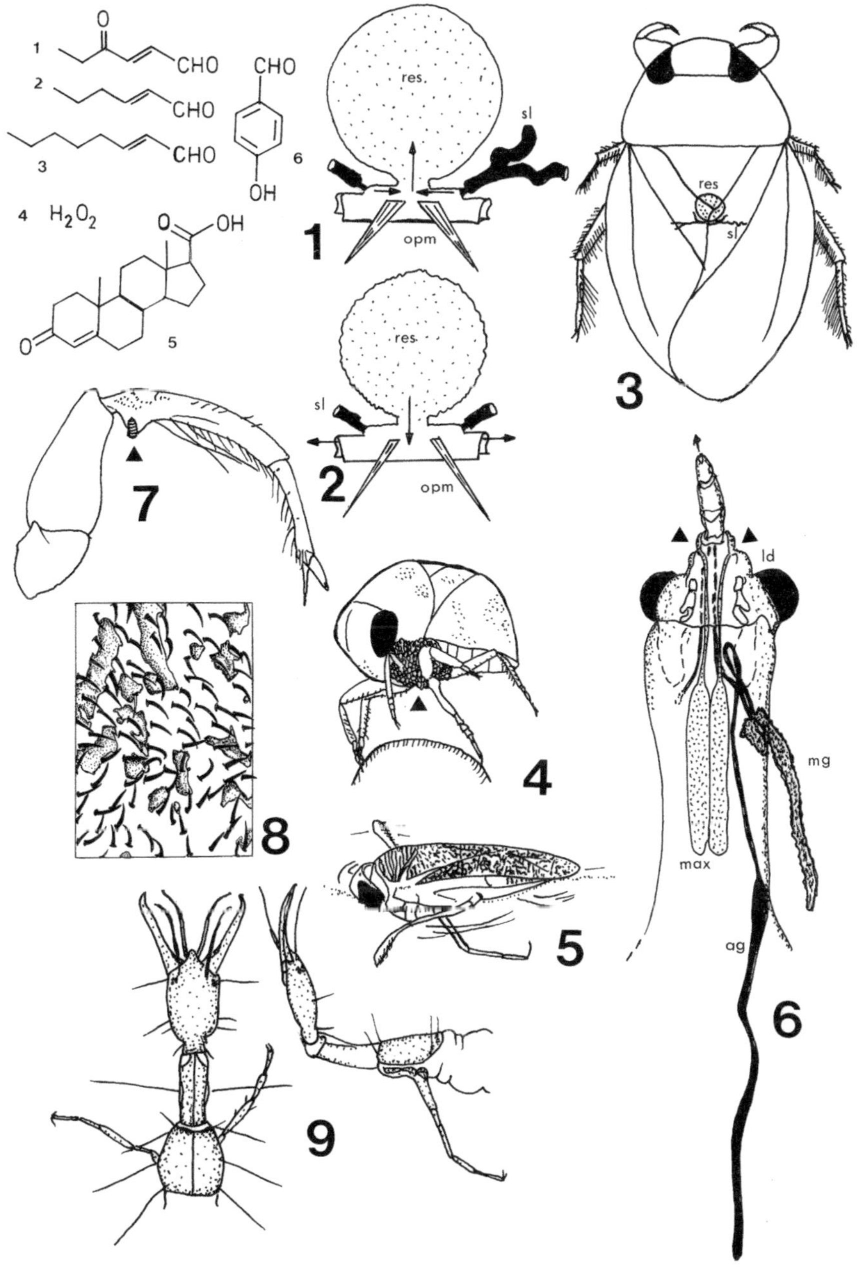


**Fig. 9.3** Defenses of Heteroptera (1–8), Neuroptera (9). 1 and 2: Probable operating of opening and closing mechanism of defensive glands in *Ilyocoris cimicoides*. 1. Opening muscles (opm) are relaxed and secretion flows from the secretory lobes (sl) into the reservoir (res). 2. Opening muscles are contracted and secretion flows from the reservoir into the vestibule and through efferent channels to the ostioles (arrows). 3. Dorsal view of adult *I. cimicoides* showing the position of metathoracic gland reservoir (res) and secretory lobes (sl). 4. Secretion grooming of *Plea minutissima*;

**Saldidae** (shore bugs; 382 sp.) and **Aepophilidae** (marine bugs; 1 sp.): When disturbed saldids run fast, leap into the air, and land some distance away after which they scurry into shelter. The metathoracic scent gland has one median reservoir and one ostiole located on the sternum of the metathorax. Pregenital eversible glands are present at the posterolateral corners of the seventh sternum (not in Aepophilidae). Schmidt (1982) reports that Saldidae probably produce venomous saliva.

**Omaniidae** (intertidal dwarf bugs; five sp.): Omaniidae are predacious and intertidal bugs. They are able to jump readily. The metathoracic scent gland has four reservoirs of a paired gland system with a single median ostiole. Chemical data are not available.

**Leptopodidae** (spiny-legged bugs; 32 sp.): Leptopodidae fly swiftly from rock to rock when disturbed. The metathoracic scent gland has two reservoirs and two ostioles. Other exocrine glands (including pregenital glands) are missing.

**Nepomorpha (aquatic bugs)**: According to Slater (1982) the group has 2404 sp. worlwide. Both aquatic and terrestrial nepomorphs inject venom to immobilize and liquefy their prey. Apart from Nepidae, Aphelocheiridae, and few Belostomatidae, Gelastocoridae, and Notonectidae within all remaining species there exists a metasternal gland (see Staddon and Thorne 1979). Absence of this type of gland is interpreted as a secondary condition. Several representatives of Nepomorpha are able to spread secretion from the metathoracic gland actively over their bodies (Andersen and Weir 2004). Therefore these species crawl out of the water (Naucoridae, Notonectidae, Pleidae) or lie on the water surface (Corixidae) in order to show secretion grooming. Usually these antimicrobic secretions serve to clean the hair pile which holds the air store (Andersen and Weir 2004). Therefore secretion grooming is primarily found in species with a physical gill. In addition also cephalic glands are present. The classification of Nepomorpha was based on Moreira et al. (2018).

**Nepoidea**: According to Moreira et al. (2018) Nepoidea (Belostomatidae + Nepidae) represent a basal sister group to other Nepomorpha. The Nepoidea species are ambush predators, which are hidden in submerged vegetation to wait on prey. They possess no abdominal scent glands.

**Nepidae** (water scorpions) 268 sp.: *Nepa rubra* and also the Australian genus *Laccotrephes* are well camouflaged due to their brown color and their mud body cover (Andersen and Weir 2004). Its dorsoventral compression is similar to dead leaf. On molestations *Nepa* and other genera show thanatosis; in some cases with its raptatorial forelegs water scorpions cling 1 h motionless to vegetation. Bites of both *Nepa* and *Ranatra* are painful for humans (Pawlowsky 1927). The poisonous secretion is produced by paired salivary or labial glands, which consist of a main gland and an accessory gland (Fig. 9.3/6). The accessory gland duct transports secretion from the accessory gland into the bipartid main gland. From there the lateral duct leads to the salivary pump and proboscis (Pawlowsky 1927). By transcriptomic studies the salivary glands of *Ranatra chinensis* revealed many transcripts encoding proteins with homology to proteases, acid phosphatases, apyrases, dipeptidylpeptidases IV, hyaluronidases, and prophenoloxidases. No metathoracic scent glands are present within Nepidae; however it is of interest that Nepidae preen only the eyes and the respiratory siphon (Andersen and Weir 2004). This indicates that they might use other secretions as from dermal or salivary glands or anal fluids (see Weiss 2006). Larvae lack dorsal abdominal scent glands. The secretion of the cephalic or maxillary glands (Fig. 9.3/6) is insecticidal and produces a burning sensation on human skin (Walker et al. 2016). Stridulations and vibrations were reported for the forelegs of *Ranatra* with a femoral plectrum and a coxal pars stridens (Aiken 1985).

**Belostomatidae** (giant water bugs) 169 sp.: They are general predators and feed on aquatic invertebrates and vertebrates (amphibians, fishes, turtles, birds), which are captured with their raptorial forelegs and quickly immobilized with the poisonous secretion from the salivary glands (de Carlo et al. 1973). The bites which are very painful for humans also represent effective defense mechanisms (Haddad et al. 2010). The belostomatid saliva reveals proteins from 5 to 55 kDa (Walker et al. 2016). Enzymic assays revealed activities of phospholipase A2, hyaluronidase, protease, amylase, esterase, a-glucosidase, glucosaminidase, invertase, lipase, nuclease, phosphatase, and phosphohydrolase (Walker et al. 2016). The saliva of the larger *Lethocerus uhleri* (Lethocerinae) which only feed on small vertebrates contains three proteolytic enzymes and no amylase, while the smaller *Belostoma lutarium* (Belostomatidae), which feeds on small invertebrates and snails, produces two proteolytic enzymes and amylase (Swart et al. 2006). The authors suggest that the presence of amylase allows to exploit the plant material already ingested by their prey. There are also reports of rich lipidic contents from the venom of *Belostoma anurum* (Walker et al. 2016). Walker et al. (2018) presented a transcriptomic and proteomic study on *Lethocerus distinctifemur* and found 132 venom proteins including putative enzymes, cytolytic toxins, and antimicrobial peptides. 73% of proteins were homologous to assassin bugs (Reduviidae), and 21% are not known from other sources (Walker et al. 2018). Metasternal scent glands of Lethocerinae and Horvathiniinae contain a median undivided lip valve, dorsoventral valve-opener muscles, and accessory glands (glands absent in Belostomatinae; Aldrich 1988). Especially in males these glands are well developed and may be detected due to their odor. (*E*)-2-hexenyl acetate was identified as main constituent in *Lethocerus indicus* males, with the corresponding butyrate as minor compound (Aldrich 1988). The functions of these esters remain unknown and might fulfill a secondary, probably sexual role (Scrimshaw and Kerfoot (1987). In Asia where especially *Lethocerus* is eaten by humans the scent gland odor and taste can be perceived (Schuh and Slater 1995). Belostomatidae larvae lack dorsal abdominal scent glands (Aldrich 1988). As defense reaction most Belostomatidae are additionally capable of ejecting a foul-smelling inky black anal liquid. Finally many families have paired so-called cephalic or maxillary glands (Fig. 9.3/6) which are saclike and open at the base of the beak (de Carlo 1961). The gland reservoirs are lined by a cuticle and bear ventrally glandular cells. In *Abedus herberti* the reservoirs are depleted after a molestation (Eisner et al. 2005). The viscous white fluid contains the four pregnanes desoxycorticosterone (Fig. 9.3/**5**), pregnenolone, progesterone, and 3a-hydroxy-pregn-5-en-20-one (Lokensgard et al. 1993). The first three pregnanes were also identified from prothoracic defensive glands of dytiscid beetles. It was evaluated that one *A. herberti* specimen contained about 0.1 mg desoxycorticosterone (in dytiscid beetles 0.03–0.4 mg/beetle). In the genus *Abedus* a soft chirping was described, which is produced by expulsion of air through the ventral spiracles (Aiken 1985).

**Corixoidea**: **Corixidae** (water boatmen; 662 sp.): Corixids are unable to use their rostrum and salivary secretion for defense. The opening of the large metathoracic scent glands is located laterally of the mesocoxae and contains a median, an undivided flap valve, and a ventral valve opener muscles. Obviously the pale yellow oil-like secretion with an aldehyde odor is depleted due to the elasticity of the gland reservoir. Larvae have dorsal abdominal scent gland openings between terga III–IV, IV–V, and V–VI. Corixids have defensive secretions composed of (*E*)-4-oxo-hex-2-enal (95% of adult secretion; Fig. 9.3/**1**) and of (*E*)-4-oxo-oct-2-enal (92% in larval secretions; Staddon et al. 1979). When attacked both adult corixids and larvae may eject secretions under water. It was found that fish guts only contain low numbers of corixids (Scrimshaw and Kerfoot 1987). The majority of corixids (Diaprepocorinae, Cymatiainae, Corixinae) rub their metathoracic gland secretions over their body parts covered with hydrofuge hairs. Within 2–3 s secretion grooming was observed while floating on the water surface (Fig. 9.3/5; Kovac and Maschwitz 1991). Secretion grooming of *Diaprepocoris* takes place on land. The behavior is released when light intensity and water temperature are raised. Due to the antibacterial properties of the aldehyde secretion, the contamination of hydrofuge hairs by microorganisms is minimized and thus loss of air store is reduced. It is peculiar that in all instars of *Micronecta* and earliest corixid instars secretion grooming is lacking: They have no hydrofuge hairs and obtain oxygen by cutaneous diffusion. Probably their strongly smelling secretions are directed against aquatic or terrestrial predators (Kovac and Maschwitz 1991).

Stridulation in Corixidae was found to be a component of the reproductive behavior and was observed in Micronectinae and Corixinae and probably occurs in Diaprepocorinae and Cymatiinae (Aiken 1985). The position and function of stridulatory organs vary considerably within Corixidae (Aiken 1985). In few cases (e.g., *Cenocorixa*) stridulation was shown to maintain individual distances between males. Corixids are able to leap out of the water into the air and take flight. Maxillary glands are present (Puchkova 1965).

**Ochteroidea**: **Ochteridae** + **Gelastocoridae**: Schmidt (1982) reports that Ochteridae and Gelastocoridae produce saliva which is probably venomous (Ochteridae) or paralytic and painful (Gelastocoridae).

**Gelastocoridae** (toad bugs; 116 sp.): Many representatives of toad bugs are cryptic and take on the color of the substrate; moreover they are characterized by their jumping ability. They usually occur near water in wet mud and vegetable debris where they usually blend well with their background. The metathoracic scent glands of Gelastocorinae contain a pair of stop valves and a dorsoventral valve with a polygonal cuticular microsculpture (scent glands are absent in Nerthrinae). The glands contain 4-oxo-2-hexenal (63%), octenal (25%; Fig. 9.3/**3**), 4-oxo-octenal (9%), and hexanal (3%; see Scrimshaw and Kerfoot 1987). Larvae lack dorsal abdominal scent glands. Maxillary glands are present (Puchkova 1965).

**Ochteridae** (velvety shore bugs; 75 sp.): Some Ochteridae species have larvae that scoop sand over the head and then push it over the dorsal surface with their front legs, thus making them closely resemble sand grains. Metathoracic scent glands are present. Adults and larvae lack dorsal abdominal scent glands. Maxillary glands are present (Puchkova 1965).

**Naucoroidea**: **Naucoridae** (creeping water bugs; 398 sp.): Naucorid bugs such as *Pelocoris* and *Ilyocoris (= Naucoris) cimicoides* should be handled with care as they can inflict a painful bite (Pawlowsky 1927) due to the venom of their salivary glands (Schmidt 1982). Their bite is characterized by Wesenberg-Lund (1943) as "worse than a bee sting." The metathoracic scent gland system *of Ilyocoris cimicoides* consists of a pair of glands, a reservoir, a closing apparatus, and an efferent apparatus (Fig. 9.3/1–3; Staddon and Thorne 1973). The complex gland system possesses a medium undivided lip valve and dorsoventral valve opener muscles (Fig. 9.3/1, 2); accessory glands are missing. In *Ilyocoris cimicoides* the aromatic compounds 4-hydroxy-benzaldehyde (Fig. 9.3/**6**) and 4-hydroxy benzoic acid methylester were identified (see Scrimshaw and Kerfoot 1987) along with minor constituents. From time to time *Ilyocoris cimicoides* leave the water in order to distribute metathoracic gland secretion over the ventral hairs (Kovac and Maschwitz 1990a). This grooming behavior prevents the contamination of these hairs with microorganisms, which would result in wetting of this region and loss of the air bubble. There is one pair of larval scent glands (gland divided) present between abdominal terga III and IV. Maxillary glands are present (Puchkova 1965). A sound production was reported from *Ilyocoris cimicoides* males; however both stridulatory structures and significance of sound production are discussed controversially (Aiken 1985). Horta et al. (2010) described a possible case of mimicry involving the naucorid Limnocoris porphyrus and an anuran tadpole of the hylid frog Scinax machadoi. Because both can sting or might be toxic it is not clear whether it is Mullerian mimicry or Bates mimicry.

## 9.3 Aphelocheiroidea: Potamocoridae + Aphelocheiridae

**Potamocoridae** (eight sp.): Tropical species which are related to Aphelocheiridae (Lytle 2015).

**Aphelocheiridae** (benthic water bugs; 78 sp.): There are reports that *Aphelocheirus* may occasionally sting or bite, which can be painful to man (Andersen and Weir 2004). No metathoracic scent glands are present in this family (Staddon and Thorne 1979) and it is of interest that *Aphelocheirus* does not show grooming behavior (Andersen and Weir 2004). Larvae possess divided dorsal abdominal scent gland openings between terga III and IV. Maxillary glands are present (Puchkova 1965).

**Notonectoidea**: **Notonectidae** (backswimmers; 422 sp.): Backswimmers swim on their backs, and often have reverse countershading. When disturbed they swim rapidly down. They are also able to leap into the air from the water and take flight.

When handled they can inject painful venomous saliva (German name "Wasserbiene," water bee; Pawlowsky 1927; Schmidt 1982). In Notonectinae the metasternal scent glands have paired stop valves, dorsoventral valve opener muscles, and a polygonal cuticular microsculpture. The glands are absent in representatives of Anisopinae, e.g., *Anisops* and *Buenoa*. Interestingly *Anisops* does not show any grooming behavior (Andersen and Weir 2004). Apart from Anisopinae in other notonectids paired accessory glands are associated with the reservoir which has a midventral opening. The odorless, brownish secretion consists of 4-hydroxybenzaldehyde (Fig. 9.3/**6**) and methyl-4-hydroxy benzoate, compounds which were also identified from pygidial glands of water beetles. It is remarkable that these antimicrobics also are deterrents (at $6 \times 10^{-7}$ moles) against cichlids as *Tilapia* (Scrimshaw and Kerfoot 1987). Larvae lack dorsal abdominal scent gland openings. Maxillary glands are present (Puchkova 1965). It is remarkable that only members of genera *Buenoa*, *Anisops,* and *Walambianisops* can stridulate, which definitely lack thoracic defensive glands. In contrast other genera of Notonectinae possess no stridulatory apparatus (Aiken 1985; Andersen and Weir 2004). The stridulatory repertoire of *Buenoa* is very complex (tibial comb/rostral prong; femur/coxal peg; femoral setae/rostral flange) but obviously and exclusively used in courtship and mating. The same type of chirping behavior as in *Buenoa* is also typical for males of *Anisops* stridulating with their forelegs (Fig. 9.3/7). It seems important to denote that *Notonecta maculata* as predators release hydrocarbons such as n-heneicosane and n-tricosane which repel oviposition by mosquitoes such as *Culiseta longiareolata* (Silberbush et al. 2010). It would be interesting to know if these hydrocarbons, which might be valuable in eliminating mosquitoes, are produced in complex glands.

## 9.4 Pleoidea: Helotrephidae + Pleidae

**Pleidae** (pygmy backswimmers; 44 sp.): Like Notonectidae, representatives of Pleidae swim in the inverted position. The predacious pygmy backswimmers have a saclike metathoracic scent gland with a single midventral opening (Aldrich 1988). Paired accessory glands are associated with the scent gland. The colorless secretion of *Plea leachi* was reported to contain a 10–15% solution of hydrogen peroxide (Fig. 9.3/**4**) together with traces of an unknown carbonyl compound (Maschwitz 1971). From time to time *Plea minutissima* leaves the water in order to distribute this secretion by its legs (secretion grooming) over its hydrofuge ventral pubescence (Fig. 9.3/4; Kovac 1993). This efficient disinfectant probably destroys epzoic microorganisms and guarantees respiration via an intact air sheath when submerged. Increase of light intensity and higher water temperatures promote secretion grooming and probably flight behavior (Kovac 1993). If adults of *Plea* are dissected very often foams are apparent, which is obviously due to a catalytical decomposition of hydrogen peroxide into water and oxygen. Scrimshaw and Kerfoot (1987) also report that the *Plea* secretion is an effective deterrent against small fishes as *Gambusia*. Pleidae larvae have dorsal abdominal scent gland openings between terga III and IV. Stridulation is known to occur in both sexes of *Plea minutissima* by contacts of a mesothoracic file with a prothoracic projection (Aiken 1985). Maxillary glands are present (Puchkova 1965). According to Schmidt (1982) the saliva of Pleidae is probably venomous.

**Helotrephidae** (164 sp.): As Notonectidae they swim with the venter up. Larvae of the tropical Helotrephidae have a single median unpaired dorsal scent gland opening located between terga II and IV. In Helotrephinae stridulatory structures were identified. Especially a serrated lateral margin of hemielytron contacts a ridge dorsally on hind femur (Schuh and Slater 1995). Finally it must be noted that the first aquatic plant-sucking insects are two **homopteran** larvae of spittlebugs of genus *Mahanarva* from Costa Rica. They live submerged in water filled flowers of various *Heliconia* species (Thompson 1997).

**MEGALOPTERA (alderflies, dobsonflies, fish flies; 100% aquatic)**: Worldwide there exist around 380 megalopteran species (Stork 2018: 354; Dijkstra et al. 2014; Ardila-Camacho and Contreras-Ramos 2018) which contain the families Sialidae (alderflies) and Corydalidae (dobsonflies and fish flies). All representatives have long-living aquatic larvae (up to 5 years; 10–12 stages) and short-lived terrestrial pupae and adults (Cover and Bogan 2015). The larvae feed on small invertebrates and vertebrates (e.g., mayfly and caddisfly larvae; small tadpoles) and also cannibalize smaller conspecifics. Few publications dealing with larval bionomy show that there exist defensive mechanisms. In the genus *Neohermes* (Corydalidae) an avoidance reaction was observed, when conspecific larvae had contacts (Smith 1970). When disturbed last-stage *Neohermes californicus* larvae (and prepupae) vomit an extremely foul-smelling, dark, greenish liquid (Smith 1970). Also adults of *Corydalus* can bite when they are disturbed (Parfin 1952). Recently defensive reactions were described for *Neoneuromus ignobilis* (Corydalidae; Cao et al. 2012). When disturbed due to the high flexibility of their bodies the larvae curl up into a ball or immediately open the palates. Cao et al. (2012) also observed the depletion of milky and water-insoluble substances from the end of the larval abdomen, a behavior which was compared with depletion of defensive secretions by whirligig beetles (Gyrinidae) or secretion grooming in aquatic *Plea* bugs. *Corydalus* females protect their egg masses with a hardening substance which appears like bird feces (Ardila-Camacho and Contreras-Ramos 2018).

**NEUROPTERA (net-winged insects; 4% aquatic)**: There exist more than 6000 neuropteran species worldwide (Stork 2018: 5868; probably 10,000). Among the 17 known familes only 2 are aquatic (Sisyridae: 60 sp.; Nevrorthidae: 19 sp.) and 1 semiaquatic (Osmylidae: 160 sp.). Neuropteran larvae are characterized by specialized sucking jaws which are formed by mandibles and distal parts of maxillae (Beutel et al. 2014). This food uptake apparatus exhibits a sucking channel and a poison channel; the paired poison and salivary glands are associated with this specialized food apparatus (Beutel et al. 2010; Beutel et al. 2014). Aquatic larvae of Sisyridae (spongillaflies) have piercing mouthparts and are specialized predators or commensales of freshwater sponges (with symbiontic Zoochlorellae) and limnic bryozoans (Heckman 2018). Their stylets curve outwards distally. For constructing their pupal cocoons by using their spinning glands, full-grown third-stage larvae (lengths 2.7–8.5 mm) leave the water (Cover and Bogan 2015). Because other larval Neuroptera such as lacewings (Chrysopidae) use their pointed jaws in piercing and sucking prey and even are capable of piercing human skins where they may cause injuries (Southcott 1991), the same situation is suggested for Sisyridae. Larvae of Nevrothidae and even their pupal stages are aquatic and larvae represent predators of benthic invertebrates. Nevrorthidae represent the sister group of all other Neuroptera, suggesting an aquatic ancestor for all Neuroptera (Cover and Bogan 2015). The slender and agile predatory larvae are usually found in mountain brooks. As described by Zwick (1967) larvae of *Nevrothus fallax* showed an unusual behavior after molestation. The larvae rapidly moved foreward and backward and then suddenly lifted the head by an extremely quick tossing movement in order to form a right angle between head with prominent sucking jaws and rest of the body (Fig. 9.3/9). This peculiar behavior is interpreted as defensive behavior. Several representatives of the Osmylidae larvae are associated with freshwater habitats and occur in riparian areas. Many species consume trichopteran eggs, and hunt on Tipulidae and Chironomidae larvae (Cover and Bogan 2015). Because chironomid prey larvae die immediately after encountering osmylid larva, it can be suggested that the larval saliva is toxic (Wesenberg-Lund 1943; Schmidt 1982). Adults of Osmylidae are characterized by defensive glands and disagreeable odors are perceivable after molestation (Dettner 2015).

**COLEOPTERA (beetles; 3% aquatic)**: Most data are from Short (2018), Jäch and Balke (2008), Arnett and Thomas (2001), Beutel and Leschen (2016), and Leschen et al. (2010). According to Jäch and Balke (2008) about 3% of the known Coleoptera species described are regarded "aquatic." In considering undescribed species it can be assumed that 18,000 water beetle species might exist among 386,500 species worldwide (Stork 2018). The following beetle families are not arranged phylogenetically but according to Jäch and Balke (2008).

## 9.5 Predominantly Aquatic Families (Jäch and Balke 2008)

**Myxophaga**: Defensive mechanisms of the four families are only known from Lepiceridae (Navarrete-Heredia 2005). **Hydroscaphidae** (skiff beetles and their larvae are hygropetric; 13 sp.). Skiff beetles live on algae covered by a thin water film.

**Lepiceridae** (two sp.): Adults are found at the edge of rivers and prefer moist environments; larvae are probably aquatic. Recently collecterd Lepicerus adults were capable to retract body appendages and show thanatosis. In addition they cover their body surface with subtrate particles, are strongly sclerotized and colored similar to substrate where they are living (Navarrete-Heredia et al. 2005). **Sphaeriusidae** (minute bog beetles; 19 sp.): Adults of many species are found along banks of streams, in stagnant waters, or in *Sphagnum* mosses. Ecological data are scarce. **Torridincolidae** (torrent beetles; Short 2018: 60 sp.): All torrent beetles have aquatic developmental stages and prefer hygropetric habitats. Even pupae are aquatic with plastron-bearing gills (Lawrence 1991).

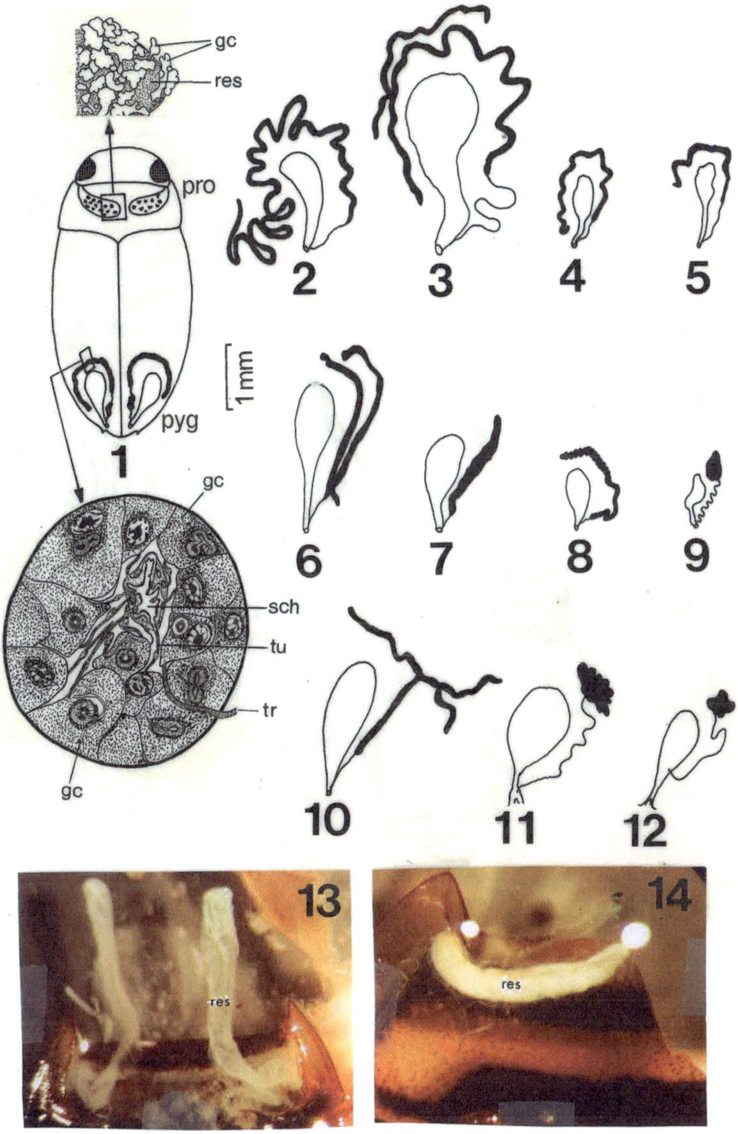


**Fig. 9.4** Site of defensive glands (prothoracic glands and pygidial glands) of Hydradephaga (1–10; according to Dettner 1985) and Geadephaga (11–12) in *Nebrioporus depressus* (1). 2. *Acilius sulcatus,* 3. *Colymbetes fuscus*, 4. *Liopterus haemorrhoidalis*, 5. *Laccophilus minutus*, 6. *Hygrobia hermanni*, 7. *Gyrinus substriatus*, 8. *Noterus clavicornis*, 9. *Haliplus heydeni*, 10. *Amphizoa lecontei*, 11. *Trachypachus gibbsi*, 12. *Omophron dentatum*. 13–14: Prothoracic gland reservoirs of *Ilybius fenestratus* (13) and *Acilius canaliculatus* (14). Abbreviation: *Pro* prothoracic glands; *Pyg* pygidial glands; *res* reservoir; *gc* glandular cells; *sch* secretory channel; *tu* tubule; *tr* trachea. 1–10: Modified according to Dettner 1985. 11–12: Modified according to Forsyth (1972). 13–14: Originals in Dettner 2014

**Adephaga**: All aquatic Adephaga (= Hydradephaga) possess paired abdominal pygidial glands (Fig. 9.4). Secretions of all Hydradephaga hitherto tested topically on the cuticles of each representative show an increase in wettability for water of treated cuticles (Fig. 9.6).

**Amphizoidae** (trout stream beetles): Larvae and adults (five species worldwide; Dettner 2016b) prefer fast-flowing rivers but are poorly adapted to aquatic environment. They share plesiomorphic features with Carabidae. Eggs are deposited submers, and mature larvae leave the water for pupation. Freshly emerged adults are often covered with mud. Larvae live gregariously, show thanatosis, and roll their bodies into balls when disturbed (Yu et al. 1993). Adults possess pygidial glands with a tripartite secretory lobe (Fig. 9.4/10; prothoracic glands absent) which are filled with a yellow fluid, which is depleted over the hind body (as in Carabidae) upon disturbance (Edwards 1953). The gland reservoir is covered by muscles; as in *Hygrobia* (Fig. 9.4/6) the gland tissue is lengthened (Fig. 9.4/10; Forsyth 1970). The secretion contains the strongly smelling dimethyl disulfide (Fig. 9.9/Formula 3), and antimicrobics and fungicides like methyl-3-indole carboxylate (Fig. 9.9/ Formula 1) and methyl-4-hydroxycarboxylate (Fig. 9.5/Formula 6; Dettner 1990).

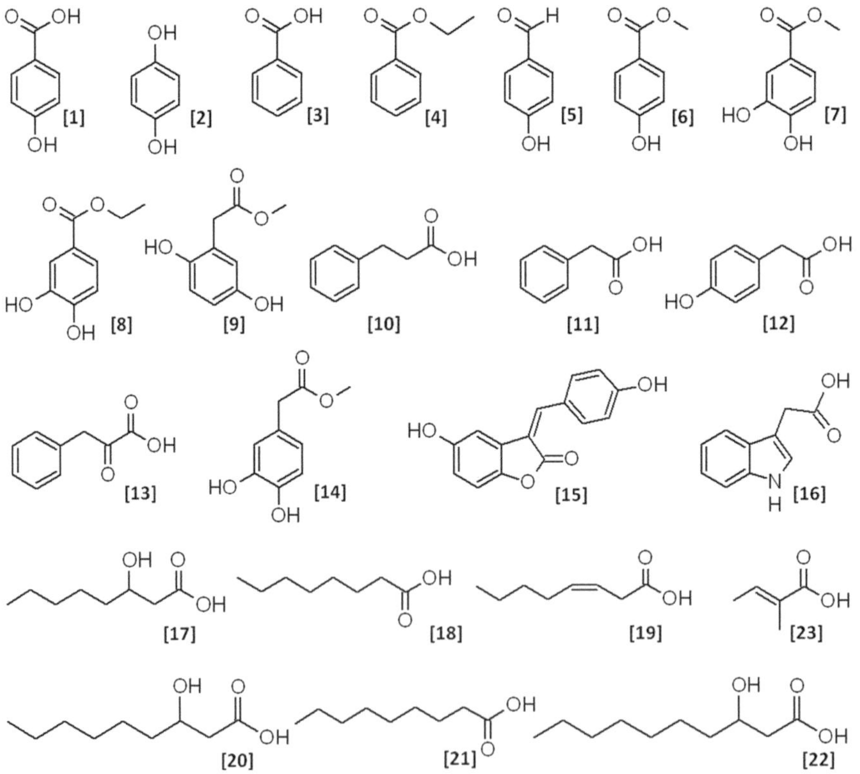


**Fig. 9.5** Pygidial gland constituents (**1–23**) of predacious diving beetles (Dytiscidae)(according to Dettner 2014). For abbreviations see Table 9.1

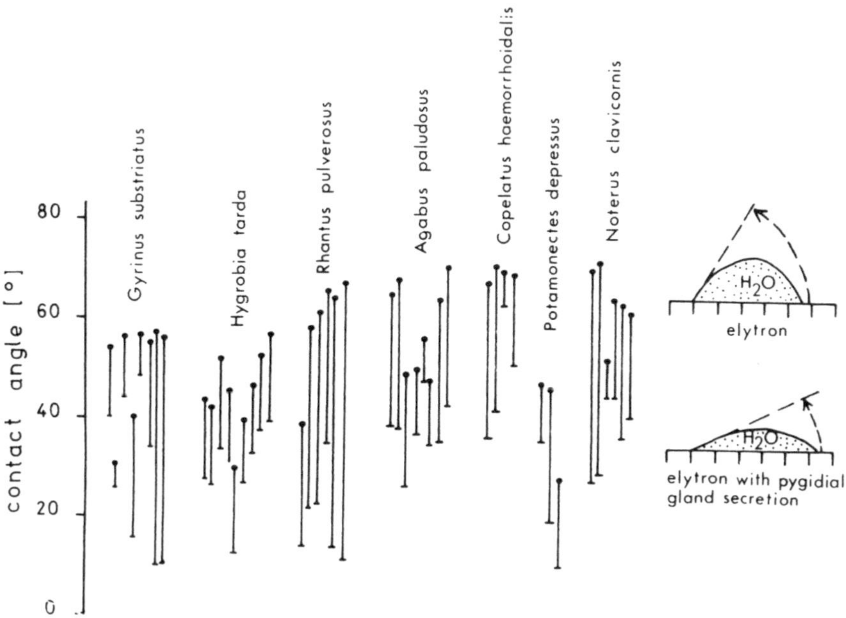


**Fig. 9.6** Contact angles of water droplets, placed on elytral surfaces of different Hydradephaga species. Vertical bars: measurements per specimen: above value = untreated control elytra; below = elytra treated with pygidial gland secretion (according to Dettner 1985)

The yellow coloration of the secretion, which according to Darlington (1929) leaves a cigarette-like stain on the fingers, is due to 4′,5-dihydroxybenzalisocumaranone (= marginalin; Fig. 9.5/Formula 15) which was first discovered from pygidial glands of *Dytiscus marginalis* (Dettner 2014).

**Aspidytidae** (cliff water beetles; two sp.): Larvae and adults of this monophyletic family (Vasilikopoulos et al. 2019) prefer hygropetric habitats and are probably predacious (Beutel et al. 2016). When disturbed adults walk rapidly. Pygidial glands must be present; until now however they are not recorded from this family , which is also true for the prothoracic glands.

**Dytiscidae** (predacious diving beetles; according to Yee 2014: 4303 sp.): Jäch and Balke (2008) estimated up to 4800 sp. of diving beetles and predacious diving beetles, which means that this water beetle family which occurs on all continents where it is found in all aquatic habitats has the first position with respect to global water beetle diversity (Balke and Hendrich 2016; Yee 2014; Miller and Bergsten 2016). Larvae and adults of nearly all species are carnivorous and aquatic. Habitats range from ponds and bogs to rivers, phytotelmata, hygropetric areas, and even subterranean waters (Yee 2014; Miller and Bergsten 2016). A lot of predacious diving beetles leave the water for secretion grooming or before flight, another fraction of species stay on land during winter or summer, and few species of the genera *Geodessus* and *Typhlodessus* are reported to be terrestrial. Thus they have reduced swimming hairs and are unable to swim (Brancucci 1985).

Dytiscidae have a lot of predators ranging from vertebrates such as fishes or amphibians to invertebrates as Odonata, Heteroptera, or other predacious water beetles. Therefore we find a lot of defensive mechanisms (Dettner 2014; Miller and Bergsten 2016). Various mechanical defenses can be observed. Larger species such as Cybistrinae or Dytiscinae can kick with their hind legs which are supplied with large spurs. In addition both larvae and adults of these larger dytiscids bite with their mandibles which is painful for humans and may result in bleeding. In addition various Laccophilinae show a characteristic jumping behavior. In contrast many dytiscid beetles may quickly escape or in contrast can hide, whereas certain larvae of Cybistrinae exhibit thanatosis when molested (Miller and Bergsten 2016). As in other water insects coloration in Dytiscidae can be variable ranging from crypsis to aposematism and resulting from structural, secretion, and pigmentary colors (Dettner 2014). Colorful taxa which are attractively marked often are found in clear water with mineral substrates (Larson 1996; Young 1960) whereas dark melanistically or green-colored species (*Laccophilus* with yellow carotenoids and blue bile pigments; Dettner 2014) often prefer habitats with dark substrates or dense vegetation (Balke et al. 1997).

Several species of Dytiscidae can produce sounds and have organs of possible stridulatory function (Larson and Pritchard 1974; Franciscolo 1979). Although detailed bioacoustical studies are missing, the significance of most stridulations seems to be courtship behavior or intraspecific communication. It is suggested that stridulations also represent alarm and stress signals and might represent defense signals (Smith 1973).

Chemical defense through allomones is widespread in Dytiscidae (Dettner 2014). Both adults and larvae possess huge rectal ampullae that means diverticuli of hindgut. These ampullae primarily represent hydrostatic organs which are normally filled with feces and water. If adults of larger Dytiscidae (e.g., Dytiscinae) are handled at first they deplete an unpleasant smelling fluid from their rectal ampulla containing also hydrogen sulfide and ammonia. During severe molestations the two complex gland secretions from pygidial and prothoracic glands are suddenly or gradually depleted. Pygidial glands are present in all Dytiscidae and contain an ovoid reservoir, covered by muscles (Fig. 9.4). The lengthened, sometimes branched, secretory lobes (Fig. 9.4/1–5) are connected to the reservoir by a shorter or longer central collecting canal (Forsyth 1968). There exist various types of glandular cells with tubules. In several hydroporine species (*Hyphydrus, Stictotarsus*) an accessory gland and an integumental gland were described (see Dettner 2014).

Pygidial gland secretion very often is like a paste and dytiscid beetles distribute the material (about 10–15% of one gland reservoir) on body surface by using their legs. This behavior was described by Dettner (1985) and Kovac and Maschwitz (1990b). The secretion represents one of the richest sources for aromatic compounds within arthropods (Figs. 9.5 and 9.11). 14 aromatic compounds were identified (Table 9.1), 7 aliphatic gland constituents, the unusual yellow pigment marginalin, the tryptophane metabolite 3-indoleacetic acid, and tiglic acid which is probably also derived from amino acids (Schildknecht et al. 1962; Dettner 2014). Now chemical data are known from 91 dytiscid species comprising 29 genera. Several aromatics can be used as chemotaxonomic markers (phenylacetic acid for Hydroporinae and Copelatinae and responsible for the strong sweetish odor of *Hydrodessus fragrans* or *Hydrotarsus*; 4-hydroxybenzaldehyde for Colymbetinae/Dytiscinae) (Dettner 2014). It is remarkable that 4-hydroxybenzaldehyde and 4-hydoxybenzoic acid methylester from pygidial glands of Dytiscidae are also present in metathoracic glands of water bugs *Notonecta glauca* and *Ilyocoris cimicoides* (see Naucoridae, Notonectidae).

Concerning the functions of pygidial glands (Fig. 9.5) both aromatics (e.g., **3**, **6**, **11**) and aliphatics (e.g.,**17**, **19**, **20**, **22**) represent excellent preservatives which maintain a clean body surface that means protection from contamination of body surface by bacteria, protozoa, or fungi. Even in the terrestrial water beetle *Geodessus besucheti* phenylacetic acid represents the main constituent of the pygidial glands as compared with its aquatic bidessine relatives (Dettner 2014; Brancucci 1985). In addition it was suggested that each of the following pygidial gland constituents, the yellow pigment marginalin (Fig. 9.5/Formula 15) acts as fixative), the glycoproteins (form coherent films), and 3-hydroxy acids (**17**, **20**, **22**; polyesters) fix and entangle epizoic microorganisms which are subsequently stripped off (Schildknecht and Bühner 1968; Barbier 1990). The significance of the plant auxin 3-indoleacetic acid (Fig. 9.5/**16**) in Hydroporinae is unknown. The pygidial gland secretion of *Acilius sulcatus* especially with compounds **5** and **6** was shown to act against ciliates. This seems important because many aquatic arthropods and especially dytiscid beetles are contaminated with many epitrichic ciliates. The effect of pygidial gland secretion on the wettability of dytiscid integument by water is highly significant. As compared with the untreated elytral control, contact angles of standard water droplets on elytra previously treated with pygidial gland secretion of various Dytiscidae species were significantly lowered (Fig. 9.6). This increase of wettability was also observed with solutions of single pygidial gland constituents (Dettner 2014). When water beetles leave the water their integument becomes hydrophobic and they have enormous difficulties to break through the water surface and reenter into the water (Brancucci 1975). In addition especially after flights the weight of the beetles is reduced due to the filled tracheal system and the empty rectal ampulla. Therefore especially small Dytiscidae such as Hydroporinae are overcompensated and are pressed out of the water (Fig. 9.7/Hydroporinae 1). In this situation the beetles start rubbing and show secretion grooming in distributing pygidial gland secretion over their bodies (Fig. 9.7/Hydroporinae 9–11). After few minutes on or below the water surface the beetles fill their rectal ampulla with water and become more and more hydrophilous and as a consequence receive a stable position within the water (Fig. 9.7/Hydroporinae 12, 13; Dettner 1985). These findings amply demonstrate that exocrine glands and their secretions have usually a multifunctional significance.

Seasonal fluctuations of pygidial gland titers in *Acilius* (Newhart and Mumma 1979) and *Agabus* (Classen and Dettner 1983) do not reflect different degrees of utilization of the secretion but are due to the different age structures of population titers. Young specimens namely after pupation have low amounts of aromatics and other compositions as compared with elder specimens (Dettner 1985).

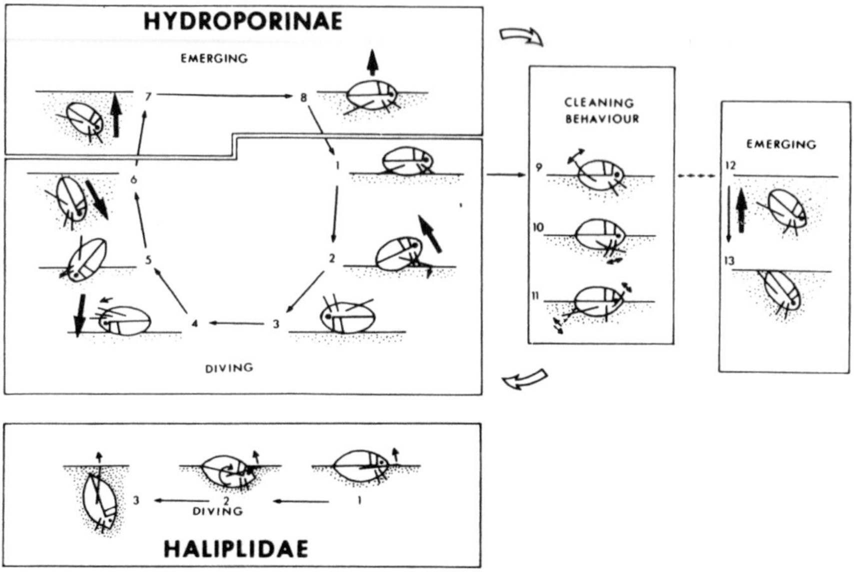


**Fig. 9.7** Diving behavior (left), subsequent cleaning behavior (center), and sloping position (right side) of Hydroporinae (Dytiscidae). Diving behavior of Haliplidae (below). According to Dettner (1985)

Prothoracic exocrine glands are present in all dytiscid beetles (Fig. 9.4/1, 13, 14). The homology of the prothoracic gland of Hygrobiidae (Fig. 9.9/10) with the gland of Dytiscidae is questionable. Prothoracic exocrine glands represent paired saclike reservoirs which open anterolaterally of prothorax. As exocrine glands they are not homologues with endocrine prothoracic glands which are present in all insects. Forsyth (1968) describes that tubule-carrying gland cells cover these paired reservoirs which have no muscles. The usually odorous prothoracic secretion often is fluid and can be exuded as a milk after a strong molestation. Several Dytiscidae such as *Ilybius fenestratus* (Fig. 9.4/13), *Hyphydrus ovatus*, or *Laccophilus minutus* possess large prothoracic gland reservoirs extending from prothorax into the abdomen (Dettner 2014).

The chemistry of the prothoracic glands was especially investigated by Schildknecht (1970). An impressive array of vertebrate steroidal hormones was identified and a lot of new beetle steroids with $C_{18}$, $C_{19}$, and $C_{21}$ skeletons (Fig. 9.8). Together with Chrysomelidae, Silphidae, Lampyridae, and Belostomatidae, Dytiscidae represent those insects, with an extraordinary biosynthetical capacity to produce these essential compounds from cholesterol precursors from their food. It was shown that labeled steroid precursors are incorporated into defensive steroids of *Acilius sulcatus* (Chapman et al. 1977). Obviously microorganisms from the crop of these beetles play an important role in transforming these steroids into prothoracic gland compounds (for details see Dettner 2014). Altogether 63 steroids are recorded (e.g., Compounds **25**, **28**, **37**, **57**, **67**, **73**, Fig. 9.8, Table 9.2). Among the

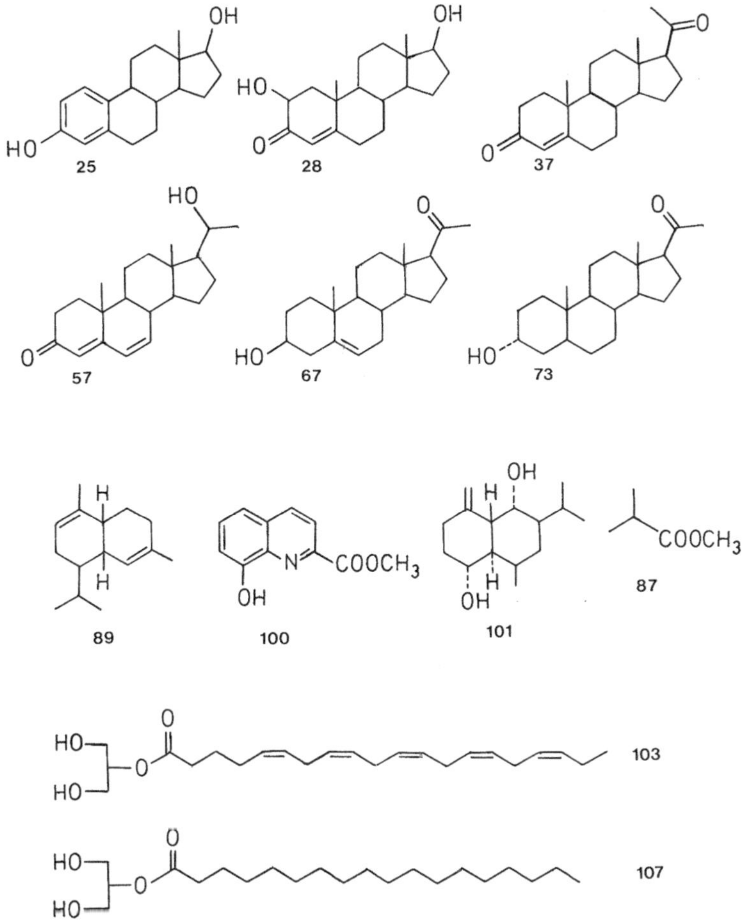


**Fig. 9.8** Prothoracic gland constituents **24–108** (for names see Table 9.2) from predacious diving beetles (Dytiscidae) (according to Dettner 2014). **25**: $C_{18}$ steroid of estratriene structure; **28**: $C_{19}$ steroid of androstene structure; **37**: $C_{21}$ steroid of pregn-4-ene structure; **57**: $C_{21}$ steroid of pregna-4,6-diene structure; **67**: $C_{21}$ steroid of pregn-5-ene structure; **73**: $C_{21}$ steroid of pregnane structure; **89** and **101**: sesquiterpenes alpha-muurolene (**89**) and platambin (**101**). Alkaloid (**100**), ester (**87**), and monoglycerides (**103**, **107**)

nonsteroidal constituents of the prothoracic glands there is methylisobutanoate (Fig. 9.8/**87**), benzoic acid (Fig. 9.5/**3**), pentadecanoic and octadecanoic acids, and monoglycerides with saturated and unsaturated side chains (Fig. 9.8/**103**, **107**; Table 9.2). In addition the yellow alkaloid methyl-8-hydroxquinoline carboxylate was found (Fig. 9.8/**100**) in *Ilybius fenestratus*, which is derived from tryptophane, and the same species together with *Dytiscus marginalis* also contains methyl esters of leucine, valine, and isoleucine. In *Ilybius fenestratus* and *Platambus maculatus* Schildknecht and his group (Schildknecht 1977) identified 13 sesquiterpenes (including for example platambin, Fig. 9.8/**101**, and alpha muurolene, Fig. 9.8/**89**)

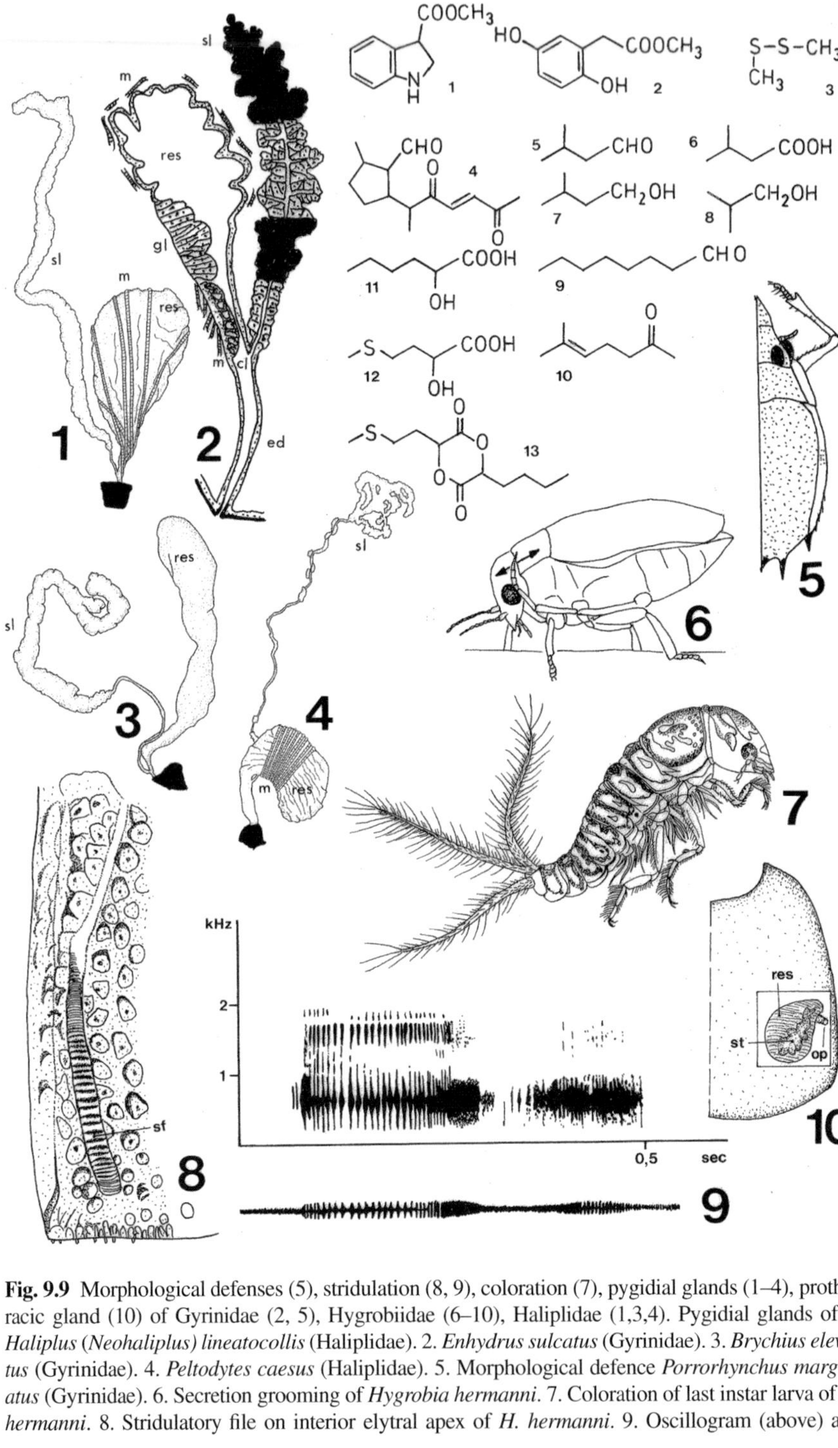


**Fig. 9.9** Morphological defenses (5), stridulation (8, 9), coloration (7), pygidial glands (1–4), prothoracic gland (10) of Gyrinidae (2, 5), Hygrobiidae (6–10), Haliplidae (1,3,4). Pygidial glands of 1. *Haliplus (Neohaliplus) lineatocollis* (Haliplidae). 2. *Enhydrus sulcatus* (Gyrinidae). 3. *Brychius elevatus* (Gyrinidae). 4. *Peltodytes caesus* (Haliplidae). 5. Morphological defence *Porrorhynchus marginatus* (Gyrinidae). 6. Secretion grooming of *Hygrobia hermanni*. 7. Coloration of last instar larva of *H. hermanni*. 8. Stridulatory file on interior elytral apex of *H. hermanni*. 9. Oscillogram (above) and

which are defensive compounds and additionally responsible for the characteristic odor of the prothoracic secretion of these two and other species. *Colymbetes fuscus* does not produce steroids but instead contains the nucleoproteid colymbetin (Table 9.2), which can lower blood pressure in small mammals.

Biological activities of prothoracic steroids which are often present in huge amounts in the secretion are primarily directed against fishes and amphibians (newts and frogs). If the beetles are ingested, the secretion and especially the steroids are absorbed via the gill membranes, partly via skin. Many of these compounds from prothoracic glands of dytiscid beetles but also constituents from pygidial glands of whirligig beetles (Gyrinidae, which lack prothoracic glands) are toxic and exhibit especially anesthetic activities, which was investigated by Miller and Mumma (1976a, b) with minnows (Fig. 9.10; *Pimephales promelas*) or by Gerhart et al. (1991) with bluegill sunfishes (*Lepomis macrochirus*). In many experiments they evaluated survival time of test fishes which was correlated with the steroid concentration and the type of steroids used in the experiments (Fig. 9.10). The activity of steroids was highly related to the degree of oxygenation (see Dettner 2014). It was suggested that the aforementioned alkaloid (Fig. 9.8/**100**) and the sesquiterpene platambin (Fig. 9.8/**101**) are directed against small mammals, because *Ilybius fenestratus* often stays on land (Schildknecht 1970, 1977). The amino acids and their methyl esters exhibit various effects on fishes and they can be both feeding stimulants and deterrents (Dettner 2014). According to Kasumyan and Doving (2003) this applies especially to the found valine, isoleucine, and leucine. In bioassays with minnows (*Phoxinus phoxinus*) it was found that two polyunsaturated monoglycerides showed a deterrent effect and may act as emulsifiers for prothoracic steroids (Schaaf and Dettner 2000).

As with pygidial gland secretion also prothoracic defensive fluid is not exclusively directed against predatory vertebrates. When peritrichic ciliates isolated from the water beetle *Agabus sturmii* were treated with prothoracic gland secretion there was a drastic negative effect that means stop of cilia movement after a short time (Dettner 2014).

When last-stage dytiscid larvae leave the water in order to construct a firmly closed pupal chamber it was noticed that within the opened chamber an aromatic lactonic odor was perceivable (Blunck 1923). It is suggested that peristigmatic glands of pupae produce these antisepticals; the same phenomenon was described in Carabidae pupal chambers (Giglio et al. 2009).

**Fig. 9.9** (continued) spectrogram (below) of stridulation noise of *H. hermanni*, 10. Half of prothorax of *H. hermanni* with prothoracic gland, covered by muscles. Abbreviation: *cl* closure; *ed* efferent duct; *op* opening of reservoir; *ogl* gland; *m* muscles; *res* reservoir; *sf* stridulatory file; *sl* secretory lobe; *st* secretory tissue. Pygidial gland constituents: *Amphizoa*: **1.** Methylindol-3-carboxylate, **2.** methyl homogentisate, **3.** dimethyl disulfide; Gyrinidae, **4.** gyrinidione, **5.** 3-methyl-1-butanal, **6.** 3-methyl-1-butanoic acid, **7.** 3-methyl-1-butanol, **8.** 2-methyl-1-propanol, **9.** octanal, **10.** 6-methyl-5-hepten-2-on, *Hygrobia*: **11.** 2-hydroxyhexanoic acid, **12.** S-methyl-2-mercaptobutanoi acid, **13.** lactide between **11** and **12.** 1, 3, 4: Dettner and Böhner 2009, 8–10: Dettner 1997, modified according to Barth 1960 (2), Miller and Bergsten 2012 (5), Kovac and Maschwitz 1990a, b (6), Klausnitzer 1996 (7)

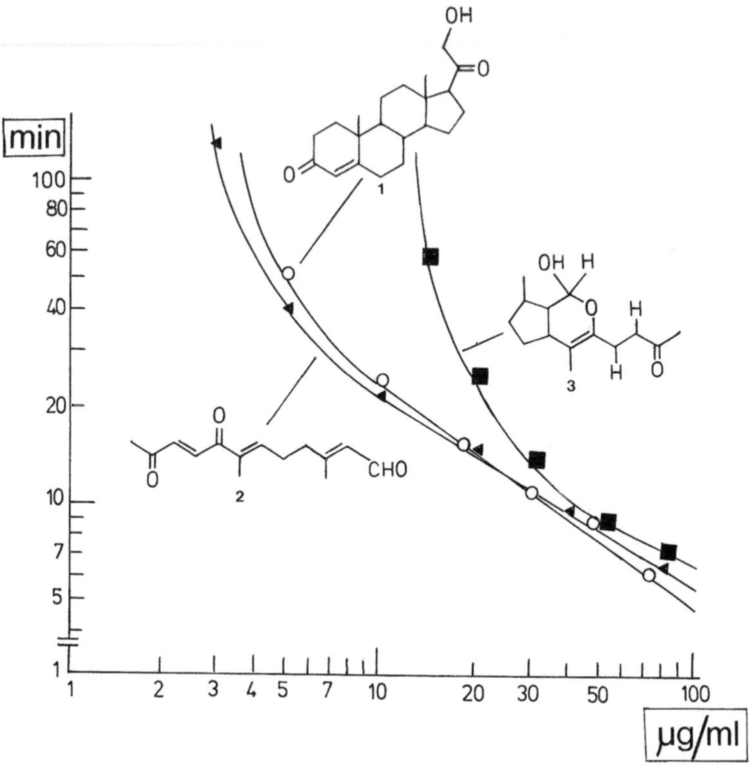


**Fig. 9.10** Structure-activity of steroid deoxycorticosterone (1), gyrinidal (2) and gyrinidone (3) from pygidial rsp. prothoracic glands of whirligig beetles (2, 3) and predacious water beetles (1). Log x log plots of the survival time of fathead minnows (*Pimephales promelas*) in 50 mL solutions of norsesquiterpenes and steroid (ug/ml). Modified according to Miller et al. (1975)

Within Dytiscidae larvae (e.g., Hydroporinae, Laccophilinae, Colymbetinae, Dytiscinae) there exist glandular organs of epipharynx. From outside many spines are found which are confined to those areas where the tips of mandibles touch the ventral surface of head (De Marzo and Nilsson 1986). Glandular cells without tubules either form a single epithelium behind spinulose areas or represent a large gland with secretory ducts (*Cybister*). The role of these secretions is discussed controversially as detergents for cleaning mandibular tips ("Organes de toilette," Bertrand 1928), poisonous secretions, or attractants for prey.

**Gyrinidae** (whirligig beetles; 850 sp.; Beutel and Roughley 2016; Jäch and Balke 2008 estimated 1000 sp.; with respect to global diversity among water beetles this family has position 6); Phylogenetically there exist three subfamilies (Spanglerogyrinae, Heterogyrinae, Gyrininae) with Orectochilini, Gyrinini, and Enhydrini based on molecular and morphological data from both adults (Miller and Bergsten 2012) and larvae (Beutel and Roughley 1993). Adults by active movements whirl on the water surface (loops, zigzags), but they can also dive and fly after leaving the water. Gyrinids are often found in aggregations with sometimes 50,000 individuals, which is interpreted as a kind of defense strategy (Yee and Kehl 2015; Vulinec and Miller 1989). Vulinec and Miller (1989) showed that adult whirligig beetles are adapted to aquatic and terrestrial life by divided compound eyes and by hind leg strokes of about 50 to 60 s with speeds of 53–144 cm/s (Yee and Kehl 2015). Larvae are strictly aquatic, need a lot of oxygen, and possess tracheal gills. Preys are digested extraintestinally. All developmental stages are carnivorous, and pupation takes place on land. If there is a danger, the heavily sclerotized adults drastically increase their movements, can immediately dive, and may move under the water surface. There they swim and must quickly sting on solid objects. Apart from these unusual neustonic adaptations whirligig beetles possess paired abdominal pygidial glands (Fig. 9.4/7; Fig. 9.9/2) with odoriferous, volatile, and fluid secretions. The paired glands of *Gyrinus caspius* were described by Forsyth (1968), and the glands of *Enhydrus sulcatus* were figured by Barth (Fig. 9.9/2; 1960). The gland reservoirs are covered by muscles, and the secretory tissue evacuates secretion into efferent ducts of reservoirs close to or far away from the reservoir opening. *Enhydrus* is unusual as its gland reservoir hosts a second type of secretory tissue, apart from the secretory lobe (Fig. 9.9/2), an analogue situation resembling the anal glands of primitive Steninae staphylinid beetles (Fig. 9.13; Betz et al. 2018). It would be interesting to know if phylogenetic data from pygidial gland morphology and chemistry are in accordance with the abovementioned phylogenies of whirligig beetles. There exist various investigations with respect to the defensive secretions which contain the norsesquiterpenes gyrinidal, isogyrinidal, gyrinidon, and gyrinidion (Fig. 9.9/Formula 4) exhibiting strong repellencies against fishes (Fig. 9.10), newts (*Notophthalmus*), birds, and small mammals (Schildknecht et al. 1972; Miller et al. 1975; Scrimshaw and Kerfoot 1987).

In all cases when the beetles were attacked all whirligig beetles were rejected and the individuals survived without damage. Eisner and Aneshansley (2000) demonstrated that the oral tolerance of the bass *Micropterus salmoides* varied, dependent on gyrinidal dosage and satiation of the fish. Most of these experiments were performed by offering topically treated mealworms to the bass in an aquarium. Moreover the bass showed an intensive oral flushing behavior during which beetle sesquiterpenes were absorbed (about 80%) by the gills of the bass. Depending on gyrinid species there are also stinking volatiles in the pygidial gland secretions (Ivarsson et al. 1996) such as 3-methyl-1-butanal (Fig. 9.9/Formula 5), 3-methyl-1-butanol (Fig. 9.9/Formula 7), 2-methyl-1-propanol (Fig. 9.9/Formula 8), 3-methyl butyric acid (Fig. 9.9/Formula 6), and octanal (Fig. 9.9/Formula 9). In few *Gyrinus* species traces of 6-methyl-5-hepten-2-one (Fig. 9.9/Formula 10) and phenylacetaldehyde were recorded (Table 9.3; Fig. 9.11). Remarkably *Macrogyrus oblongus* did not contain the aforementioned four norsesquiterpenes, but instead other unidentified terpenes. Finally pygidial glands of *Orectochilus villosus* contained $C_{17}$, $C_{18}$, $C_{21}$, and $C_{22}$ alkanes; however the chemistry of the strong violet odor as described by Lucht (1974) remains unknown (Table 9.3). In addition the pygidial gland secretions are applied to the body surface by using their legs in order to keep it free from contaminating bacteria, protozoa, or fungi. In gyrinid beetles this secretion groom-

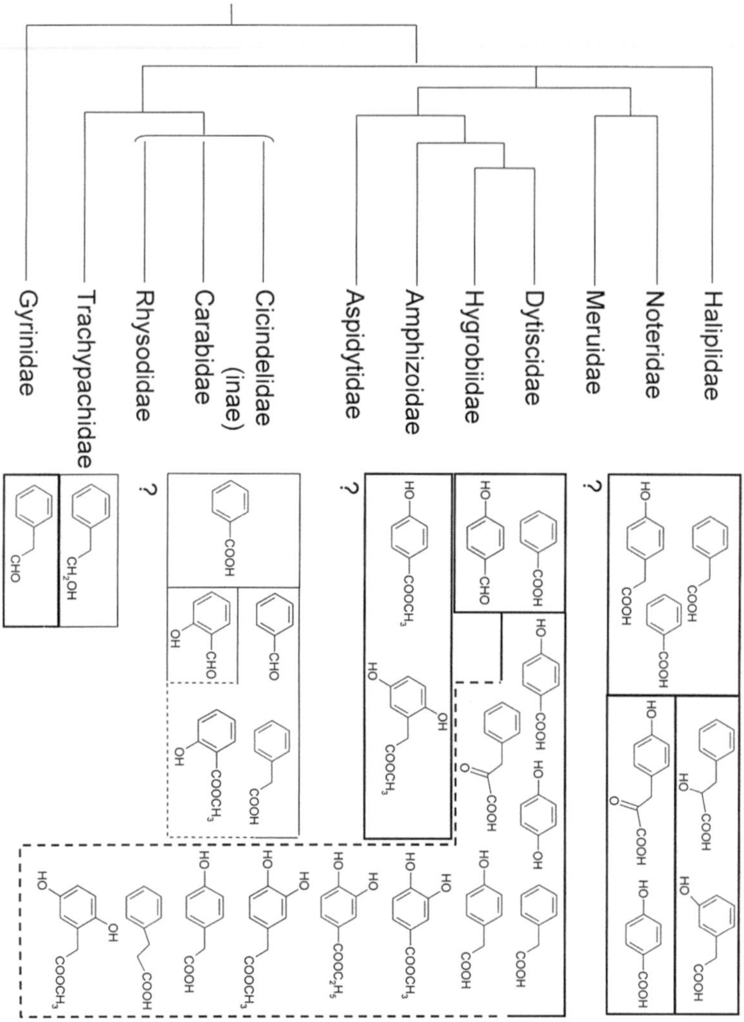


**Fig. 9.11** Aromatic pygidial gland constituents in aquatic and terrestrial adephagan beetles (Dettner and Böhner 2009). Fat boxes: Aromatics are present in nearly all members of appropriate families. Interrupted boxes: Compounds only present in these taxa, especially large rectangular box of Dytiscidae. Thin boxes: Compounds only in minor amounts in single species per taxon. Phylogeny according to Beutel et al. 2006. Terrestrial families: Cicindelidae, Carabidae, Rhysodidae, Trachypachidae

ing was described by Kovac and Maschwitz (1990b). In addition pygidial gland secretion of *Gyrinus substriatus* increased wettability of the beetle integument (Fig. 9.6). As compared with the untreated elytral control, contact angles of standard water droplets on elytra previously treated with pygidial gland secretion were significantly lowered.

As demonstrated by fossil Permian Gyrinidae this family became adapted to swimming on the water surface long before the mesozoic invasion of the aquatic environment by Dytiscoidea took place (Yan et al. 2018). In the primitive Spanglerogyrinae the dorsal and ventral eyes are only narrowly separated as compared with the more derived Gyrininae. Molested *Spanglerogyrus albiventris* shows a peculiar jumping behavior (Steiner and Anderson 1981) which is also typical for some other gyrinid beetles. It is remarkable that *Gyrinus* larvae have a non-wettable integument and therefore are captured at the water surface (Bott 1928). *Gyrinus* larvae also gather debris which is mixed with mucopolysaccharides from midgut cells (Rupprecht 1971) in order to construct a pupal chamber on the top of emergent plants.

Various studies demonstrated the role of the aggregations of gyrinid beetles. It was found that they aggregate in the daytime into single- and multispecies groups (rafting of 100 to 1000 individuals). With increasing temperature the beetles dispersed from these rafts to shallow edges of the waters (Fitzgerald 1987; Heinrich and Vogt 1980). Naive fishes ate single beetles; however they learned to avoid these beetles, indicating that rafting places represent safe places for gyrinids (Heinrich and Vogt 1980). Henrikson and Stenson (1993) finally demonstrated that water prepared with gyrinid elicited evasive behavior in the beetles. They concluded that aggregations of the dark, highly contrasting beetles in exposed areas advertise their bad taste, which represents a case of aposematism. Probably, the spreading activity of the Gyrinidae secretion on the water surface has no relevance for living beetles (Vulinec 1987), since this activity is typically found in defensive secretions of many arthropods. It is important to denote that most adult Gyrinidae have a hard integument. Very often the upper body sideboard is brightly colored and elytra exhibit large and small spines at elytral apices as found in *Porrorhynchus marginatus* (Fig. 9.9/5; Gustafson and Miller 2016), which might be interpreted as defense against fishes.

**Haliplidae** (crawling water beetles; 240 sp.): Haliplidae hold the 11th position with respect to global water beetle diversity according to Jäch and Balke (2008). Larvae and adults are strictly aquatic and prefer stagnant or slow-running waters often with filamentous algae and characeans (van Vondel 2016). Adults are characterized by metacoxal plates and move their hind legs alternately. The three herbivorous larval stages have tracheal gills, move very slowly, and are often hidden within water plants. Haliplid larvae have a slender body shape and show a hard integument. Elder larvae are often camouflaged and covered by diatoms rendering a black body coloration; however they are fed by fishes, frogs, or carnivorous insect larvae (Hickman 1931). According to Falkenström (1926) pupae of *Haliplus* and in contrast to Dytiscidae pupae exhibit thanatosis. Mousseau and Roughley (2007) discuss that characteristic elytral maculations in Haliplidae and especially in *Brychius* can have deceptive character or may represent a warning coloration. Adult haliplids are consumed by fishes, frogs, ducks, or carnivorous insects (larvae of Dytiscidae and Odonata, adult dytiscid beetles, Hickman 1931).

The paired pygidial glands of adults of *Haliplus ruficollis* were described by Forsyth (1968), and the glands of other species from the genera *Peltodytes*, *Brychius*, and *Haliplus* were figured by Dettner and Böhner (2009). The gland reservoirs are covered by muscles, which either are confined to the reservoir in the primitive genus *Peltodytes* (Fig. 9.9/4) or extend to the reservoir opening in other genera (Fig. 9.9/1). The more primitive *Peltodytes* representatives exhibit compact glandular tissues (Fig. 9.9/4) as in Carabidae (Fig. 9.4/12) and show longer collecting channels which unite with the gland reservoir far away from the gland opening. In *Brychius* and *Haliplus* the glandular tissue is lengthened; however *Brychius* and *Haliplus* (subgenus *Neohaliplus*) exhibit short collecting channels and have reservoirs which gradually narrow towards opening (Fig. 9.9/1, 3, 4). Phylogenetic data from pygidial gland morphology are in accordance with phylogenies based on adult characters (Beutel and Ruhnau 1990). Phenylacetic acid (Fig. 9.5/**11**) which is also found in hydroporine Dytiscidae (and *Copelatus*) and Noteridae was recorded in all haliplid species investigated (Table 9.3). In all genera also further antimicrobials from selected Dytiscidae (*Hydrovatus, Hygrotus, Laccophilus*) such as 3-hydroxy octanoic, 3-hydroxy decanoic acid (Fig. 9.5/**17**, **22**), benzoic acid (Fig. 9.5/**3**; Dytiscidae: Colymbetinae, Dytiscinae), or 4-hydroxyphenylacetic acid (Fig. 9.5/**12**; Dytiscidae: Hydroporinae; Noteridae) were recorded (Dettner and Böhner 2009). Exclusively few Haliplidae species secrete 3-hydroxyphenylacetic acid and phenyllactic acid (Fig. 9.11). All aliphatic and aromatic recorded gland constituents from Haliplidae are bactericidals, fungicidals, and disinfectants and are distributed on the beetle surfaces by secretion grooming (Dettner 1985; Kovac and Maschwitz 1990b). On land and on water surface the beetles distributed secretions with their legs on their body surface in order to dive in a characteristic way typical for Haliplidae (Fig. 9.7/Haliplidae 1–3). In addition phenylacetic acid and other components increase wettability of the haliplid body surface when the beetles hang on the water surface and show their characteristic cleaning behavior in order to dive (Fig. 9.6/Haliplidae; Dettner 1985). Whether the 3-hydroxyesters are transformed to polyesters on the body surface in order to enclose, kill, and eliminate epizoic microorganisms has to be investigated (Dettner and Böhner 2009). Stridulation was reported from three *Haliplus* species (Beier 1929; Seeger 1971; Aiken 1985). Sclerotized teeth at the underside of the elytra represent the pars stridens, whereas the plectrum is found in the pleural fold of abdominal segments 2–4. Stridulation behavior which is achieved by stretching of abdomen was described before and during mating and probably is not a defense mechanism.

**Hygrobiidae** (squeak beetles; six sp.): It is a relict family (also named Paelobiidae) with species in China, Australia, and Europe. Larvae and adults live in mud, silt, and detritus of ponds (Dettner 1997b; Dettner 2016c). Eggs are deposited on submerged plants. There exist three larval instars with tubular gills and characteristic strong but variable brownish pigmentation (Fig. 9.9/7). They pupate in a pupal chamber on land. Adults possess pygidial glands consisting of a gland reservoir (covered by muscles) with efferent duct and bifurcated secretory lobe (Fig. 9.4/6; Forsyth 1970) filled with colorless to white fluid which is depleted on land and distributed on the body surface of adult beetles (Kovac and Maschwitz 1990b). *Hygrobia* produces a series of unusual 2-hydroxyacids such as 2-hydroxyhexanoic acid (Fig. 9.9/Formula 11) and S-methyl-2-hydroxy mercaptobutanoic acid (Fig. 9.9/Formula 12), which may form diverse lactides (Fig. 9.9/ Formula 13). Traces of benzoic acid and 4-hydroxybenzaldehyde, which also represent main gland constituents in Dytiscidae, were also recorded (Fig. 9.11; see Dettner 1985, 1987). *Hygrobia hermanni* also showed a typical behavior which was described as secretion grooming (Fig. 9.9/6, Kovac and Maschwitz 1990b). In addition pygidial gland secretion of *H. hermanni* increased wettability of the beetle integument (Fig. 9.6). As compared with the untreated elytral control, contact angles of standard water droplets on elytra previously treated with pygidial gland secretion were significantly lowered (Fig. 9.6). As in Dytiscidae the sister group of Hygrobiidae (Fig. 9.11) *Hygrobia hermanni* is characterized by paired prothoracic glands which are covered by secretory cells and a muscle cover (Fig. 9.9/10; Forsyth 1970). In contrast to Dytiscidae the prothoracic glands of *Hygrobia* are situated posterolaterally within the prothorax wherefore homology seems questionable. Recent molecular data indicate that there is a sistergroup relationship of Hygrobiidae to a clade comprising Amphizoidae, Aspidytidae and Dytiscidae (Vasilikopoulos et al. 2019). The separate evolution of a prothoracic gland in Hygrobiidae therefore seems possible. Since there are no chemical data available, significance of the secretion remains unknown. When molested both sexes of the beetles ("squeak beetles") produce audible stridulations (Fig. 9.9/9) by rubbing abdominal apex on a file at the underside of an elytron (Fig. 9.9/8; Beutel 1986; Aiken 1985; Wichard et al. 1995). Probably *Hygrobia* also stridulates during contacts with conspecifics and in courtship.

**Meruidae** (comb-clawed cascade beetles; one sp.): The single small-sized species (adult body length 0.9 mm) was described from Venezuela and lives in the margins of cascades and mountain streams (Balke et al. 2018). Its larva has several characteristics of *Noterus* larvae. In order to submerge, adults turnover with the head directed downward and kick with their alternating leg motion until finding an underwater foothold. Similar behaviors are known from Haliplidae or small Dytiscidae. Pygidial glands were identified morphologically, and prothoracic glands are absent (Beutel et al. 2006).

**Noteridae** (burrowing water beetles; 280 sp., according to Jäch and Balke (2008)): Noteridae hold the 9th position with respect to global water beetle diversity. Noteridae are especially found burrowing in muddy substrate and in root mats in ponds and marshes, there exist few subterranean species. Adults and larvae are mainly carnivorous, and larvae (only few species have been described) have a siphon at abdominal tip in order to expire oxygen from aquatic plant aerenchyma. Air-filled pupae of some species are attached to the aerenchyma of water plants. Prothoracic glands are absent, and typical paired pygidial glands are present. They consist of a reservoir covered with muscles and a bipartite secretory lobe with two types of glandular cells (Fig. 9.4/**8**; Forsyth 1968). As in Dytiscidae main gland constituents are antimicrobics as phenylacetic (sweet odor; Fig. 9.5/**11**; Fig. 9.11), 4-hydroxybenzoic (Fig. 9.5/**5**), benzoic (Fig. 9.5/**3**), and 4-hydroxyphenylacetic acids (Fig. 9.5/**12**). Also 3-indole-acetic acid (Fig. 9.5/**16**) of unknown function is present. Noteridae also show secretion grooming behavior outside the water in order to distribute pygidial gland secretions on their bodies (Kovac and Maschwitz

1990b). In addition pygidial gland secretion of *Noterus clavicornis* increased wettability of the beetle integument (Fig. 9.6; Dettner 1985, 1997a, 2016a). As compared with the untreated elytral control, contact angles of standard water droplets on elytra previously treated with pygidial gland secretion were significantly lowered.

## *9.5.1 Polyphaga: Hydrophiloidea*

**Hydrophilidae** (water scavenger beetles or silver water beetles; 2932 sp.; Short: 2950 sp.; according to Jäch and Balke 2008: 3rd position with respect to global water beetle diversity): Hydrophilids and their larvae are mainly found in stagnant and also running waters and hygropetric habitats; however their ecologic habits are extremely broad. Adults prefer plants and decaying organic matters as food whereas larvae are usually predacious. There are no complex glands present. Nevertheless Kovac and Maschwitz (2000) observed terrestrial and submerged grooming in various Hydrophilidae (*Hydrophilus, Laccobius, Berosus, Enochrus, Spercheus*: Spercheidae) apart from the terrestrial *Helophorus nubilus* (see Helophoridae). It is remarkable that *Hydrochus*, *Hydrobius*, and *Berosus* use anal droplets for grooming behavior.

Stridulation is widely distributed in adult Hydrophilidae and was obviously evolved several times (Aiken 1985). In many cases hydrophilid beetles stridulate when they are irritated, and sometimes sonograms indicate that sounds are important during courtship (Scheloske 1975). In several cases stridulatory structures in hydrophilid beetles are controversially discussed (Hansen 1991; Scheloske 1975).

Within Berosini (*Berosus*) and Oocyclini (*Laccobius*; Scheloske 1975; Fig. 9.12/2) both sexes stridulate. In both genera laterosternite 3, which is concealed under the elytra, exhibits a ribbed area representing a stridulatory organ together with protuberances on the underside of elytra (Hansen 1991). Within Hydrophilini stridulatory organs are present in genera *Hydrophilus*, *Hydrobius, Limnoxenus*, and *Tropisternus* (Wilson et al. 2015) and seem to be reduced in *Helochares* and *Enochrus* (Hansen 1991). In *Tropisternus* this behavior was named disturbance stridulation and acoustic aposematism which deters predators such as wolf spiders (Masters 1979). In Chaetarthriini, Sperchopsini, and Sphaeridiinae stridulatory structures are missing (Hansen 1991).

**Fig. 9.12** (continued) and metatibia and metatarsus (d). 4. Ventral aspect of head and prothorax of *Ochthebius quadricollis* (Hydraenidae) with part of exocrine secretion delivery system (e: end apparatus of gland cells, su: sulcus for collecting secretion, ssd: setae for secretion delivery). 5. Venter of left side of prothorax of *Hydraena riparia* with part of secretion delivery system (*e*: end apparatus of gland cells; *hh* hydrofuge hypomeron; *su* sulcus; *wh* wet hypomeron). 6 and 7. Macerated larvae of *Luciola leii* (Lampyridae) with everted, fork-like reservoirs of defensive glands (arrows in enlarged right figure). 8. *Georyssus* spec. (Georyssidae) with camouflage (covered by sand/mud; left), clean surface (right). 9. Aposematically colored *Ancyronyx* spec. (Elmidae). Gland constituents of lampyrid larva *L. leii*: **1.** Terpinolene, **2.** gamma-terpinene. Originals (1, 6, 7). Modified according to Scheloske (2), Perkins 1997 (3–5), Merritt and Cummins 1984 (8), 1996 (9).

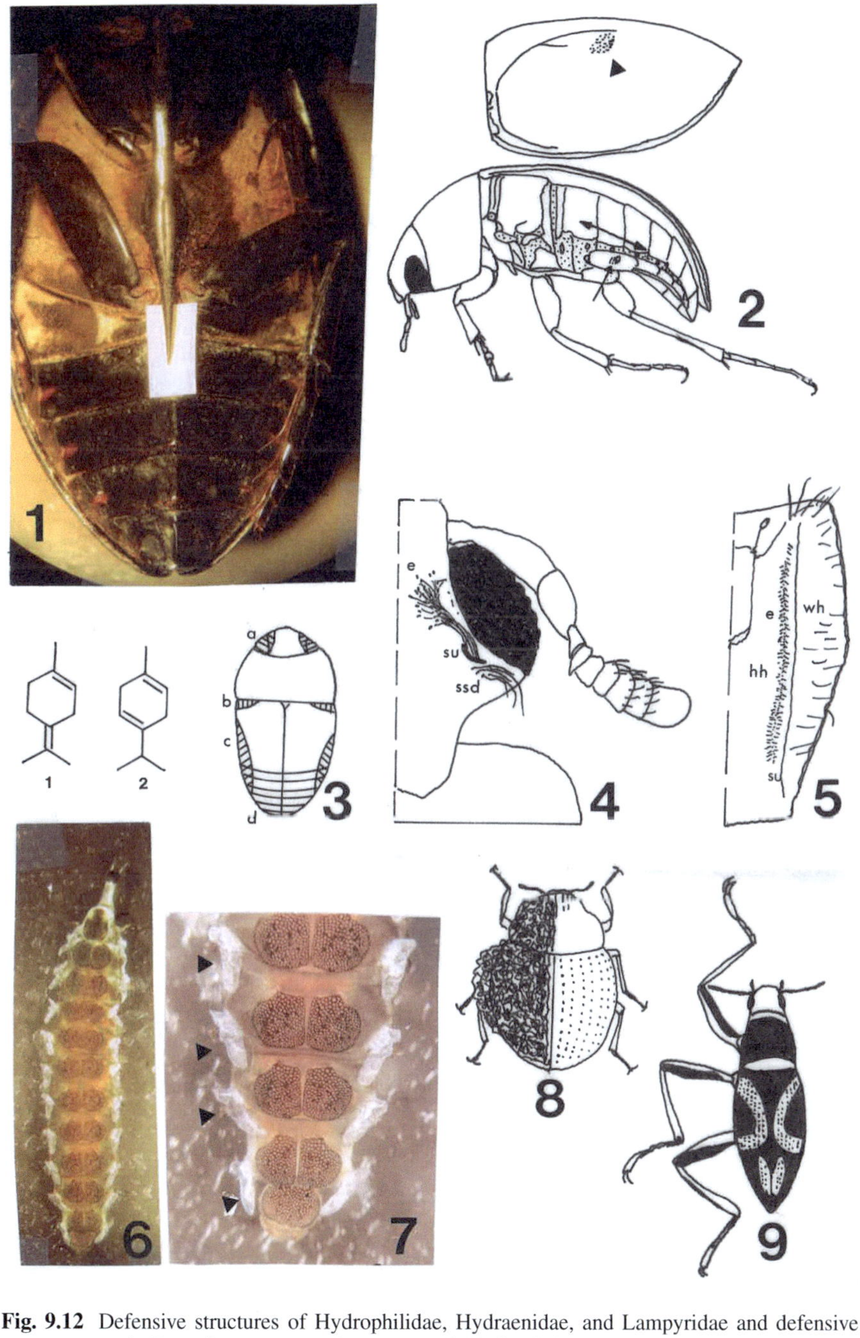


**Fig. 9.12** Defensive structures of Hydrophilidae, Hydraenidae, and Lampyridae and defensive components. 1. Posterior metasternal process of *Hydrophilus aterrimus* (Hydrophilidae). 2. Stridulation organs of *Laccobius minutus* (Hydrophilidae) with protuberances on underside of elytron and pleural area (pars stridens) of third abdominal segment. 3. Dorsal side of *Limnebius piceus* (Hydraenidae) with areas groomed by protibia and protarsus (a), protibia (b), mesotibia (c),

The posterior metasternal process of *Hydrophilus*-species (Fig. 9.12/1) probably represents a defensive structure. In addition this process serves for keeping the ventral air bubble. Predator avoidance in Hydrophilidae also may include crypsis. For example *Derallus* larvae bear setiferous projections with a lot of algae and detritus. When molested larvae of *Hydrophilus* may immediately defecate like an explosion and suddenly lift their head backwards (Pavlovsky 1922).

**Helophoridae** (192 sp.): Helophoridae hold the 12th position with respect to global water beetle diversity according to Jäch and Balke (2008). Due to sticky dermal glands, adults are often dirty or covered by sand and must be cleaned before determination of these brownish, metallic, or black specimens. Adult Helophoridae are found especially in shallow water, and their larvae are found among vegetation an in soil near water (Angus 1992). The laterosternite 3 of adult *Helophorus* has no ribbed area (which would represent a stridulatory organ; Hansen 1991) and no stridulation has been recorded (Oliva 1992).

**Epimetopidae** (56 sp.; Short 2018: 66 sp.): The beetles seem to be riparian and live within sand substrates. Adults may be covered by soil particles, and they possess no abdominal stridulatory organ (Hansen 1991).

**Hydrochidae** (181 sp.): Hydrochidae hold the 11th position with respect to global water beetle diversity according to Jäch and Balke (2008). Hydrochid beetles are found in stagnant to slowly flowing water. The dark or metallic adults are often covered by a secretion and crust of vegetable matter (Oliva 1992). Adults feign death when disturbed and possess no abdominal stridulatory organ (Hansen 1991).

**Spercheidae** (filter-feeding water scavenger beetles; 20 sp., Yee and Kehl 2015): Representatives of these filter-feeding beetle family are found in stagnant waters and in mud. When molested on land or within water *Spercheus* adults of both sexes are able to stridulate in a different manner. Obviously stridulation is primarily a defensive reaction but can also be important in sexual communication (Von Frankenberg 1937). Laterosternites 3, which are concealed under the elytra, exhibit transversely ribbed areas which are moved against the ridges under each elytra in order to stridulate. Remarkably *Spercheus* adults also exhibit thanatosis (longer lasting) together with the defensive stridulation (Von Frankenberg 1937).

## *9.5.2 Polyphaga; Staphylinoidea*

**Hydraenidae** (minute moss beetles; about 1300 sp.; Short 2018: 1600 described species; according to Jäch and Balke 2008 there was an estimation of 2500 sp. which means that Hydraenidae occupy the second position among water beetles with respect to global water beetle diversity): Many species of these true water beetles (few of them are terrestrial) possess a microplastron that means a ventrally situated thin layer of air, which is held by hydrofuge setae (Jäch et al. 2016). An exocrine secretion delivery system with various glands especially located at head and prothorax, e.g., in *Ochthebius* or *Hydraena* species and various cuticular structures such as grooves, ridges, and setae (Fig. 9.12/4, 5), plays an enormous role in secretion grooming (Jäch et al. 2016; Kovac and Maschwitz 2000; Perkins 1997). Perkins (1997) presented a detailed morphological investigation on these secretion delivery systems, the wetted and hydrofuge integumental body areas, and studied grooming behavior of selected species such as *Limnebius* (Fig. 9.12/3); however due to the small body size of these hydraenid beetles no chemical data are available. Integument of terrestrial and coastal species is covered with encrustations. Adult hydraenids cannot swim but walk on underside of surface film. Stridulation is quite common and two main stridulatory areas are known: (a) abdomen/elytra (all species) and (b) head/pronotum (only few species). Since stridulatory organs are sexually dimorphic, a role in courtship and mating was suggested (Aiken 1985). The similar grooming patterns of Hydraenidae within water and on land were described by Kovac and Maschwitz (2000).

**Scirtidae** (marsh beetles; Yee and Kehl 2015: 1000 sp.; over 1800 sp.): Scirtidae hold the 11th position with respect to global water beetle diversity according to Jäch and Balke (2008). They prefer various habitats such as streams, pools, phytotelmata, and groundwater. Marsh beetle larvae are usually aquatic, and ecology and bionomy of the family are poorly studied. If molested larvae are very sensitive and immediately stop feeding and may even escape (Klausnitzer 1996).

**Elmidae** (riffle beetles; 1500 sp.; Yee and Kehl: 1300 sp.): Elmidae hold the 11th position with respect to global water beetle diversity according to Jäch and Balke (2008). Aquatic Elmidae are herbivores or detrivores and are usually found in running water with high oxygen content, and seldom lakes and ponds. Adults possess an efficient plastron that means a thin air film coats the body surface. Interestingly secretion grooming by brushing legs over the ventral air layer is important but can be only observed under water (Kovac and Maschwitz 2000). White (1989) offered elmid specimens to a turtle and predatory fishes and showed that aposematically colored (red to yellow markings on elytra and thorax) beetles were almost always rejected (Fig. 9.12/9). Larvae do not have aposematic colorations and were never rejected or ignored by fishes. It was further proven that elmid adults have a chemical defense restricted to the elytra; until now active principles have not been elucidated. Elliott (2008) showed that lanceolate larvae of *Elmis*, *Esolus*, or *Oulimnius* rolled up like an armadillo when disturbed. Both larvae and adults usually have hard integuments.

**Dryopidae** (long-toed water beetles; 280 sp.; according to Jäch and Balke 2008/ Yee and Kehl 2015: 300 sp.): Dryopidae hold the 8th position with respect to global water beetle diversity according to Jäch and Balke (2008). Adults are found in aquatic or other habitats, and larvae are terrestrial or under water surface. Beetles and larvae are slowly moving and generally lethargic insects. Adults are purely aquatic and are characterized by a plastron. For *Dryops* secretion grooming by using their legs was reported (Kovac and Maschwitz 2000). Ecology of Dryopidae is only purely known. Adults and larvae do not have aposematic colorations and are accepted by predators (Scrimshaw and Kerfoot 1987). According to Lawrence (1991) Dryopidae pupae have unpaired gin traps, which means sclerotized and often armed portions of two adjacent sclerites (Lawrence 1991).

**Psephenidae** (water pennies; 287 sp.; Jäch and Balke 2008: 272 sp.): Psephenidae hold the 11th position with respect to global water beetle diversity according to Jäch and Balke (2008): The flattened larvae with their powerful claws are usually in running waters. They can hug the rocks as tenaciously as limpets, so they are safe from predators. Their coloration varies and often they match with their background substrate (Lawrence 1991). Aquatic psephenid pupae are characterized by plastron-bearing gills; in addition they have gin traps (Lawrence 1991), and adults are terrestrial. The nocturnal feeding behavoir of the larvae was reported to be a defense mechanism against predators. In addition the larvae can form a ball when being swept away (Beutel and Leschen 2016).

**Cneoglossidae** (eight sp.; Yee and Kehl 2015: ten sp.): Beetles are found in submerged bushwood in small streams. Larvae are aquatic, and adults are terrestrial. The life history of this small family is unknown.

**Eulichadidae** (forest stream beetles; 42 sp.): Larvae are aquatic; the beetles are in sand or litter under rocks in streams.

**Lutrochidae** (travertine beetles; Jäch and Balke 2008/Yee and Kehl 2015: 15 sp.): Travertine beetles occur submerged in decaying wood. Their life history is unknown and their taxonomy has to be revised. Adults which inhabit calcareous water streams may be encrusted with this material (Yee and Kehl 2015).

## 9.6 Not Predominantly Aquatic Families According to Jäch and Balke (2008)

**Nitidulidae** (sap beetles; according to Jäch and Balke 2008 about 3000 sp.): One species *Amphicrossus japonicus* is aquatic and can be found in in water-filled stems of bamboo.

**Carabidae** (ground beetles; 37,600 sp.; Beutel and Leschen 2016): Among these Geadephaga there are a lot of species which are riparian, found near freshwater habitats, or even stay underwater for feeding or concealing (*Carabus clathratus, Oodes, Nebria, Bembidion, Chlaenius, Brachygnathus*). Representatives of the genus *Omophron* are found within wet sand and even resemble dytiscid water beetles. However phylogenetically there is no relation between *Omophron* and Hydradephaga. All Geadephaga share the same pygidial glands with Hydradephaga. However the gland morphology of *Carabus* or *Omophron* (Fig. 9.4/12) is not similar with Hydradephaga, because secretory tissues are compact and not lengthened. In addition they produce other pygidial gland constituents (e.g., *Omophron*: 2-methylpropanoic acid, 3-methylbutyric acid; *Carabus*: 2-methylpropenoic acid, (E)-2-methyl-but-2-enoic acid; also in water beetle *Ilybius*; Dazzini Valcurone and Pavan 1980). In *Omophron* stridulation was investigated and was named disturbance stridulation and acoustic aposematism which deters predators such as wolf spiders (Masters 1979).

**Staphylinidae** (rove beetles; 55,244 sp. Beutel and Leschen 2016): Rove beetles from many subfamilies (e.g., *Aleochara, Omalium, Paederus, Philonthus, Thinobius*) are semiaquatic and can be found on beaches, along rivers, ponds, bogs, and nearly every kind of freshwater. Due to their short elytra and their unprotected abdomina most staphylinid species have defensive glands or toxic hemolymph in order to defend against predators (Dettner 1993). Especially two taxa have special relationships with water. Many species of *Bledius* (Oxytelinae) live subsocially within wet sand of shores (freshwater; sea) where they feed on algae. Females feed and defend their larvae and use their quinoid abdominal gland secretions against aggressors (Dettner 1993; Thayer 2016). In many cases the formulation of these mixtures of defensive secretions is optimized against special target organisms (Steidle and Dettner 1993, 1995). Another rove beetle group, representatives of Steninae (worldwide with 3300 sp.), is typical for the water surface of all kind of freshwaters (Betz et al. 2018). The beetles are predators of aquatic springtails and other small arthropods and are found in every wet place. Many species walk on the water surface and depending on species show a unique skimming behavior after a molestation. They secrete terpenes and alkaloids such as stenusin or cicindeloin from their anal glands (Fig. 9.13; Dettner 1987, 1993; Betz et al. 2018). Comparable to *Enhydrus* glands

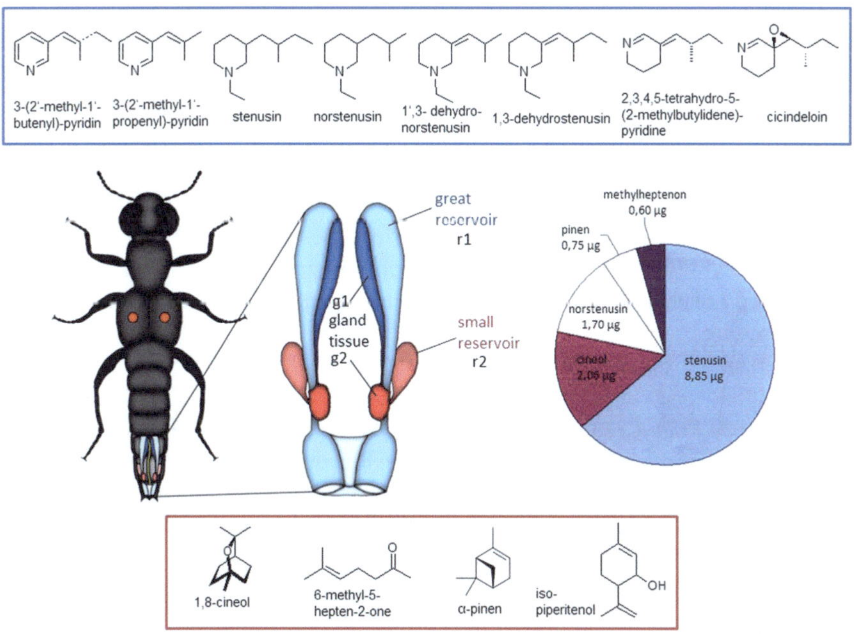


**Fig. 9.13** Anal glands and secretions of *Stenus* rove beetles (Staphylinidae). Paired anal glands with large reservoirs (gland tissue g1 and gland reservoir r1; blue) and small gland reservoirs (r2) and gland tissue (g2; red). Alkaloids from the large gland are figured above (blue box), terpenes from the small gland are figured below (red box). Circle demonstrates quantitative amounts of alkaloids stenusin and norstenusine from large gland as compared with remaining terpenes from smaller glands of *Stenus comma* (according to Betz et al. 2018)

(Gyrinidae), *Stenus* abdominal glands have two types of secretory tissues (gland 1 and gland 2) which are morphologically separated and synthesize different compounds. The large gland 1 delivers alkaloids; in contrast the small gland produces terpenes. All alkaloids are biosynthetically derived from amino acids from food (e.g., springtails, aphids). Together with molecular data, the Steninae alkaloids are excellent chemotaxonomic markers (Betz et al. 2018).

Both special alkaloids (formerly named "spreading alkaloids") and selected terpenes may have considerable spreading activity. However the spreading ability and velocity of a *Stenus* species on the water surface vary. These differences are due to gland composition and the ability of the beetle species to deplete their outer gland reservoir on the water surface (Fig. 9.13; Lang et al. 2012). These compounds have multiple functions as antimicrobics, fungicides, and deterrents against predatory arthropods and fishes (Schierling et al. 2013). In addition the rove beetles exhibit grooming behavior or deplete parts of their anal glands, when molested. Spreading abilities are unusual phenomena in nature; however it is also observed in semiaquatic Veliidae (Heteroptera) which in contrast spray saliva on the water surface. As mentioned by Bush et al. (2008) even pine needles floating on the water surface are propelled across the surface since the resin at its base decreases the surface tension.

**Leiodidae** (round fungus beetles): *Platypsyllus castoris* lives on beavers, and *Silphopsyllus desmanae* lives on water moles. These two species represent parasitic water beetles (Jäch and Balke 2008).

**Scarabaeidae** (scarab beetles; 27,000 sp.): Among this large cosmopolitan family there exists the aquatic rice beetle *Dyscinetus morator*. This species is often found underwater, especially while escaping disturbances (Jäch and Balke 2008).

**Ptilodactylidae** (toe-winged beetles; Yee and Kehl 2015: approximately 500 sp.): They are found in aquatic, especially riparian, habitats. Larvae feed on submerged decaying wood. As defense ptilodactylid pupae have abdominal gin traps (Beutel and Leschen 2016). Not all species are aquatic. Ecological information is mostly lacking.

**Lampyridae** (lightning bugs/fireflies; approx. 2000 sp. according to Jäch and Balke (2008) and Yee and Kehl (2015)): Larvae of few species (*Luciola, Pyractomena, Pristolycus*) are secondarily aquatic (sometimes with tracheal gills) and are found in running water (lotic systems) or ditches (lentic systems). As terrestrial larvae which feed on land snails, the aquatic larvae feed on aquatic molluscs such as mussels and snails. Elder larvae may be good swimmers in order to catch prey. Since the snail preys in Japan sometimes represent hosts for human parasites these larvae are important for biocontrol (Yee and Kehl 2015). Aquatic and semiaquatic lampyrid larvae leave the water before pupation in order to construct terrestrial mud cells (Branham 2010). All lampyrid larvae aposematically produce light via paired photic organs which are located ventrally at the eighth abdominal segment (Branham 2010). Fu et al. (2007, 2009) reported that the glow intensity of three aquatic *Luciola* species is weak. Okada (1928) and Fu et al. (2007, 2009) described defensive glands in larvae of the genus *Luciola*. After molestation from meso- and metathorax and every abdominal segment (8) paired forked organs are everted from slit-like crevices which are situated above the gills (Fig. 9.12/6, 7). The anterior branch of these glandular sacs is longer than the posterior one. Fu et al. (2007) described the odor of *Luciola leii* secretion (present in all six instars) as pine oil-like, whereas Okada (1928) reported an odor in a Japanese *Luciola*-species resembling resin and peppermint. Until now terpinolene (Fig. 9.12/Formula 1) and gamma-terpinene (Fig. 9.12/Formula 2) could be identified in the defensive secretion of *L. leii* (Fu et al. 2007). When aquatic larvae are under attack they may run away, and exhibit thanatosis, glowing and with release of defensive secretion. Preliminary bioassays showed that *L. leii* larvae were refused by various predatory fishes or dragonfly larvae (Fu et al. 2007). However one has to take into consideration that most lampyrid beetles and their larvae synthesize toxic cardiac glycosides, the lucibufagins (Eisner et al. 2005).

**Chrysomelidae** (leaf beetles; about 46,000 species in many subfamilies are known): From several subfamilies such as Alticinae, Chrysomelinae, Galerucinae, or Hispinae there are known members living on emergent aquatic plants or even within phytotelmata. Adults of several species can live underwater or at the water surface (Jäch and Balke 2008). Representatives of subfamily Donaciinae with genera such as *Donacia, Macroplea*, or *Plateumaris* are especially adapted to freshwater. Their larvae and pupae are generally submerged, but also adults, which may carry an extensive plastron, are often found underwater. Larvae have caudal spines in order to penetrate tissues of aquatic plants to take up air. Cocoons which are spun from larval silk glands are fixed underwater and are water- and airtight. Donaciinae usually feed on monocotyledonous host plants or Nymphaeaceae; adults also prefer pollen. Defense mechanisms in Donaciinae are not known; however the abovementioned subfamilies of Chrysomelidae with semiaquatic species represent interesting taxa to investigate, since they exhibit various mechanisms of mechanical (Hispinae) and chemical defenses including sequestration of plant toxins (e.g., Burse and Boland 2015; Dettner 1987, 2015; Hilker and Meiners 2002; Opitz and Müller 2009; Pasteels et al. 1984, 1990).

**Nanophyidae** (about 300 species Jäch and Balke 2008): Several species of *Nanophyes* represent phytophilous water beetles feeding on *Alternanthera* and *Ludwigia.*

**Erirhinidae** (about 300 species, Jäch and Balke 2008) with some phytophilous water beetles.

**Curculionidae** (weevils; largest family of animals; more than 60,000 species, Jäch and Balke 2008): Few species from various subfamilies have invaded aquatic environments. Most species belong to Bagoinae but also those from Cleoninae, Notarinae, Stenopelminae, Tanysphyrinae, or Ceutorhynchinae aquatic representatives are known. Many species (larvae and adults) are mono- and oligophageous (e.g., *Alisma, Azolla, Equisetum, Butomus, Eichhornia, Glyceria, Lemna, Myriophyllum, Pistia, Potamogeton, Ranunculus, Ricciocarpus, Salvalina, Stratiotes*) and few species are used as biological control agents of introduced nuisance water plants (Merritt and Cummings 1996).

Most aquatic weevils swim and dive underwater and use plastron respiration. Larvae are mining but usually slowly crawl on host plants and pupate within air-filled cocoons located on roots of plants (Klausnitzer 1996). Adults of most species are mechanically defended by hard integuments; sometimes as in *Dicranthus elegans* they have additionally long and hard elytral spines. In several genera such as *Helodytes, Eubrychius*, or *Hydronomus* terrestrial and aquatic grooming pattern was described (Kovac and Maschwitz 2000). For grooming dark and later clear fecal droplets were repeatedly released from the abdominal tips.

**Trachypachidae** (false ground beetles; six sp.): The enigmatic Trachypachidae (Adephaga) have many morphological similarities with Dytiscoidea (e.g., Arndt and Beutel 1995); however they are not found in wet habitats and therefore not covered by Jäch and Balke (2008). In contrast they are xerophilous (Beutel and Arndt 2016) and molecular data place them phylogenetically within Geadephaga (Maddison et al. 2008). Although biology of this family is nearly unknown, the pygidial glands resemble those of carabids (Fig. 9.4/11, 12). The defensive chemicals of two *Trachypachus* species (Attygalle et al. 2004) with methacrylic acid and tiglic acid also resemble those secretions of terrestrial carabids. In contrast, trachypachids secrete octanoic acid, unknown sulfur compounds, or an aromatic (Fig. 9.11) which is typical for Hydradephaga (Dettner 1985; Dettner and Böhner 2009). Therefore further research is necessary in order to evaluate the chemotaxonomic value of these micrometabolites.

## 9.7 Shore Beetles According to Jäch and Balke (2009)

**Georissidae** (minute mud-loving beetles; 80 sp.): These beetles are usually riparian. By using buccal secretions adults are often covered by sand or mud (Fig. 9.12/8) and it is difficult to observe the melanized or metallic surfaces. Bameul (1989) described the camouflage of *G. crenulatus* which even prevents flight (Shepard 2003). In spring when georissid beetles are very active they exhibit a short thanatosis before moving away. Adults have no stridulatory organs (Hansen 1991).

**Limnichidae** (minute mars-loving beetles; 387 sp., Beutel and Leschen 2016; according to Yee and Kehl 2015: 335 sp.): These beetles feed on algae, are semiaquatic, and can be found on sandy beaches or coralline formations. Adults of Thaumastodinae (jumping shore beetles) can jump. Beetles of this family are accepted by fishes (Scrimshaw and Kerfoot 1987).

**Heteroceridae** (variegated mud-loving beetles; 300 sp.; Yee and Kehl: over 250 sp.): Adults and larvae live in tunnels in wet sand at water edges, in riparian areas. Adult heterocerid larvae can use secretion from dermal or salivary glands in order to stabilize their pupal chamber. After molestations beetles break through the mud and fly away (Klausnitzer 1996).

**HYMENOPTERA (sawflies, wood wasps, bees, wasps, ants; 0.1% aquatic)**: Among worldwide 132,000 hymenoptera species (Stork 2018: 116,861) only about 150 from 11 families represent mostly internal parasitoids of aquatic stages (eggs, larvae) of various water insects and hygrophilic spiders (Cover and Bogan 2015; Morse 2017; Wesenberg-Lund 1943). In contrast selected Agriotypinae and Pompilidae (e.g., *Anoplius*) represent external parasitoids of Trichoptera pupae (*Goera, Silo*) of, respectively, *Dolomedes* or *Pardosa* spiders (Morse 2017).

Interestingly by abdominal silk glands *Agriotypus* larvae produce a helical outgrowth from the trichopteran pupa which serves as a breathing device (Wesenberg-Lund 1943). Wesenberg-Lund (1943) and Heckman (2018) summarized typical aquatic hymenopterans from Chalcididae (e.g., *Pleurotropis, Trichogramma, Prestwichia*; hosts: cocoons and eggs of Gyrinidae, Dytiscidae, *Sialis*, Odonata, water bugs), Proctotrupidae (*Limnodytes, Litus, Anteris, Anagrus, Anaphes* = *Polynema*; hosts: eggs of water beetles, and water bugs, Odonata), Ichneumonidae (*Hemiteles, Apsilops, Atractodes, Trichocryptus*; hosts: cocoons of Gyrinidae, larvae, and pupae of various Diptera, Trichoptera, Lepidoptera), Braconidae (e.g., *Dacnusa, Chaenusa, Liposcia, Gyrocampa Hydroplitis*; hosts: larvae of *Hydrellia, Hydrocampa*), and Agriotypinae (see above). Several biological aspects of aquatic parasitoid species were discussed by Heckman (2018).

As in other insects, parasitic wasps are attacked by various predators ranging from vertebrates to invertebrates such as spiders and other insects. Because natural histories of aquatic hymenopteran species are mainly unknown, here only selected defensive mechanisms within those families are presented where aquatic representatives were reported. As known from various hymenopteran parasitoids they may be brightly colored, which indicates their unpalatability (Quicke 1997). Other larger species such as ichneumonids and braconids defend against vertebrates and invertebrates by stinging (Quicke 1997), which sometimes might be really painful. Certainly only females are effective, since males lack ovipositors for venom injection. Certain parasitoids in addition produce distinct, sometimes unpleasant, odors which originate from exocrine glands. In female *Alloxysta* species mandibular glands produce deterrents such as 6-methyl-5-hepten-2-one, actinidin, and various iridoids (Völkl et al. 1994) which are directed against predators. Other species produce also perillene-isomers or iridoids such as nepetalactone and iridomyrmecin (Hübner et al. 2002). Many ichneumonids but also braconids and certain Oppiinae are odoriferous which may also originate from mandibular glands (own observations, Quicke 1997) and are probably directed against wasps. Unfortunately only few data exist on exocrinology of hymenopteran parasitoid wasps; however some pheromones, kairomones, and venom proteins are compiled by Keeling et al. (2004), Dettner (2015), and Asgari and Rivers (2010). Other parasitic Hymenoptera show thanatosis (death feigning). Interestingly some braconids are polymorphic for this behavior and in *Habrobracon* the gene controlling thanatotic behavior is inherited as a single dominant allele (Grosch 1988; Quicke 1997). Parasitic wasps may act as models in mimicry systems or may be mimics themselves (for theory see Dettner and Liepert 1994). In addition especially ichneumonids also might construct cryptic or mimetic cocoons (Quicke 1997). Finally larger braconids may be protected from predators by heavily sclerotized thoracic spines (Quicke 1997). In larger ichneumonids there occur even poison claws and urticating hairs (Hanson and Gauld 1995).

**DIPTERA (true flies; about 50% aquatic)**: Worldwide there exist about 154,000 species of Diptera (Stork 2018: 155,477), most of which have aquatic or semiaquatic eggs and larvae. There exist various reviews where aquatic and semiaquatic dipteran larvae are compiled (e.g., Italian and European species: Rivosecchi 1984; Faasch 2015; neotropical species: Fusari et al. 2018; world: Hennig 1968, Morse 2017). Especially the following aquatic and subaquatic families should be denoted: Ceratopogonidae,

Chironomidae, Deuterophlebiidae, Simuliidae, Thaumaleidae, Canacidae, Culicidae, Dixidae, Corethrellidae, Chaoboridae, Blephariceridae, Nymphomyiidae, Psychodidae, Stratiomyidae, Dryomyzidae, Syrphidae, Periscelididae, and Ephydridae. In addition further families may also be of importance: Cylindrotomidae, Tipulidae, Limoniidae, Tanyderidae, Ptychopteridae, Scatopsidae, Athericidae, Rhagionidae, Tabanidae, Pelecorhynchidae, Empididae, Dolichopodidae, Lonchopteridae, Phoridae, Muscidae, Scathophagidae, Sarcophagidae, and Sciomyzidae. Since bionomics of many Diptera are unknown (Oosterbroek 2006), especially at the larval stage (Hennig 1968), the knowledge on defenses of dipteran larvae is deficient. As demonstrated by Crespo (2011) *Chaoborus* larvae (Chaoboridae) can sense fish kairomones which affect their vertical migration. Also chironomids can assess different concentrations of infochemicals. For the predatory *Toxorhynchites* larvae it was shown that not chemical cues but vibrations (detected by thoracic and abdominal setae) were necessary to detect prey organisms (Crespo 2011). As a whole Crespo (2011) stressed that studies on predator avoidance have not received the same attention in Diptera as in other aquatic insects.

One fascinating aspect of mosquito perceiving of kairomones of larval predators was presented by Crespo (2011). Recent molecular data on olfactory binding proteins in larvae and adult mosquitoes showed that several of them are present in both immature and mature stages. Therefore it seems conceivable that adult mosquitoes may utilize information which is relevant during their larval stages. That female *Culiseta* mosquitoes only oviposit in those waters where no *Notonecta* kairomones (such as n-heneicosane), the predators of the *Culiseta*-larvae, are present was discussed in chapter Notonectidae.

Apart from flight reactions (often achieved by peristaltic movements) which might be abundant in dipteran larvae, secondary defenses such as chemical defense or thanatosis obviously are rare (Dettner 2015). However primary defense mechanisms such as larval shelters ( Chironomidae, Simuliidae), crypsis, translucence, or stable and elastic cuticles may be more abundant in spite of the fact that larval body colors are not abundant (Dettner 2015). Tergites and sternites of fly larvae usually are reduced in order to obtain optimal mobility.

Larval integuments of Diptera may sometimes possess micro- and macrotrichia, spines, and dermal glands but no complex glands. Often larvae show fleshy projections, papillae, pseudopods, or suctorial discs. Even pupae of aquatic Diptera often have abdominal spines. That the abovementioned structural characters such as hairs and/or body sizes of the larvae can protect chironomid larvae from different kind of predators is amply illustrated by the study of Hershey and Dodson (1987). Large larvae of *Cricotopus bicinctus* were less susceptible to predation by *Hydra* and small *Ischnura* larvae. In contrast long hairs protected *Cricotopus sylvestris* from *Hydra* but not from *Ischnura.*

In the following examples of defense in aquatic or semiaquatic dipteran larvae are given.

**Ceratopogonidae** (biting midges): According to Ronderos et al. (2018) there exist worldwide 6268 sp. In Ceratopogonidae it was proved that female canthariphilous species such as *Atrichopogon trifasciatus* are able to transfer toxic cantharidin into their eggs and larvae (Dettner 1997). Larvae of *Forcipomyia* possess secretory setae dorsally on their bodies (Fig. 9.14/2), where various antibacterial and hygroscopic

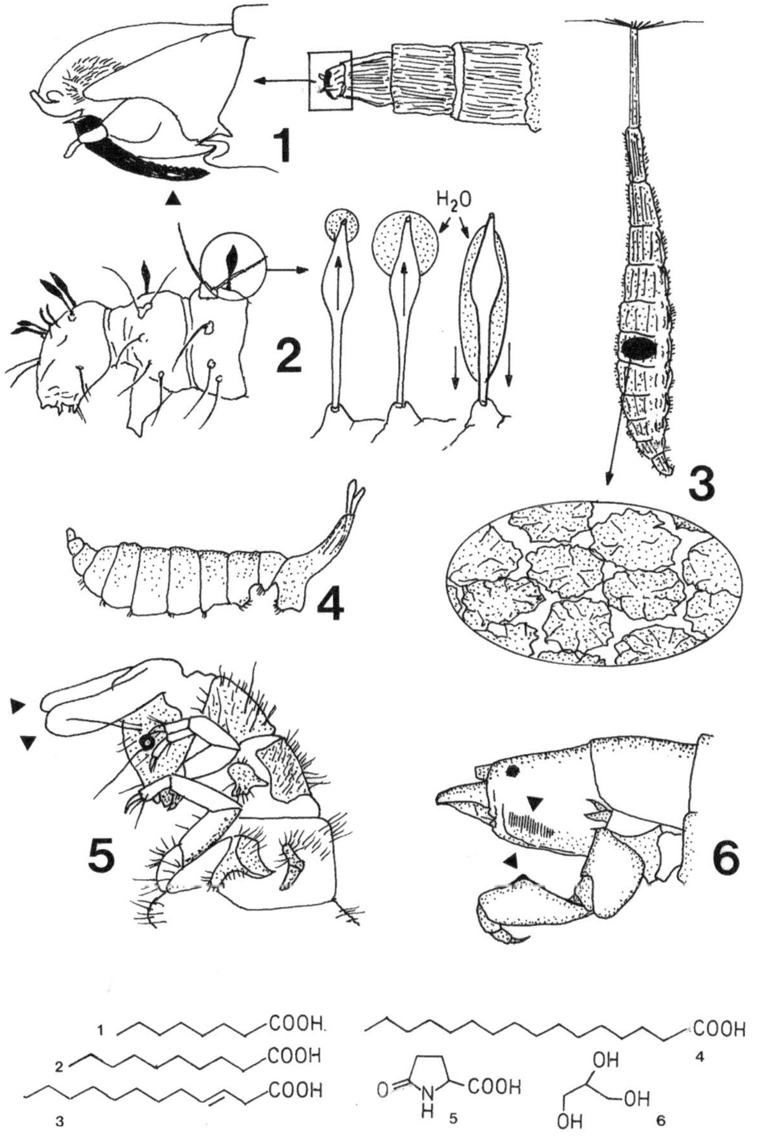


**Fig. 9.14** Morphological defense, defensive glands, and stridulatory organ of dipteran (1–4) and trichopteran larvae (5, 6). 1. *Tabanus* spec. larva (Tabanidae; right) and head capsule with pierced mandible (left). 2. Fore-end of *Forcipomyia* larva (Ceratopogonidae) with secretory setae (black; left side). Enlargement of secretory setae (right) showing depletion of secretion and its hydroscopic effects. 3. Larva of *Stratiomys* spec. (Stratiomyidae) with hard integument by calcareous crystals (enlargement). 4. Pupa of *Ephydra bivittata* (Ephydridae) with gin trap between segments 6 and 8. 5. Head and thorax of *Apataniana hellenica* (Limnephilidae) with everted prothoracic glands (arrows). 6. Stridulatory organ of *Hydropsyche pellucidula* (Hydropsychidae) with file on head underside and scraper on profemur (arrows). Defensive secretions: **1.** Caprylic acid (*Apatania, Forcipomyia*), **2**. capric acid (*Apatania, Forcipomyia*), **3**. 3-dodecenoic acid (*Apatania*), **4**. palmitic acid (*Forcipomyia*), **5**. pyroglutamic acid (*Forcipomyia*), **6.** glycerol (*Forcipomyia*).
Modified according to Jacobs and Renner 1974 (1, 3), Urbanek et al. 2011 (2), Rivosecchi 1984 (4), Wichard et al. 1995 (5), Aiken (1985)

compounds such as fatty acids (Fig. 9.14/**1**, **2**, **4**), glycerol (Fig. 9.14/**6**), and pyroglutamic acid (Fig. 9.14/**5**) are produced (Urbanek et al. 2012). *Atrichopogon* larvae and pupae of various genera bear abdominal spines which may represent a kind of mechanical defense (Ronderos et al. 2018).

**Lonchopteridae** (spur-winged flies): 58 species worldwide (Fusari et al. 2018): Lonchopterid larvae live on shores of water. They are often dark colored and have broad dorsal and smaller lateral sclerotized plates which may have a protective value.

**Tabanidae** (horse and deer flies): According to Fusari et al. (2018) there exist worldwide about 4500 species. Larvae are predators and often have a thick integument (Fig. 9.14/1). From each segment 4–10 of the often colored larvae there exists a thick ring, each supplied with various parapodia and sometimes with additional spines. Larvae have a retractile head capsule with pierced mandibles, where toxins from the salivary gland are applied (Fig. 9.14/1). Bites of some *Tabanus* larvae are known to be as painful as bee stings. Injections of tabanid venoms into insects produce moveless animals after one or two spasms (Schmidt 1982). In the field even tadpoles are killed immediately.

**Stratiomyidae** (soldier flies): According to Pitaluga de Godoi and Pujol-Luz (2018) there exist worlwide 2800 species (928 species are aquatic, Morse 2017). The integument of armored, dorsoventrally flattened larvae is leatherlike, with a lot of calcareous crystals which are produced within malpighian tubules (Fig. 9.14/3). The hard and flexible integument protects the larvae from predators but also represents a protective device against hymenopteroid parasitoids (Wesenberg-Lund 1943).

**Syrphidae** (syrphid flies; worldwide about 6000 sp.; 1341 aquatic species according to Morse 2017): Certain larvae of usually saprophagous aquatic Eristalinae possess a long anal segment ("rat-tailed larvae") which acts as a protecting sheath for the breathing tube. Since *Eristalis tenax* and various other syrphid species are able to survive in extremely polluted aquatic habitats various antibacterial peptides could be identified within the hemolymph of these larvae (Altincicek and Vilcincskas 2007). In contrast cuticles of syrphid larvae with both short and long breathing tubes have impervious cuticles which prevent entering of many toxic substances from the water (Heckman 2018). Similar microbiological results and unusual products were achieved by analysis of *Helaeomyia petrolei* larvae (**Ephydridae**, shore flies; 1251 species Morse 2017) from extreme habitats (Kadavy et al. 2000). Pupal representatives of this family such as *Ephydra bivittata* also possess gin traps as defense against parasitoids (Fig. 9.14/4). Predatory syrphid larvae of *Ocyptamus* live in tanks of bromeliads and subdue their prey with venoms (Miranda and Rotheray 2018).

**TRICHOPTERA (caddisflies; more than 99% aquatic):** Few defensive mechanisms are known from Trichoptera, the sister group of Lepidoptera. They have worldwide about 15,000 species (Stork 2018: 14,391) and are found in nearly all freshwater habitats (Heckman 2018; Pes et al. 2018). Nearly all species have long-living aquatic larvae with about five larval stages. In 1984 Otto showed that trout in daylight preferably attacked caddis-larvae which possessed head capsules with contrasting color patterns compared to uniformly colored specimens. However this preference was reversed under reduced illuminations (Otto 1984). Many Trichoptera larvae produce transportable case buildings from silk glands as primary defensive behavior against both vertebrate and invertebrate predators (Morse 2003). A wide range of plant and mineral materials are used and are orientated longitudinally or horizontally. Such caddisfly cases may be classified as tube cases (most Integripalpia: Limnephilidae, Goeridae, Phryganeidae, Brachycentridae, Lepidostomatidae, Beraeidae, Sericostomatidae, Odontoceridae, Molannidae, Leptoceridae), saddle or tortoise cases (Glossosomatidae), or purse cases (Hydroptilidae). Some cases are only made from silk, and may contain shells of freshwater molluscs, freshwater sponges, or pieces of corals (marine species). In contrast about half the caddisfly species do not build cases, but spin silken retreats which are fixed on the substrates and serve for capturing food (e.g., Philopotamidae, Psychomyiidae, Ecnomidae, Polycentropodidae, Hydropsychidae, Arctopsychidae). There are also free-living larvae moving on substrates (e.g., Rhyacophilidae, Hydrobiosidae). Efficiency of different trichopteran larvae with various types of cases was tested against predatory *Anax* dragonfly larvae (Ferry et al. 2013). It was astonishing that the presence of any case, regardless of the material used, offers survival benefits against dragonfly larvae. In contrast caddisflies removed from their case were attacked and captured. Larval cases of *Limnephilus* larvae surviving exposure to dragonfly larvae and predacious minnows were significantly stronger (greater mineral fraction), longer, and wider than cases of victims (Nislow and Molles Jr 1993). Finally Boyero et al. (2006) showed that *Potamophylax* larvae can discriminate predators such as dragonfly larvae, salamander larvae, and trouts by chemical cues and alter their choice of case type according to risks of predation. As mentioned by Crespo (2011) long-range identification of food was not dependent on intact antennae of trichopteran larvae but may be achieved by sensilla in maxillary and galeal palps. He also stressed that predator avoidance responses had not been studied so far in trichopteran larvae.

However one exceptional example is described below. When disturbed aquatic larvae of the genera *Apatania* and *Apataniana* (Limnephilidae) release a defensive fluid (1–2 μg secretion per specimen) from an eversible prothoracic gland (Fig. 9.14/5; Wichard et al. 1995), which shows paralyzing effects against small invertebrates (other trichopteran larvae such as *Rhyacophila*). When *Rhyacophila* penetrates the *Apatania* case, at first the larvae withdraw, and then release droplets of the secretion. In contrast the secretion is ineffective against larger targets such as plecopteran larvae or fishes (Wagner et al. 1990). The secretion contains about 40 $C_7$- to $C_{14}$-carboxylic acids (Fig. 9.14/**1–3**) with the main constituents 5-octenoic-, octanoic-, decanoic-, 3-dodecenoic- (Fig. 9.14/**3**), 3,5-dodecadienoic-, and 3,5,7,11-tetradecatetrenoic acids (Wagner et al. 1990). These eversible larval limnephilid defensive glands resemble the osmeteria which are found in various caterpillars such as Notodontidae or Papilionidae. It is noteworthy that a single irritation does not result in a complete loss of defensive secretion. Another defensive behavior was observed in larvae of Hydropsychidae. In order to defend their retreats against intruders they communicate and produce stridulatory signals in rubbing the scraper of the forefemur against a file on the underside of the head (Fig. 9.14/6; Aiken

1985). Also *Plectrocnemia* larvae will defend their retreats and fight with open mandibles against intruders (Solem and Gullefors 1996).

**LEPIDOPTERA (moths, butterflies; 0.5% aquatic)**: According to Graca and Solis (2018) there exist 157,000 lepidopteran species worldwide (Stork 2018: 157,338). Most of the species are terrestrial; however several families contain few or various semiaquatic or aquatic species at least caterpillars. Females of aquatic species usually deposit eggs in underwater vegetation. In few species females cover their egg masses with thick secretion. In *Acentria ephemerella* there exist even females which have no wings and are living completely underwater (Reichholf 1970; Vallenduuk and Cuppen 2004; Wesenberg-Lund 1943). Newly hatched larvae may attach silken threads to the leaf as anchor. As a kind of protection from enemies especially early larval instars of several species feed as miners or boreres within stems or leaves of aquatic or semiaquatic plants. By using their salivary glands later instars or larvae from other species may construct cases from their food plants like trichopteran larvae and therefore to some extent might be protected. Oxygen intake by larvae may be achieved through tracheal gills, plastron-like devices, and use of bubbles in interstitial places or in cases made with the host plant. Depending on species and larval stages hydrophobicity and wettability of larval integuments may vary considerably. Pupation of aquatic lepidopteran larvae takes place in or out of water inside silken cocoons.

Examples of aquatic lepidopteran families, some genera involved, and food plant families are given below: **Cosmopterigidae** (1792 sp.): *Cosmopterix atte* (Cyperaceae); **Momphidae** (114 sp.): *Monpha* (Onagraceae); **Tortricidae** (10,387 sp.): *Bactra* (Cyperaceae); *Clepsis* more than 60 species associated with (Cyperaceae) *Choristoneura* and *Archips* (Typhaceae); **Pyralidae** (5921 sp.): *Acola* (Amaranthaceae); and **Crambidae** (9655 sp.) within subfamily Acentropinae according to Cover and Bogan (2015) most of the 730 species worldwide are aquatic: *Acentria, Argyractis, Cataclysta, Donacaula, Elophila, Hygraula, Neargyractis, Niphograpta, Nomophila, Nymphula, Ostrinia, Oxyelophia, Parapoynx, Petrophila, Samea, Sameodes, Schoenobius, Thalia, Usingeriessa,* and *Xubida*. Their food plant families are Alismataceae, Amaranthaceae, Araceae, Callitrichaceae, Cannaceae, Ceratophyllaceae, Characeae, Cyperaceae, Gramineae, Haloragaceae, Hydrocharitaceae, Lemnaceae, Marantaceae, Nelumbonaceae, Nymphaeaceae, Poaceae, Polygonaceae, Pontederiaceae, Potamogetonaceae, Salviniaceae, Sparganiaceae, Typhaceae, and Zosteraceae.

**Sphingidae** (1463 sp.): *Eumorpha* (Onagraceae), **Erebidae** (24,600 sp. including Arctiidae and Lymantriidae): *Paracles*, and **Noctuidae**: *Bellura* (Cambombaceae, Gramineae, Hydrocharitaceae, Lentibulariaceae, Nymphaceae, Pontederiaceae, Typhaceae).

As a whole there are not known defensive mechanisms of aquatic Lepidoptera. Generally Pyralidae larvae are known to regurgitate as a response to disturbance (Dettner 2015). A lot of terrestrial caterpillars feed on toxic plants and are able to tolerate and to enrich these toxins as defense mechanisms (Nishida 2002; Opitz and Müller 2009). In comparing the abovementioned aquatic lepidopteran families and their food plants (plant family level) with defensive mechanisms and toxin sequestration in terrestrial Lepidoptera there was no correspondence at all.

**MECOPTERA (scorpionflies, hangingflies; 1% aquatic)**: There is one relict family (with 8 species) in Australia and South America among 9 families worldwide (600 species; Stork 2018: 757) which is characterized by long, slender aquatic larvae, which resemble Corydalidae or Trichoptera larvae. These representatives of Nannochoristidae have four instars and live in stream benthos and prey upon larvae of Chironomidae and other dipterans (Machado 2018; Cover and Bogan 2015; Pilgrim 1972). Adults of many Mecoptera fall to the ground and show thanatosis when disturbed (Dettner 2015). Very often they emit a malodorous digestive fluid from the mouth when they are irritated. Until now larval defense in Mecoptera is unknown.

## 9.8 Conclusions

After surveying defense mechanisms of aquatic and semiaquatic water insects and after comparing the results with antipredatory data of Witz (1990) there are some interesting results. Generally there is a high significance of thanatosis, escape, withdrawal, and especially stabile elastic or hard exoskeleton. Comparable to all arthropods (Witz 1990) defensive stridulation is more abundant in water insects and especially distributed in adults of different water beetles or Heteroptera, but also uniquely in larvae of Odonata and Trichoptera. Following result is astonishing: Exclusively notonectid bugs which have secondarily lost their complex thoracic defensive gland exhibit stridulatory organs. Such stridulatory organs are not found in those notonectids with intact thoracic glands. Generally stridulation in water insects seems primarily defensive stridulation, before it is used intraspecifically for alarm or communication.

The presence of hard exoskeletons and spines as observed in larvae of Odonata or Ephemeroptera should not only represent defenses against fishes but also might be due to kairomonal effects which seem to be abundant in freshwater ecosystems (Brönmark and Hansson 2012). Many phenomena concerning such mechanisms of defense have to be investigated in the future and especially in tropical water insects (especially Indo-Malayan or Afrotropical regions, see Stork 2018). Moreover behavior and bionomy of aquatic dipteran larvae are mostly unknown.

Exocrine glands and chemical defense in aquatic insects are not so widespread and abundant as in terrestrial arthropods (Witz 1990) or in marine systems. The only widespread complex gland in adephagan beetles is present within all Adephaga. In addition the thoracic gland of water bugs is primarily present in all Heteroptera. The same applies to the diversity and biosynthetic variations of the constituents. It is remarkable that the ubiquitous defensive system in terrestrial habitats, the quinones, is absolutely absent in water insects (Blum 1981; Dettner and Peters 2010). The extensive use of aromatic compounds and antimicrobics in freshwater in contrast to terrestrial systems is amply demonstrated by comparing adephagan land and water beetles. Since insects do not possess the metabolic capacity to generate the steroid skeleton de novo, steroids are seldom, because precursors must be taken from outside. Therefore it is remarkable that steroids are abundant in prothoracic glands of larger Dytiscidae and the maxillary glands of belostomatid bugs. Three of four steroids from the maxillary gland of belostomatids are also present in prothoracic beetle glands and this convergent evolution might indicate that these qualitatively identical compounds in extremely high concentrations are targeted against larger predators such as fishes or amphibians. Typical defensive mechanisms of terrestrial taxa such as reflex bleeding of toxic hemolymph or sequestration of toxic plant compounds and use for own defense are absolutely missing in water insects. The only case of reflex bleeding in Ephemeroptera seems to be a mechanical defense and has nothing to do with toxic hemolymph. Moreover there are no toxic spines or urticating hairs in aquatic insects. Apart from tabanid larvae and several predatory bugs also venoms and stings in freshwater insects evolved only few times as compared with terrestrial arthropods.

Highly astonishing is the scarcity of defensive mechanisms in aquatic larval stages. Chemical defense by exocrine glands in larval stages is only present in aquatic lampyrids and one trichopteran group. In contrast larvae of Odonata, Plecoptera, Ephemeroptera, and most of Trichoptera lack any defensive chemicals. Also defensive stridulation is rare within larvae; however larval mechanical defense seems widespread. Many larvae of Heteroptera possess abdominal defensive glands; however there exist no chemical or behavioral data. In contrast to Heteroptera larvae the larval dytiscid cephalic gland seems not to be defensive. What are the reasons for missing larval defenses in aquatic freshwater habitats? Larvae are often short-lived (not in Odonata, etc.); therefore biosynthesis of allomones or development of stridulation structures might be too expensive.

As compared with marine or terrestrial systems, in freshwater arthropods we cannot observe any uptake of toxic compounds from plants in order to generate and optimize own chemical defense and increase of fitness. Obviously there exist no trichopteran or aquatic lepidopteran larvae, which would incorporate toxic plant structures in their case or alternatively would incorporate these compounds into their body. In terrestrial systems incorporation of toxic plant constituents is widely distributed and overwhelming at least in lepidopteran larvae (Nishida 2002; Opitz and Müller 2009). In addition there was no record that toxic invertebrates from freshwater such as aposematically colored and toxic water mites or flatworms (Dettner 2010) were fed by carnivorous water insects in order to incorporate toxic chemicals. Finally urticating hairs or allergic reactions which are widespread in terrestrial arthropods are not known in aquatic insects apart from allergic reactions to haemoglobin of certain Chironomidae.

There is another chapter with only few data in aquatic entomology. The significance of microorganisms as producers of behavioral modifying compounds such as allomones or their role as symbionts or parasites of their insect hosts is mainly unknown. In Dytiscidae a rich aerobic, heterotrophic bacterial flora was identified inside the foregut which shows elements of an autochthonous bacterial flora in the foregut (Schaaf and Dettner 1997). Later it was shown that actinomycete species within the crops of beetles play an important role in transformation of food steroid precursors into prothoracic defensive steroids (Dettner 2014). Furthermore the role and significance of intracellular microorganisms in water insects have to be investigated. However their phylogenetic relationships seem interesting: *Rickettsia* species from Dytiscidae were shown to be related to rickettsial isolates from aquatic Limoniidae (Diptera), *Lutzomyia apache* (Diptera: Psychodidae), or *Hemiclepsis* or *Torix* leeches (Küchler et al. 2009).

As was demonstrated by various publications of Kovac and Maschwitz (1990b, 1991, 2000) secretion grooming is an abundant behavior especially in adult freshwater insects. This behavior prevents contamination of the hydrofuge respiratory regions by bacteria or peritrichic ciliates, which would otherwise lead to loss of the air bubble, wetting of these areas, and finally drowning of the insect. Usually the antimicrobial secretions of complex glands such as pygidial glands of water beetles or thoracic glands of aquatic bugs increase wettability of the hydrophobous integuments of aquatic insects (Dettner 1985). In order to decrease this wettability there must be certain dermal glands which can rapidly produce (maybe through oil-secreting) waterproofing surface areas (Holdgate 1955; Beament 1976). There is considerable knowledge on physicochemical properties of integuments of water-walking arthropods (Bush et al. 2008) in contrast to true water insects which may also show amphibious habits. Certainly the qualitative and quantitative distribution, chemistry and functional role of these dermal glands must be investigated in selected water insects especially with respect to the passage from terrestrial into aquatic environment and vice versa (a), a stable and balanced position within the water (b), and an optimal function of breathing (c). In order to fulfill these various demands as change of terrestrial/aquatic habitats, breathing, care of body, and predator defenses it is conceivable that most exocrine secretions of water insects have a multifunctional significance.

**Acknowledgments** In order to prepare this manuscript the help of the following collaborators and colleagues is highly acknowledged: A. Böttcher (Bayreuth), B. Dettner (Bayreuth), E. Helldörfer (Bayreuth), H. Luthardt (Bayreuth), S. Wagner (Bayreuth), V. Lavanya (Springer), Prof. Dr. Kleber del Claro (Fed. Univ. Uberlandia, Brazil) and Prof. Dr. R. Guillermo Ferreira (Sao Carlos, Brazil). During my Ernst Bresslau guest professorship at the Zoological Institute (chemical ecology in marine and freshwater systems) of University of Cologne I was highly supported by Prof. Dr. E. von Elert and PD Dr. P.Fink.

## References

Aiken RB (1985) Sound production by aquatic insects. Biol Rev 60:163–211

Aldrich JR (1988) Chemical ecology of the Heteroptera. Annu Rev Entomol 33:211–238

Altincicek B, Vilcincskas A (2007) Analysis of the immune-inducible transcriptome from microbial stress resistant, rat-tailed maggots of the drone fly *Eristalis tenax*. BMC Genomics 8:326

Andersen NM (1982) The semiaquatic bugs (Hemiptera, Gerromorpha). Phylogeny, adaptations, biogeography and classification. Entomonograph 3:1–455

Andersen NM (1996) Heteroptera: Gerromorpha. Semiaquatic bugs. In: Nilsson A (ed) Aquatic insects of North Europe. Apollo Books, Stenstrup, pp 77–90

Andersen NM, Weir TA (2004) Australian water bugs. Apollo Books, Stenstrup

Angus R (1992) Insecta Coleoptera Hydrophilidae Helophorinae. In: Schwoerbel J, Zwick P (eds) Süßwasserfauna von Mitteleuropa 20/10-2. Fischer, Stuttgart

Anjos-Santos D, Neiss UG, Pessacq P (2018) Superfamily Calopterygidae. In: Hamada N, Thorp JH, Rogers DC (eds) Keys to neotropical hexapoda, Thorp and Covich's freshwater invertebrates, vol 3. Academic Press, London, pp 449–468

Ardila-Camacho A, Contreras-Ramos A (2018) Order Megaloptera. In: Hamada N, Thorp JH, Rogers DC (eds) Keys to neotropical hexapoda, Thorp and Covich's freshwater invertebrates, vol III. Academic Press, London, pp 217–227

Armisen D, Refki PN, Crumiere AJJ, Viala S, Toubiana W, Khila A (2015) Predator strike shapes antipredator phenotype through new genetic interactions in water striders. Nat Commun 6:8153

Arndt E, Beutel RG (1995) Larval morphology of *Systolosoma* solier and *Trachypachus* motschulsky (Coleoptera: Trachypachidae) with phylogenetic considerations. Ent Scand 26:439–446

Arnett RH, Thomas MC (2001) American Beetles, vol 1. CRC, Boca Raton, FL

Asahina S (1950) On the life history of *Epiophlebia superstes* (Odonata, Anisozygoptera). Proceedings of the VIIIth Congress of Entomology, Stockholm, pp 337–341

Asgari S, Rivers DB (2010) Venom proteins from endoparsitoid wasps and their role in host-parasite interactions. Annu Rev Entomol 56:313–335

Attygalle AB, Wu X, Ruzicka J, Rao S, Garcia S, Herath K, Meinwald J, Maddison DR, Will KW (2004) Defensive chemicals of two species of *Trachypachus motschulski*. J Chem Ecol 30:577–588

Avelino-Capistrano F, Pessacq P, Barbosa LS (2018) Order Plecoptera. In: Hamada N, Thorp JH, Rogers DC (eds) Keys to neotropical hexapoda, Thorp and Covich's freshwater invertebrates, vol 3. Academic Press, London, pp 119–141

Balke M, Hendrich L (2016) 7.6 Dytiscidae Leach, 1815. In: Beutel RG, Kristensen NP (eds) Handbook of zoology, Arthropoda: Insecta. Coleoptera, Beetles, vol 1, 2nd edn. De Gruyter, Berlin, pp 118–140

Balke M, Larson DJ, Hendrich L (1997) A review of the New Guinea species of *Laccophilus* Leach 1815 with notes on regional melanism (Coleoptera Dytiscidae). Trop Zool 10:295–320

Balke M, Alarie Y, Beutel RG (2018) 7.8 Meruidae Spangler & Steiner, 2005. In: Beutel RG, Kristensen NP (eds) Handbook of zoology, Arthropoda: Insecta. Coleoptera, Beetles, vol 1, 2nd edn. De Gruyter, Berlin, pp 149–158

Bameul F (1989) Description du comportement de camouflage d'un Coléoptère: le déguisement actif de *Georissus crenulatus* (Coleoptera Georissidae), et proposition d'une nouvelle classification des déguisements chez les invertébrés. Compt Rend Acad Sciences Paris 309(série III):351–356

Barbier M (1990) Marginalin, a substance from the pygidial glands of *Dytiscus marginalis* (Coleoptera): molecular associations with polyamines in vitro. Z Naturfoschg 45b:1455–1456

Barth R (1960) Ueber die Pygidialdrüse von *Enhydrus sulcatus* (Wied., 1821) (Coleoptera, Gyrinidae). Mem Inst Oswaldo Cruz 58(2):135–147

Bauernfeind E, Soldan T (2012) The mayflies of Europe (Ephemeroptera). Brill Academic Publishers, Leiden

Bay EC (1974) Predator-prey relationships among aquatic insects. Annu Rev Entomol 19:441–453

Barthlott W, Riede K, Wolter M (1994) Mimicry and ultrastructural analogy between the semi-aquatic grasshopper Paulinia acuminata (Orthoptera: Pauliniidae) and ist foodplant, the water-fern Salvinia auriculata (Filicatae: Salviniaceae). Amazoniana 12:47-58

Beament JWL (1976) The ecology of cuticle. In: Hepburn HR (ed) The insect integument. Elsevier, Amsterdam, pp 359–374

Beaty SR (2015) The Plecoptera of North Carolina: a biologist's handbook for the identification of stonefly nymphs with standard taxonomic effort levels. Version 4.0. North Carolina

Department of Environmental Quality, Division of water resources, Biological Assessment Branch, Raleigh, NC
Beier M (1929) Zur Kenntnis der Lebensweise von *Haliplus wehnckei* Gerh. Z f Morph Ökol Tiere 14:191–233
Bell WJ, Roth LM, Nalepa CA (2007) Cockroaches: ecology, behavior, and natural history. Hopkins University Press, Baltimore, MD
Benfield EF (1974) Autohemorrhage in two stoneflies (Plecoptera) and its effectiveness as a defense mechanism. Ann Entomol Soc Am 67:739–742
Benton MJ, Pritchard G (1990) Mayfly locomotory Responses to endoparasitic infection and predator presence: the effects on predator Encounter rate. Freshw Biol 23:363-371
Bertrand H (1928) Les Larves et Nymphes des Dytiscides, Hygrobiides, Haliplides. Encyclopédie Entomologique X, Lechevalier, Paris
Betz O (2010) Adhesive exocrine glands in insects: morphology, ultrastructure, and adhesive secretion. In: Byern JV, Grunwald I (eds) Biological adhesive systems. Springer, Wien, pp 111–152
Betz O, Kölsch G (2004) The role of adhesion in prey capture and predator defence in arthropods. Arthropod Struct Dev 33:3–30
Betz O, Koerner L, Dettner K (2018) The Biology of Steninae. In: Betz O, Irmler U, Klimaszewski J (eds) Biology of Rove Beetles (Staphylinidae). Springer, Cham, pp 229–283
Beutel R (1986) Skelet und Muskulatur des Kopfes und Thorax von *Hygrobia tarda* (Herbst). Ein Beitrag zur Klärung der phylogenetischen Beziehungen der Hydradephaga (Insect: Coleoptera). Stuttg Beitr Naturk Ser A 7 Nr 388:1–54
Beutel RG, Arndt E (2016) Trachypachidae C. G. Thomson, 1857. In: Beutel RG, Kristensen NP (eds) Handbook of zoology, Arthropoda: Insecta. Coleoptera, Beetles, vol 1, 2nd edn. De Gruyter, Berlin, pp 159–162
Beutel RG, Leschen AB (eds) (2016) Handbook of zoology, Arthropoda: Insecta. Coleoptera, Beetles, vol 1, 2nd edn. De Gruyter, Berlin
Beutel RG, Roughley RE (1993) Phylogenetic analysis of Gyrinidae based on characters of the larval head (Coleoptera: Adephaga). Entomo Scand 24:459–468
Beutel RG, Roughley RE (2016) 7.1 Gyrinidae Latreille, 1810. In: Beutel RG, Kristensen NP (eds) Handbook of zoology, Arthropoda: Insecta. Coleoptera, Beetles, vol 1, 2nd edn. De Gruyter, Berlin, pp 80–89
Beutel RG, Ruhnau S (1990) Phylogenetic analysis of the genera of Haliplidae (Coleoptera) based on characters of adults. Aquat Insects 12:1–17
Beutel RG, Balke M, Steiner ES Jr (2006) The systematic position of Meruidae (Coleoptera, Adephaga) and the phylogeny of the smaller aquatic adephagan beetle families. Cladistics 22:102–131
Beutel RG, Friedrich F, Aspöck U (2010) The larval head of Nevrothidae and the phylogeny of Neuroptera (Insecta). Zool J Linnean Soc 158:533–562
Beutel RG, Friedrich F, Ge SQ, Yang XK (2014) Insect morphology and phylogeny. De Gruyiter, Berlin
Beutel RG, Balke M, Ribera I (2016) 3.1. Aspidytidae Ribera, beutel, Balke and Vogler, 2002. In: Beutel RG, Kristensen NP (eds) Handbook of zoology, Arthropoda: Insecta. Coleoptera, Beetles, vol 1, 2nd edn. De Gruyter, Berlin, pp 21–28
Blum MS (1981) Chemical defenses of Arthropods. Academic Press, New York
Blunck H (1923) Die Entwicklung des *Dytiscus marginalis* L. vom Ei bis zur Imago. 2.Teil. Die Metamorphose (B Das Larven und das Puppenleben). Z Wiss Zool Abt A 121:172–392
Bott HR (1928) Beiträge zur Kenntnis von *Gyrinus natator substriatus* Steph. Z Morph Ökol Tiere 10:207–306
Boyero L, Rincon PA, Bosch J (2006) Case selection by a limnephilid caddisfly (*Potamophylax latipennis* (Curtis)) in response to different predators. Behav Ecol Sociobiol 59:364–372
Brancucci M (1975) Die Dytisciden und die Oberflächenspannung des Wassers (Coleoptera). Deutsche Entomol Z NF 24:423–424

Brancucci M (1985) A Review of the biology and structure of *Geodessus besucheti* Brancucci (Coleoptera, Dytiscidae). Proc Acad Nat Sci Phil 137:29–32

Branham MA (2010) 4.15 Lampyridae. In: Kristenasen NP, Beutel RG (eds) Handbook of zoology, Coleoptera, Beetles, vol 2, 2nd edn. De Gruyter, Berlin, pp 141–149

Brittain JE (1990) Life history strategies in Ephemeroptera and Plecoptera. In: Campbell IC (ed) Mayflies and stoneflies. Kluwer, Amsterdam, pp 1–12

Brönmark C, Hansson LA (2012) Chemical ecology in aquatic systems. Oxford University Press, Oxford

Brossut R (1993) Allomonal secretions in cockroaches. J Chem Ecol 9:143–158

Burks RL, Lodge DM (2002) Cued in: advances and opportunities in freshwater chemical ecology. J Chem Ecol 28:1901–1917

Burse A, Boland W (2015) RNAi based functional analysis of biosynthetic enzymes and transport proteins involved in the chemical defense of juvenile leaf beetles. In: Hoffmann KH (ed) Insect molecular biology and ecology. CRC, Boca Raton, FL, pp 350–375

Bush JWM, Hu DL, Prakash M (2008) The integument of water-walking arthropods: form and function. Adv Insect Physiol 34:117–192

Cao C-Q, Liu ZW, Chen SZ, Tong C (2012) Swimming behavior of the aquatic larva of *Neoneuromus ignobilis* (Megaloptera: Corydalidae: Corydalinae). Acta Entomol Sin 55:133–138

Chang SI (1966) Some physiological observations on two aquatic Collembola. Trans Am Microsc Soc 85:359–371

Chapman JC, Lockley WJS, Rees HH, Goodwin TW (1977) Stereochemistry of olefine bond formation in defensive steroids of *Acilius sulcatus* (Dytiscidae). Eur J Biochem 81:293–298

Chivers DP, Smith RJF (1998) Chemical alarm signalling in aquatic predator-prey systems: a review and prospectus. Ecoscience 5:338–352

Cipola NG, Silva DD, Bellini BC (2018) Class Collembola. In: Hamada N, Thorp JH, Rogers DC (eds) Keys to neotropical hexapoda, Thorp and Covich's freshwater invertebrates, vol 3. Academic Press, London, pp 11–55

Classen R, Dettner K (1983) Pygidial defensive titer and population structure of *Agabus bipustulatus* L. and *Agabus paludosus* F (Coleoptera, Dytiscidae). J Chem Ecol 9:201–209

Corbet PS (1999) Dragonflies: behaviour and ecology of odonata. Harley Books, Colchester

Cordoba-Aguilar A (2008) Dragonflies and damselflies – model organisms for ecological and evolutionary research. Oxford University Press, Oxford

Cover MR, Bogan MT (2015) Minor Insect orders. In: Ecology and general biology: Thorp and Covich's freshwater invertebrates, vol I, 4th edn. Academis Press/Elsevier, Amsterdam, pp 1059–1072

Crespo FG (2011) A review of chemosensation and related behavior in aquatic insects. J Insect Sci 11:62; insectscience.org/11.62

Crowell HH (1946) Notes on an amphibious cockroach from the republic of Panama. Entomol News 75:171–172

Curio E (1976) The ethology of predation. Springer, Berlin

Darlington PJ Jr (1929) Notes on the habits of *Amphizoa*. Psyche 36:383–385

Dazzini Valcurone M, Pavan M (1980) Glandole pigidiali e secrezione difensive dei Carabidae (Insecta Coleoptera). Pubbl Ist Entomol Univ Pavia 12:1–36

Dazzini-Valcurone M, Pavan M (1978) Scent glands and defensive secretions of Rhynchota. Pubbl Ist Entomol Univ Pavia 5:1–46

De Carlo JM (1961) Anatomia e histologia de las glandulas maxilares o cefalicas de *Belostoma elegans* (Mayr). Actas Trabajos Congr Sudamericano LaPlata Zool 5:127–141

De Carlo JM, Pellerano GN, Maggese MC (1973) Anatomia microscopica del tracto digestivo y glandulas salivales de *Lethocerus mazzai* (Hemiptera, Belostomatidae). Physis Buenos Aires Sect C 32(85):295–314

De Marzo L, Nilsson AN (1986) Morphological variation and fine structure of some head structures of Dytiscidae (Coleoptera). Entomol Basil 11:29–42

Dettner K (1985) Ecological and phylogenetic significance of defensive compounds from pygidial glands of hydradephaga (Coleoptera). Proc Acad Nat Sci Phil 137:156–171
Dettner K (1987) Chemosystematics and evolution of beetle chemical defenses. Annu Rev Entomol 32:17–48
Dettner K (1990) Chemische Abwehr bei der ursprünglichen Käferfamilie der Amphizoidae- ein beitrag zur Evolution der Pygidialdrüse der Hydradephaga. Mitt D Ges Allg Angew Entomol 7:519–526
Dettner K (1993) Defensive secretions and exocrine glands in free-living staphylinid beetles-Theit bearing on phylogeny (Coleoptera: Staphylinidae). Biochem Syst Ecol 21:143–162
Dettner K (1997) Inter- and intraspecific transfer of toxic Insect compound cantharidin. Ecol Stud 130:115–145
Dettner K (1997a) Insecta: Coleoptera: Noteridae. In: Schwoerbel J, Zwick P (eds) Süßwasserfauna von Mitteleuropa 20/2,3,4. Fischer, Stuttgart, pp 97–126
Dettner K (1997b) Insecta: Coleoptera: Hygrobiidae. In: Schwoerbel J, Zwick P (eds) Süßwasserfauna von Mitteleuropa 20/2,3,4. Fischer, Stuttgart, pp 127–147
Dettner K (2010) Chemical defense and toxins of lower terrestrial and freshwater animals. In: Mori K (ed) Comprehensive natural products II Chemistry and Biology, Chemical Ecology, vol 4. Elsevier, Amsterdam, pp 387–410
Dettner K (2014) Chapter 6 Chemical Ecology and Biochemistry of Dytiscidae. In: Ecology, systematics, and the natural history of predacious diving beetles (Coleoptera: Dytiscidae). Springer, Dordrecht, pp 235–306
Dettner K (2015) Toxins, defensive compounds and drugs from Insects. In: Hoffmann KH (ed) Insect molecular biology and ecology. CRC, Boca Raton, FL, pp 39–93
Dettner K (2016a) 7.3 Noteridae Thomson, 1857. In: Beutel RG, Kristensen NP (eds) Handbook of zoology, Arthropoda: Insecta. Coleoptera, Beetles, vol 1, 2nd edn. De Gruyter, Berlin, pp 98–107
Dettner K (2016b) 7.4 Amphizoidae LeConte, 1853. In: Beutel RG, Kristensen NP (eds) Handbook of zoology, Arthropoda: Insecta. Coleoptera, Beetles, vol 1, 2nd edn. De Gruyter, Berlin, pp 107–112
Dettner K (2016c) 7.5 Hygrobiidae Régimbart, 1879, 112-118. In: Beutel RG, Kristensen NP (eds) Handbook of zoology, Arthropoda: Insecta. Coleoptera, Beetles, vol 1, 2nd edn. De Gruyter, Berlin
Dettner K, Böhner M (2009) Dier Pygidialdrüse der Wassertreter (Coleoptera: Haliplidae): Morphologie, Chemie, Funktion und phylogenetische Bedeutung. Contr Nat Hist 12:437–460
Dettner K, Liepert C (1994) Chemical mimicry and camouflage. Annu Rev Entomol 39:129–154
Dettner K, Peters W (2010) Lehrbuch der Entomologie Teil 1 & 2. Spektrum, Heidelberg
DeWalt RE, Kondratieff B, Sandberg JB (2015) Order Plecoptera. In: Ecology and general biology: Thorp and Covich's freshwater invertebrates, vol 1, 4th edn. Academic Press/Elsevier, Amsterdam, pp 933–949
Dijkstra KDB, Monaghan MT, Pauls SU (2014) Freshwater biodiversity and aquatic insect diversification. Annu Rev Entomol 59:143–163
Edmunds M (1974) Defence in animals. Longman, Harlow
Edwards JG (1953) The real source of *Amphizoa*-secretions. Coleopt Bull 7:4
Eisner T, Aneshansley DJ (2000) Chemical defense: aquatic beetle (*Dineutes hornii*) vs. fish (*Micropterus salmoides*). PNAS 97:11313–11318
Eisner T, Eisner M, Siegler M (2005) Secret weapons. Belknap Press, Cambridge
Elliott JM (2008) The ecology of riffle beetles (Coleoptera: Elmidae). Fr Rev 1:189–203
Evans DL, Schmidt JO (1990) Insect defenses. Adaptive mechanisms and strategies of prey and predators. State University of New York Press, Albany
Faasch H (2015) Identification guide to aquatic and semiaquatic Diptera larvae. DGL Arbeitshilfe 1-2015. DGL, Hardegesen
Falkenström G (1926) Beiträge zur Kenntnis der Biologie der Haliplidae und der Metamorphose von Haliplus immaculatus Gerh. Entomol Tidskr Arg 47(1):1–28

Ferrari MCO, Wisenden BD, Chivers DP (2010) Chemical ecology of predator-prey interactions in aquatic ecosystems: a review and prospectus. Can J Zool 88:698–724

Ferry EE, Hopkins GR, Stokes AN, Mohammadi S, Brodie ED Jr, Gall BG (2013) Do all portable cases constructed by caddisfly larvae function in defense ? J Insect Sci 13:5

Fitzgerald VJ (1987) Social behavior of adult whirligig beetles (*Dineutes nigrior* and *D. discolor* Coleoptera: Gyrinidae). Am Midl Nat 118:439–447

Fleck G, Hamada N, Carvalho AL (2009) A remarkable new genus and species of dragonfly (Odonata: Anisoptera: Libellulidae) from Brazil and notes on its bionomics and phylogenetic affinities. Ann Soc Entomol Fr (ns) 45:275–284

Forbes MR, Robb T (2008) Testing hypotheses about parasite-mediated selection using odonate hosts. In: Cordoba-Aguilar A (ed) Dragonflies and damselflies – model organisms for ecological and evolutionary research. Oxford University Press, Oxford, pp 175–188

Forsyth DJ (1968) The structure of the defence glands in the Dytiscidae, Noteridae, Haliplidae and Gyrinidae (Coleoptera). Trans R Entomol Soc Lond 120:159–181

Forsyth DJ (1970) The structure of the defence glands of Cicindelidae, Amphizoidae, and Hygrobiidae (Insecta: Coleoptera). J Zool Lond 160:51–59

Forsyth DJ (1972) The structure of the pygidial defence glands of Carabidae (Coleoptera). Transact Zool Soc Lond 32:249–309

Franciscolo ME (1979) Coleoptera Haliplidae, Hygrobiidae, Gyrinidae, Dytiscidae, Fauna d'Italia, vol 14. Calderini, Bologna

Fu X, Vencl FC, Nobuyoshi O, Meyer-Rochow VB, Lei C, Zhang Z (2007) Structure and function of the eversible glands of the aquatic firefly *Luciola lei* (Coleoptera: Lampyridae). Chemoecology 17:117–124

Fu X, Meyer-Rochow VB, Tylwer J, Suzuki H, de Cock R (2009) Structure and function of the eversible organs of several genera of larval firefly (Coleoptera: Lampyridae). Chemoecology 19:155–168

Fusari LM, Dantas GPS, Pinho LC (2018) Order Diptera. In: Hamada N, Thorp JH, Rogers DC (eds) Keys to neotropical hexapoda, Thorp and Covich's freshwater invertebrates, vol 3. Academic Press, London, pp 607–623

Gerhart DJ, Bondura ME, Commito JA (1991) Inhibition of sunfish feeding by defensive steroids from aquatic beetles: structure-activity relationships. J Chem Ecol 17:1363–1370

Giglio A, Brandmayr P, Dalpozzo R, Sindona G, Tagarelli A, Talarico F, Brandmayr TZ, Ferrero EA (2009) The defensive secretion of *Carabus lefebrei* Dejean 1826 pupa (Coleoptera, Carabidae): gland ultrastructure and chemical identification. Microsc Res Tech 72:352–361

Gorb SN (2007) Smooth attachment devices in insects: functional morphology and biomechanics. Adv Insect Physiol 34:81–115

Graca MB, Solis MA (2018) Order Lepidoptera. In: Hamada N, Thorp JH, Rogers DC (eds) Keys to neotropical hexapoda, Thorp and Covich's freshwater invertebrates, vol 3. Academic Press, London, pp 325–338

Grant PM (2001) Mayflies as food. In: Dominguez E (ed) Trends in research in Ephemeroptera and Plecoptera. Kluwer, Amsterdam, pp 107–124

Grosch DS (1988) Genetic research on braconid wasps. Adv Genet 35:109–184

Gustafson GT, Miller KB (2016) Revision of the Southeast Asian whirligig beetles genus *Porrorhynchus* Laporte, 1835 (Coleoptera: Gyrinidae: Gyrininae: Dineutini). Colept Bull 70:675–714

Gutjahr ALN, Braga CES (2018) Order Orthoptera. In: Hamada N, Thorp JH, Rogers DC (eds) Keys to neotropical hexapoda, Thorp and Covich's freshwater invertebrates, vol 3. Academic Press, London, pp 143–168

Haddad V, Schwartz EF, Schwartz CA (2010) Bites caused by giant water bugs belonging to Belastomatidae family (Hemiptera, Heteroptera) in humans: a report of seven cases. Wilderness Environ Med 21:130–133

Hamada N, Thorp JH, Rogers DC (2018) Keys to neotropical Hexapoda, Thorp and Covich's freshwater invertebrates, vol 3. Academic Press, London

Hansen M (1991) The hydrophiloid beetles. Biolog Skrifter 40:1–368
Hanson P, Gauld LD (eds) (1995) The hymenoptera of Costa Rica. Oxford University Press, Oxford
Hauser R (1985) Ein Diapausesekret bei Wasserläufern (Hemiptera, Gerridae). Mitt Schweiz Entomol Ges 85:511–525
Heckman CW (2018) Ecological strategies of aquatic insects. CRC, Baton Rouge
Heinrich B, Vogt FD (1980) Aggregation and foraging behavior of whirligig beetles (Gyrinidae). Behav Ecol Sociobiol 7:179–186
Hennig W (1968) Die Larvenformen der Dipteren. Akademieverlag, Berlin
Henrikson BI, Stenson JAE (1993) Alarm substance in *Gyrinus aeratus* (Coleoptera, Gyrinidae). Oecologia 93:191–194
Hershey AE, Dodson SI (1987) Predator avoidance by *Cricotopus* cyclomorphosis and the importance of being big and hairy. Ecology 68:913–920
Hickman JR (1931) Contribution to the biology of the Haliplidae (Coleoptera). Ann Entomol Soc Am 24:129–142
Hilker M, Meiners T (2002) Chemoecology of Insect eggs and egg deposition. Blackwell, Berlin
Hinton HE (1981) Biology of insect eggs, vol 1-3. Pergamon Press, Oxford
Holdgate MW (1955) The wetting of insect cuticles by water. J Exper Biol 32:591–617
Honkavaara J, Rantala MJ, Suhonen J (2009) Mating status, immune defence, and multi-parasite burden in the damselfly *Coenagrion armatum*. Entomol Exp Appl 132:165–171
Honma A, Oku S, Nishida T (2006) Adaptive significance of death feigning posture as a specialized inducible defense against gape-limited predators. Proc Biol Sci 273:1631–1636
Hopkin SP (1997) Biology of the Springtails (Insecta: Collembola). Oxford University Press, Oxford
Horta MAP, de Melo AL, Bertoluci J (2010) A possible case of mimicry involving a heteropteran insect and an anuran tadpole. Herpetol Bull 114:1-7
Hübner G, Völkl W, Francke W, Dettner K (2002) Mandibular gland secretions in alloxystine wasps (Hymenoptera, Cynipoidea, Charipidae): do ecological or phylogenetic constraints influence occurrence or composition ? Biochem Syst Ecol 30:505–523
Ivarsson P, Henrikson BI, Stenson JAE (1996) Volatile substances in the pygidial secretion of gyrinid beetles (Coleoptera: Gyrinidae). Chemoecology 7:191–193
Jäch MA, Balke M (2008) Global diversity of water beetles (Coleoptera) in freshwater. Hydrobiologia 595:419–442
Jäch MA, Beutel RG, Delgado JA, Diaz JA (2016) 14.1 Hydraenidae Mulsant, 1844. In: Beutel RG, Kristensen NP (eds) Handbook of zoology, Arthropoda: Insecta. Coleoptera, Beetles, vol 1, 2nd edn. De Gruyter, Berlin, pp 316–345
Jacobs W, Renner M (1974) Taschenlexikon zur Biologie der Insekten. Fischer, Stuttgart
Johansson F, Mikolajewski DJ (2008) Evolution of morphological defenses. In: Cordoba-Aguilar A (ed) Dragonflies and damselflies- Model organisms for ecological and evolutionary Research. Oxford University. Press, Oxford, pp 70–74
Johnson MD (1991) Behavioural ecology of larval dragonflies and damselflies. Trends Ecol Evol 6:8–13
Kadavy DR, Hornby JM, Haverkost T, Nickerson KW (2000) Natural antibiotic resistance of bacteria isolated from larvae of the oil fly *Helaeomyia petrolei*. Appl Environ Microbiol 66:4615–4619
Kasumyan AO, Doving KB (2003) Taste preferences in fishes. Fish Fish 4:289–347
Keeling CI, Plettner E, Slessor KN (2004) Hymenopteran Semiochemicals. In: Schulz S (ed) Topics in current chemistry, vol 1. Springer, Berlin, pp 133–177
Kerfoot WC, Sih A (1987) Predation-direct and indirect impacts on aquatic communities. University Press of New England, Hanover
Kicklighter C (2012) Chemical defences against predators. Oxford University Press, Oxford, pp 236–249
Klausnitzer B (1996) Käfer im und am Wasser. Westarp, Magdeburg & Spektrum, Heidelberg

Kovac D (1993) A quantitative analysis of secretion-grooming behaviour in the water bug *Plea minutissima* Leach (Heteroptera, Pleidae): Control by abiotic factors. Ethology 93:41–61

Kovac D, Maschwitz U (1990a) Sekretputzen bei *Ilyocoris cimicoides* (Heteroptera, Naucoridae). Nachr Entomol Ver Apollo Frankfurt NF 11:155–164

Kovac D, Maschwitz U (1990b) Secretion-grooming in aquatic beetles (Hydradephaga): a chemical protection against contamination of the hydrofuge respiratory region. Chemoecology 1:131–138

Kovac D, Maschwitz U (1991) The function of the metathoracic scent gland in corixid bugs (Hemiptera, Corixidae): Secretion-grooming on the water surface. J Nat Hist 25:331–340

Kovac D, Maschwitz U (2000) Protection of hydrofuge respiratory structures against detrimental microbic growth by terrestrial grooming water beetles (Coleoptera: Hydrophilidae, Hydraenidae, Dryopidae, Elmidae, Curculionidae). Entomol Gener 24:277–292

Kovac D, Maschwitz U, Hirschel K (1991) *Plea minutissima* (Pleidae)-Sekretputzen Secretion-grooming. Film E 3101. Publ Wiss Film Biol 21:13–20

Küchler SM, Kehl S, Dettner K (2009) Characterization and localization of *Rickettsia* sp. in water beetles of genus *Deronectes* (Coleoptera; Dytiscidae). FEMS Microbiol Ecol 68:201–211

Lampert W, Sommer U (2007) Limnoecology, 2nd edn. Oxford University Press, Oxford

Lancaster J, Downes BJ (2013) Aquatic entomology. Oxford University Press, Oxford

Lang C, Seifert K, Dettner K (2012) Skimming behaviour and spreading potential of *Stenus* species and *Dianous coerulescens* (Coleoptera: Staphylinidae). Naturwissenschaften 99:937–947

Larson DJ (1996) Color patterns of dytiscine water beetles (Coleoptera: Dytiscidae, Dytiscinae) of arroyos, billabongs and wadis. Coleopt Bull 50:231–235

Larson DJ, Pritchard G (1974) Organs of possible stridulatory function in water-beetles (Coleoptera: Dytiscidae). Coleopt Bull 28:53–63

Lawrence JW (1991) Order Coleoptera. In: Stehr FW (ed) Immature insects, vol 2. Kendall & Hunt, Dubuque, pp 144–185

Leschen RAB, Beutel RG, Lawrence JF (eds) (2010) Handbook of zoology, Arthropoda: Insecta. Coleoptera, Beetles, vol 2. De Gruyter, Berlin

Lokensgard J, Smith RL, Eisner T, Meinwald J (1993) Pregnanes from defensive glands of a belostomatid bug. Experientia 49:175–176

Lucht W (1974) Studien über die Biologie und Ethologie des *Orectochilus villosus* Müll. (Col., Gyrinidae). Entomol Blätter 70:12–34

Lytle DA (2015) Order Hemiptera. In: Ecology and general biology: Thorp and Covich's freshwater invertebrates, vol 1, 4th edn. Academic Press/Elsevier, Amsterdam, pp 951–963

Machado RJP (2018) Order Mecoptera. In: Hamada N, Thorp JH, Rogers DC (eds) Keys to neotropical hexapoda, Thorp and Covich's freshwater invertebrates, vol 3. Academic Press, London, pp 349–351

Maddison DR, Moore W, Baker MD, Ellis TM, Ober KA, Cannone JJ, Gutell RR (2008) Monophyly of terrestrial adephagan beetles as indicated by three nuclear genes I (Coleoptera: Carabidae and Trachypachidae). Zool Scr 38:43–62

Marcussen BM, Axelsen JA, Toft S (1999) The value of two Collembola species as food for a linyphiid spider. Entomol Exp Appl 92:29–36

Maschwitz U (1971) Wasserstoffperoxid als Antiseptikum bei einer Wasserwanze. Naturwissenschaften 68:572

Masters WM (1979) Insect disturbance stridulation: its defensive role. Behav Ecol Sociobiol 5:187–200

McCormick S, Polis GA (1982) Arthropods that prey on vertebrates. Biol Rev 57:29–58

McPeek MA (2000) Predisposed to adapt? Clade-level differences in characters affecting swimming performance in damselflies. Evolution 54:2072–2080

Merritt R, Cummins KW (1984) An introduction to the aquatic insects of North America, 1. edn. Kandall/Hunt, Dubuque

Merritt R, Cummins KW (1996) An introduction to the aquatic insects of North America, 3. edn. Kandall/Hunt, Dubuque

Messner B (1969) Zur Morphologie und Histologie einer neuen Drüse in den Mitteltibien von *Tridactylus variegatus* Taltr. 1809 (Orthoptera). Wiss. Z. Ernst-Moritz-Arndt-Universität Greifswald XVIII. Math Naturwiss 18:93–98

Millar JG (2005) Pheromones of true bugs. Top Curr Chem 240:37–84

Miller KB, Bergsten J (2012) Phylogeny and classification of whirligig beetles (Coleoptera: Gyrinidae): relaxed-clock model outperforms parsimony and time-free Bayesian analyses. Syst Entomol 37:706–746

Miller KB, Bergsten J (2016) Diving beetles of the world. Johns Hopkins University Press, Baltimore, MD

Miller JR, Mumma RO (1976a) Physiological activity of water beetle defensive agents. I Toxicity and anesthetic activity of steroids and norsesquiterpenes administered in solution in the minnow *Pimephales promelas* Raf. J Chem Ecol 2:115–130

Miller JR, Mumma RO (1976b) Physiological activity of water beetle defensive agents. II Absorption of selected anesthetic steroids and norsesquiterpenes across gill membranes of the minnow *Pimephales promelas* Raf. J Chem Ecol 2:131–146

Miller JR, Hendry LB, Mumma RO (1975) Norsesquiterpenes as defensive toxins of whirligig beetles (Coleoptera: Gyrinidae). J Chem Ecol 1:59–82

Miranda GFG, Rotheray G (2018) Family Syrphidae. In: Hamada N, Thorp JH, Rogers DC (eds) Keys to neotropical hexapoda, Thorp and Covich's freshwater invertebrates, vol 3. Academic Press, London, pp 779–783

Moore KA, Williams DD (1990) Novel strategies in the complex defense repertoire of a stonefly (*Pteronarcys dorsata*) nymph. Oikos 57:49–56

Moreira FFF, Rodrigues HDD, Sites RW, Cordeiro IRS, Magalhaes OM (2018) Order Hemiptera. In: Hamada N, Thorp JH, Rogers DC (eds) Keys to neotropical hexapoda, Thorp and Covich's freshwater invertebrates, vol 3. Academic Press, London, pp 175–216

Moriya N (1989) Morphology and histology of the scent glands of the pigmy mole cricket, *Tridactylus japonicus* de Haan (Orthoptera: Tridactylidae). Appl Entomol Zool 24:161–168

Moriya N, Ichinose T (1988) Function of the scent from the pigmy mole crickets, *Tridactylus japonicus* de Haan (Orthoptera: Tridactylidae). Appl Entomol Zool 23:321–328

Morse JC (2003) Trichoptera (Caddisflies). In: Resh VC, Cardé RT (eds) Encyclopedia of insects. Academic Press, Amsterdam, pp 1154–1151

Morse JC (2017) Biodiversity of aquatic insects. In: Foottit RG, Adler PH (eds) Insect biodiversity, vol 1. Wiley-Blackwell, Oxford, pp 205–227

Mousseau T, Roughley RE (2007) Taxonomy, classification, reconstructed phylogeny and biogeography of Nearctic species of *Brychius* Thomson (Coleoptera: Haliplidae). Coleopt Bull 61:351–397

Navarrete-Heredia JL, Cortés-Aguilar J, Beutel RG (2005) New findings on the enigmatic beetle family lepiceridae (Coleoptera: Myxophaga). Entomol Abh 62:193–201

Neiss UG, Fleck G, Pessacq P, Tennessen KJ (2018) Odonata: Superfamily Libelluloidea. In: Hamada N, Thorp JH, Rogers DC (eds) Keys to neotropical hexapoda, Thorp and Covich's freshwater invertebrates, vol 3. Academic Press, London, pp 399–447

Nesemann H, Shah RDT, Shah DN, Sharma S (2010) First records of *Rhicnoda natatrix* and *Ricnoda rugosa* (Blattodea: Blaberidae) from Nepal and India (Maharashtra) with notes on habitat quality. J Threat Taxa 2:648–652

Newhart AT, Mumma RO (1979) Defensive secretions of three species of *Acilius* (Coleoptera, Dytiscidae) and their seasonal variations as determined by high-pressure liquid chromatography. J Chem Ecol 5:643–652

Nishida R (2002) Sequestration of defensive substances from plants by Lepidoptera. Annu Rev Entomol 47:57–92

Nislow KH, Molles MC Jr (1993) The influence of larval case design on vulnerability of *Limnephilus frijole* (Trichoptera) to predation. Freshw Biol 29:411–417

Okada YK (1928) Two Japanese aquatic glowworms. Trans R Entomol Soc Lond 76:101–108

Oliva A (1992) Cuticular microstructure in some genera of Hydrophilidae (Coleoptera) and their phylogenetic significance. Bull Inst R Sci Nat Belg Entomol 62:33–56
Oosterbroek P (2006) The European families of the Diptera. KNNV Publishing, Utrecht
Opitz SEW, Müller C (2009) Plant chemistry and insect sequestration. Chemoecology 19:117–154
Otto, C. (1984) Coloração de cabeça adaptativa em larvas de caddis produtoras de casos. Freshw Biol 14:317–321.
Parfin S (1952) Notes on the Bionomics of *Corydalus cornutus* (Linné), *Chauliodes rastricorn* is Rambur. *C. pectinicornis* (Linné) and *Neohermes* spec. The American Midl. Naturalist 47:426-434.
Pasteels JM, Rowell-Rahier M, Braekman JC, Daloze D (1984) Chemical defences in leaf beetles and their larvae: the ecological, evolutionary and taxonomic significance. Biochem Syst Ecol 12:395–406
Pasteels JM, Duffey S, Rowell-Rahier M (1990) Toxins in chrysomelid beetles: Possible evolutionary sequence from de novo synthesis to derivation from food-plant chemicals. J Chem Ecol 16:211–222
Pavlovsky EN (1922) On the biology and structure of the larvae of Hydrophilus caraboides L. J Cell Sci S2-66:627-655
Pawlowsky EN (1927) Gifttiere und ihre Giftigkeit. Gustav Fischer, Jena
Peckarsky B (1980) Predator-prey interactions between stoneflies and mayflies: Behavioral observations. Ecology 61:932–943
Peckarsky BL (1984) Predator-Prey interactions among aquatic insects. In: Resh VH, Rosenberg DM (eds) The ecology of aquatic insects. Praeger, New York, pp 196–253
Peckarsky B (1987) Mayfly cerci as defense against stonefly predation: deflection and detection. Oikos 48:161–170
Peckarsky B (1996) Alternative predator avoidance syndromes of stream-dwelling mayfly larvae. Ecology 77:1888–1905
Peddle SM, Larson DJ (1999) Cuticular evidence of traumatic experiences of water beetles (Coleoptera: Dytiscidae, Hydrophilidae). Coleopt Bull 53:42–51
Perkins PD (1997) Life on the effective bubble: Exocrine secretion delivery systems (ESDS) and the evolution and classification of beetles in the family Hydraenidae (Insecta, Coleoptera). Ann Carnegie Museum 66:89–207
Pes AM, Holzenthal RW, Sganga JV, Santos APM, Barcelos Silva P, Camargos LM (2018) Order Trichoptera. In: Hamada N, Thorp JH, Rogers DC (eds) Keys to neotropical hexapoda, Thorp and Covich's freshwater invertebrates, vol 3. Academic Press, London, pp 237–324
Pescador ML, Peters WL (1974) The life history and ecology of *Baetisca rogersi* Berner (Ephemeroptera: Baetiscidae). Bull Fla State Mus Biol Sci 17:151
Pescador ML, Richard BA, Hubbard MD, Staniczek AH (2009) Evolution of Baetiscidae (Ephemeroptera): current state of knowledge of the family. Aquat Insects 31(Suppl 1):137–147
Pessacq P, Muzon J, Neiss UG (2018) Order Odonata. In: Hamada N, Thorp JH, Rogers DC (eds) Keys to neotropical Hexapoda, Thorp and Covich's freshwater invertebrates, vol 3. Academic Press, London, pp 355–494
Pierce CL (1988) Predator avoidance, microhabitat shift, and risk-sensitive foraging in larval dragonflies. Oecologia 77:81–90
Pilgrim RL (1972) The aquatic larva and the pupa of *Choristella philpotti* Tillyard 1917 (Mecoptera: Nannochoristidae). Pac Insects 14:151–168
Pitaluga di Godoi FS, Pujol-Luz JR (2018) Family Stratiomyidae. In: Hamada N, Thorp JH, Rogers DC (eds) Keys to neotropical hexapoda, Thorp and Covich's freshwater invertebrates, vol 3. Academic Press, London, pp 717–777
Puchkova LV (1965) Maxillary glands of Heteroptera, their function and origin. Zool Zh Akad Nauk SSSR 44:1801–1808
Quicke DLJ (1997) Parasitic wasps. Chapman & Hall, London
Reichholf J (1970) Untersuchungen zur Biologie des Wasserschmetterlings Nymphula nymphaeata L. (Lepidoptera, Pyralidae). Int Rev ges Hydrobiol 55:687-728

Rentz DCF, Su YN (2003) Orthoptera. In: Resh VH, Cardé RT (eds) Encyclopedia of insects. Academic Press, Amsterdam, pp 827–839
Resh VH, Rosenberg DM (1984) The ecology of aquatic insects. Praeger, New York
Rivosecchi L (1984) Ditteri (Diptera) in Guide per il riconoscimento delle specie animali delle aque interne Italiane. 177 pp, Consiglio Nazionale delle Ricerche AQ/1/206
Ronderos MM, Diaz F, Marino PI, Fereira-Keppler RL (2018) Family Ceratopogonidae. In: Hamada N, Thorp JH, Rogers DC (eds) Keys to neotropical hexapoda, Thorp and Covich's freshwater invertebrates, vol 3. Academic Press, London, pp 625–659
Rupprecht R (1971) Der Kokonbau der Taumelkäferlarve (*Gyrinus substriatus* Stephens). MZ Naturwiss Arch 10:195–202
Ruxton GD, Sherratt TN, Speed MP (2004) Avoiding attack. The evolutionary ecology of crypsis, warning signals, and mimicry. Oxford University Press, Oxford
Salles FF, Dominguez E, Molineri C, Boldrini R, Nieto C, Dias LG (2018) Order Ephemeroptera. In: Hamada N, Thorp JH, Rogers DC (eds) Keys to neotropical hexapoda, Thorp and Covich's freshwater invertebrates, vol 3. Academic Press, London, pp 61–117
Sartori M, Brittain JE (2015a) Order Ephemeroptera. In: Ecology and general biology: Thorp and Covich's freshwater invertebrates, vol 1, 4th edn. Academic Press/Elsevier, Amsterdam, pp 873–891
Sartori M, Brittain JE (2015b) Order Ephemeroptera 873–891, in: Ecology and General Biology: Thorp and Covich's Freshwater Invertebrates. Vol I, 4th edition. Academic Press/Elsevier, Amsterdam. In defensive glands of the water beetle *Agabus affinis*. Lipids 35:543–550
Schaaf O, Dettner K (1997) Microbial diversity of aerobic heterotrophic bacteria inside the foregut of two tyrphophilous water beetle species (Coleoptera: Dytiscidae). Microbiol Res 152:57–64
Schaaf O, Dettner K (2000) Polyunsaturated monoglycerides and a pregnadiene in defensive glands of water beetles *Agabus affinis*. Lipids 35:543–550
Schal C, Frase J, Bell WJ (1982) Disturbance stridulation and chemical defence in nymphs of the tropical cockroach *Megaloblatta blaberoides*. J Insect Physiol 28:541–552
Scheloske HW (1975) Fortpflanzungsverhalten und Lauterzeugung bei *Laccobius minutus* (L.) (Coleoptera, Hydrophilidae). Verh Dtsch Zool Ges 1974:329–334
Schierling A, Seifert K, Sinterhauf SR, Rieß JB, Rupprecht JC, Dettner K (2013) The multifunctional pygidial gland secretion of the Steninae (Coleoptera: Staphylinidae): ecological significance and evolution. Chemoecology 23:45–57
Schildknecht H (1970) Die Wehrchemie von Land- und Wasserkäfern. Angew Chemie 82:17–25
Schildknecht H (1977) Protective substances of arthropods and plants. Pontif Acad Sci Scr Varia 41:1–49
Schildknecht H, Bühner R (1968) Über ein Glykoproteid in den Pygidialwehrblasen des Gelbrandkäfers. Z Naturforsch 23b:1209–1213
Schildknecht H, Holoubek K, Wolkenstörfer M (1962) Über einen Inhaltsstoff in den Pygidialblasen vom Gelbrandkäfer. Z Naturforsch 17b:81–83
Schildknecht H, Neumaier H, Tauscher B (1972) Gyrinal, die Pygidialdrüsensubstanz der Taumelkäfer (Coleoptera: Carabidae). Liebigs Ann Chem 756:155–161
Schmidt JO (1982) Biochemistry of Insect venoms. Ann Rev Ent 27:339–368
Scholtz CH, Holm E (1986) Insects of Southern Africa. Butterworths, Durban
Schuh RT, Slater JA (1995) True bugs of the world. Comstock Publishing Associates/Cornell University Press, Ithaca
Schulz S, Messer C, Dettner K (1997) Poduran, an unusual tetraterpen from the springtail *Podura aquatica*. Tetrahedron Lett 38:2077–2080
Schulze P (1924) Biologie der Tiere Deutschlands. Apterygota Teil 25.2, 25–56. Borntraeger, Berlin
Scrimshaw S, Kerfoot WC (1987) 16. Chemical defenses of freshwater organisms: beetles and bugs. In: Kerfoot WC, Sih A (eds) Predation. University Press New England, Hanover, pp 240–262

Seeger W (1971) Morphologie, Bionomie und Ethologie von Halipliden, unter besonderer berücksichtigung funktionsmorphologischer Gesichtspunkte (Haliplidae: Coleoptera). Arch Hydrobiol 68:400–435

Shal C, Gautier JY, Bell WJ (1984) Behavioural ecology of cockroaches. Biol Rev 59:209–254

Shepard WD (2003) The biology of *Georissus californicus* LeConte (Coleoptera, Hydrophilidae, Georissinae). Spec Bull Jpn Soc Coleopt 6:121–128

Short AEZ (2018) Systematics of aquatic beetles (Coleoptera): current state and future directions. Syst Entomol 43:1–18

Silberbush A, Markman S, Lewinsohn E, Bar E, Cohen JE, Blaustein L (2010) Predator-released hydrocarbons repel oviposition by a mosquito. Ecol Lett 13:1129–1138. https://doi.org/10.1111/j.1461-0248.2010.01501.x

Slater JA (1982) Hemiptera. In: Parker S (ed) Synopsis and classification of living organisms, vol 2. McGraw-Hill, New York, pp 417–447

Smith EL (1970) Biology and structure of the Dobsonfly, *Neohermes californicus* (Walker). Pan Pac Entomol 46:142–150

Smith RL (1973) Aspects of the biology of three species of the genus *Rhantus* (Coleoptera: Dytiscidae) with special reference to the acoustical behavior of two. Can Entomol 105:909–919

Solem JO, Gullefors B (1996) Trichoptera, caddisflies. In: Nilsson A (ed) Aquatic Insects of North Europe. Apollo Books, Stenstrup, pp 223–255

Sotka EE, Forbey J, Horn M, Poore AGB, Raubenheimer D, Whalen KE (2009) The emerging role of pharmacology in understanding consumer-prey interactions in marine and freshwater systems. Integr Comp Biol 49:291–313

Southcott RV (1991) Injuries from larval Neuroptera. Med J Aust 154:329–332

St. Quentin D, Beier M (1968) V. Überordnung Libelluloidea und 6. Ordnung Odonata (Libellen). 39 pp, Handbuch der Zoologie IV, 2-2/6

Staddon BW, Thorne MJ (1973) The structure of the metathoracic scent gland system of the water bug *Ilyocoris cimicoides* (L.) (Heteroptera: Naucoridae). Trans R Entomol Soc Lond 124:343–363

Staddon BW, Thorne MJ (1979) The metathoracic gland system in Hydrocorisae (Heteroptera: Nepomorpha). Syst Entomol 4:239–250

Staddon BW, Everton IJ, Games DE (1979) Organ specificity and scent constitution in Corixidae (Heteroptera: Hydrocorisae). Comp Biochem Physiol 62B:259–262

Steidle JLM, Dettner K (1993) Quantitative composition of the defensive secretion of *Bledius* species (Coleoptera: Staphylinidae: Oxytelinae) is adapted to naturally occurring predators. Chemoecology 4:63–71

Steidle JLM, Dettner K (1995) Abdominal gland secretion of *Bledius* rove beetles as an effective defence against predators. Entomol Exp Appl 76:211–216

Steiner WE, Anderson JJ (1981) Notes on the Natural history of *Spanglerogyrus albiventris* Folkerts, with a new Distribution record (Coleoptera: Gyrinidae). Pan Pac Entomol 57:124–132

Stocks I (2004) Reflex bleeding (autohemorrhage). In: Capinera JP (ed) Encyclopedia of entomology. Springer, Berlin, pp 3132–3238

Stoks R (1999) Effect of lamellae on survival and foraging success of the damsel fly *Lestes sponsa* (Odonata: Lestidae). Oecologia 117:443–448

Stork NE (2018) How many species of insects and other terrestrial arthropods are there on earth? Annu Rev Entomol 63:31–45

Suhling F, Müller O (1996) Die Flußjungfern Europas (Gomphidae). Westarp, Magdeburg & Spektrum, Heidelberg

Suhling F, Sahlen G, Gorb S, Kalkman VJ, Dijkstra K-D B, van Tol J (2015) Order Odonata. In: Ecology and general biology: Thorp and Covich's freshwater invertebrates, vol 1, 4th edn. Academic Press/Elsevier, Amsterdam, pp 893–932

Swart CC, Deaton LE, Felgenhauer BE (2006) The salivary gland and salivary enzymes of the giant water bugs (Heteroptera; Belostomatidae). Comp Biochem Physiol Part A 145:114–122

Thayer MK (2016) Staphylinidae Latreille, 1802. In: Kristenasen NP, Beutel RG (eds) Handbook of zoology, Coleoptera, Beetles, vol 2, 2nd edn. De Gruyter, Berlin, pp 394–442

Thompson DJ (1987) Regulation of damselfly populations: the effects of weed on larval mortality due to predation. Frehw Biol 17:367–371

Thompson V (1997) Spittlebug nymphs (Homoptera: Cercopidae) in Heliconia Flowers (Zingiberales: Heliconiaceae): Preadaptation and Evolution of the first aquatic Homoptera. Rev Biol Trop 45:905-912

Thorp JH, O'Neill BJ (2015) Hexapoda-Introduction to Insects and Collembola. In: Ecology and general biology: Thorp and Covich's freshwater invertebrates, vol 1, 4th edn. Academic Press/ Elsevier, Amsterdam, pp 849–871

Thorp JH, Rogers DD (2015) Ecology and general biology: Thorp and Covich's freshwater invertebrates, vol 1, 4th edn. Academic Press/Elsevier, Amsterdam

Thorp JH, Rogers DD (2016) Keys to Nearctic fauna: Thorp and Covich's freshwater invertebrates, vol 2, 4th edn. Academic Press/Elsevier, Amsterdam

Urbanek A, Richert M, Gilka W, Szadziewski R (2011) Morphology and histology of secretory setae in terrestrial larvae of biting midges of the genus *Forcipomyia* (Diptera: Ceratopogonidae). Arthropod Struct Dev 40:485–494

Urbanek A, Szadziewski R, Stepnowski P, Boros-Majewska J, Dawgul M, Kamysz W, Sosnowska D, Golebiowski M (2012) Composition and antimicrobial activity of fatty acids detected in the hygroscopic secretion collected from the secretory setae of larvae of the biting midge *Forcipomyia nigra* (Diptera: Ceratopogonidae). J Insect Physiol 58:1265–1276

Vasilikopoulos A, Balke M, Beutel RG, Donath A, Podsiadlowski L, Pflug JM, Waterhouse RM, Meusemann K, Peters RS, Escalona HE, Mayer C, Liu S, Hendrich L, Alarie Y, Bilton DT, Jia F, Zhou X, Maddison DR, Niehuis O, Misof B (2019) Phylogenomics of the superfamily Dytiscoidea (Coleoptera: Adephaga) with an Evaluation of the phylogenetic conflict and systematic error. Mol Phyl and Evol 135:270-285

Vallenduuk HJ, Cuppen HMJ (2004) The aquatic living caterpillars (Lepidoptera: Pyraloidea: Crambidae) of Central Europe. A key to the larvae and autecology. Lauterbornia 49:1–17

Van Vondel BJ (2016) 7.2 Haliplidae Aubé, 1836. In: Beutel RG, Kristensen NP (eds) Handbook of zoology, Arthropoda: Insecta. Coleoptera, Beetles, vol 1, 2nd edn. De Gruyter, Berlin, pp 89–98

Völkl W, Hübner G, Dettner K (1994) Interactions between *Alloxysta brevis* (Hymenoptera, Cynipoidea, Alloxystidae) and honeydew collecting ants: how an aphid hyperparasitoid overcomes ant aggression by chemical defense. J Chem Ecol 20:2621–2635

Von Frankenberg G (1937) Das Zirporgan des Wasserkäfers *Spercheus*. Aus der Heimat 50:47–50

Vulinec K (1987) Swimming in whirligig beetles (Coleoptera: Gyrinidae): a possible role of the pygidial gland secretion. Coleopt Bull 41:151–153

Vulinec K, Miller MC (1989) Aggregation and predator avoidance in whirligig beetles (Coleoptera: Gyrinidae). J N Y Entomol Soc 97:483–447

Wagner R, Aurich M, Reder E, Veith HJ (1990) Defensive secretions from the larvae of *Apatania fimbriata* (Pictet) (Trichoptera: Limnephilidae). Chemoecology 1:96–104

Walker AA, Weirauch C, Fry BG, King GF (2016) Venoms of Heteropteran insects: a treasure Trove of diverse pharmacological toolkits. Toxins 8:1–32

Walker AA, Vargas MJH, Corzo G, Frau BG, King GG (2018) Giant fish-killing water bug reveals ancient and dynamic venom evolution in Heteroptera. Cell Mol Life Sci 75:3215–3229

Weatherston J, Percy JE (1978) Venoms of rhynchota (Hemiptera). In: Bettini S (ed) Arthropod venoms. Springer, Berlin, pp 489–509

Weiss MR (2006) Defecation behavior and ecology of insects. Annu Rev Entomol 51:635–661

Wesenberg-Lund C (1943) Biologie der Süßwasserinsekten. Nordisk Vorlag, Kopenhagen & Springer Berlin

White DS (1989) Defense mechanisms in riffle beetles (Coleoptera: Dryopoidea). Ann Entomol Soc Am 82:237–241

Wichard W, Arens W, Eisenbeis G (1995) Atlas zur Biologie der Wasserinsekten. Gustav Fischer, Stuttgart

Wildermuth H (2000) Totstellreflex bei Großlibellenlarven (Odonata). Libellula 19:17–39

Williams DD, Feltmate BW (1992) Aquatic insects. CAB International, Wallingford
Wilson N, Flinn MB, West B, Hereford J (2015) Identification of sound-producing hydrophilid beetles (Coleoptera: Hydrophilidae) in underwater recordings using digital signal processing. Coleopt Bull 69:305–315
Witz BW (1990) Antipredator mechanisms in arthropods: a twenty year literature survey. Florida Entomol 73:71–99
Wohlfahrt B, Mikolajewski DJ, Joop G, Suhling F (2005) Are behavioural traits in prey sensitive to the risk imposed by predatory fish? Freshw Biol 51:76–84
Yan EV, Beutel RG, Lawrence JF (2018) Whirling in the late Permian: ancestral Gyrinidae show early radiation of beetles before Permian-Triassic mass extinction. BMC Evol Biol 18:33 (1-10)
Yee DA (2014) Ecology, systematics and the natural history of predacious diving beetles (Coleoptera: Dytiscidae). Springer, Dordrecht
Yee DA, Kehl S (2015) Order Coleoptera. In: Ecology and general biology: Thorp and Covich's freshwater invertebrates, vol 1, 4th edn. Academic Press/Elsevier, Amsterdam, pp 1003–1042
Young FN (1960) The colors of desert water beetles - environmental effect or protective coloration? Ann Entomol Soc Am 53:422–425
Yu P, Xie W, Fengqin L (1993) Bionomics and morphology of the larvae of *Amphizoa sinica* Yu & Stork (Coleoptera: Amphizoidae). Scient treatises Syst Evol Zool 2:107–114
Zwick P (1967) Beschreibung der aquatischen Larve von *Neurorthus fallax* (Rambur) und Errichtung der neuen Planipennierfamilie Neurorthidae fam. nov. Gewässer Abwasser 44:65–86
Zwick P (1980) Handbuch der Zoologie. IV. Band Arthropoda; 2. Hälfte Insecta, 2. Teil: 7. Plecoptera (Steinfliegen). De Gruyter, Berlin
Heckman CW (2018) Ecological strategies of aquatic insects. CRC, Boca Raton
Heckman CW (2018) Ecological strategies of aquatic insects. CRC, Boca Raton

# Chapter 10
# Reproductive Behavior and Sexual Selection

**Diogo Silva Vilela and Iago Sanmartín-Villar**

**Abstract** Aquatic insects present diverse reproductive strategies, tactics, and complex life cycles which require distinct amounts of space and different food sources. And to achieve the reproductive goals they make use of many other strategies. Such reproductive repertoire allowed these insects to evolve the most intricate and fantastic mating systems, often involving conspicuous ornaments. This chapter treats some of these subjects, providing examples from studies carried out worldwide with aquatic insect orders, approaching differences among males and females, sex ratio, sex maturation, reproductive strategies, and all aspects that are involved in success or failure of the competitors. In the first part of this chapter, we treat all insect orders that are aquatic (complete cycle under the water) and semiaquatic (larval stage under the water), mentioning the life cycles and some reproductive tactics. In the second part, we discuss about the theoretical approach concerning insect reproduction such as sexual selection, trade-offs, and roles of polymorphism.

**Keywords** Behavior · Evolution · Mating systems · Territoriality · Sexual ornaments

## 10.1 Introduction

Among aquatic insects, a great array of reproductive strategies, tactics, and behaviors are known to exist, as well as complex life cycles which require distinct space and food sources. For instance, there are the territorial strategies, where males (in most of the cases) defend a small portion of the habitat from other males (i.e., competitors). In counterpoint, there are also the sneakers, males that are incapable to hold a territory and sneak behind the "landlord" to gain access to females. There are

D. S. Vilela (✉)
Graduate Program in Entomology, Department of Biology, University of São Paulo (USP), Ribeirão Preto, Brazil

I. Sanmartín-Villar
IVEP Team, CRCA – CNRS, Université Toulouse III - Paul Sabatier, Toulouse, France



K. Del-Claro, R. Guillermo (eds.), *Aquatic Insects*,
https://doi.org/10.1007/978-3-030-16327-3_10

many other ways in which insects reach its final goal: to reproduce. This reproductive repertoire allowed these insects to evolve the most intricate and fantastic mating systems, often involving conspicuous ornaments such as morphological structures and colorful patterns, which were, in 1871, called by Charles Darwin as secondary sexual characters, closely related to the reproductive success, but not part of the sex per se. Finally, when put together variation in morphological and colorful ornaments, different strategies from what to make use of those ornaments, and a limited resource (often females), a war is started.

In this chapter we treat some of these issues providing some interesting examples from studies carried out worldwide with aquatic insect orders, differences among males and females, sex ratio, sex maturation, reproductive strategies, and all aspects that are involved in success or failure of the competitors, which can lead to an arms race between them. In the first part of this chapter, we treat all insect orders that have aquatic (complete cycle under the water) and semiaquatic (larval stage under the water) developments separately, approaching the life cycles and reproductive tactics. In the second part, we treat the theoretical approach concerning insect reproduction such as sexual selection, trade-offs, and roles of polymorphism.

## 10.2 Aquatic Insects with Terrestrial Adult Life

In this part is treated orders in which immatures are developed inside water or closely related to water environments, and the adults have a terrestrial life. Most of the insect orders that spend some of their time in aquatic habitats (11 out of 13) have only the immature stage completely water dependent. Despite remaining close to water as adults (e.g., for oviposition, territorial defense), they all emerge to a terrestrial life.

### *10.2.1 Blattodea*

Blattodea is a well-distributed and successful group, inhabiting several terrestrial environments such as forests, urban areas, deserts, and also aquatic habitats (Roth 1970; Bell et al. 2007; Nessimian 2014). Freshwater representatives of the order are proven to be found in the suborder Blaberidae and range from semiaquatic (e.g., Blaberidae: Epilamprinae) to aquatic (e.g., Blaberidae: *Rhicnoda*) habitats of development (Bell et al. 2007; Nesemann et al. 2010; Lopes 2018).

The aquatic habitats inhabited by cockroaches vary from the hitherto little studied species from phytotelmata like bromeliads to streams, ponds, and other small aquatic environments (Nessimian 2014). Cockroaches are great organic matter decomposers in terrestrial environments (Bell et al. 2007). This is also true for the aquatic species, being among the main agents responsible for disintegrated organic debris in streams and pools inside forests (Irmler and Furch 1979; Nessimian 2014).

Although spending most of the time in the water, aquatic cockroaches did not have any specialized modifications to swim like modified legs (Bell et al. 2007; Nessimian 2014). Despite that, they can easily swim (including terrestrial species such as *Periplaneta americana* L., Bell et al. 2007) below water surface to either feed or flee from potential predators and also under rocks (Shelford 1909; Bell et al. 2007; Nesemann et al. 2010; Lopes 2018).

The reproductive system of aquatic cockroaches (Blaberidae) frequently involves ovoviviparity and mating is poorly reported, mainly due to its crepuscular and nocturnal habits (Roth 1970; Bell et al. 2007). Therefore, the females are more often observed and hence have greater behavioral repertoire recorded (Roth 1970). After mating, females extrude its fertilized eggs into a structure called brood sac or uterus. There the eggs will remain until mature and the female finally makes the posture of the ootheca (Roth 1970). After hatching from the ootheca, nymphs remain sheltered by its mother, in a space between her abdomen and forewings (Pruthi 1933; Roth 1970, 1981; Bell et al. 2007; Fig. 10.1), until ready to obtain food and shelter by themselves, which may not happen until they reach second instar (Roth 1981).

**Fig. 10.1** Blaberid female with many newly hatched nymphs on her back. Photography of Natasha Mhatre

### 10.2.2 Diptera

The order Diptera is one of the megadiverse holometabolous groups of hexapoda, containing over 150,000 described species of flies, of which about 50% are aquatic (Carvalho et al. 2012; Fusari et al. 2018). Among aquatic flies are species with an important financial and public health appeal, vectors of tropical diseases like dengue and yellow fever by the genus *Aedes* and malaria by *Anopheles* species. Aquatic immature dipterans are also important from the ecological point of view, being part of the regular diet of fishes and other insects that controls their numerous populations (Fusari et al. 2018). Dipterans are found in all continents including Antarctica and the immature stages of aquatic species occupy all freshwater habitats and even marine zones, except for the open sea (Pinho 2008).

The reproductive rendezvous for most of the aquatic flies consists of a swarm composed by males who try to get the best spots of the swarm to grab a passing female, similarly to other ancient lineages such as Ephemeroptera (Marchand 1984; Gullan and Cranston 2014). Yuval et al. (1993) found that the numerous fly aggregations of the culicid *Anopheles freeborni* Aitken may favor some males to the detriment of other, due to physical characteristics. In early evening (5–10 min after sunset), they observed and sampled over 6000 mosquito swarms (97% composed of males), in which males showed significant body and wing size differences. Resting mosquitoes were, on average, 0.2 mm smaller than swarming mosquitoes and its mating success was related to wing size (Fig. 10.2), thus indicating that size played a role in reproduction in this species.

After mating, females remain to the task of ovipositing, a delicate time for mosquitoes that may influence the chances for the larvae to thrive until emergence to adult phase and even determine the timing of disease transmission to humans (Reiskind and Wilson 2004). To mitigate the risks of mortality, at the time of oviposition females may choose to lay the eggs in sites according to the amount of competition with other larvae. Reiskind and Wilson (2004) found that in the culicid *Culex restuans* Theobald females maintained under artificial conditions laid more eggs in sites with increased amount of nutrients, and fewer egg clutches from conspecifics, and the number of eggs deposited per day increased with the availability of oviposition sites, or plots (Fig. 10.3a–b). These results indicate that in natural conditions, females of *Cx. restuans* may be attracted to oviposit in sites with greater amount of nutrients and less competition to ensure larvae development and mitigate predation. Furthermore, the oviposition behavior and habitat preferences of this culicid seem to be common for mosquitoes and its understanding may also help control programs in cases of disease transmission risk (Reiskind and Wilson 2004).

### 10.2.3 Ephemeroptera

The mayflies, insects of the order Ephemeroptera (Fig. 10.4), compose a small group of about 3500 species which occupies freshwater habitats worldwide, except Antarctica (Sartori and Brittain 2015). An interesting feature about mayflies is

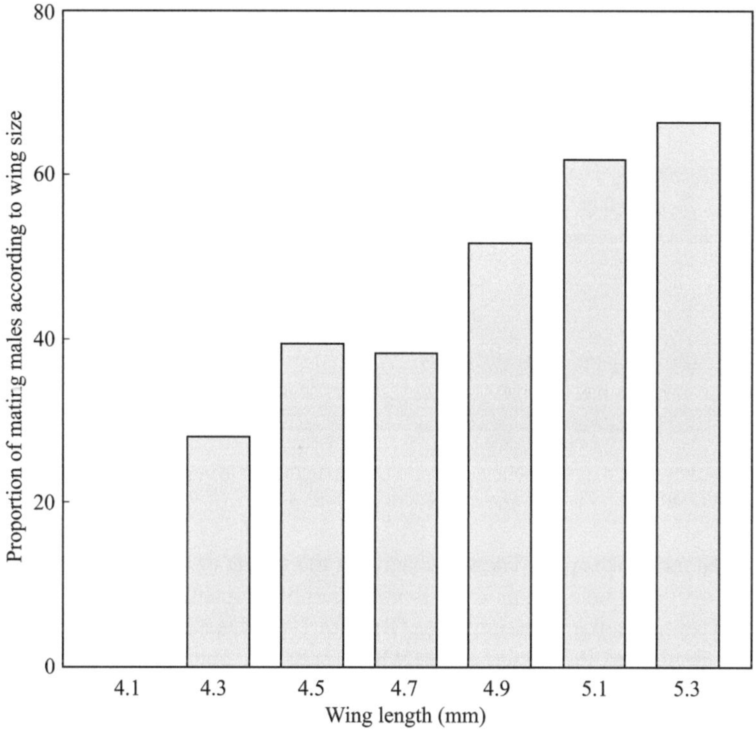


**Fig. 10.2** Correlations between wing size and proportion of matings by the culicid *Anopheles freeborni* Aitken. Adapted from Yuval et al. 1993

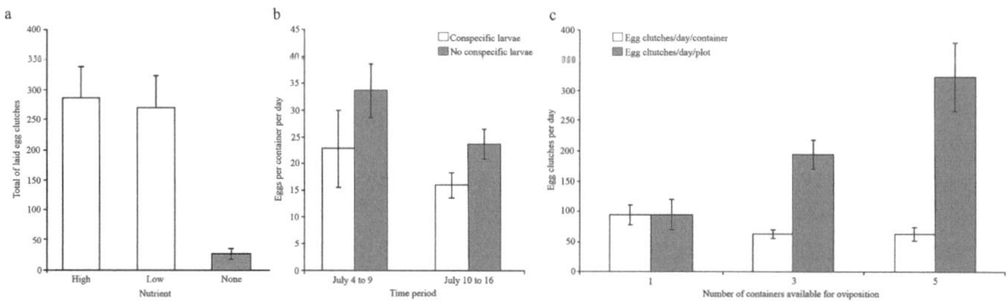


**Fig. 10.3** Graphics showing: **a**—the total eggs laid in proportion to the amount of provided nutrients, **b**—eggs laid in the presence or absence of conspecific larvae, and **c**—number of eggs laid according to the number of available containers. Adapted from Reiskind and Wilson 2004

present as soon as the aquatic nymph emerges to adult stage: Ephemeroptera is the only insect order in which a winged form passes through a molting phase (Burks 1975; Brittain 1982). When the nymph emerges, it remains for some hours (or even days, depending on the species) in a stage known as subimago. Recently emerged insect adults are known as imagos and, thus, the subimago stands for a stage before

**Fig. 10.4** Adult male mayfly (Ephemeroptera: Leptophlebiidae: *Miroculis yulieae* Raimundi and Salles) on the substrate. Photography of Frederico F. Salles

the winged molt takes part. The subimago has the aspect of an immature adult with the wings yet opaque and this form is not generally capable to mate, as its sexual structures are in final maturation stage (Brittain 1982; Da-Silva and Salles 2012).

After molting to the imago stage, which now has functional wings and fully developed reproductive organs, it begins its ephemeral journey to find a sexual partner. Mayfly lives are short; they usually live no more than few hours (Clifford 1982) and therefore its adult life is totally devoted to sex (Brittain 1982). In fact, most of the Ephemeroptera groups have stunted mouthparts and some species even present vestigial legs (Da-Silva and Salles 2012).

In order to reproduce, males have to be able to recognize the females within a "cloud" (i.e., swarm) of mayflies that are mostly formed by males at the emergence place. To do so, males make use of their large compound eyes that gives them greater possibilities in recognizing the females even in low-light conditions, while females fly towards the "male cloud" to mate, in the air (Brodskiy 1973; Allan and Flecker 1989). It is important to point out that those swarms have complex behavioral patterns, serving as a specific mechanism to isolate the species, such as the drumming behavior of stoneflies (Brittain 1990). In fact, the whole wing apparatus of mayflies is specialized to perform the most important behavior of their lives (Brodskiy 1973). The nuptial dance precedes the mating and consists of an intricate pattern of display by males, which can assume the right position to mate, only achieved during flight (Brodskiy 1973; Allan and Flecker 1989). It consists of low horizontal flights, mostly according to wind direction.

During mating, males grasp the females with the aid of its forelegs and then the mating takes place, for just a few seconds to some minutes. After mating ends, the work of the males is done. To the females, the last task is to lay the eggs (often thousands of them) dipping the abdomen in the water, preferably away from predators, to ensure that they will hatch (Brittain 1982). In some species, this dipping behavior causes female abdomen to break as it touches the water, releasing the eggs (Da-Silva and Salles 2012).

As exceptions to the regular egg development, reproductive system, and swarming behaviors of Ephemeroptera, some species of mayflies (Baetidae) can be ovoviviparous (Gibbs 1979; Brittain 1990; Gaino and Rebora 2005) and even parthenogenetic (Degrange 1960; Bergman and Hilsenhoff 1978; Kazuki and Tojo 2010). Ovoviviparous species are only found in baetids and has been reported in many species of that family (Gibbs 1979; Gaino and Rebora 2005). Gibbs (1979) first reported ovoviviparity in two species of *Callibaetis*, in which imagos presented longevity up to 14 days, which is very rare among mayflies, associating this longevity with the ovoviviparity. An interesting feature about ovoviviparous species is shown by Gaino and Rebora (2005), where the authors demonstrated that the egg structure of oviparous species *Baetis rhodani* Pictet is far more complex and thicker than in the ovoviviparous *Cloeon dipterum* L. This is due to the fact that the ovoviviparous egg hatches almost simultaneously to the contact with the water, and the oviparous egg has to provide a certain security for the embryo to complete its development.

Regarding the parthenogenetic species of mayflies, some are obligatory, as in *Baetis hageni* Eaton, and others are facultative, as in *B. spinosus* McDunnough and some *Cloeon* species (Bergman and Hilsenhoff 1978; Harker 1997). In fact, Harker (1997) demonstrated that in the European species *Cloeon dipterum*, if mature nymphs and recently emerged subimagos are somehow injured (e.g., by predators), it should release eggs which thrive and develop to adults. Further evidence to this facultative parthenogenesis is that oviposition was recorded for both virgin and swarming imagos (Harker 1997).

### 10.2.4 Hymenoptera

With near 150,000 described species, Hymenoptera is one of the most diverse and cosmopolitan insect orders, inhabiting both terrestrial and aquatic environments (Melo et al. 2012; Fernandes et al. 2018). Most of the success can be attributed to the diversity if niches that hymenopterans occupy, ranging from plant dependents such as pollinators and seed dispersers to predators and parasitoids (Fernandes et al. 2018). Although being very speciose, the amount of aquatic Hymenoptera represents less than 0.2% (150 taxa of 11 families) of all species of the order, and all of them are parasitoids (Bennett 2007).

Being strictly parasitoids, aquatic wasps often depend on aquatic hosts to complete the life cycle. For example, in the species *Caraphractus cinctus* Walker (Mymaridae), females oviposit submerged and place its eggs inside the eggs of other insects, preferably of Dytiscidae beetles. However, when lacking aquatic hosts, females may leave the water to search for potential organisms and lay the eggs flying towards the marginal vegetation (Jackson 1958, 1966). Jackson (1961, 1963, 1966) made a thorough description on the biology of this mymarid. He demonstrated that mating (lasts about 10 s) occurs under water in the majority of times; males (which emerge first) wait for females to hatch from the eggs of the host, then raise the wings vibrating it, and then mount the female remaining on the top, curving its abdomen to reach female genitalia. Males may wait for several females to emerge and then mate with them subsequently. Females can also be parthenogenetic, laying eggs almost immediately after emergence (Jackson 1966). After mating, females actively search for the eggs of potential hosts under water, often placed under leaves (Jackson 1963, 1966).

Other species may paralyze their hosts and take them away from the water to lay the eggs (Guarisco 2010). This is the case of the pompilid *Anoplius depressipes* Banks, which are specialized to lay eggs on spiders, often from the genus *Dolomedes*. Roble (1985) documented the oviposition behavior of the pompilid on the spider, remarking that this is the only aquatic wasp with parasitoid habits, unlike other species that are parasites. These pompilids choose only female spiders standing at the water surface and place just one egg per spider. The wasp followed the fleeing spider, which submerged in an attempt to escape from the parasitoid, and stung its ovipositor on its abdomen, paralyzing the spider in a few seconds. Then, the wasp took the spider to a burrow, a safe place for the larvae to feed on its paralyzed host (Roble 1985).

### *10.2.5 Lepidoptera*

Another megadiverse insect order, Lepidoptera have almost 160,000 described species, occurring in all continents and occupying nearly all habitats, in both land and freshwater. There are no aquatic butterflies, though. All freshwater lepidopterans are moths, which are represented in eight families, with Crambidae being the family with most aquatic representatives (Duarte et al. 2012; Graça and Solis 2018). Aquatic moths are also part of the trophic chain of freshwater environments by acting as leaf miners and root eaters and also acting as predators of small insects and being predated by small vertebrates (Gorayeb and Pinger 1978; Adis 1983; Solis 2008). For instance, the larvae of the erebid *Palustra laboulbeni* Bar were found in lentic and slow waters of French Guiana and North Brazil feeding on the leaves and roots of water plants and also algae, and being predated by frogs (Adis 1983).

The biology of aquatic moths is poorly reported and consequently their reproductive habits. Odor plume seems to be an important agent in helping adult moths to find potential mates; however, this subject is unexplored for aquatic species (Vickers 2005). Nevertheless, there are studies on species with biological control importance (Bennett 1970; Denoth et al. 2002), such as the pyralid *Sameodes albiguttalis* Warren, that damage leaves of invasive aquatic plants. Females take advantage of small cuts and other deformities that the leaves may have to lay the eggs inside plant cells. Then, the larvae will develop eating the plant tissues, eventually leading to its decay (DeLoach and Cordo 1978). Another reported issue concerning aquatic moth larvae is the accidental introduction of species, which may lead to the decreasing of native aquatic plants (Buckingham and Bennett 1989). Finally, one of the greatest challenges of studying aquatic moths, leading to a poor knowledge of its natural history, seems to be the aquatic habitat per se, often located in areas hard to access which makes it difficult to find the small larvae (Graça and Solis 2018).

### 10.2.6 Neuroptera

With more than 4000 described species, Neuroptera (i.e., antlions and lacewings) has a great arrange of morphological variation and also habitat preferences Freitas and Penny 2012). Being predominantly terrestrial, neuropterans are represented by only 3 of the 17 families in aquatic environments. Of these three aquatic families, Sisyridae and Osmylidae have worldwide distribution and Nevrorthidae are restricted to Asia and Oceania (Martins and Ardila-Camacho 2018). Sisyrids are commonly known as parasites of freshwater sponges (e.g., *Spongilla* genus) and they rely on these hosts as food source and substrate until they leave to pupate and emerge as adults (Poirrier 1969). On the other hand, the sisyrids *Climacia areolaris* Hagen and *Sisyra vicaria* Walker were reported to feed on a wider range of food, such as aphid nymphs and adults, alderfly eggs, and even pollen grains (Pupedis 1987).

Most of the freshwater-related species of Neuroptera have nocturnal habits; thus, mating, oviposition, and larvae migration are present in the hours between dusk and dawn (Pupedis 1987). Although remaining in land, after emergence adults forage and rest near the aquatic environments in vegetal substrate like shrubs and small plants. Searching for food, adult sisyrids would check actively on the borders of the leaves on the shrubs, touching it with their antennae, until anything edible is found (Pupedis 1987). Aquatic neuropterans are also predated, and have specialized taxa to be careful with: *Sisyridivora* (i.e., eater of sisyrids) females are parasitoids that construct a "feeding tube" on the sisyrid's cocoon with its ovipositor (Fig. 10.5). The tube is filled with fluids of the host, and once the female finds the tip of the tube, it starts to feed moving its mouthparts, in a movement similar to licking (Pupedis 1978).

Reproductive behavior on aquatic neuropterans often involves the use of pheromones in a courtship. In the genus *Osmylus*, during the courtship males would signalize with their wings and evert structures similar to glands located at the end of the abdomen, and the female touches it with antennae and palpi. This occurs before mating takes part and can last for over an hour, before the male grabs the female with his front legs and also with the aid of his mouthparts (Withycombe 2009; Henry 1997). After mating, males transfer a spermatophore to the female, which they often feed upon. To avoid that energetic loss, males may rest with the females for several hours after mating, to ensure egg fertilization. In a couple days, females lay the eggs side by side, where the larvae will hatch few weeks later (Withycombe 2009).

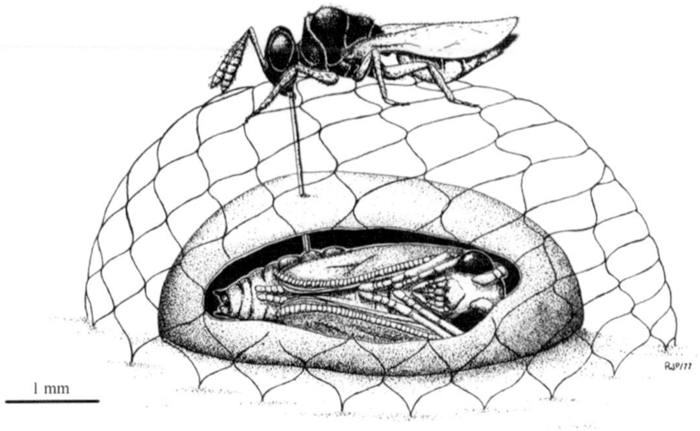


**Fig. 10.5** The parasitoid wasp *Sisyridivora cavigena* Gahan feeding on the pupa of the sisyrid *Climacia areolaris* Hagen. Adapted from Pupedis 1978

## *10.2.7 Odonata*

The order Odonata comprises the group of dragonflies and damselflies (Fig. 10.6a–d). Odonata is, today, composed by over 6000 species described, which occurs in all continents except Antarctica (Dijkstra et al. 2013). Nearly 1800 of those species are from the neotropical region (Neiss and Hamada 2014). The individuals of this order inhabit ponds, lakes, streams, and rivers, being one of the most important organisms to the freshwater ecosystems which they occur due to its highly predatory behavior, which controls the population of other invertebrates in the larval stages (Corbet 1999; Mandal et al. 2008; Saha et al. 2012), and also bioindicators of pristine environments (Monteiro-Júnior et al. 2013; Nasirian and Irvine 2017). The larval stage is aquatic for almost all species. In few species larvae seem to be semiaquatic.

The reproductive system in Odonata often involves courtship, elaborated behavioral contest between males, and sperm competition; the latter is treated further in this chapter (Cordero-Rivera et al. 2004; Córdoba-Aguilar and Cordero-Rivera 2005; Guillermo-Ferreira and Bispo 2012). Recently, Guillermo-Ferreira and collaborators have been investigating the case of the neotropical calopterygid *Mnesarete pudica* Hagen in Selys (Guillermo-Ferreira and Bispo 2012; Guillermo-Ferreira et al. 2014, 2015), in which males exhibit elaborated courtship displays and territorial disputes prior to mating. They found that, unlike other calopterygids, males also perform a cross-display with their red-colored wings perched in front of females, while other calopterygid species often perform circular hovering flights only (Guillermo-Ferreira and Bispo 2012; Fig. 10.7). Furthermore, they found that the red coloration of the wings can trigger an aggressive response in other males, which are territorial and fight over patches near oviposition sites (Guillermo-Ferreira et al. 2014). Those fights are usually settled when *M. pudica* males assess rival quality by the opaqueness and size of wing red spots: the larger, the better (Guillermo-Ferreira

**Fig. 10.6** Dragonflies (**a**—larvae and **b**—adult) and damselflies (**d**—larvae and **c**—adult). Photographies by Frederico Falcão Salles (**a**, **d**), Adolfo Cordero Rivera (**c**), and Diogo Vilela (**b**)

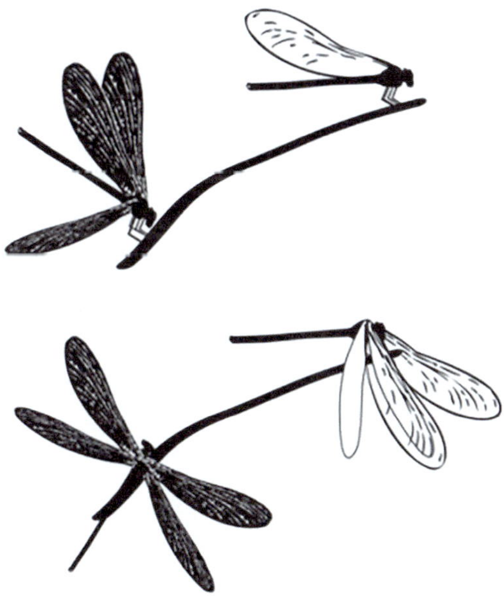

**Fig. 10.7** The wing display by *Mnesarete pudica* Hagen in Selys male. Adapted from Guillermo-Ferreira and Bispo 2012

et al. 2015). On the other hand, the same visual signal incites a sexual response by females: spreading the wings as a signal of refusal to mate, leaves flying to a distant perch, or the wing-flapping behavior, which allows the male to carry on with the courtship display (Guillermo-Ferreira and Bispo 2012).

Another interesting feature about the group, which increases its behavioral peculiarities, is that males have two different organs responsible for reproduction. One effectively produces gametes (i.e., primary gonopore) and is located on the ventral portion of the ninth segment and the other, the genital ligula or penis, is responsible to place the sperm on the females and often to remove sperm from other males, located on the ventral side of second abdominal segment (Waage 1979; Siva-Jothy 1987; Corbet 1999). This condition leads the males to an important behavioral task: prior to mating, they have to ensure that its penis is "loaded" in order to fertilize the eggs; to do so, they curve the abdomen so that its ninth segment (which contains the primary gonopore) is placed facing the second segment (which contains the penis) and then make the sperm transfer. The sperm transfer can occur in distinct moments: after grabbing the female, in tandem or alone, immediately after one mating event.

Mating in odonates can be as fast as 10 s long (as observed in *Orthetrum coerulescens* Fabricius, Rehfeldt 1991) or last for some hours (as in *Ceriagrion tenellum* de Villers, Andres and Cordero-Rivera 2000, and *Ischnura graellsii* Rambur, Cordero 1992). Despite the time spent on mating, it is important to complete successfully the sperm displacement and sperm transfer to female. After this transfer is ended, males can remain in tandem position to guard females as they oviposit, as observed in *Acanthagrion truncatum* Selys; guard females with no contact, as observed in *Diastatops obscura* Fabricius; or even leave the female to oviposit alone as in the coenagrionids *Cyanallagma nigrinuchale* Selys and *Argia reclusa* Selys (Vilela et al. 2016).

### *10.2.8 Orthoptera*

Only 0.16% of all valid species of Orthoptera are semiaquatic. This order has over 28,000 described species in all continents, with Amazon being the most diverse region (Gutjahr and Braga 2018). Semiaquatic orthopterans are only found in the family of the true locusts, Acrididae, which complete its development associated and feeding to aquatic plants such as *Eichhornia* and *Paspalum* (Nunes 1989; Vieira and Santos 2003; Braga et al. 2011). In this sense, species such as *Paulinia acuminata* De Geer are even used as biological control to mitigate populations of the macrophyte *Salvinia molesta* Mitchell, a weed that disperses faster in areas altered by deposition of organic waste, causing a hydric disbalance (De Olivera 2014).

Together with their ecological importance, the life cycles of some semiaquatic orthopterans were also well studied in the Brazilian Amazon (Nunes 1989; Nunes et al. 1992; Nunes and Adis 1994; Nunes-Gutjahr and Braga 2012; Vieira and Santos 2003; Braga et al. 2011; De Olivera 2014). The species *Cornops aquaticum* Bruner, one of the most studied, is a biological controller of the *Eichhornia* macrophytes, where they feed. Matings occur on the plant and may last from 40 to 120 min, with the females ovipositing on the petiole of the leaves. Eggs are protected with a cement-like substance and will hatch within 30–40 days where the nymphs will consume the host and adjacent plants (Vieira and Santos 2003).

### 10.2.9 Megaloptera

The small order Megaloptera (alderflies, dobsonflies, and fish flies) is comprised by almost 400 species present in all continents except Antarctica. Larval stage of all species is aquatic, which may inhabit lakes, streams, and turbulent waters of rivers (Hamada and Azevêdo 2012; Ardila-Camacho and Contreras-Ramos 2018).

Mating occurs near water, on the adjacent vegetation. Males of Sialidae may use acoustic signals in a courtship behavior. Similarly to stoneflies, some Megaloptera species may use abdominal vibrations to find potential mates and, as the females approach, males may tap wings and abdomen on the ground (Rupprecht 1975; Theischinger 2000). After mating, females aggregate with others to place the eggs on rocks, vegetation, bridges, and some other substrate near the water. After hatching, larvae have to immediately search for water, to prevent dehydration from heat and sunlight exposure (Theischinger 2000; Hamada and Azevêdo 2012). Megalopteran larvae are commonly found under rocks of rapid streams, where they feed on other small invertebrates. After completing the development, larvae leave the water and build small chambers on the soil adjacent to the water body to pupate for few weeks until they emerge as an adult (Theischinger 2000). Adult lives are short (10 days) and almost entirely devoted to find a partner to mate. They feed only on liquids, as their intestines are atrophied (Hamada and Azevêdo 2012).

### 10.2.10 Plecoptera

Insects of the order Plecoptera (Fig. 10.8), also commonly named stoneflies, are hemimetabolous insects whose nymphs inhabit pristine and well-oxygenated habitats, occurring in the freshwaters of all continents except Antarctica (Hynes 1976; Fochetti et al. 2009). These insects are closely related to freshwater habitats even though its adults are terrestrial, due to the poor ability of the adults to disperse for long distances. This limited flight ability combined with the strict preferences of the

**Fig. 10.8** Adult Plecoptera (stonefly, *Anacroneuria debilis* Pictet) in vegetation near a water body. Photography of Frederico F. Salles

nymphs is probably the main reason that makes this group highly endemic (Fochetti and Figueroa 2008). In its turn, nymphs are aquatic and depend on a combination of ecological traits to develop to the adult stage, which are mainly well-oxygenated and preserved running water bodies (Hynes 1976). After emerging to adults, stoneflies pass through a maturation period, which can last from 3 to 5 days (Harper 1973). Males are, in general, smaller than females (Froelich 2012), and its sexing can be difficult to determine without examining through a microscope, although some females can be spotted with visible eggs on the abdomen.

An interesting feature about mature male stoneflies is their drumming behavior. The drumming behavior of stoneflies consists of a complex communication system in which males use his abdominal hammer to produce substrate-borne vibrations, to communicate with females (Stewart 1997; Sandberg and Stewart 2003). Males and females, then, engage in a duet-like communication mode (Abbott and Stewart 1993; Fig. 10.9), which can help individuals to find potential mates, help in species recognition, and even help females assess male fitness (Stewart 1997).

After finding its sexual partner, at the copulation the males stay on top of females, curving its abdomen to reach female reproductive organs. Copulations can occur several times a day and after insemination females can lay batches of dozens to hundreds of eggs (Harper 1973). After the egg development, which varies between species and usually takes from 30 to 60 days to hatch, the nymphs begin to rapidly grow, passing through several molts until they emerge to the adult form.

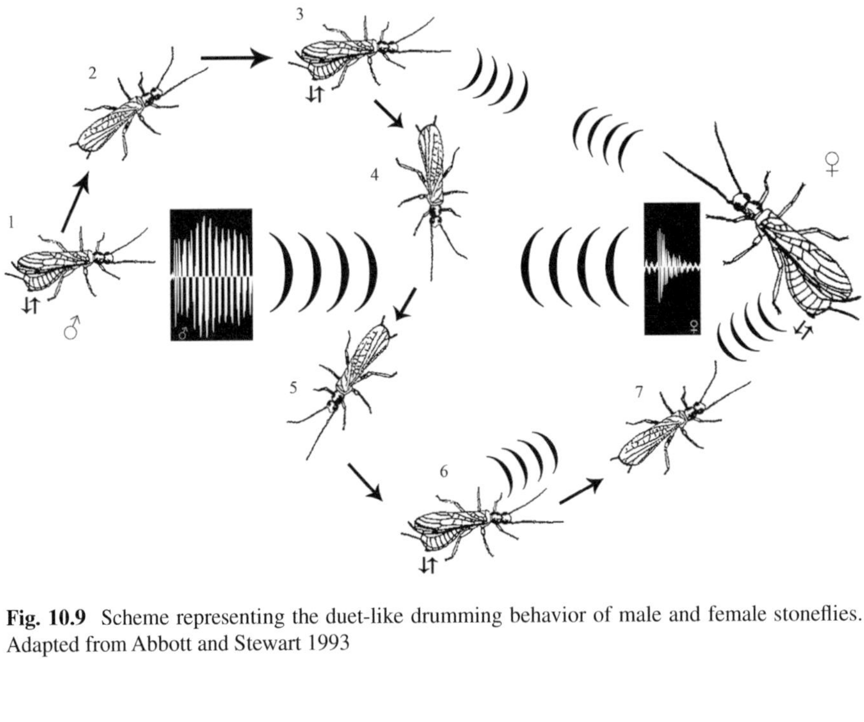


**Fig. 10.9** Scheme representing the duet-like drumming behavior of male and female stoneflies. Adapted from Abbott and Stewart 1993

### 10.2.11 Trichoptera

The Trichoptera, or caddisflies (Fig. 10.10), are cosmopolitan, present in freshwater habitats of every continent, except Antarctica. With over 15,000 described species, Trichoptera is one of the most successful aquatic orders in terms of diversity (Paprocki 2012; Pes et al. 2018). Larval and pupal stages are all aquatic, depending, in its large majority, on running waters (Paprocki 2012), except for two genera of Limnephilinae (Mackay and Wiggins 1979).

As in its sister group, Lepidoptera, trichopterans also rely on pheromones to perform its sex activities of searching and locating potential mates (Resh et al. 1987). First thought as being of defensive use, sex pheromones were evidenced first by Wood and Resh (1984). They found in a field experiment that, in cages with females *Gumaga griseola* McLachlan the number of approaching males was significantly higher than in control cages without females. Furthermore, in cages with the pheromone extract only, males were less abundant, indicating that the visual clues are also important to find a mate (Wood and Resh 1984). After pheromones reach their targets and many individuals are gathered, they would swarm near the water face forming a column, where mating takes part (Paprocki 2012).

After mating, females may lay the eggs on top of rocks, leaves, directly on water surface, and diving or even evolve in association with other organisms such as the starfish *Patiriella regularis* Verrill (Winterbourn and Anderson 1980; Reich et al. 2011; Paprocki 2012). Caddisfly larvae can explore a great deal of food resources, as different groups may present different adaptations to scrap, predate other organisms, and feed on plant fluids (Paprocki 2012). An important ecological and taxonomic feature of trichopteran larvae is their capacity to build portable cases (Fig. 10.11). Using silk, they put together tiny grains of sand, twigs, and leaves which shelters the larvae until the moment of pupate (Paprocki 2012; Pes et al. 2018).

**Fig. 10.10** An adult male Trichoptera (Odontoceridae: *Marilia sp.*). Photography of Frederico F. Salles

**Fig. 10.11** A larvae of Odontoceridae (*Marilia sp.*) carrying its case. Photography of Frederico F. Salles

## 10.3 Aquatic Insects with Aquatic Adult Life

The two following orders have totally aquatic representatives, that is, both immature and mature stages remain on freshwater environments. Living so close to the water has pressured these organisms to develop into structures, mechanisms which have allowed them to explore different niches and habitats that, therefore, have demanded different life strategies and behavioral traits.

### *10.3.1 Coleoptera*

By far the most speciose insect order, with over 400,000 described species, Coleoptera (i.e., beetles) is present throughout all regions of the planet except Antarctica. Beetles are extremely diverse in terms of morphology, habitats, and niche occupancy and it may be the main reason for its radiation success (Grimaldi et al. 2005; Jäch and Balke 2007; Casari and Ide 2012; Benetti et al. 2018). In this sense, the aquatic Coleoptera is expected to be numerous. They indeed are with almost 20,000 species distributed in 37 families; Coleoptera has also the largest number of aquatic representatives among insects, and more speciose than many entire orders (Benetti et al. 2018). With regard to so many differences within the order, there are also different types of aquatic beetles. Some are true water beetles—submerges for most of its life in adult stage, leaving water just for dispersal (e.g., Dytiscidae, Gyrinidae, and Elmidae)—or false water beetles—terrestrial adult life (Scirtidae, Psephenidae). There are other four categories of water-related beetles: phytophilous, parasitic, facultative, and shore beetles (Jäch 1998). Here, we treat the true water beetles.

One of the most diverse true water beetle families, Dytiscidae or diving beetles have its bodies shaped perfectly for the aquatic life (Fig. 10.12). Their middle and hind legs are flattened to help in the swimming, in which they are very skillful. Another modified trait is the dilated tarsomeres on the forelegs, which helps to grab females during mating (Casari and Ide 2012). Mating behavior was reported

**Fig. 10.12** A water beetle (Coleoptera: Dytiscidae). Photography of Frederico F. Salles

for the Japanese water beetle *Dytiscus alaskanus* Balfour-Browne (Aiken 1992). At the rendezvous, while females are swimming around, the male mounts on top of female without apparent courtship behavior, grasping her with his modified foreleg tarsomeres (see Kolmes 1983 for similar precopulatory behavior in Gyrinidae). If the mounting is successful, that is, the male was able to be in the right position so that female does not escape, the mating is completed (Aiken 1992). To fit in the optimal position, after mounting males make lateral shakes to prevent the escaping attempts from females, which have drawn their legs, and the mating takes place. When the males succeed to transfer its spermatophores, the males build a plug on female's gonopore, preventing them from extruding his gametes. After mating, females would oviposit mainly in aquatic plants, such as *Oenanthe javanica* Blume (Inoda 2010).

## *10.3.2 Hemiptera*

Hemiptera (true bugs) is the largest and most diverse order among hemimetabolous insects with almost 90,000 species described worldwide. Members of the family Heteroptera (Fig. 10.13) are the only actual aquatic, although in some other families as Gerridae the species are found close to freshwater environments, skating on the water surface (Grazia et al. 2012; Gullan and Cranston 2014; Moreira et al. 2018). Among the most remarkable water bugs are the representatives of Belostomatidae family. Their large size is intimidating; they are predators of invertebrates and even small vertebrates of ponds, lakes, and streams (Aldrich 1988). Also, if attacked, these bugs may squirt a black secretion and flee to avoid predation (Leston 1979).

Still in belostomatidae, parental care is essential to ensure that the eggs will hatch safely. Males are the ones who take care of the eggs, either by watching them on the sites of oviposition (often marginal vegetation) or by caring for the eggs placed on its back, preventing them from desiccation and drowning. Females of the subfamily Lethocerinae place the eggs on marginal vegetation and those of Belostomatinae oviposit directly on males' backs (Tallamy 2001). In the lethocerin

**Fig. 10.13** A water bug on the substrate (Hemiptera: Naucoridae: *Ambrysus lamprus* Nieser, Pelli, and Melo). Photography of Frederico F. Salles

*Lethocerus deyrollei* Vuillefroy, females mate repeatedly (over 20 times) with the same male which, in its turn, is convinced of the paternity (see Smith 1979) and perform the postcopulatory paternal care (Ichikawa 1989). The repeated mating also occurs in belostomatines. Females can store sperm from previous males and fertilize the eggs. So, to ensure that the majority (if not all) of the eggs placed on its back belong to them, males repeatedly mate with females, because the last male to copulate ensures the paternity of the majority of fertilized eggs (Smith 1979).

## 10.4 Sexual Selection in Aquatic Insects

### *10.4.1 Sexual Selection as Evolutionary Pressure*

By walking across a stream or taking a look in a pond we are able to perceive in a fast look the diversity of shapes, colors, and behaviors that present the aquatic (complete cycle under the water) and semiaquatic (larval stage under the water) insects. Under the Darwinian interpretation (Darwin 1859), this variability is produced by the effect of natural selection. This concept explains the evolutionary pressure that promotes the reproductive success of individuals better adapted to their contemporary environmental conditions. Individuals that possess advantageous characters are able to better survive until maturity, compete with their conspecifics, being selected by their mates, and produce better adapted offspring. In the case of insects, this specialization was produced by at least more than 400 millions of years (around 10,000 times more than the apparition of the human species).

Back to our excursion along the water mass and after discriminating individuals by species with the help of a book guide or a taxonomic key, we still observe variability among individuals of the same species. Males and females show different morphological and behavioral traits. In general, females are larger and males present more complex structures. The existence of this sexual dimorphism (two forms/phenotypes among sexes) was interpreted as the product of another different evolutionary pressure: the sexual selection (Darwin 1871). This concept explains how traits are maintained across generations due to the mate preferences of selection more than the adaptation of the rest of environmental factors. It is assumed that the individual preference for a trait is genetically determined because it is correlated with the expression of the trait (Fisher 1958) or because of the inheritance of genes from parents that already possessed the preference (Dawkins 1996). The selected traits can be morphological characteristics as ornaments or armaments or behavioral displays or abilities. For instance, color-winged males that perform costly courtships in territorial damselflies are preferred by females (e.g., Orr and Hämäläinen 2007; Watanabe and Taguchi 1990); thicopteran females are selected among males according to the pheromones produced (e.g., Jackson and Resh 1991); in the stonefly *Pteronarcella badia* Hagen (Plecoptera: Pteronarcyidae), males that perform drum duets (timing vibrations) find females faster than those that do not drum or perform anomalous drums (Abbott and Stewart 1993). Traits try to stimulate the perception of possible future partners according to their sensorial system. For instance, dragonflies use colors and movements because they possess complex eyes, thicopterans use pheromones due to the chemical receptors of the adult antennae, and stoneflies use drums because of their developed mechanoreceptors. According to the runaway hypothesis (Fisher 1958), attractive traits are those that allow to win competitions and/or are related with the male fitness. For instance, bold-aggressive behaviors are selected in several species (even in humans) because they allow to obtain more resources as food and territories (see Schuett et al. 2010). *Ephemera danica* Müller (Ephemeroptera: Ephemeridae) females prefer courtships performed by long-winged males, who are better jostle competitors and obtain more reproductive success in at least four other ephemeropteran species (see Harker 1992). In the water boatman *Sigara falleni* Fieber (Heteroptera: Corixidae), females select males according to the size of their body and foreleg, a trait developed during the juvenile stage that allows the competition against other males.

The selection of traits across generations produces by positive feedback the exaggeration of characters, a fact increased in species in which females can determine the phenotype of their offspring (see Mousseau and Fox 1998). However, the investment of resources in attractive traits possesses consequences when natural and sexual selections follow opposed directions. The boundaries of the trait exaggeration appear when they decrease the probabilities to survive or reproduce. Instead small individuals achieve more matings due to their aerobatic behavior in the midge *Chironomus plumosus* L. (Diptera: Chironomidae), and very small individuals are not able to perform longer flights (Neems et al. 1990). Based on this phenomenon, the handicap selection hypothesis (Zahavi 1975) explains that traits are selected by the cost that they represent in production, maintenance, and ability to survive with a trait that disrupts with the efficiency to survive. For instance, colorful, noisy, or bolder individuals increase the encounter with predators, so individuals must be healthy and competitive if they are able to survive under these handicapped conditions.

After identifying the sexual variability in different species in our excursion we perceive that in general females are larger than males while males show more exaggerated traits (morphology and behavior) than females. Due to the conflict arisen by the different conditions and costs of each sex (Chapter *Sexual Conflict*), the probability to mate is commonly ensured for females but it is more variable for males (Bateman 1948). This produces an ecological scenario in which (a) the ratio of male:female searching for copulas is frequently biased towards the formers (e.g., in ephemeropteran, thicopteran, and dipteran swarms; Sullivan 1981; Svensson and Petersson 1987; also see Janicke and Morrow 2018); (b) males possess higher variability (Fairbairn and Preziosi 1994); and (c) wrong matings produce higher costs in females due to the investment in reproduction (Thornhill 1983). Therefore, the selection of mates is generally performed by females. As this reduces the probabilities to mate by males, the latter employ different strategies to compete between them for the access of copula. The firefly *Luciola lei* males compete against other males with the emission of flash lights but stop to produce them at the end of the copula to reduce the conspicuousness of the female (Fu et al. 2006). The increase of competition and the variability produces intrasexual selection and enrichment of the species genome (Janicke et al. 2018). Therefore, females suffer the harassment of territorial and sneaker males. The competition among males increases the differences towards females, emphasizing sexual dimorphism and producing intersexual selection. However, females can also be subject of selection. In the case of the long-tailed dance fly *Rhamphomyia longicauda* Loew (Diptera: Empididae), males select and give nuptial gifts to females concentrated in the same location (lek) according to the size of their abdomen, which represents a signal of fertility (Funk and Tallamy 2000).

### *10.4.2 Reproductive Trade-Off*

The copula formation constitutes a costly and risky phenomenon for both sexes: it is time and resource consuming (e.g., Martens and Rehfeldt 1989), produces injuries performed by the genitalia interactions (Chapman et al. 1995), increases the exposure to diseases and parasites (Daly 1978; Hamilton and Zuk 1982) and the conspicuousness towards predators (e.g., Yuval and Bouskila 1993), and decreases individual locomotion and general life span (in mosquitoes, South et al. 2009; in stoneflies, Taylor and Peckarsky 1998). However, it is considered that copulation process is more costly for females. In several species, females can fertilize all their eggs with a single ejaculation of a male (compare Cordero et al. 1995, 1998). Therefore, female fitness is based on the quality of the mate. However, the advantages of the polyandry should be taken into account (see Fincke 1984; Zeh and Zeh 2003), especially if the first copula was performed by a poor-quality male. For instance, damselfly females of *Mnais costalis* Selys (Odonata: Calopterygidae) mate with a dominant male after a copula with a subordinate (Watanabe and Taguchi 1990). On the other side, the sperm of males can be extracted from the sperm chamber of the female by other males that mate after or can be removed by the female (see sperm competition and female cryptic selection in Chapter *Sexual Conflict*). To ensure their mating success, male reproductive strategy is focused on the quantity of matings. Under this scenario, the previous selection of mates described turns into an intersexual arms race based on the female harassment and the male avoidance (by both sexes). For example, in the water strider *Aquarius remigis* Say (Heteroptera: Gerridae), males are able to increase the length, mobility, strength, and rigidity of the phallus to fix the female during copula, while females increase the length of their abdominal spines to avoid the male assemblage (Adriaenssens and Johnsson 2010; Arnqvist and Rowe 1995; Fairbairn et al. 2003). It was proposed that in the damselfly *Platycypha caligata* Selys, females perform a collective defense mechanism against male reproductive attempts by grasping their wings in groups (Martens and Rehfeldt 1989). As male and female reproductive structures and strategies were coevolved under millions of years to reach opposed goals (see Rice 1998) complex interactions aroused during all stages of the reproductive events to promote the selection of mates. As we saw, the selection starts before copula by the presentation of morphological traits (ornaments/armaments) or resources (e.g., nuptial gifts, territories). However, the strength of the male harassment can suppose drastic consequences for females, especially in non-territorial species and/or those that do not possess mechanisms or signals to be selected. Therefore, female reduces the cost of the male attempts by the acceptance of the copula, especially if they are able to modify the outcome of the mate. The morphological assemblage of the genitalia and the secondary sexual structures were proposed as mechanisms of interspecific recognition and sexual selection by fertilization success (Simmons 2014). Due to the species-specific evolution of the genital structures, these traits constitute nowadays the best key to identify insect species. Even if a male was accepted, the lock-and-key compatibility between individuals will determine the success of the copula and the selection of the best partner. The selection of males is also dependent on their behavioral tactics during copula, meaning by their copulatory courtship (Eberhard 1991). This behavior is based on the capability of males to stimulate females with the intromission and movement of their genitalia and secondary sexual characters. The last selection is produced after copulation. Females judge the male behavior and determine the use of their sperm or the inversion in the eggs fertilized with this sperm. For instance, long guarding events could be interpreted as a postmating courtship to condition the female investment in the next oviposition (Eberhard 1996). Damselflies of the genus Coenagrionidae (Odonata: Zygoptera) perform pre- and postcopulatory tandems for hours, even until the end of the day (e.g., Andrés and Cordero-Rivera 2000; Cordero 1989). In the case of the water strider *Gerris lateralis* (Hemiptera: Gerridae) the sperm recovery increases with the time of the copula, a factor positively related with the fertilization success (Arnqvist and Danielsson 1999). In this case, the multiplicity of mates reduces the amount of sperm transfer in mates performed closer in time. *Mystacides azurea* L. (Trichoptera: Leptoceridae) old males prolong the duration of mating due to their reduced possibilities to reach a mate (Petersson 1990). After all the processes, females of several species possess the last word by the use or the rejection of the sperm (e.g., Córdoba-Aguilar and Cordero-Rivera 2008). In some cases, females that expulsed the sperm of the last male exploit the male guarding to profit the protection towards new male attempts (personal observation in *Polythore derivata* McLachlan and *Archineura incarnata* Karsch).

### *10.4.3 Established Variability: Polymorphism*

As we saw, sexual selection could increase the intrasexual variability until creation of different phenotypes (polymorphism). The maintenance of polymorphism is assumed by the dependence of mating success on environmental factors (e.g., see Cordero 1992). The female color di/polymorphism of damselflies (reviewed in Sanmartín-Villar 2017) is currently interpreted by a mechanism of male harassment avoidance dependent on the frequency of phenotypes in the population (Van Gossum et al. 2008). At lower population frequencies of color male-like females (those that mimic males in color and behavior; Robertson 1985), males identify them as other males and do not try to develop sexual interactions (Cordero and Andrés 1996; Sanmartín-Villar and Cordero-Rivera 2016). In the case of the territorial damselflies of the genus *Mnais* (Zygoptera: Calopterygidae), dominant color winged males are preferred by females. However, color female-like male phenotypes can infiltrate the territories of the dominant phenotype and sneak their protected females when the territorial males fight with intruders (e.g., Watanabe and Taguchi 1997). In at least five *Dytiscus* species (Coleoptera: Dytiscidae), female dimorphism prevents male grasping. In these species, the number and position of the male tarsal discs (setae) varied according to each female phenotype (smooth or structured dorsum) (Bergsten et al. 2001; Bilton et al. 2008; Inoda et al. 2012). One of the most extreme dimorphisms is represented by the aquatic moth *Acentria ephemerella* Denis and Schiffermüller (Lepidoptera: Crambidae). If the preference or the mating success is not balanced between phenotypes (e.g., phenotypes of one sex prefer different phenotypes of the other sex or only phenotype is always selected and the other does not reach enough fitness), the outcome of sexual selection could be the radiation of species by divergent evolution or the loss of variability by the extinction of phenotypes.

### *10.4.4 Plasticity: Counteracting Sexual Selection*

To finish, the fact that the outcome of sexual selection is not only a question of genes should be taken into account. The evolutionary pressure can punctually be modified by plasticity and/or learning processes. For instance, territorial males become sneakers with the age or when the population frequency of territorial males is elevated in damselflies of the genus *Mnais* (Odonata: Calopterygidae), even when they still present the territorial wing coloration (Tsubaki et al. 1997; Watanabe and Taguchi 1990). The perception and posterior selection of females are also conditioned by factors as the female status. Water strider females of the species *Gerris buenoi* Kirkaldy (Heteroptera: Gerridae) prefer large males only when they are hungry (Ortigosa and Rowe 2002). Under lower frequencies of heteromorphic females (different color than males) and after trying to mate with all kind of conspecific phenotypes (even with other males), damselfly males of the genus *Enallagma* (Zygoptera: Coenagrionidae) identify the mimesis of color male-like females as individuals of the opposed sex (Miller and Fincke 1999). Instead the capability of inflating the abdomen is commonly related with the maturity of the eggs in female long-tailed dance flies *Rhamphomyia longicauda* Loew (Diptera: Empididae); females not ready to mate can cheat increasing their abdomen to obtain nuptial gifts from males (Funk and Tallamy 2000).

Despite their apparent simplicity, the insect world hides sophisticated structures and strategies shaped by long-term evolutionary processes. Their protection and study will continue to give us a source of knowledge and will teach us about the complexity of life and how to modify our technology in order to solve problems as they do. Performing this chapter we evidenced the elevated lack of information referent to the reproductive behavior and the evolutionary pressures that they concern in many aquatic insect orders, so we would like to encourage teams and students to continue discovering the incredible mysterious mechanisms that underlie the biology, ethology, and ecology of the aquatic insects.

**Acknowledgments** We would like to thank the organizers of the book for the opportunity to write this chapter. We also thank Natasha Mhatre (Fig. 10.1), Adolfo Cordero Rivera (Fig. 10.6c), and Frederico Falcão Salles (Figs. 10.4, 10.6a, d, 10.8, 10.10, 10.11, and 10.13) for the use of photographs.

## References

Abbott JC, Stewart KW (1993) Male search behavior of the stonefly, *Pteronarcella badia* Hagen (Plecoptera: Pteronarcyidae), in relation to drumming. J Insect Behav 6(4):467–481

Adis J (1983) Eco-entomological observations from the Amazon. IV. Occurrence and feeding habits of the aquatic caterpillar *Palustra laboulbeni* Bar, 1873 (Arctiidae: Lepidoptera) in the vicinity of Manaus, Brazil. Acta Amazon 13(1):31–36

Adriaenssens B, Johnsson JI (2010) Shy trout grow faster: exploring links between personality and fitness-related traits in the wild. Behav Ecol 22(1):135–143. https://doi.org/10.1093/beheco/arq185

Aiken RB (1992) The mating behaviour of a boreal water beetle, *Dytiscus alaskanus* (Coleoptera Dytiscidae). Ethol Ecol Evol 4(3):245–254. https://doi.org/10.1080/08927014.1992.9523136

Aldrich J (1988) Chemical ecology of the Heteroptera. Annu Rev Entomol 33(1):211–238. https://doi.org/10.1146/annurev.en.33.010188.001235

Allan JD, Flecker AS (1989) The mating biology of a mass-swarming mayfly. Anim Behav 37:361–371

Andrés JA, Cordero-Rivera A (2000) Copulation duration and fertilization success in a damselfly: an example of cryptic female choice? Anim Behav 59(4):695–703. https://doi.org/10.1006/anbe.1999.1372

Ardila-Camacho A, Contreras-Ramos A (2018) Order Megaloptera. In: Thorp and Covich's freshwater invertebrates. Academic Press, pp 217–227

Arnqvist G, Danielsson I (1999) Postmating sexual selection: the effects of male body size and recovery period on paternity and egg production rate in a water strider. Behav Ecol 10(4):358–365. https://doi.org/10.1093/beheco/10.4.358

Arnqvist G, Rowe L (1995) Sexual conflict and arms races between the sexes: A morphological adaptation for control of mating in a female insect. Proc R Soc B Biol Sci 261(1360):123–127. https://doi.org/10.1098/rspb.1995.0126

Bateman AJ (1948) Intra-sexual selection in drosophila. Heredity 2(Pt. 3):349–368

Bell WJ, Roth LM, Nalepa CA (2007) Cockroaches: ecology, behavior, and natural history. JHU Press

Benetti CJ, Michat MC, Archangelsky M (2018) Order Coleoptera: introduction. In: Thorp and Covich's freshwater invertebrates. Academic Press, Cambridge, pp 497–517

Bennett FD (1970) Insects attacking water hyacinth in the West Indies, British Honduras and the USA. Hyacinth Control Journal 8(2):10–13

Bennett AM (2007) Global diversity of hymenopterans (Hymenoptera; Insecta) in freshwater. In: Freshwater animal diversity assessment. Springer, Dordrecht, pp 529–534

Bergman EA, Hilsenhoff WL (1978) Parthenogenesis in the mayfly genus Baetis (Ephemeroptera: Baetidae). Ann Entomol Soc Am 71(2):167–168

Bergsten J, Töyrä A, Nilsson AN (2001) Intraspecific variation and intersexual correlation in secondary sexual characters of three diving beetles (Coleoptera: Dytiscidae). Biol J Linn Soc 73(2):221–232. https://doi.org/10.1006/bijl.2001.0540

Bilton DT, Thompson A, Foster GN (2008) Inter- and intrasexual dimorphism in the diving beetle *Hydroporus memnonius* Nicolai (Coleoptera: Dytiscidae). Biol J Linn Soc 94(4):685–697. https://doi.org/10.1111/j.1095-8312.2008.01029.x

Braga CE, Nunes-Gutjahr AL, de Morais JW, Adis J (2011) Fenologia de *Cornops aquaticum* (Orthoptera: Acrididae) associado a *Eichhornia crassipes* (Pontederiaceae) em um Lago de Várzea na Amazônia Central, Brasil. Rev. Soc. Entomol. Argent. Mendoza 70(3–4):185–196

Brittain JE (1982) Biology of mayflies. Annu Rev Entomol 27(1):119–147

Brittain JE (1990) Life history strategies in Ephemeroptera and Plecoptera. In: Mayflies and stoneflies: Life histories and biology. Springer, Dordrecht, pp 1–12

Brodskiy AK (1973) The swarming behavior of mayflies (Ephemeroptera). Entomol Rev 52:33–39

Buckingham GR, Bennett CA (1989) Laboratory host range of *Parapoynx diminutalis* (Lepidoptera: Pyralidae), an Asian aquatic moth adventive in Florida and Panama on *Hydrilla verticillata* (Hydrocharitaceae). Environ Entomol 18(3):526–530

Burks BD (1975) The mayflies, or Ephemeroptera, of Illinois. Entomological Reprint Specialists, Los Angeles, p 216

Carvalho CD, Rafael JA, Couri MS, Silva VC (2012) Diptera. In: Rafael JAA, Melo GAR, Carvalho CJB, Casari SA, Constantino R (eds) Insetos do Brasil: Diversidade e Taxonomia. Ribeirão Preto, Holos Editora, pp 701–743

Casari SA, Ide S (2012) Coleoptera Linnaeus, 1758. In: Rafael JAA, Melo GAR, Carvalho CJB, Casari SA, Constantino R (eds) Insetos do Brasil: Diversidade e Taxonomia. Ribeirão Preto, Holos Editora, pp 453–536

Chapman T, Liddle LF, Kalb JM, Wolfner MF, Partridge L (1995) Cost of mating in *Drosophila melanogaster* females is mediated by male accessory gland products. Nature 373(6511):241. https://doi.org/10.1038/373241a0

Clifford HF (1982) Life cycles of mayflies(Ephemeroptera), with special reference to voltinism. Quaest Entomol 18(1):15–90

Corbet PS (1999) Dragonflies: behaviour and ecology of Odonata. Harley Books

Cordero A (1989) Reproductive behaviour of *Ischnura graellsii* (Rambur) (Zygoptera: Coenagrionidae). Odonatologica 18(3):237–244
Cordero A (1992) Density-dependent mating success and colour polymorphism in females of the damselfly *Ischnura graellsii* (Odonata: Coenagrionidae). J Anim Ecol 61:769–780
Cordero A, Andrés JA (1996) Colour polymorphism in odonates: females that mimic males? J Br Dragonfly Soc 12(2):50–60
Cordero A, Santolamazza-Carbone S, Utzeri C (1995) Male disturbance, repeated insemination and sperm competition in the damselfly *Coenagrion scitulum* (Zygoptera: Coenagrionidae). Anim Behav 49:437–449. https://doi.org/10.1006/anbe.1995.0057
Cordero A, Santolamazza-Carbone S, Utzeri C (1998) Mating opportunities and mating costs are reduced in androchrome female damselflies, *Ischnura elegans* (Odonata). Anim Behav 55(1):185–197. https://doi.org/10.1006/anbe.1997.0603
Cordero-Rivera A, Andrés JA, Córdoba-Aguilar A, Utzeri C (2004) Postmating sexual selection: allopatric evolution of sperm competition mechanisms and genital morphology in calopterygid damselflies (Insecta: Odonata). Evolution 58(2):349–359
Córdoba-Aguilar A, Cordero-Rivera A (2005) Evolution and ecology of Calopterygidae (Zygoptera: Odonata): status of knowledge and research perspectives. Neotrop Entomol 34(6):861–879
Córdoba-Aguilar A, Cordero-Rivera A (2008) Cryptic female choice and sexual conflict. In Dragonflies: model organisms for ecological and evolutionary studies. vol. 15, p 189–202. https://doi.org/10.1093/acprof
Daly M (1978) The cost of mating. Am Nat 112(948):771–774
Darwin CR (1859) On the origin of the species. In: Murray J ed., New York
Darwin CR (1871) The descent of man, and sexual selection in relation to sex
Da-Silva ER, Salles FF (2012) Ephemeroptera Hyatt and Arms, 1891. In: Insetos do Brasil: Diversidade e taxonomia. Ribeirão Preto, Holos, pp 231–244
Dawkins R (1996) The blind watchmaker: why the evidence of evolution reveals a universe without design (Reissued in 2006 and 1996). Norton
De Olivera AF (2014) Avaliação do potencial de *Cornops aquaticum* (Bruner, 1906) (Orthoptera: Acrididae) para o controle biológico de *Eichhornia crassipes* (Pontederiaceae) no Parque Ecológico do Utinga, Belém, Pará. University of Pará State, Masters Dissertation, p 58
Degrange C (1960) Recherches sur la reproduction des Ephéméroptères. Trav Lab Hydrobiol Piscic Univ Grenoble 51:7–193
DeLoach CJ, Cordo HA (1978) Life history and ecology of the moth Sameodes albiguttalis, a candidate for biological control of water hyacinth. Environ Entomol 7(2):309–321
Denoth M, Frid L, Myers JH (2002) Multiple agents in biological control: improving the odds? Biol Control 24:20–30
Dijkstra KDB, Bechly G, Bybee SM, Dow RA, Dumont HJ, Fleck GW, Garrison R, Hamalainen M, Kalkman VJ, Karube H, May ML, Orr A, Paulson DR, Rehn AC, Theischinger G, Trueman JWH, Van Tol J, Von Ellenrieder N, Ware J (2013) The classification and diversity of dragonflies and damselflies (Odonata). Zootaxa 3703(1):36–45
Duarte M, Marconato G, Specht A, Casagrande MM (2012) Lepidoptera. In: Insetos do Brasil: Diversidade e Taxonomia. Ribeirão Preto, Holos, pp 625–682
Eberhard WG (1991) Copulatory courtship and cryptic female choice in insects. Biol Rev 66(1):1–31. https://doi.org/10.1111/j.1469-185X.1991.tb01133.x
Eberhard WG (1996) Female control: sexual selection by cryptic female choice. Princeton
Fairbairn DJ, Preziosi RF (1994) Sexual selection and the evolution of allometry for sexual size dimorphism in the water strider, *Aquarius remigis*. Am Nat 144(1):101–108. https://doi.org/10.1086/285663
Fairbairn DJ, Vermette R, Kapoor NN, Zahiri N (2003) Functional morphology of sexually selected genitalia in the water strider *Aquarius remigis*. Can J Zool 81:400–413. https://doi.org/10.1139/Z03-021

Fernandes DRR, Querino RB, Hamada N (2018) Order hymenoptera. In: Thorp and Covich's freshwater invertebrates. Academic Press, Cambridge, pp 339–347

Fincke O (1984) Sperm competition in the damselfly *Enallagma hageni* Walsh (Odonata: Coenagrionidae): benefits of multiple mating to males and females. Behav Ecol Sociobiol 14(3):235–240. https://doi.org/10.1007/BF00299623

Fisher RA (1958) The genetical theory of natural selection. Dover Publ, New York. https://doi.org/10.1111/jeb.12566

Fochetti R, Figueroa JMT (2008) Global diversity of stoneflies (Plecoptera; Insecta) in freshwater. Hydrobiologia 595:365–377. http://dx.doi.org/10.1007/978-1-4020-8259-7_39.

Fochetti R, Ravizza CA, Figueroa JMT (2009) Plecoptera, vol Vol. 43. Fauna d'Italia, Calderini, Bologna, p 536

Freitas S, Penny ND (2012) Neuroptera Linnaeus, 1758. In: Rafael JA, GAR M, CJB d C, Casari AS (eds) Insetos do Brasil, pp 537–546

Froelich CG (2012) Plecoptera. In: Rafael JA, GAR M, CJB d C, Casari AS (eds) Insetos do Brasil, pp 257–261

Fu X, Nobuyoshi O, Vend FV, Lei C (2006) Life cycle and behaviour of the aquatic firefly Luciola leii (Coleoptera: Lampyridae) from Mainland China. Can Entomol 138(6):860–870. https://doi.org/10.4039/n05-093

Funk DH, Tallamy DW (2000) Courtship role reversal and deceptive signals in the long-tailed dance fly, *Rhamphomyia longicauda*. Anim Behav 59(2):411–421. https://doi.org/10.1006/anbe.1999.1310

Fusari LM, Dantas GP, Pinho LC (2018) Order Diptera. In: Thorp and Covich's Freshwater Invertebrates. Academic Press, Cambridge, pp 607–623

Gaino E, Rebora M (2005) Egg envelopes of Baetis rhodani and *Cloeon dipterum* (Ephemeroptera, Baetidae): a comparative analysis between an oviparous and an ovoviviparous species. Acta Zool 86(1):63–69

Gibbs KE (1979) Ovoviviparity and nymphal seasonal movements of *Callibaetis* spp. (Ephemeroptera: Baetidae) in a pond in southwestern Quebec. Can Entomol 111(8):927–931

Gorayeb IS, Pinger RR (1978) Detecção de predadores naturais das larvas de *Simulium fulvinotum* Cerq. e Mello, 1968 (Diptera, Nematocera). Acta Amazon 8(4):629–637

Graça MB, Solis MA (2018) Order Lepidoptera. In: Thorp and Covich's freshwater invertebrates. Academic Press, Cambridge, pp 325–338

Grazia J, Cavichioli RR, Wolff VRS, Fernandes JAM, Takiya DM (2012) Hemiptera. In: Rafael JA, GAR M, CJB d C, Casari AS (eds) Insetos do Brasil, pp 347–406

Grimaldi D, Engel MS, Engel MS (2005) Evolution of the insects. University Press, Cambridge

Guarisco H (2010) The Fishing Spider genus *Dolomedes* (Araneae: Pisauridae) in Kansas. Trans Kans Acad Sci 113(1/2):35–43

Guillermo-Ferreira R, Bispo PC (2012) Male and female interactions during courtship of the Neotropical damselfly *Mnesarete pudica* (Odonata: Calopterygidae). Acta Ethologica 15(2):173–178

Guillermo-Ferreira R, Therézio EM, Gehlen MH, Bispo PC, Marletta A (2014) The role of wing pigmentation, UV and fluorescence as signals in a neotropical damselfly. J Insect Behav 27(1):67–80

Guillermo-Ferreira R, Gorb SN, Appel E, Kovalev A, Bispo PC (2015) Variable assessment of wing colouration in aerial contests of the red-winged damselfly *Mnesarete pudica* (Zygoptera, Calopterygidae). Sci Nat 102(3–4):13

Gullan PJ, Cranston PS (2014) The insects: an outline of entomology. John Wiley and Sons, Hoboken

Gutjahr ALN, Braga CES (2018) Order Orthoptera. In: Thorp and Covich's freshwater invertebrates. Academic Press, Cambridge, pp 143–168

Hamada N, Azevêdo CAS (2012) Megaloptera. In: Rafael JA, GAR M, CJB d C, Casari AS (eds) Insetos do Brasil, pp 547–552

Hamilton WD, Zuk M (1982) Heritable true fitness and bright birds: a role for parasites? Science 218(4570):384–387

Harker JE (1992) Swarm behaviour and mate competition in mayflies (Ephemeroptera). Journal of Zoology, 228(4):571–587.

Harker J (1997) The role of parthenogenesis in the biology of two species of mayfly (Ephemeroptera). Freshw Biol 37(2):287–297

Harper PP (1973) Emergence, reproduction, and growth of setipalpian Plecoptera in southern Ontario. Oikos 24:94–107

Henry CS (1997) 11-Modern mating systems in archaic Holometabola: sexuality in neuropterid Insects. The evolution of mating systems in insects and arachnids, p 193–210

Hynes HBN (1976) Biology of Plecoptera. Annu Rev Entomol 21(1):135–153

Ichikawa N (1989) Repeated copulations benefit of the female in *Lethocerus deyrollei* Vuillefroy (Heteroptera: Belostomatidae). J Ethol 7(2):113–117. https://doi.org/10.1007/bf02350032

Inoda T (2010) Preference of oviposition plant and hatchability of the diving beetle, *Dytiscus sharpi* (Coleoptera: Dytiscidae) in the laboratory. Entomol Sci 14(1):13–19. https://doi.org/10.1111/j.1479-8298.2010.00407.x

Inoda T, Suzuki G, Ohta M, Kubota S (2012) Female dimorphism in Japanese diving beetle *Dytiscus marginalis czerskii* (Coleoptera: Dytiscidae) evidenced by mitochondrial gene sequence analysis. Entomol Sci 15(3):357–360. https://doi.org/10.1111/j.1479-8298.2012.00524.x

Irmler U, Furch K (1979) Production, energy, and nutrient turnover of the cockroach *Epilampra irmleri* Rocha e Siva and Aguiar in a Central-Amazonian inundation forest. Amazoniana 6:497–520

Jäch MA (1998) Annotated check list of aquatic and riparian/littoral beetle families of the world (Coleoptera). Water Beetl China 2:25–42

Jäch MA, Balke M (2007) Global diversity of water beetles (Coleoptera) in freshwater. Hydrobiologia 595(1):419–442. https://doi.org/10.1007/s10750-007-9117-y

Jackson DJ (1958) Observations on the biology of *Caraphractus cinctus* Walker (Hymenoptera: Mymaridae), a parasitoid of the eggs of Dytiscidae: I. Methods of Rearing and Numbers Bred on Different Host Eggs. Trans R Ent Soc Lond 110(17):533–554

Jackson DJ (1961) Observations on the biology of *Caraphractus cinctus* Walker (Hymenoptera: Mymaridae), a parasitoid of the eggs of Dytiscidae (Coleoptera). 2. Immature stages and seasonal history with a review of mymarid larvae. Parasitology 51(3–4):269–294

Jackson DJ (1963) Diapause in *Caraphractus cinctus* Walker (Hymenoptera: Mymaridae), a parasitoid of the eggs of Dytiscidae (Coleoptera). Parasitology 53(1–2):225–251

Jackson DJ (1966) Observations on the biology of *Caraphractus cinctus* Walker (Hymenoptera: Mymaridae), a parasitoid of the eggs of Dytiscidae (Coleoptera) III. The adult life and sex ratio. Trans R Ent Soc Lond 118(2):23–49

Jackson JK, Resh VH (1991) Periodicity in mate attraction and flight activity of three species of caddisflies (Trichoptera). Journal of the North American Benthological Society, 10(2):198–209.

Janicke T, Morrow EH (2018) Operational sex ratio predicts the opportunity and direction of sexual selection across animals. Ecol Lett 21(3):384–391. https://doi.org/10.1111/ele.12907

Janicke T, Ritchie MG, Morrow EH, Marie-Orleach L (2018) Sexual selection predicts species richness across the animal kingdom. Proc Biol Sci 285(1878):20180173. https://doi.org/10.1098/rspb.2018.0173

Kazuki S, Tojo K (2010) Automictic parthenogenesis of a geographically parthenogenetic mayfly, Ephoron shigae (Insecta: Ephemeroptera, Polymitarcyidae). Biol J Linn Soc 99(2):335–343

Kolmes SA (1983) Precopulatory behavior of the whirligig beetle *Dineutes discolor* (Coleoptera: Gyrinidae). J NY Entomol Soc 91:273–279

Leston D (1979) The eversible rectal organ of certain Miridae (Hemiptera) and its function. Flor Entomol 62(4):409–411

Lopes SM (2018) Order Blattodea. In: Thorp and Covich's freshwater invertebrates. Academic Press, Cambridge, pp 169–173

Mackay RJ, Wiggins GB (1979) Ecological diversity in Trichoptera. Annu Rev Entomol 24(1):185–208

Mandal SK, Ghosh A, Bhattacharjee I, Chandra G (2008) Biocontrol efficiency of odonate nymphs against larvae of the mosquito, *Culex quinquefasciatus* Say, 1823. Acta Trop 106(2):109–114

Marchand RP (1984) Field observations on swarming and mating in *Anopheles gambiae* mosquitoes in Tanzania. Nether J Zool 34:367–387

Martens A, Rehfeldt G (1989) Female aggregation in Platycypha caligata (Odonata: Chlorocyphidae): a tactic to evade male interference during oviposition. Anim Behav 38:369–374. https://doi.org/10.1016/S0003-3472(89)80029-6

Martins CC, Ardila-Camacho A (2018) Order Neuroptera. In: Thorp and Covich's freshwater invertebrates. Academic Press, Cambridge, pp 229–236

Melo GAR, Aguiar AP, Garcete-Barrett BR (2012) Hymenoptera. Insetos do Brasil: diversidade e taxonomia. p 553–612

Miller MN, Fincke O (1999) Cues for mate recognition and the effect of prior experience on mater recognition in Enallagma damselflies. J Insect Behav 12(6):801–814

Monteiro-Júnior CS, Couceiro SRM, Hamada N, Juen L (2013) Effect of vegetation removal for road building on richness and composition of Odonata communities in Amazonia, Brazil. Int J Odonatol 16(2):135–144

Moreira FFF, Rodrigues HDD, Sites RW, Cordeiro IRS, Maglhães OM (2018) In Thorp and Covich's Freshwater Invertebrates. Academic Press, Cambridge, pp 175–216

Mousseau T, Fox C (1998) The adaptive significance of maternal effects. Trends Ecol Evol 13(10):403–407. https://doi.org/10.1016/S0169-5347(98)01472-4

Nasirian H, Irvine KN (2017) Odonata larvae as a bioindicator of metal contamination in aquatic environments: application to ecologically important wetlands in Iran. Environ Monit Assess 189(9):436

Neems RM, McLachlan AJ, Chambers R (1990) Body size and lifetime mating success of male midges (Diptera: Chironomidae). Anim Behav 40(4):648–652. https://doi.org/10.1016/S0003-3472(05)80694-3

Neiss UG, Hamada N (2014) Ordem Odonata (odous = dente; gnatha = mandíbula). In: Hamada N, Nessimian JL, Querino RB (eds) Insetos Aquáticos na Amazônia Brasileira: taxonomia, biologia e ecologia, pp 217–282

Nesemann H, Shah RDT, Shah DN, Sharma S (2010) First records of Rhicnoda natatrix and Rhicnoda rugosa (Blattodea: Blaberidae) from Nepal and India (Maharashtra) with notes on habitat quality. J Threat Taxa 2(1):648–652

Nessimian JL (2014) Ordem Blattodea. In: Hamada N, Nessimian JL, Querino RB (eds) Insetos Aquáticos na Amazônia brasileira: taxonomia, biologia e ecologia. Ed. INPA, Manaus, pp 288–295

Nunes AL 1989 Estudo sobre o ciclo de vida e fenologia de *Stenacris fissicauda* fissicauda (Bruner, 1908) (Orthoptera-Acrididae) em um lago de várzea da Amazônia Central, Brasil. M.Sc. Thesis, National Institute of Amazon Researches, Manaus, Brazil, pp 122

Nunes AL, Adis J (1994) Comportamento populacional de *Tucayaca gracilis* (Giglio-Tos, 1897) (Orthoptera-Acrididae) frente à oscilação do nível d'água na várzea da Amazônia Central. Boletim do Museu Paraense Emílio Goeldi 10:211–224

Nunes AL, Adis J, Nunes de Mello JA (1992) Estudo sobre o ciclo de vida e fenologia de *Stenacris fissicauda fissicauda* (Bruner, 1908) (Orthoptera: Acrididae) em um lago de várzea da Amazônia Central, Brasil. Boletim do Museu Paraense Emílio Goeldi 8:349–374

Nunes-Gutjahr AL, Braga CE (2012) Gafanhotos da Flona Caxiuanã: Guia prático. Museu Paraense Emilio Goeldi, Belém, PA, p 62

Ortigosa A, Rowe L (2002) The effect of hunger on mating behaviour and sexual selection for male body size in *Gerris buenoi*. Anim Behav 64(3):369–375. https://doi.org/10.1006/anbe.2002.3065

Orr AG, Hämäläinen M (2007) The metalwing demoiselles of the eastern tropics. Kota Kinabalu, Malaysia: Natural History Publications (Borneo). J. R. Soc. Interface, 7. 130 pp

Paprocki H (2012) Trichoptera. In: Rafael JA, GAR M, CJB d C, Casari AS (eds) Insetos do Brasil, pp 613–623

Pes AM, Holzenthal RW, Sganga JV, Santos AP, Barcelos-Silva P, Camargos LM (2018) Order Trichoptera. In: Thorp and Covich's freshwater invertebrates. Academic Press, Cambridge, pp 237–324

Petersson E (1990) Male age, copulation duration and insemination success in Mystacides azurea (Leptoceridae, Trichoptera). Ethology 85:156–162. https://doi.org/10.1111/j.1439-0310.1990.tb00395.x

Pinho LC 2008 Diptera. In: Guia on-line: Identificação de larvas de Insetos Aquáticos do Estado de São Paulo. Froehlich, C.G. (org.). http://sites.ffclrp.usp.br/aguadoce/Guia_online/Guia_Diptera.pdf

Poirrier MA (1969) Some Fresh-water Sponge Hosts of Louisiana and Texas Spongilla-Flies, with New Locality Records. Am Midl Nat 81(2):573. https://doi.org/10.2307/2423990

Pruthi HS (1933) An interesting case of maternal care in an aquatic cockroach, Phlebonotus pallens Serv.(Epilamprinae). Curr Sci 1(9):273–273

Pupedis RJ (1978) Tube feeding by *Sisyridivora cavigena* (Hymenoptera: Pteromalidae) on *Climacia areolaris* (Neuroptera: Sisyridae). Ann Entomol Soc Am 71(5):773–775. https://doi.org/10.1093/aesa/71.5.773

Pupedis RJ (1987) Foraging behavior and food of adult spongila-flies (Neuroptera: Sisyridae). Ann Entomol Soc Am 80(6):758–760. https://doi.org/10.1093/aesa/80.6.758

Rehfeldt GE (1991) Site-specific mate-finding strategies and oviposition behavior in *Crocothemis erythraea* (Brullé)(Odonata: Libellulidae). J Insect Behav 4(3):293–303

Reich P, Hale R, Downes BJ, Lancaster J (2011) Environmental cues or conspecific attraction as causes for egg mass aggregation in hydrobiosid caddisflies. Hydrobiologia 661(1):351–362

Reiskind MH, Wilson ML (2004) *Culex restuans* (Diptera: Culicidae) oviposition behavior determined by larval habitat quality and quantity in southeastern Michigan. J Med Entomol 41(2):179–186

Resh VH, Jackson JK, Wood JR (1987) Techniques for demonstrating sex pheromones in Trichoptera. In: Proceedings of the Fifth International Symposium on Trichoptera. Springer, Dordrecht, pp 161–164

Rice WR (1998) Male fitness increases when females are eliminated from gene pool: Implications for the Y chromosome. Proc Natl Acad Sci 95(11):6217–6221. https://doi.org/10.1073/pnas.95.11.6217

Robertson HM (1985) Female dimorphism and mating behaviour in a damselfly, Ischnura ramburi: females mimicking males. Anim Behav 33:805–809

Roble SM (1985) Submergent capture of *Dolomedes triton* (Araneae, Pisauridae) by *Anoplius depressipes* (Hymenoptera, Pompilidae). J Arachnol

Roth LM (1970) Evolution and taxonomic significance of reproduction in Blattaria. Annu Rev Entomol 15(1):75–96

Roth LM (1981). The mother-offspring relationship of some blaberid cockroaches (Dictyoptera: Blattaria: Blaberidae). Proceedings-Entomological Society of Washington (USA)

Rupprecht R (1975) Die Kommunikation von Sialis (Megaloptera) durch Vibrationssignale. J Insect Physiol 21(2):305–320

Saha N, Aditya G, Banerjee S, Saha GK (2012) Predation potential of odonates on mosquito larvae: Implications for biological control. Biol Control 63(1):1–8

Sandberg JB, Stewart KW (2003) Continued studies of drumming in North American Plecoptera; Evolutionary implications. Research update on Ephemeroptera and Plecoptera. University of Perugia, Perugia, pp 73–81

Sanmartín-Villar I (2017) Eco-etoloxía de cabaliños do demo con polimorfismo de cor. Universidade de Vigo, Vigo

Sanmartín-Villar I, Cordero-Rivera A (2016) The inheritance of female colour polymorphism in Ischnura genei (Zygoptera: Coenagrionidae), with observations on melanism under laboratory conditions. PeerJ 4:e2380. https://doi.org/10.7717/peerj.2380

Sartori M, Brittain JE (2015) Order Ephemeroptera. In: Thorp and Covich's freshwater invertebrates, 4th edn. Academic Press, Cambridge, pp 873–891

Schuett W, Tregenza T, Dall SRX (2010) Sexual selection and animal personality. Biol Rev Camb Philos Soc 85(2):217–246. https://doi.org/10.1111/j.1469-185X.2009.00101.x

Shelford R (1909) Note on some amphibious cockroaches. Rec Ind Mus 3:125–127

Simmons LW (2014) Sexual selection and genital evolution. Aust Entomol 53:1–17. https://doi.org/10.1111/aen.12053

Siva-Jothy MT (1987) Variation in copulation duration and the resultant degree of sperm removal in *Orthetrum cancellatum* L. (Libellulidae: Odonata). Behav Ecol Sociobiol 20(2):147–151

Smith RL (1979) Paternity assurance and altered roles in the mating behaviour of a giant water bug, *Abedus herberti* (Heteroptera Belostomatidae). Anim Behav 27:716–725

Solis MA (2008) Aquatic and semiaquatic Lepidoptera. In: An introduction to the aquatic insects of North America, 4th edn, revised printing. Kendall/Hunt Publishing Company, Dubuque

South SH, Steiner D, Arnqvist G (2009) Male mating costs in a polygynous mosquito with ornaments expressed in both sexes. Proc R Soc B Biol Sci 276(1673):3671–3678. https://doi.org/10.1098/rspb.2009.0991

Stewart KW (1997) Insect life: vibrational communication in insects. Am Entomol 43(2):81–91

Sullivan RT (1981) Insect swarming and mating. Fla Entomol 64(1):44–65

Svensson BG, Petersson E (1987) Sex-role reversed courtship behaviour, sexual dimorphism and nuptial gifts in the dance fly, *Empis borealis* (L.). In: Annales Zoologici Fennici. Finnish Ac, vol. 24, p 323–334

Tallamy DW (2001) Evolution of exclusive paternal care in arthropods. Annu Rev Entomol 46(1):139–165

Taylor BW, Peckarsky BL (1998) Effects of size at metamorphosis on stonefly fecundity, longevity, and reproductive success. Int Rev Hydrobiol 114:494–502

Theischinger G (2000) Australian alderfly larvae and adults (Insecta: Megaloptera): a preliminary guide to the identification of larvae and survey of adults of Australian alderflies: presented at the taxonomy workshop held at the Murray Darling Freshwater Research Center, Albury, 8–9 February, 2000. Cooperative Research Centre for Freshwater Ecology

Thornhill R (1983) Cryptic female choice and its implications in the Scorpionfly *Harpobittacus nigriceps*. Am Nat 122(6):765–788

Tsubaki Y, Hooper RE, Siva-Jothy MT (1997) Differences in adult and reproductive lifespan in the two male forms of Mnais pruinosa costalis Selis (Odonata: Calopterygidae). Res Polol Ecol 39(2):149–155. https://doi.org/10.1007/BF02765260

Van Gossum H, Sherratt TN, Cordero-Rivera A (2008) The evolution of sex-limited colour polymorphisms. In: Córdoba-Aguilar A (ed) Dragonflies: model organisms for ecological and evolutionary studies. Oxford University Press, Oxford, pp 219–229

Vickers NJ (2005) Winging it: moth flight behavior and responses of olfactory neurons are shaped by pheromone plume dynamics. Chem Senses 31(2):155–166

Vieira MF, Santos AC (2003) Duração do ciclo de vida de *Cornops aquaticum* (Bruner, 1906) (Orthoptera: Acrididae: Leptysminae) e aspectos de seu comportamento alimentar na Amazônia central. Acta Amazonica, Manaus 33(4):711–714

Vilela DS, Guillermo-Ferreira R, Del-Claro K (2016) The Odonata community of a Brazilian vereda: seasonal patterns, species diversity and rarity in a palm swamp environment. Biosci J 32:486–495. https://doi.org/10.14393/BJ-v32n2a2016-30491

Waage JK (1979) Reproductive character displacement in *Calopteryx* (Odonata: Calopterygidae). Evolution 33(1Part1):104–116

Watanabe M, Taguchi M (1990) Mating tactics and male wing dimorphism in the damselfly, Mnais pruinosa costalis Selys (Odonata: Calopterygidae). J Ethol 8(2):129–137. https://doi.org/10.1007/BF02350283

Watanabe M, Taguchi M (1997) Competition for perching sites in the hyaline-winged males of the damselfly Mnais pruinosa costalis Selys that use sneaky mata-securing tactics (Zygoptera: Calopterygidae). Odonatologica 26(2):183–191

Winterbourn MJ, Anderson NH (1980) The life history of *Philanisus plebeius* Walker (Trichoptera: Chathamiidae), a caddisfly whose eggs were found in a starfish. Ecol Entomol 5(3):293–304
Withycombe CL (2009) XX. Notes on the biology of some British neuroptera (planipennia). Trans R Entomol Soc Lond 70(3–4):501–594. https://doi.org/10.1111/j.1365-2311.1923.tb02844.x
Wood JR, Resh VH (1984) Demonstration of sex pheromones in caddisflies (Trichoptera). J Chem Ecol 10(1):171–175
Yuval B, Bouskila A (1993) Temporal dynamics of mating and predation in mosquito swarms. Oecologia 95(1):65–69. https://doi.org/10.1007/BF00649508
Yuval B, Wekesa JW, Washino RK (1993) Effect of body size on swarming behavior and mating success of male *Anopheles freeborni* (Diptera: Culicidae). J Insect Behav 6(3):333–342
Zahavi A (1975) Mate selection-A selection for a handicap. J Theor Biol 53:205–214. https://doi.org/10.1016/0022-5193(75)90111-3
Zeh JA, Zeh DW (2003) Toward a new sexual selection paradigm: Polyandry, conflict and incompatibility (invited article). Ethology 109(12):929–950. https://doi.org/10.1046/j.1439-0310.2003.00945.x

# Chapter 11
# Sexual Conflict in Water Striders, Dragonflies and Diving Beetles

**Adolfo Cordero-Rivera and Anais Rivas-Torres**

**Abstract** The field of sexual selection has been historically dominated by a stereotyped view of the sexual roles, with competing males and selective females, but in recent decades there has been a paradigm switch, with the emergence and dominance of the concept of sexual conflict. Put simply, there is sexual conflict when the optimum value for a trait (or a group of traits) is different for females and males. Although the recent literature mainly considers sexual conflict as a process separate from the other well-known processes in the field of postcopulatory sexual selection (sperm competition and cryptic female choice), our approach is that sexual conflict is the consequence of several pressures related to natural and sexual selection, and not a process by itself. Therefore, here we consider sexual conflict as a part of a continuum of sexual selection mechanisms. We concentrate on the effects of sexual conflict on reproductive behaviour of three groups of aquatic insects, whose habitats differ markedly, water striders, odonates and diving beetles, but also include some examples of studies addressing sexual conflict in other groups of aquatic insects. Our hypothesis is that the dimensional structure of the habitat will affect the intensity of sexual conflict over mating rate, copulation duration and postcopulatory guarding. There is abundant evidence and comprehensive reviews of the conflict over mating rates in water striders, odonates and, to a lesser degree, diving beetles. The bi-dimensionality of the water surface allows an easy monopolisation of females by males in this microhabitat, and water striders conform to this rule, so that the commonest mating system is characterised by strong conflicts and struggles before and after copulation. For animals like odonates, which are fast fliers and can use diverse terrestrial microhabitats, the opportunities for males to force females to copulate are certainly limited. In the case of diving beetles, the situation seems more favourable for the females, as they could control male approaches by hiding easily in the vegetation, or even in the case of extreme male density they could fly away and move to a different water body. The sexual conflict over mating duration is also intense in water striders, and also relevant in the other reviewed groups. In the field of postcopulatory conflicts, odonates have offered the best examples of male

A. Cordero-Rivera (✉) · A. Rivas-Torres
ECOEVO Lab, EE Forestal, Campus Universitario, Pontevedra, Spain
e-mail: adolfo.cordero@uvigo.es



K. Del-Claro, R. Guillermo (eds.), *Aquatic Insects*,
https://doi.org/10.1007/978-3-030-16327-3_11

adaptations and female counter-adaptations, which are even more elaborated when studying the evolution of genitalia. We end by summarising our main conclusions and propose some ideas for future work. We stress that a comprehensive understanding of sexual conflicts in animals requires the study of both male and female anatomies, as well as their behaviours, avoiding assumptions or gender stereotypes, which have historically biased research to a male-view approach.

**Keywords** Behaviour · Equity · Cryptic female choice · Sperm competition

## 11.1 What Is Sexual Conflict

Observing a pair of animals contributing to the rearing of the next generation may seem at first sight a beautiful example of cooperation for a common goal. However, as many other widespread layman misconceptions about animal behaviour, the two members of a pair cooperate but are at the same time in permanent conflict (Arnqvist and Rowe 2005). This is especially strong in species with high parental investment, where both partners will have more to gain than to lose if "convincing" the other progenitor to invest more in the offspring. Sexual selection will therefore favour cheating the partner. This "sexual conflict" is only one of the many evolutionary conflicts of interests that can be found in all levels of biological complexity (Queller and Strassmann 2018).

George C. Williams was the first to use the term "sexual conflict" in his book "Adaptation and Natural Selection" (Williams 1966, p. 184), and although it received some attention the dominant view at that time was cooperation between males and females. After some years of silent, the concept of sexual conflict was brought to life again thanks to the chapter on "Parental Investment and Sexual Selection" by Trivers (1972).

In 1983, as a second-year biology student, the first author was "gently obliged" by his genetics professor to read a recent book (at that moment) with the curious title "The Selfish Gene" (Dawkins 1976). This book, that has sold more than a million copies, clearly proposed the idea of conflicts between genes as one of the main forces in evolution. There is a chapter entitled "The Battle of the Sexes", where the idea of sexual conflict is introduced, using these terms:

> "As a convenient approximation, we can once again assume that each individual body is a selfish machine, trying to do the best for all its genes. The best policy for such a selfish machine will often be one thing if it is male, and quite a different thing if it is female."

However, the credit for popularising the concept of sexual conflict among evolutionary biologists is for Geoff Parker, who in 1970 published a review of mating behaviour in insects, and defined the term "sperm competition". He introduced the idea that intrasexual selection between ejaculates continues after mating, inside the female (Parker 1970), and a few years after studied the evolutionary consequences of the conflict between the sexes (Parker 1979). Parker (2006) defined the term as

> "[…] a conflict between the evolutionary interests of individuals of the two sexes".

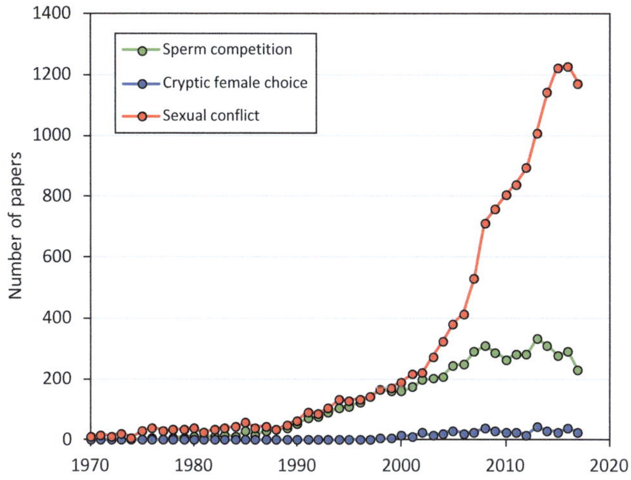


**Fig. 11.1** The diffusion of the idea of sexual conflict is an example of a paradigm switch in evolutionary ecology. The figure shows the trends in the number of papers that cite each concept in their title or abstract since 1970 to 2017. Note that "cryptic female choice" does not appear in the titles of papers until 1995. Source: Web of Sciences, accessed October 2018

The picture was completed when the female perspective was also explicitly considered in models of sexual selection, and was denominated "cryptic female choice", because it was not apparent to an observer (Thornhill 1983; Eberhard 1996). Put it simply, there is sexual conflict when the optimum value for a trait (or a group of traits) is different for females and males. The "trait" can also be different for both sexes. The recent literature tends to use the term "sexual conflict" (Fig. 11.1), as a process separate from the other well-known processes in the field of postcopulatory sexual selection (sperm competition and cryptic female choice), but sexual conflict is in fact the consequence of several pressures related to natural and sexual selection (Cordero Rivera and Córdoba-Aguilar 2016), and not a process by itself (Shuker 2014). Therefore we here consider sexual conflict as a part of a continuum of sexual selection mechanisms (Simmons 2014). Some have argued that sexual conflict is so intrinsically linked to sexual selection, that both terms can be synonymous or that sexual selection is better described by the term "sex-dependent selection" (Carranza 2009). The important question for us is that sexual conflict can occur before, during and/or after mating, and therefore it can affect almost all the behavioural repertories of males and females related to reproduction. Sexual conflict can even occur in monogamous species (for instance for mating rate), while sexual selection will be weak.

Although sexual conflict can be obvious in animals with parental care, particularly vertebrates, the first conceptual and experimental studies were mainly inspired by insects, where parental care is extremely rare (Machado and Trumbo 2018). If parents die before the offspring hatches, as it occurs in many insects, parental care cannot evolve. Insects are therefore good models to study conflicts over mating rate and postcopulatory conflicts over fertilisation. For a general review of what is known about sexual conflict in animals see Arnqvist and Rowe (2005).

## 11.2 The Ecology of Sexual Conflict in Aquatic Insects

We here concentrate on the effects of sexual conflict on reproductive behaviour of three groups of aquatic insects, whose habitats differ markedly, water striders, odonates and diving beetles, but also include some examples of studies addressing sexual conflict in other groups of aquatic insects. Our hypothesis is that the dimensional structure of the habitat will affect the intensity of sexual conflict over mating rate, copulation duration and postcopulatory guarding. There is ample evidence for the relevance of local ecological settings in the evolution of sexual conflict (Perry and Rowe 2018).

Water striders have been at the core of sexual conflict studies over mating rate, due to the prolific studies of Locke Rowe and Göran Arnqvist, who pioneered the field (e.g. Rowe et al. 1994). Water striders live in a world essentially bi-dimensional, on the surface of rivers, ponds, lakes and even the sea (Fig. 11.2). This fact conditions the ability of females to avoid unwanted male mating attempts, and likely is a factor contributing to the extreme adaptations exhibited by some species. For a review of their ecology and behaviour see Spence and Anderson (1994).

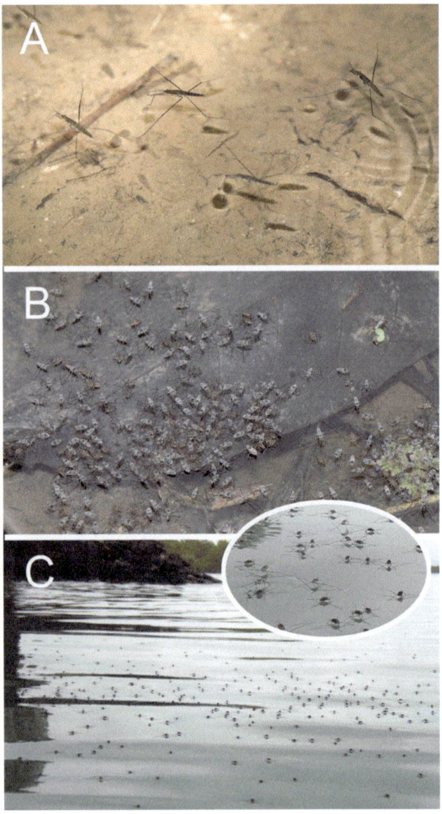


**Fig. 11.2** The bi-dimensionality of the water surface, and the high densities of many populations of water striders around the world, conditions the ability of females to avoid unwanted matings, and therefore sexual conflict over mating rate is usually very high. (**a**) A group of *Aquarius najas* in an Italian stream; (**b**) an aggregation of water striders and veliids (likely *Microvelia*) in a pond in the National Park Pacaya-Samiria in Peru; (**c**) an aggregation ("flotilla") of *Halobates robustus* in the coast of Santa Cruz island (Galapagos), consisting of males, with a detail in the insert. Pictures by ACR

The second group that we will review is totally different, in the sense that the adults live in a complex tri-dimensional world, are highly mobile and are fast flyers: the odonates. Since the early studies of postcopulatory sexual selection, damselflies and dragonflies have been a textbook example for sperm competition, particularly after the clever experiments of Jonathan Waage (1979) showing that the male intromittent organ can be used to remove sperm from rival males before insemination (Cordero-Rivera and Córdoba-Aguilar 2010). This fact conditioned for long time our view of odonate mating behaviour, but recent evidence has shown the relevance of other mechanisms apart from sperm competition (Córdoba-Aguilar and Cordero-Rivera 2008).

Finally, we review the literature on reproductive behaviour and sexual conflict in a group of aquatic insects whose adults live in the water, but also have the ability to fly, the diving beetles (the family Dytiscidae), and therefore have a multidimensional world available: the aquatic and the aerial habitats. We predict that this multidimensional space will increase the ability of females to escape from males and therefore this fact will diminish the intensity of sexual selection and particularly of sexual conflict.

### *11.2.1 Sexual Conflict Over Mating Rate*

There is ample evidence that the optimum number of matings is different for males and females. The common pattern is the so-called Bateman's principle that states that male reproductive success increases in a direct way with the number of matings achieved (i.e. males are limited by the number of females they mate with), whereas in females the limitation is the amount of energy and time available for egg production and maturation (i.e. females are limited by resources and less by the number of mates) (Bateman 1948). This asymmetry predicts that males will be more indiscriminate and try to maximise the number of matings, whereas females are predicted to be more selective and maximise mate quality and resources. Some studies have shown that increasing the number of matings over the optimum can be negative for females (Arnqvist and Nilsson 2000). Therefore, there is sexual conflict over the number of matings.

If the optimum number of matings is different for males and females, then another prediction of sexual conflict theory is that selection would favour the evolution of adaptations in males that allow them to hold females and even force copulation (a diverse array of grasping appendages—see Fig. 11.6—but also behavioural adaptations) (Clutton and Parker 1995). The same hypothesis predicts the evolution of mechanisms in females to avoid being grasped by males or to make this extremely difficult or even dangerous. For instance, females can evolve anti-grasping mechanisms (like in *Gerris*; Arnqvist and Rowe 2002). Sexual cannibalism (Fig. 11.3b) could also be a way for females to solve the conflict, and at the same time get extra resources for future reproduction (Schneider 2014).

**Fig. 11.3** Odonates are fast fliers and live in a complex tri-dimensional habitat. Female odonates have to collaborate to mate (like in **a**, *Ischnura hastata* at the Galapagos). Sexual conflict over mating rate and mating duration can be very intense, and sometimes females show sexual cannibalism (**b**, a female *I. hastata* cannibalising a mature male; Zarabanda, Cuba). Males cannot force copulations under normal densities, but can harass females until a convenience polyandry is the best of a bad job for them (**c**, an ovipositing female of *Calopteryx haemorrhoidalis* being taken forcibly in tandem by two males; Pontecorvo, Italy). Pictures by ACR

The bi-dimensionality of the water surface allows an easy monopolisation of females by males in this microhabitat, and water striders conform to this rule, so that the commonest mating system is characterised by strong conflicts and struggles before and after copulation (type I sensu Arnqvist 1997). For example, in *Gerris buenoi* females are frequently harassed by males in search of mating, and the only way to escape for females is to fight with the male (pre-mating struggles; Fig. 11.4a), which has the negative effect of increasing predation rate by backswimmers (Notonectidae) (Rowe 1994). In fact, the mating frequency of females in *Aquarius remigis* is positively related to male density (Lauer et al. 1996). Mating is specially costly for females, because they have to carry the males on their back for long periods of time (Fig. 11.4b, c), decreasing their skating efficiency (Rowe et al. 1994), and likely increasing predation. A review of 15 species of the genus *Gerris* indicates that conflict over mating rate is widespread in the group, with species where males apparently have gained control over mating rate, and others where females retain a relative advantage (Rowe and Arnqvist 2002). When harassment rate is experimentally increased, females reduce their resistance and therefore mating rates increase (Fig. 11.5), a fact that suggests that this mating system can be characterised as "convenience polyandry" (Rowe 1992). However, water striders can exhibit also

**Fig. 11.4** Sexual conflict over mating rate in water striders is evident by pre- and postcopulatory mating struggles. (**a**) Struggle between two *Gerris* individuals; (**b**) copulation (*Gerris* sp., Pontecorvo, Italy) can be brief or very long, and in many species males remain in the back of females for hours, days or even weeks, increasing predation risk, like this couple of *Aquarius najas*, heavily parasitised by water mites (**c**; Tarragona, Spain). Pictures by ACR

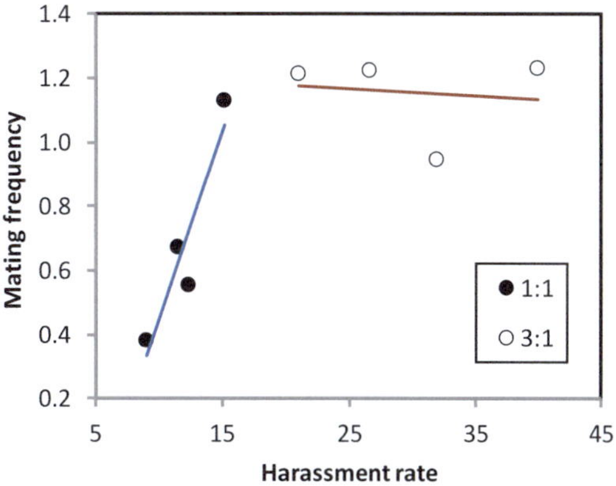


**Fig. 11.5** The effect of harassment rate of male *Gerris buenoi* over female mating frequency, at two different sex ratios. Harassment was measured as the number of male attempts to copulate with females per single female hour. Lines represent regressions. Source: modified from Rowe (1992)

a second mating system, with little apparent conflict (type II sensu Arnqvist 1997). In the commonest case, males have apparently "won" the battle for controlling mating rate, but in the type II mating system males have become territorial and court females, so that females have higher control over mating rates (Arnqvist 1997).

The above discussion can be extended to other groups of aquatic insects that live on the water surface, like Mesoveliidae, Veliidae and Hydrometridae (Heteroptera). However, very little research has been done using these other families as models. Searching for sexual conflict and the family name in the Web of Sciences gives no papers for Mesoveliidae and Hydrometridae and only two papers for Veliidae. One of such studies for the subfamily Microveliinae found several examples of putative grasping and anti-grasping adaptations, suggesting that they have evolved due to sexual conflict over mating rate (Cassis et al. 2018). Some Veliidae show males that remain for long periods over their females (Polhemus 1974), a situation paralleling a common behaviour in water striders, which is known to generate costs on females (see above). There is even one case where females have evolved a gland on their dorsal surface that produces a secretion used by males to feed, a clear example of a counter-adaptation of females to reduce kleptoparasitism by their mates, which remain on the back of females for long periods of time (Arnqvist et al. 2006).

For animals like odonates, which are fast fliers and can use diverse terrestrial microhabitats, the opportunities for males to force females to copulate are certainly limited. Female odonates have to blend up their abdomen and cooperate with the male for a successful copulation (Figs. 11.3a and 11.8a). This has led some authors to assume that female odonates cannot be forced to mate (e.g. Fincke 1997). This view is probably true in most ecological settings. However, males have evolved the ability to concentrate in places that females have to visit to lay their eggs, monopolising the area; that is, they behave as territorial (Baker 1983). Under these circumstances, males can be so insistent as to force copulation (Fig. 11.3c), again in a case of "convenience polyandry" (Cordero 1999; Cordero and Andrés 2002).

The odonates are also particular because males have a secondary genitalia, and therefore both individuals contact in two body regions to mate. The typical pattern is for males to search for females either actively or waiting at the shore of the water bodies. When a female is found, the male tries to grasp her in tandem, using his anal appendages (Cordero-Rivera 2002), and at this moment conflict over mating rate can be intense. Females can avoid being grasped by escape flights, by refusal displays (Utzeri 1988) or by visiting the water bodies when males are absent (Cordero-Rivera et al. 1999). Once grasped, they can also refuse to bring the abdomen in contact with male genitalia, even if males can be insistent (Cordero et al. 1992). In *Phylolestes ethelae*, and endemic species of the mountains of the island of Hispaniola, males commonly grasp females in tandem during the morning, and remain in this position until the night, when they mate and release the female (ACR unpubl. obs.). This is probably the longest precopulatory tandem in the order. In general, however, the tri-dimensional habitat of adult odonates allows females to have greater control over their mating rate. In an extreme case, females of some species have been observed practising sexual cannibalism (Cordero 1992), which is the most dramatic (for the male) way of solving the conflict (Fig. 11.3b). On the other hand, the existence of conflict over mating rate seems to explain the widespread occurrence of female colour polymorphism in odonates (Cordero et al. 1998; Fincke 2004), a topic out of the scope of this chapter (for a review see Van Gossum et al. 2008).

Males of odonates are selected to be efficient in grasping females in tandem, and the structure, size and orientation of their anal appendages seem the result of conflict over mating rate, particularly among the Gomphidae, a family whose males have conspicuous forceps as anal appendages (Fig. 11.6). However, the only experimental study that has addressed this question did not find any pattern of hyperallometry in the grasping apparatus of five species, a prediction from the sexual conflict hypothesis (Córdoba-Aguilar et al. 2015). This topic needs to be studied in a large number of species, and in a phylogenetic context. Other possible ways for males to control mating frequency derive from ethological adaptations. For example, males could search for newly emerged females and force mating, because these females cannot refuse or escape with efficiency. This has been observed in *Ischnura aurora*, whose New Zealand populations consist of males that mate with teneral females (Rowe 1978).

**Fig. 11.6** Sexual conflict over mating rates may select for complex and strong "grasping devices" in males, and for "anti-grasping devices" in females. Among the odonates, the family Gomphidae shows the most elaborated anal appendages, whose function as potential sexual conflict adaptations merits further study. (**a**) Male of *Paragomphus genei* (Cáceres, Spain); (**b**) male of *Leptogomphus intermedius* (Nankunshan, China); and (**c**) male of *Lamelligomphus annakarlorum* (Yunnan, China). Pictures by ACR

**Fig. 11.7** Sexual conflict is evident in the sexually dimorphic forelegs of dytiscid beetles. (**a**) Male of *Dytiscus* in a pond. The arrow indicates the enlarged foreleg, used as a grasping device during mating, and a detail of the adhesive setae of a male *Cybister* (ventral view; the size of the sucker is 2.4 × 3.3 mm). (**b**) Lateral view of a female *Cybister*. Note the absence of modifications in the forelegs. Pictures by ACR

In the case of diving beetles, the situation seems more favourable for the females, as they could control male approaches by hiding easily in the vegetation, or even in the case of extreme male density they could fly away and move to a different water body. Apparently, in some species, there is intense conflict for mating rates, because grasping and anti-grasping mechanisms have been found (Miller 2003). Although little or no courtship behaviour is known, it seems that males could perform acoustic and chemical signals to attract and detect females. Diving beetles produce chemicals for defence and it could be that males also use their chemical senses to detect or attract females (Dettner and Schwinger 1980). In some species, males use stridulatory devices to produce acoustic signals to attract receptive females (Miller and Bergsten 2014).

From the scarce evidence available, we can say that the Dytiscidae show three phases during their reproduction, some divided into subphases. The first phase would be divided into attachment and female resistance, the second in copulation and postcopulatory guarding (see below in the next sections), and finally the release phase, end of copula (Cleavall 2009). In this section, we are concerned with the first phase, attachment and female resistance. These two subphases are the result of the arms war derived from sexual conflict. The males develop sucker-shaped adhesive setae on the pro-tarsomeres, principally (Fig. 11.7). These setae vary in density, number or size, and this fact is not random; the complexity of these sucker setae will go hand in hand with the mechanisms of resistance that the females possess, and these go from modifications of the cuticular surface of the pronotum and the elytra, by dense punctuations, striations or furrows (Miller 2003). In fact, it is possible to establish a parallel classification joining the adaptations of the setae in the males and the surface of the elytra in the females. Sucker setae with three large and smaller cups are related to dense and large punctures on the surface of females' elytra, and furrows on females' elytra co-evolved with setae that have three medium and minute cups or one large and two medium cups (Miller and Bergsten 2014).

### 11.2.2 *Sexual Conflict Over Mating Duration*

During copulation, the ability of the pair to move, fly and escape from predators is evidently reduced, and these are only part of the costs of mating (see Daly 1978). Therefore, if there is sexual conflict for copulation duration, the prediction is that both sexes will be continuously under antagonistic coevolution.

Natural selection should favour brief copulations, enough to complete insemination, but sexual selection often favours the opposite. From a natural selection perspective, copulation should be efficient in sperm transfer, and should be completed as soon as possible. However, when copulatory processes are studied in detail, often we discover that insemination occurs at the end of copulation, after several minutes (Cordero-Rivera and Zhang 2018), or even hours (Andrés and Cordero-Rivera 2000). Prolonged copulations can be the result of sperm competition, cryptic female choice or both (Andrés and Cordero-Rivera 2000). In the case of water striders, copulation duration can be highly variable among species. In most cases, copulations are brief, between 5 and 20 min (Arnqvist 1997). In the American water strider *Aquarius remigis*, it lasts for 2–3 h in the laboratory, and insemination occurs at the end (Campbell and Fairbairn 2001). In *A. remigis*, copulations lasting less than 15 min did not produce any viable egg, suggesting that no sperm was transferred during that period of time, whereas "once sperm transfer began, however, virtually the entire sperm load was delivered instantaneously to the female" (Rubenstein 1989). Surprisingly, the European *Aquarius najas* (see Fig. 11.4c) mates for only 1–4 min, but males remain over their mates for weeks (Sattler 1957). It is clear that insemination can be completed in a brief copulation, and the experimental evidence in water striders indicates that copulation duration has high relevance for paternity (Arnqvist and Danielsson 1999).

Some odonates are known for their long copulations (even 7 h; Fig. 11.8) which may be a way for males to guard their mates while still in copula, acting as a living mating plug (Cordero 1990). However, long copulations can be costly for both partners, and conflict is expected. Copulatory processes in odonates are well studied. Copulation starts with a stage I, very variable in duration, which is employed by the male to remove rivals' sperm from the sperm storage organs of the female, and only the last phase is employed for insemination (stage II) (Cordero-Rivera and Córdoba-Aguilar 2010). The damselfly *Ischnura graellsii* mates during the afternoon, and males of this species have apparently evolved the ability to retain their females in copula until the end of reproductive activity (Fig. 11.8), so that females are forced to lay their eggs the day after, at noon (Cordero 1989). Interestingly, its sister species, *Ischnura elegans*, usually mates early in the morning, also for several hours,

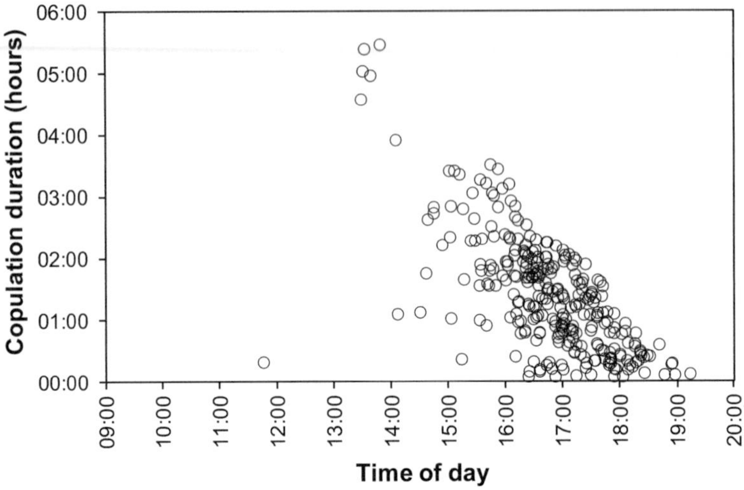


**Fig. 11.8** The relationship between time of day and copulation duration in a high-density population of the damselfly *Ischnura graellsii* at Cangas (Spain). Copulation duration ($N = 282$) was estimated as the time between the first and the last observation of marked mating pairs, and in most cases is therefore an underestimation of the whole copulation duration. Note that there is a limit to copulation duration, which in this case coincides with the end of reproductive activity. In this species males control copulation duration, and females are released too late, and therefore are forced to lay eggs the day after. The maximum density of ovipositing females occurs between 12 and 14 h. Source: unpublished data from Cordero-Rivera and Egido Pérez (1998)

and ends copulations by noon, seemingly the best moment for females to lay eggs (Miller 1987a). These two species exemplify contrasting solutions to the same problem by closely related species, by means of interspecific ethodiversity (Cordero-Rivera 2017a). As a by-result, this time separation between the activity of both species contributes also to minimise hybridisation when they are sympatric (Sánchez-Guillén et al. 2014).

Copulation duration has been investigated in detail in many species of odonates, and has been classified as short (less than 1 min), medium (1–5 min) and long (more than 5 min to several hours) (Corbet 1962). The fact that very brief copulations are as effective as copulations that last for hours clearly indicates that copulation duration in many species is not optimised by natural selection. Postcopulatory sexual selection affects copulation duration in odonates, by means of sperm competition and cryptic female choice mechanisms, which usually determine sexual conflict (reviewed by Córdoba-Aguilar and Cordero-Rivera 2008). Sexual selection also affects copulation duration by its interaction with territorial behaviour, and as a consequence territorial species of odonates copulate on average for shorter times than non-territorial ones (Córdoba-Aguilar et al. 2009).

Experimental manipulations in damselflies have shown that copulation duration is under male control (Miller 1987a). Males have sensilla in their genital ligula and are able to detect female mating status, prolonging copulation with mated females compared to virgin females (Andrés and Cordero-Rivera 2000; Uhía and

Cordero-Rivera 2005). Only the pre-insemination phase of copulation is prolonged, suggesting that this is the result of postcopulatory sexual selection (cryptic female choice) rather than natural selection (Andrés and Cordero-Rivera 2000). Previous studies have concluded that the source of conflict is the sperm stored in the spermatheca (Cordero-Rivera et al. 2004), the organ used by females for long-term storage of sperm (see below).

In diving beetles, as we mentioned above, their mating behaviour includes three phases, and in this section we are interested in the second phase, copulation and the transfer of the spermatophore. Copula durations range from 1 to 10 min in species such as *Thermonectus nigrofasciatus* and *Rhantus binotatus* (Cleavall 2009). In contrast, in *Dytiscus alaskanus* copulation varies between 5.4 and 8.3 h (Aiken 1992). In other species, the only information available includes the complete duration of the mating, which includes the three phases discussed above. In this case, a mating event duration would last between 4 and 45 min in species of the genera *Agabus*, *Colymbetes*, *Acilius* and *Rhantus* (Cleavall 2009). Due to the large duration of the mating event, especially in the Dysticinae subfamily, it could be that the females have evolved to a smaller size and thus avoid the risk to be drowned by the male, since these would be able to breathe directly from the water (Madsen 2012) and discard asphyxiation by the male as a coercive strategy.

### 11.2.3 Postcopulatory Sexual Conflicts

There is a large variation in the association between males and females after copulation. In a classic paper, Parker (1970) showed that the competition between males continues after copulation, by battles between the ejaculates inside the female, a process he named "sperm competition". He also showed that many postcopulatory behaviours of males could be interpreted as adaptations to reduce sperm competition. In fact, postcopulatory associations between the male and the female are widespread in the insects (Alcock 1994).

Again, if costs and benefits are taken into account, the fact that males remain in close contact with the females, grasping them or in the females' back, is likely to be costly for females (Figs. 11.3a, 11.4b, c), who probably would benefit by ending the association immediately after copulation. However, in some circumstances, the presence of the male could be beneficial for the female, if the male helps in finding suitable oviposition sites (Byers and Eason 2009), defends the female from the interference of other males (e.g. type II mating of Gerridae, sensu Arnqvist 1997) or helps to detect predators. The duration of postcopulatory associations can be therefore a matter of sexual conflict, and it has been shown to be highly variable among and inside species (Alcock 1994).

As occurred with mating rate, the dimensionality of the habitat can theoretically affect the outcome of this conflict. Many water striders (Gerridae), but also other insects of the water surface, are typical examples of males kleptoparasiting their females, by remaining over their backs for weeks, feeding on the prey that the female has secured. To dislodge their mates, females of water striders use vigorous struggles (Fig. 11.4a), which in some species invariably dislodge the male, but in others do not (Arnqvist 1997). These struggles are clear indicators of conflict over (post)mating duration, but are a risky behaviour: predators are five times more successful capturing struggling pairs compared to single individuals (Rowe 1994). On the other hand, the costs related to carrying the males are likely to be density dependent, because, at high density, other males will try to mount the females if they are successful in dislodging the first male. In fact, in *A. remigis* it has been found that paired females can be more efficient in foraging (85% of efficiency) compared to lone females (32%) precisely because they receive less disturbance by mate-searching males (Wilcox 1984).

In the case of odonates, there is a large variation in the extent of male-female associations after copulation. In a classical paper, Waage (1984) argued that most postcopulatory behaviours could easily be interpreted in the context of sperm competition theory, but also included a cautionary note about other possible functions of postcopulatory mate associations. For instance, males in tandem sometimes adopt a vertical position, which could allow them a better vision over the oviposition substrates and help in detecting potential predators. In this case, both individuals would benefit from the behaviour. The possible behavioural alternatives were recently reviewed as an example of the concept of ethodiversity (Cordero-Rivera 2017a), and are here presented in Table 11.1.

One example of a female behaviour which could be related to cryptic female choice due to sexual conflict is the ejection of sperm after some copulations, provided that the sperm ejected belongs to the last male to copulate with her (Fig. 11.9). This has not been demonstrated with genetic markers, and is a priority for future research. However, the fact that in the laboratory, where female mating history is controlled, females sometimes eject sperm after their first mating agrees with a sexual conflict interpretation via cryptic female choice. The presence and duration of postcopulatory tandem have been associated to male density and disturbance (e.g. Utzeri and Ercoli 2004), and understood as an example of male adaptation to reduce sperm competition. In agreement with this, in some species, males show plasticity in this behaviour, remaining in tandem after copula when rivals' density is high but switching to non-contact guarding at low density (Latty 2006).

The mode of oviposition, including site selection and duration, can also be a source of sexual conflict (Table 11.1). There is apparently a trend from females that are never guarded by males, to others that always lay their eggs in tandem, and this trend seems related to the intensity of male disturbance. In the Libellulidae, the level of male harassment affects female oviposition behaviour, but inconsistently among species (Koch 2006), and there is no evidence for a special female strategy to cope with male disturbance. Perhaps the most extreme case of female behaviour apparently evolved to avoid the negative effects of male disturbance in underwater oviposition (Fig. 11.10), a behaviour which is the rule in some *Enallagma* species (Fincke 1986). If this behaviour has the function to reduce sexual conflict, the expectation is that females would submerge more often at higher male densities. To our knowledge this topic has not been experimentally addressed. However, females

**Table 11.1** Alternative behaviours after copulation in the Odonata, with representative examples

| Behaviour | Alternative | Example | Reference |
|---|---|---|---|
| Postcopulatory sperm ejection | Never | *Hemiphlebia mirabilis* | Cordero-Rivera (2016) |
| | Occasional | *Ischnura graellsii* | Pers. obs. |
| | Common, almost immediately after copulation | *Enallagma praevarum* | Córdoba-Aguilar (2006) |
| Postcopulatory tandem duration | Absent | *Pseudolestes mirabilis* | Cordero-Rivera and Zhang (2018) |
| | Short (<1 min) | *Ischnura graellsii* | Cordero (1989) |
| | Medium (1–10 min) | *Sympetrum internum* | Latty (2006) |
| | Long (>10 min) | *Lestes virens* | Utzeri and Ercoli (2004) |
| Postcopulatory courtship | Absent | *Ischnura graellsii* | Cordero (1989) |
| | Depending on male strategy | *Diastatops obscura* | Bañuelos Irusta and Araújo (2007) |
| | Common, almost always after copulation | *Calopteryx haemorrhoidalis* | Córdoba-Aguilar (2000) |
| Oviposition mode | Female alone, not guarded | *Anax imperator* | Corbet (1957) |
| | Female alone, non-contact guarding | *Calopteryx maculata* | Alcock (1979) |
| | In tandem | *Lestes sponsa* | Stoks et al. (1997) |
| | Male and female submerge | *Pseudagrion pilidorsum* | Matsubara and Hironaka (2005) |
| | Submerged, male guarding out of water | *Hetaerina vulnerata* | Alcock (1982) |

Some alternatives are clearly more favourable for males' interests (e.g. prolonged postcopulatory tandem), and others for females (e.g. sperm ejection), suggesting that postcopulatory sexual conflict is common in the order

of some species do submerge to oviposit, even at very low male density, or in the complete absence of males, like parthenogenetic females of *Ischnura hastata* at the Azores (Fig. 11.10b, c; Lorenzo-Carballa et al. 2009). Furthermore, in some species both the male and the female submerge to oviposit (e.g. Matsubara and Hironaka 2005). It seems therefore premature to interpret that underwater oviposition is a result of postcopulatory sexual conflicts. Alternative explanations are to avoid the risk of eggs drying due to changing water levels, or even to escape from egg parasitoids (Santolamazza et al. 2011).

In diving beetles, there is a high complexity in the female reproductive tract and a diverse array of sperm morphology (Miller and Bergsten 2014). As expected by sexual conflict theories, both the modifications of the reproductive tract and the great variety of morphology in the sperm seem to have evolved due to an arms race. Therefore, for the female these modifications of the tract would allow long-term sperm storage and then make a selection of sperm, and consequently a high control over paternity (i.e. cryptic female choice). The common disposition of the

**Fig. 11.9** Postcopulatory sexual conflicts can be subtle, like the ejection of sperm after copulation. In odonates this behaviour is known in several species and can be labelled as "cryptic female choice" (Córdoba-Aguilar 2006). (**a**) The figure shows a mating pair of *Pseudolestes mirabilis* (Hainan, China), and (**b**) the female perched after copulation, ejecting sperm (see detail in the insert). Source: Cordero-Rivera and Zhang (2018). Pictures by ACR

**Fig. 11.10** Male disturbance is a source of conflict after copulation, if females are accessible to mate-searching males during oviposition. Underwater oviposition (**a**; *Calopteryx haemorrhoidalis* from Pontevedra, Spain) could be a female adaptation to cope with unwanted male mating attempts, but other explanations, not related to sexual conflict, are also possible. For instance, parthenogenetic females of *Ischnura hastata* from the Azores islands frequently oviposit underwater, in the absence of males (**b** and **c**), although sometimes are harassed by males of *Ischnura pumilio*

reproductive tract is in loop with two genital openings (Heberdey 1931). One of the openings receives the spermatophore and is named *bursa copulatrix*. From this bursa, a spermathecal duct leads to the spermatheca and from this a fertilisation duct to the vagina near the oviduct (see Miller and Bergsten 2014). Therefore, we are facing a decoupling between the site of fertilisation and oviposition, allowing the female this cryptic choice. Miller (2001) was the first to establish a Dytiscidae female internal genital morphology classification, based principally on the number of genital openings, "Amphizoid type" and "Hydroporine type" with two genital openings, and "Dytiscinae type" with a single genital opening. Therefore, the main differences between Dytiscidae subfamilies, Hydroporinae and Dytiscinae, might be associated with the importance that pre-insemination or post-insemination systems would acquire, the latter being the most complex in the Hydroporinae subfamily, both in females and males, as we will see below.

Then, as mentioned above, the diving beetles are within an arms race and therefore females seem to counteract the sperm competition mechanisms by their cryptic female choice abilities. These adaptations are reflected in both the external and internal genitalia of the male. In external genitalia several species have spines or sawlike structures (Sharp and Muir 1912). On the other hand, it has been found conjugation (Higginson and Pitnick 2011) and/or heteromorphism on sperm morphology (Higginson et al. 2012a). Dytiscidae diving beetles have different degrees of conjugation (i.e. two or more spermatozoa physically unite for transport in a female reproductive tract): simple conjugation (two sperm heads unite), medium conjugation (aggregation of multiple sperm heads) and complex conjugation or "rouleaux". However, in some cases sperm is found solely too, named as singleton (Higginson et al. 2012b). Apart from co-occurrence with conjugation, diving beetles also show sperm heteromorphism (i.e. the simultaneous production of more than one type of sperm) that is manifested by head shape and sperm length (Higginson et al. 2012a), which in other organisms is associated to sperm competition mechanisms (for a review, see Cordero Rivera and Córdoba-Aguilar 2016).

In addition to the adaptations concerning the internal and external genitalia of both sexes, it has been found that some *Dytiscus* species produce a mating plug in the apex of the female abdomen (Aiken 1992) that remains attached to the female during long periods after fertilisation, a strategy that is believed to be used in male-male competition for paternity.

### *11.2.4 Sexual Conflict and Genitalia*

The change in paradigm that occurred in sexual selection studies in the last decades (Fig. 11.1) was in fact initiated when the relationship between animal genitalia and sexual selection became evident, with the pioneering work by William Eberhard (1985). The recent review of the field has found extensive support for the sexual selection hypothesis as the main driver of genital evolution (Leonard and Córdoba-Aguilar 2010). However, the relative importance of sperm competition and cryptic female choice, and the associated sexual conflicts, remains controversial, and particularly understudied in vertebrates (Brennan and Prum 2015).

One prediction from the sexual conflict hypothesis is that it leads to sexually antagonistic co-evolution at the genital level, where each sex is selected to advance control over mating and fertilisation, and as a consequence there is a co-evolutionary arms race (Brennan and Prum 2015).

The relationship between sexual conflict and genital evolution has been reviewed recently in insects (Simmons 2014), and in-depth discussions exist for several taxa, including water striders (Arnqvist 1997), odonates (Córdoba-Aguilar et al. 2003; Cordero-Rivera and Córdoba-Aguilar 2010) and water beetles (Miller and Bergsten 2014). Therefore, here we concentrate in a few clear examples, and suggestions for future experiments.

Distinguishing between hypotheses of genitalic sexual selection is extremely difficult, due to the interactions between males and females and the fact that the same prediction can be derived from different hypotheses, like sexually antagonistic co-evolution and cryptic female choice (Eberhard 2010). In the context of this review, the most relevant pattern is the diversity and rate of evolution of genitalia when compared with other traits. Genitalia evolves much faster (Eberhard 1985), and this needs an explanation. In the Gerridae several studies have shown that genitalia is under sexual selection, and that the form of particular male sclerites is related to fertilisation success, easily interpreted as the result of variable ability among males in sperm competition, mediated by female phenotype (i.e. cryptic female choice) (Arnqvist and Danielsson 1999). There is also evidence that genitalia may evolve by sexually antagonistic co-evolution (Perry et al. 2017).

In the odonates, there is good knowledge of functional morphology of copulation, especially after the work of Peter Miller (1987a, b). The form of the genital ligula is characterised by the presence of apical and lateral lobes and horns, and by a variable amount of backwards-oriented spines, which are used to trap and remove sperm from the previous mates of the female (reviewed by Córdoba-Aguilar and Cordero-Rivera 2008; Cordero-Rivera and Córdoba-Aguilar 2010). These spines in the intromittent organ could also function as copulatory weapons, known in other insects (Reinhardt et al. 2014), and merit further study. If during copulation males damage the internal genitalia of females, there could be an arms race and sexually antagonistic co-evolution.

In the European genus *Calopteryx*, common in rivers and streams (Fig. 11.3c), this sexual conflict over control of sperm reserves can be experimentally studied by using "hand-pairing" (Opphenheimer and Waage 1987), a technique which allows obtaining even interspecific matings. If females co-evolve with conspecific males at the genital level, heterospecific males should have an advantage when interspecific matings occur because females will show less resistance to them than to conspecific males. This is what Cordero-Rivera (2017b) found when *Calopteryx haemorrhoidalis* and *C. splendens* were hand-paired, and it is the first experimental evidence in odonates for such a sexually antagonistic co-evolution in the genitalia (Fig. 11.11).

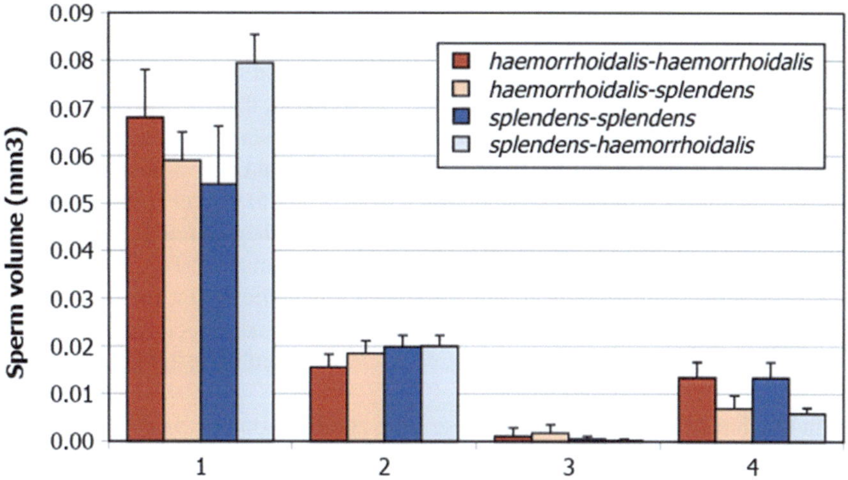


**Fig. 11.11** The volume of sperm in the *bursa copulatrix* and spermatheca of double-mated females after conspecific and heterospecific matings (postcopula treatment; the male species is indicated first in the legend), compared to the volume stored by females mated to a conspecific male and then to a conspecific or heterospecific male, but interrupted after 60 movements of stage I, before insemination. The volume of sperm in the *bursa* was reduced by 98–100% in all cases. The volume of the spermatheca in intraspecific matings was reduced by 14% in *Calopteryx haemorrhoidalis* and by 26% in *C. splendens*. In contrast, volume reduction was 63–69% in heterospecific matings. These results indicate that females have more control over spermathecal sperm when mating with a conspecific male (i.e. there is sexually antagonistic co-evolution). Source: Cordero-Rivera (2017b)

## 11.3 Conclusions and Suggestions for Future Work

Our review of the literature about three groups of aquatic insects is consistent with our working hypothesis about the dimensionality of the habitat as a modulator of male ability to control reproductive decisions. The existence of complex grasping apparatus in some species suggests that sexual conflict is behind the evolution of these characters. We propose that, in the odonates, the family Gomphidae is a priority for future work, given their elaborated male anal appendages (Fig. 11.6). Unfortunately, the study of gomphid behaviour is lagging behind other odonate families, probably because they are rarely seen in mating. In diving beetles, some subfamilies are more affected by the modifications that derive from the pre-copula phase (i.e. Cybistrinae and Dytiscinae), others from the post-copula phase (i.e. Hydroporinae) and other subfamilies that are at a morphological and behavioural intermediate point (Miller and Bergsten 2014). Thanks to this fact, the Dytiscidae beetle family gives us the opportunity to study the sexual conflict and its gradient, in the different phases of sexual behaviour. However, nearly all the phases of reproductive behaviour, and particularly of copulation in this group, require further investigations (Miller and Bergsten 2014).

Other aquatic insects are promising for future work in this field. For instance, in the water bugs of the family Nepidae, male genitalia has a pair of parameres with a hook (Keffer 2004), and we hypothesise that these could have evolved as adaptations to grasp females for copulation. In the Megaloptera, there is good evidence for a trade-off between elaborated weapons and size of nuptial gifts (Liu et al. 2015), and they are another group worth studying from a sexual conflict perspective. The whirling beetles (Gyrinidae) live, as water striders, on the water surface, and are also constrained to a bi-dimensional habitat (although they can also dive). Sexual selection can therefore be intense, at least on some organs (e.g. Fairn et al. 2007). More studies of morphological and ethological adaptations that can be used by males to try to monopolise females and also a detailed analysis of female counter-adaptations are needed to obtain a balanced view of conflict resolution (Ah-King et al. 2014).

In relation to copulation duration, we have good information from the Gerridae and many families of odonates, but for diving beetles information is scarce. Morphological studies have been completed in many genera, but without an ethological perspective the picture is clearly incomplete. The possibility that sexual conflict over mating duration has selected for smaller body size in females, allowing them to breathe directly from the water (Madsen 2012), is a suggestion for future investigations. It must be noted that this hypothesis predicts that females should be smaller at increasing intensity of sexual conflict over mating duration. Species with very long copulations (Aiken 1992) are particularly interesting in this context. Other aquatic insects, like Corixidae, Notonectidae and Aphelocheiridae, some of which obtain the oxygen from the water, should also be studied to detect sexual conflicts.

In some species, sexual conflict has apparently resulted in the evolution of aggressive copulation by males, whose intromittent organ has evolved as a wounding device, resulting in extreme sexual conflict (Reinhardt et al. 2014). The genital ligula in male odonates has a diverse form and commonly is covered by spines, which are thought to be devices to trap and remove rivals' sperm (Córdoba-Aguilar et al. 2003). However, its function as wounding devices has to be studied, and the same can be applied to other aquatic insects (i.e. the aedeagus of *Dytiscus*; see Guignot 1933). Although not an aquatic taxon, the genus *Neotrogla* (Psocodea: Prionoglarididae) from Brazilian caves is an excellent model of study because it shows reversed genitalia, with female penis, male vagina and females that mount males and that control copulations that may last for 40–70 h (Yoshizawa et al. 2014), likely resulting in extreme sexual conflict. The ecological pressures that are behind the evolution of this sex-role reversal might also apply to other taxa.

The field of postcopulatory sexual conflicts, derived from sperm competition and cryptic female choice mechanisms, is another priority for future research. It is known that some odonate females eject sperm after copulation (Fig. 11.9), but this behaviour is so cryptic that probably it is much more common than previously thought. A comprehensive review of this behaviour is needed, and an experimental approach with genetic markers is an absolute priority to know if this behaviour is a consequence of sperm competition (the sperm ejected is the amount removed by the last copulating male; e.g. Lindeboom 1998) or is cryptic female choice in action (Córdoba-Aguilar 2006). Furthermore, do Gerridae females or diving beetles eject sperm after copulation? We could not find any reference to this behaviour in the surveyed literature.

The behaviour of males and females after copulation can also be affected by sexual conflict. We have discussed the strategy of underwater oviposition in female odonates, and the same could be applicable to insects of the water surface. Gerridae females sometimes lay eggs underwater. Is this behaviour related to male disturbance, as expected from sexual conflicts? Or other natural selection pressures, like avoiding egg parasites (e.g. Amano et al. 2008), or the presence of a helping male in tandem (Hirayama and Kasuya 2008), are more relevant? In diving beetles, there is evidence for arms races between male and female genitalia (Miller and Bergsten 2014), and the existence of mating plugs (Aiken 1992) seems easily interpretable as an example of a male adaptation to control fertilisation, which can generate a sexual conflict. If so, mating plugs should be more common in species with high-density populations, where the encounter rate between sexes is high.

The study on sexual selection, sexual conflict and mating behaviour, among others, was based on beliefs and stereotypes that are far from a feminist vision (i.e. equality). Since Darwin proposed that females were timid per se, that idea further fuelled a male-centred stereotype, which even today persists in many minds both in men and women (Ah-King et al. 2014). The influential work of Darwin was used as a justification by contemporary researchers to maintain stereotyped thinking about female behaviour, to such an extent that they came to question the mental abilities of females to choose their mates (Milam 2010). Like the snake that bites its own tail, the assumption that the role of the female in sexual selection or in sexual conflict is merely the one of a secondary protagonist, or not even that, but as a simple spectator leads to the belief that it is difficult to study female genitalia. But that belief loses all its validity with the modern technologies, or just with an unbiased review of the taxonomic literature (e.g. Eberhard 2006). In fact, there are no biological reasons for this lack of studies of female genitalia and behaviour, and the bias does not depend on the gender of the investigator (Ah-King et al. 2014).

We conclude that a comprehensive understanding of sexual conflicts in animals requires the study of both male and female anatomies, as well as their behaviours, avoiding previous assumptions, or stereotypes. Conflict appears when two individuals are involved, a conclusion that a priori seems obvious, since the same word conflict implies two or more individuals, so it does not make sense to study only one part. A balanced study of this field requires a more egalitarian approach, i.e. a feminist view of the science (Ah-King et al. 2014).

**Acknowledgements** Funding was provided by a grant from the Spanish Ministry of Economy and Competitiveness, including FEDER funds (CGL2014-53140-P). ART was supported by a FPI grant (BES-2015-071965).

## References


Ah-King M, Barron AB, Herberstein ME (2014) Genital evolution: why are females still understudied? PLoS Biol 12:1–7. https://doi.org/10.1371/journal.pbio.1001851

Aiken RB (1992) The mating behaviour of a boreal water beetle, *Dytiscus alaskanus* (Coleoptera Dytiscidae). Ethol Ecol Evol 4:245–254. https://doi.org/10.1080/08927014.1992.9523136

Alcock J (1979) Multiple mating in *Calopteryx maculata* (Odonata: Calopterygidae) and the advantage of non-contact guarding by males. J Nat Hist 13:439–446

Alcock J (1982) Postcopulatory mate guarding by males of the damselfly *Hetaerina vulnerata* Selys (Odonata: Calopterygidae). Anim Behav 30:99–107

Alcock J (1994) Postinsemination associations between males and females in insects: The mate guarding hypothesis. Annu Rev Entomol 39:1–21

Amano H, Hayashi K, Kasuya E (2008) Avoidance of egg parasitism through submerged oviposition by tandem pairs in the water strider, Aquarius paludum insularis (Heteroptera: Gerridae). Ecol Entomol 33:560–563. https://doi.org/10.1111/j.1365-2311.2008.00988.x

Andrés JA, Cordero-Rivera A (2000) Copulation duration and fertilization success in a damselfly: An example of cryptic female choice? Anim Behav 59:695–703

Arnqvist G (1997) The evolution of water strider mating systems: causes and consequences of sexual conflicts. In: Choe JC, Crespi BJ (eds) The evolution of mating systems in insects and arachnids. Cambridge University Press, Cambridge, pp 146–163

Arnqvist G, Danielsson I (1999) Copulatory behavior, genital morphology, and male fertilization success in water striders. Evolution 53:147–156

Arnqvist G, Nilsson T (2000) The evolution of polyandry: multiple mating and female fitness in insects. Anim Behav 60:145–164

Arnqvist G, Rowe L (2002) Correlated evolution of male and female morphologies in water striders. Evolution 56:936–947. https://doi.org/10.1554/0014-3820(2002)056[0936:CEOMAF]2.0.CO;2

Arnqvist G, Rowe L (2005) Sexual conflict. Princeton University Press, Princeton

Arnqvist G, Jones TM, Elgar MA (2006) Sex-role reversed nuptial feeding reduces male kleptoparasitism of females in Zeus bugs (Heteroptera; Veliidae). Biol Lett 2:491–493. https://doi.org/10.1098/rsbl.2006.0545

Baker RR (1983) Insect territoriality. Annu Rev Entomol 28:65–89

Bañuelos Irusta J, Araújo A (2007) Reproductive tactics of sexes and fitness in the dragonfly, *Diastatops obscura*. J Insect Sci 7:24

Bateman AJ (1948) Intra-sexual selection in Drosophila. Heredity 2:349–368

Brennan PLR, Prum RO (2015) Mechanisms and evidence of genital coevolution: the role of natural selection, mate choice, and sexual conflict. In: Rice WR, Gavrilets S (eds) The genetics and biology of sexual conflict. Cold Spring Harbor Laboratory Press, Cold Spring Harbor, pp 385–413

Byers CJ, Eason PK (2009) Conspecifics and their posture influence site choice and oviposition in the damselfly *Argia moesta*. Ethology 115:721–730

Campbell V, Fairbairn DJ (2001) Prolonged copulation and the internal dynamics of sperm transfer in the water strider *Aquarius remigis*. Can J Zool 79:1801–1812. https://doi.org/10.1139/cjz-79-10-1801

Carranza J (2009) Defining sexual selection as sex-dependent selection. Anim Behav 77:749–751

Cassis G, Hodgins M, Weir TA, Tatarnic NJ (2018) Phylogenetic reclassification and genitalic morphology of the small water strider genus Nesidovelia Andersen & Weir and allied Microveliinae (Hemiptera:Veliidae). Aust Entomol 57:92–106. https://doi.org/10.1111/aen.12273

Cleavall L (2009) Description of Thermonectus nigrofasciatus and *Rhantus binotatus* (Coleoptera: Dytiscidae) mating behavior. University of New Mexico, Albuquerque

Clutton BTH, Parker G (1995) Sexual coercion in animal societies. Anim Behav 49:1345–1365

Corbet PS (1957) The life-history of the Emperor dragonfly Anax imperator Leach (Odonata: Aeshnidae). J Anim Ecol 26:1–69

Corbet PS (1962) A biology of dragonflies. Classey (fac-simile 1983), Faringdom
Cordero A (1989) Reproductive behaviour of Ischnura graellsii (Rambur) (Zygoptera: Coenagrionidae). Odonatologica 18:237–244
Cordero A (1990) The adaptive significance of the prolonged copulations of the damselfly, Ischnura graellsii (Odonata: Coenagrionidae). Anim Behav 40:43–48
Cordero A (1992) Sexual cannibalism in the damselfly species Ischnura graellsii (Odonata: Coenagrionidae). Entomol Gen 17:17–20
Cordero A (1999) Forced copulations and female contact guarding at a high male density in a Calopterygid damselfly. J Insect Behav 12:27–37
Cordero A, Andrés JA (2002) Male coercion and convenience polyandry in a Calopterygid damselfly (Odonata). J Insect Sci 2:14. Available online: insectscience.org/2.14
Cordero Rivera A, Córdoba-Aguilar A (2016) Selección postcópula: competencia espermática. In: Etología adaptativa: el comportamiento como producto de la selección natural. p 479–504
Cordero A, Santolamazza-Carbone S, Utzeri C (1992) A twenty-four-hours-lasting tandem in Coenagrion scitulum (Ramb.) in the laboratory (Zygoptera: Coenagrionidae). Not Odonatol 3:166–167
Cordero A, Santolamazza-Carbone S, Utzeri C (1998) Mating opportunities and mating costs are reduced in androchrome female damselflies, Ischnura elegans (Odonata). Anim Behav 55:185–197
Cordero-Rivera A (2002) Influencia de la selección sexual sobre el comportamiento reproductor de los odonatos. In: Evolución: la base de la biología. p 497–507
Cordero-Rivera A (2016) Sperm removal during copulation confirmed in the oldest extant damselfly, Hemiphlebia mirabilis. Peer J 4:e2077. https://doi.org/10.7717/peerj.2077
Cordero-Rivera A (2017a) Behavioral diversity (ethodiversity): a neglected level in the study of biodiversity. Front Ecol Evol 5:1–7. https://doi.org/10.3389/fevo.2017.00007
Cordero-Rivera A (2017b) Sexual conflict and the evolution of genitalia: male damselflies remove more sperm when mating with a heterospecific female. Sci Rep 7:7844. https://doi.org/10.1038/s41598-017-08390-3
Cordero-Rivera A, Córdoba-Aguilar A (2010) Selective forces propelling genitalic evolution in Odonata. In: Leonard J, Córdoba-Aguilar A (eds) The evolution of primary sexual characters in animals. Oxford University Press, Oxford, pp 332–352
Cordero-Rivera A, Egido Pérez FJ (1998) Mating frequency, population density and female polychromatism in the damselfly _Ischnura graellsii_: an analysis of four natural populations. Etología 6:61–67
Cordero-Rivera A, Zhang H (2018) Ethological uniqueness of a damselfly with no near relatives: the relevance of behaviour as part of biodiversity. Anim Biodivers Conserv 41:161–174
Cordero-Rivera A, Utzeri C, Santolamazza-Carbone S (1999) Emergence and adult behaviour of Macromia splendens (Pictet) in Galicia, northwestern Spain (Anisoptera: Corduliidae). Odonatologica 28:333–342
Cordero-Rivera A, Andrés JA, Córdoba-Aguilar A, Utzeri C (2004) Postmating sexual selection: allopatric evolution of sperm competition mechanisms and genital morphology in calopterygid damselflies (Insecta: Odonata). Evolution 58:349–359
Córdoba-Aguilar A (2000) Reproductive behaviour of the territorial damselfly Calopteryx haemorrhoidalis asturica Ocharan (Zygoptera: Calopterygidae). Odonatologica 29:295–305
Córdoba-Aguilar A (2006) Sperm ejection as a possible cryptic female choice mechanism in Odonata (Insecta). Physiol Entomol 31:146–153
Córdoba-Aguilar A, Cordero-Rivera A (2008) Cryptic female choice and sexual conflict. In: Córdoba-Aguilar A (ed) Dragonflies and damselflies. Model organisms for ecological and evolutionary research. Oxford University Press, Oxford, pp 189–202
Córdoba-Aguilar A, Uhía E, Cordero Rivera A (2003) Sperm competition in Odonata (Insecta): the evolution of female sperm storage and rivals' sperm displacement. J Zool 261:381–398. https://doi.org/10.1017/S0952836903004357

Córdoba-Aguilar A, Serrano-Meneses MA, Cordero-Rivera A (2009) Copulation duration in nonterritorial odonate species lasts longer than in territorial species. Ann Entomol Soc Am 102:694–701. https://doi.org/10.1603/008.102.0414
Córdoba-Aguilar A, Vrech DE, Rivas M et al (2015) Allometry of male grasping apparatus in odonates does not suggest physical coercion of females. J Insect Behav 28:15–25
Daly M (1978) The cost of mating. Am Nat 112:771–774
Dawkins R (1976) The selfish gene. Oxford University Press, Oxford
Dettner K, Schwinger G (1980) Defensive substances from pygidial glands of water beetles. Biochem Syst Ecol 8:89–95. https://doi.org/10.1016/0305-1978(80)90032-0
Eberhard WG (1985) Sexual selection and animal genitalia. Harvard University Press, Cambridge
Eberhard WG (1996) Female control: sexual selection by cryptic female choice. Princeton University Press, Princeton
Eberhard WG (2006) Sexually antagonistic coevolution in insects is associated with only limited morphological diversity. J Evol Biol 19:657–681
Eberhard WG (2010) Evolution of genitalia: theories, evidence, and new directions. Genetica 138:5–18
Fairn ER, Schulte-Hostedde AI, Alarie Y (2007) Sexual selection on accessory glands, genitalia and protarsal pads in the whirligig beetle *Dineutus nigrior* Roberts (Coleoptera: Gyrinidae). Ethology 113:257–266
Fincke OM (1986) Underwater oviposition in a damselfly (Odonata: Coenagrionidae) favors male vigilance, and multiple mating by females. Behav Ecol Sociobiol 18:405–412
Fincke OM (1997) Conflict resolution in the Odonata: implications for understanding female mating patterns and female choice. Biol J Linn Soc 60:201–220
Fincke OM (2004) Polymorphic signals of harassed female odonates and the males that learn them support a novel frequency-dependent model. Anim Behav 67:833–845
Guignot F (1933) Les Hydrocanthares de France. Les Frères Douladoure, Toulouse
Heberdey RF (1931) Zur Entwicklungsgeschichte, Vergleichenden Anatomie und Physiologie der Weiblichen Geschlechtsausführwege der Insekten. Zeitschrift für Morphol und Ökologie der Tiere 22:416–586
Higginson DM, Pitnick S (2011) Evolution of intra-ejaculate sperm interactions: do sperm cooperate? Biol Rev 86:249–270. https://doi.org/10.1111/j.1469-185X.2010.00147.x
Higginson DM, Miller KB, Segraves KA, Pitnick S (2012a) Female reproductive tract form drives the evolution of complex sperm morphology. Proc Natl Acad Sci 109:4538–4543. https://doi.org/10.1073/pnas.1111474109
Higginson DM, Miller KB, Segraves KA, Pitnick S (2012b) Convergence, recurrence and diversification of complex sperm traits in diving beetles (Dytiscidae). Evolution 66:1650–1661. https://doi.org/10.1111/j.1558-5646.2011.01532.x
Hirayama H, Kasuya E (2008) Factors affecting submerged oviposition in a water strider: level of dissolved oxygen and male presence. Anim Behav 76:1919–1926. https://doi.org/10.1016/j.anbehav.2008.08.013
Keffer SL (2004) Morphology and evolution of water scorpion male genitalia (Heteroptera: Nepidae). Syst Entomol 29:142–172. https://doi.org/10.1111/j.0307-6970.2004.00236.x
Koch K (2006) Effects of male harassment on females' oviposition behaviour in Libellulidae (Odonata). Int J Odonatol 9:71–80. https://doi.org/10.1080/13887890.2006.9748264
Latty TM (2006) Flexible mate guarding tactics in the dragonfly Sympetrum internum (Odonata: Libellulidae). J Insect Behav 19:469–477. https://doi.org/10.1007/s10905-006-9037-0
Lauer MJ, Sih A, Krupa JJ (1996) Male density, female density and inter-sexual conflict in a stream-dwelling insect. Anim Behav 52:929–939. https://doi.org/10.1006/anbe.1996.0241
Leonard JL, Córdoba-Aguilar A (2010) The evolution of primary sexual characters in animals. Oxford University Press, Oxford
Lindeboom M (1998) Post-copulatory behaviour in Calopteryx females (Insecta, Odonata, Calopterygidae). Int J Odonatol 1:175–184

Liu X, Hayashi F, Lavine LC, Yang D (2015) Is diversification in male reproductive traits driven by evolutionary trade-offs between weapons and nuptial gifts? Proc R Soc B Biol Sci. https://doi.org/10.1098/rspb.2015.0247
Lorenzo-Carballa MO, Beatty CD, Utzeri C et al (2009) Parthenogenetic *Ischnura hastata* revisited: present status and notes on population ecology and behaviour (Odonata: Coenagrionidae). Int J Odonatol 12:395–411
Machado G, Trumbo ST (2018) Parental care. In: Córdoba-Aguilar A, González-Tokman D, González-Santoyo I (eds) Insect behavior. Oxford University Press, Oxford, pp 203–2018
Madsen BL (2012) Submersion respiration in small diving beetles (Dytiscidae). Aquat Insects 34:57–76. https://doi.org/10.1080/01650424.2012.643026
Matsubara K, Hironaka M (2005) Postcopulatory guarding behaviour in a territorial damselfly, Pseudagrion p. pilidorsum (Brauer), for submerged ovipositing females (Zygoptera: Coenagrionidae). Odonatologica 34:387–396
Milam EL (2010) Looking for a few good males, female choice in evolutionary biology. John Hopkins University Press, Baltimore
Miller PL (1987a) An examination of the prolonged copulations of Ischnura elegans (Vander Linden) (Zygoptera: Coenagrionidae). Odonatologica 16:37–56
Miller PL (1987b) Sperm competition in Ischnura elegans (Vander Linden) (Zygoptera: Coenagrionidae). Odonatologica 16:201–207
Miller KB (2001) On the phylogeny of the Dytiscidae (Insecta: Coleoptera) with emphasis on the morphology of the female reproductive system. Insect Syst Evol 32:45–92
Miller KB (2003) The phylogeny of diving beetles (Coleoptera: Dytiscidae) and the evolution of sexual conflict. Biol J Linn Soc 79:359–388
Miller KB, Bergsten J (2014) Predaceous diving beetle sexual systems. In: Yee DA (ed) Ecology, systematics, and the natural history of predaceous diving beetles (Coleoptera: Dytiscidae). Springer, Dordrecht, pp 199–233
Opphenheimer SD, Waage JK (1987) Hand-pairing: a new technique for obtaining copulations within and between Calopteryx species (Zygoptera: Calopterygidae). Odonatologica 16:291–296
Parker GA (1970) Sperm competition and its evolutionary consequences in the insects. Biol Rev 45:535–567. https://doi.org/10.1111/j.1469-185X.1970.tb01176.x
Parker GA (1979) Sexual selection and sexual conflict. In: Blum MS, Blum NA (eds) Sexual selection and reproductive competition in insects. Academic Press, New York, pp 123–166
Parker GA (2006) Sexual conflict over mating and fertilization: an overview. Philos Trans R Soc Lond B Biol Sci 361:235–259
Perry JC, Rowe L (2018) Sexual conflict in its ecological setting. Philos Trans R Soc Lond B Biol Sci 373:20170418
Perry JC, Garroway CJ, Rowe L (2017) The role of ecology, neutral processes and antagonistic coevolution in an apparent sexual arms race. Ecol Lett 20:1107–1117. https://doi.org/10.1111/ele.12806
Polhemus JT (1974) The austrina group of the genus Microvelia (Hemiptera; Veliidae). Gt Basin Nat 34:207–217
Queller DC, Strassmann JE (2018) Evolutionary conflict. Annu Rev Ecol Evol Syst 49:73–93. https://doi.org/10.1146/annurev-ecolsys-110617-062527
Reinhardt K, Anthes N, Lange R (2014) Copulatory wounding and traumatic insemination. In: Rice WR, Gavrilets S (eds) The genetics and biology of sexual conflict. Cold Spring Harbor Laboratory Press, Cold Spring Harbor, pp 115–140
Rowe RJ (1978) Ischnura aurora (Brauer), a dragonfly with unusual mating behaviour (Zygoptera: Coenagrionidae). Odonatologica 7:375–383
Rowe L (1992) Convenience polyandry in a water strider: foraging conflicts and female control of copulation frequency and guarding duration. Anim Behav 44:189–202
Rowe L (1994) The costs of mating and mate choice in water striders. Anim Behav 48:1049–1056

Rowe L, Arnqvist G (2002) Sexually antagonistic coevolution in a mating system: combining experimental and comparative approaches to address evolutionary processes. Evolution 56:754–767

Rowe L, Arnqvist G, Sih A, Krupa J (1994) Sexual conflict and the evolutionary ecology of mating patterns: water striders as a model system. Trends Ecol Evol 9:289–293

Rubenstein DI (1989) Sperm competition in the water strider, *Gerris remigis*. Anim Behav 38:631–636. https://doi.org/10.1016/S0003-3472(89)80008-9

Sánchez-Guillén RA, Córdoba-Aguilar A, Cordero-Rivera A, Wellenreuther M (2014) Rapid evolution of prezygotic barriers in non-territorial damselflies. Biol J Linn Soc 113:485–496

Santolamazza S, Baquero E, Cordero-Rivera A (2011) Incidence of Anagrus obscurus (Hymenoptera: Mymaridae) egg parasitism on Calopteryx haemorrhoidalis and Platycnemis pennipes (Odonata: Calopterygidae: Platycnemididae) in Italy. Entomol Sci 14:366–369. https://doi.org/10.1111/j.1479-8298.2011.00454.x

Sattler W (1957) Beobachtungen zur Fortpflanzung von Gerris najas De Geer (Heteroptera). Zeitschrift für Morphol und Ökologie der Tiere 45:411–428

Schneider JM (2014) Sexual Cannibalism as a Manifestation of Sexual Conflict. Cold Spring Harb Perspect Biol 6(11):a017731. https://doi.org/10.1101/cshperspect.a017731

Sharp D, Muir F (1912) The comparative anatomy of the male genital tube in Coleoptera. Trans R Entomol Soc London 60:477–532

Shuker DM (2014) Sexual selection theory. In: Shuker DM, Simmons LW (eds) The evolution of insect mating systems. Oxford University Press, Oxford, pp 20–41

Simmons LW (2014) Sexual selection and genital evolution. Aust J Entomol 53:1–7

Spence JR, Anderson NM (1994) Biology of water striders: interactions between systematics and ecology. Annu Rev Entomol 39:101–128. https://doi.org/10.1146/annurev.en.39.010194.000533

Stoks R, De Bruyn L, Matthysen E (1997) The adaptiveness of intense contact mate guarding by males of the Emerald Damselfly, Lestes sponsa (Odonata, Lestidae): The male's perspective. J Insect Behav 10:289–298

Thornhill R (1983) Cryptic female choice in the scorpionfly Harpobittacus nigriceps and its implications. Am Nat 122:765–788

Trivers RL (1972) Parental investment and sexual selection. In: Campbell B (ed) Sexual selection and the descent of man 1871-1971. Aldine, Chicago, pp 136–179

Uhía E, Cordero-Rivera A (2005) Male damselflies detect female mating status: importance for postcopulatory sexual selection. Anim Behav 69:797–804

Utzeri C (1988) Female "refusal display" versus male "threat display" in Zygoptera: is it a case of intraspecific imitation? Odonatologica 17:45–54

Utzeri C, Ercoli C (2004) Disturbance by unpaired males prolongs postcopulatory guarding duration in the damselfly Lestes virens (Charpentier) (Zygoptera: Lestidae). Odonatologica 33:291–301

Van Gossum H, Sherratt TN, Cordero-Rivera A (2008) The evolution of sex-limited colour polymorphisms. In: Córdoba-Aguilar A (ed) Dragonflies and damselflies. Model organisms for ecological and evolutionary research. Oxford University Press, Oxford, pp 219–229

Waage JK (1979) Dual function of the damselfly penis: sperm removal and transfer. Science 203:916–918

Waage JK (1984) Sperm competition and the evolution of odonate mating systems. In: Smith RL (ed) Sperm competition and the evolution of animal mating systems. Academic Press, Orlando, pp 251–290

Wilcox RS (1984) Male copulatory guarding enhances female foraging in a water strider. Behav Ecol Sociobiol 15:171–174. https://doi.org/10.1007/BF00292971

Williams GC (1966) Adaptation and natural selection. Princeton University Press, Princeton

Yoshizawa K, Ferreira RL, Kamimura Y, Lienhard C (2014) Female penis, male vagina, and their correlated evolution in a cave insect. Curr Biol 24:1006–1010. https://doi.org/10.1016/j.cub.2014.03.022

# Chapter 12
# Paternal Care in Giant Water Bugs

**Shin-ya Ohba, José Ricardo I. Ribeiro, and Melania Santer**

**Abstract** We describe general biology of giant water bugs (Heteroptera, Belostomatidae), exclusive paternal care (back-brooding behavior in Belostomatinae and emergent-brooding behavior in Lethocerinae), and recent topics in sexual selection in this family. In the general biology, we introduced phylogenetic relationships of Belostomatidae within Nepomorpha and among Belostomatidae genera, distribution, food, behavior, and general egg morphology. After Smith's evolutionary hypothesis (Smith RL, The evolution of social behavior in insects and arachnids. Cambridge University Press, Cambridge, 1997), feeding habit of Belostomatidae was revealed and the result agreed with his hypothesis. In the exclusive paternal care, we reviewed the role of paternal care, oviposition site selection, and evolution of sexual size dimorphism and its relation to the presence of paternal care. After controlling for phylogenetic signal, it seems that the infanticide phenomenon does not necessarily explain the origin of sexual size dimorphism biased toward females in Belostomatidae, since the ancestor of all Belostomatidae was most likely a non-brooder. Finally, we show the future direction of the study in giant water bugs: coevolution between male and female genitalia in Belostomatidae, allometry for sexual size dimorphism and its disagreement with Rensch's rule, and phylogenetic test of predictions provided by Smith's evolutionary scenario assuming alternative conditions (*Horvathinia* is an emergent-brooder, a back-brooder, or a non-brooder).

**Keywords** Belostoma · Eggs · Behavior · Freshwater · Evolution

---

S.-y. Ohba (✉)
Biological Laboratory, Faculty of Education, Nagasaki University, Nagasaki, Japan
e-mail: ooba@nagasaki-u.ac.jp

J. R. I. Ribeiro · M. Santer
Laboratório de Estudos da Biodiversidade do Pampa (LEBIP), Universidade Federal do Pampa (UNIPAMPA), São Gabriel, Rio Grande do Sul, Brazil
e-mail: joseribeiro@unipampa.edu.br



K. Del-Claro, R. Guillermo (eds.), *Aquatic Insects*,
https://doi.org/10.1007/978-3-030-16327-3_12

## 12.1 General Biology of Belostomatidae

Giant water bugs (Belostomatidae, Heteroptera, Insecta) are classic examples of paternal care systems in arthropods, being shown them in some textbooks (Costa 2006; Royle et al. 2012). They include ca. 150 species distributed throughout the world's subtropical and tropical areas (Lauck and Menke 1961), consisting of eight genera in the subfamily Belostomatinae (*Horvathinia* Montandon, *Hydrocyrius* Spinola, *Limnogeton* Mayr, *Abedus* Stål, *Weberiella* De Carlo, *Belostoma* Latreille, *Appasus* Amyot & Serville, and *Diplonychus* Laporte) and three genera in the subfamily Lethocerinae (*Benacus* Stål, *Kirkaldyia* Montandon, and *Lethocerus* Mayr) (Polhemus 1995; Perez-Goodwyn 2006; Estévez and Ribeiro 2011; Ribeiro et al. 2018). They are found in freshwater habitats such as rice fields, marsh, ponds, lakes, and slow-moving rivers (Cullen 1969; Menke 1979; Mukai et al. 2005). Besides their large bodies, their reproductive biology is unique and well known (Clutton-Brock 1991; Shuster and Wade 2003; Arnqvist and Rowe 2005; Smiseth 2014), which includes female competition and male parental care (Ichikawa 1988, 1990, 1991; Smith and Larsen 1993; Thrasher et al. 2015). In this chapter, we review current knowledge on the function and evolution of paternal care in the giant water bugs. Additional study of giant water bug paternal care in particular, and their mating systems in general, will be an important advance in behavioral ecology. Controlling for body size in males and females via multiple regression, a phylogenetic perspective is given as well as a behavioral approach to discuss the evolution of paternal care. Using a phylogenetic approach is essential because closely related genera could have similar body sizes; that is, the traits could exhibit phylogenetic signal (Morand 1997).

## 12.2 Distribution and Phylogenetic Relationships of Belostomatidae Within Nepomorpha and Among Belostomatidae Genera

As mentioned, Belostomatinae is a widespread subfamily with the highest number of described genera among the family Belostomatidae. *Abedus* is restricted to the Nearctic Region, Mexico, where there is the largest number of representatives, and to Central America. *Belostoma*, the most diverse genus, occurs in the New World, with the largest number of representatives in South America (Lauck and Menke 1961; Moreira et al. 2011). *Appasus* and *Diplonychus* are distributed throughout Southeast Asia, Africa, India, and Australia (Estévez and Ribeiro 2011). *Hydrocyrius* is restricted to the African continent, and little is known about its biology. *Limnogeton*, restricted to north-eastern Africa, comprises apparently strict predators of snails. Its members bear smooth profemora, lacking sulci, and are thought to be the only representatives of Belostomatidae which do not have middle and posterior legs modified for swimming (Ribeiro et al. 2014). *Weberiella* is monobasic, constituted by *Weberiella rhomboides* (Menke), and is recorded from the states of Amazonas, Mato Grosso and Rondônia in Brazil, as well as French Guyana and Venezuela (Menke 1965; De Carlo 1966; Estévez and Ribeiro 2011). Its members inhabit the surface film of freshwaters exclusively (*kinon* by Fittkau 1977). *Horvathinia* is also monobasic, and it is recorded from southeast and central south of South America (Lauck and Menke 1961; Moreira et al. 2011). The genus comprises only two species, *Horvathinia lenti* De Carlo, and *H. pelocoroides* Montandon (Schnack and Estévez 2005).

Lethocerinae, now reformulated as a subfamily comprising the genera *Benacus* and *Kirkaldyia* (both being monobasic genera), and *Lethocerus* (according to Perez-Goodwyn 2006 and Ribeiro et al. 2018), is a cosmopolitan subfamily, being better represented in the Neotropical region (with about ten recorded species according to Ribeiro et al. 2009 and Perez-Goodwyn 2006). *Kirkaldyia* is distributed throughout Japan, southeastern Asia, China, Taiwan, and Korea, whereas *Benacus* extends from Oriental North America to southward, reaching Honduras, Cuba, and Jamaica (Perez-Goodwyn 2006).

Nepoidea (including Belostomatidae and Nepidae) has been generally considered the most basal branch of Nepomorpha (Mahner 1993; Hebsgaard et al. 2004; Hua et al. 2009), and Spooner (1938) was the first author who formally and intuitively proposed that Belostomatidae and Nepidae are closely related, indicating some morphological similarities. Later on, the conspicuous phallus of representatives of these two families, with the apex fully sclerotized, was highlighted as evidence of such a relationship by Dupuis (1955).

By including modifications or clarifications of both somatic and genitalic characters (including some spermatheca traits), as well as multiple genes (COI, 18S rDNA, and 16S rDNA), Ribeiro et al.'s (2018) comparative study of these genera yielded putative homology hypotheses coded as 104 morphological characters and 1829 aligned characters (Fig. 12.1a). Accordingly, the authors have corroborated the monophyly of Belostomatidae based on the retractable condition of the air straps into the abdomen, eyes somewhat projected laterally with angled borders, and tarsi 2- or 3-segmented, rarely 1-segmented. Lauck and Menke (1961) intuitively subdivided Belostomatidae into three subfamilies based on aspects of male genitalia morphology and in a more critical analysis of other characteristics, such as the condition of the abdominal sternites divided or not by a suture, and the degree of development of tarsal claws of the first pair of legs. Lee (1991) concurred with the decision made by Lauck and Menke (1961) of separating Belostomatidae into the Lethocerinae, Horvathiniinae, and Belostomatinae. However, the restriction of Belostomatidae to such three subfamilies has not been corroborated, and Ribeiro et al. (2018) have suggested the inclusion of *H. pelocoroides* within the subfamily Belostomatinae instead of comprising the monobasic subfamily Horvathiniinae related to the subfamily Belostomatinae (Fig. 12.1a).

Lethocerinae was recovered as a monophyletic group, despite having low support (Ribeiro et al. 2018). The authors suggest that the lateral lobes of segment 7 of abdomen with proximal portions subdivided into sublateral and lateral plates, and the presence of a conspicuous pouch below spermatheca, support Lethocerinae.

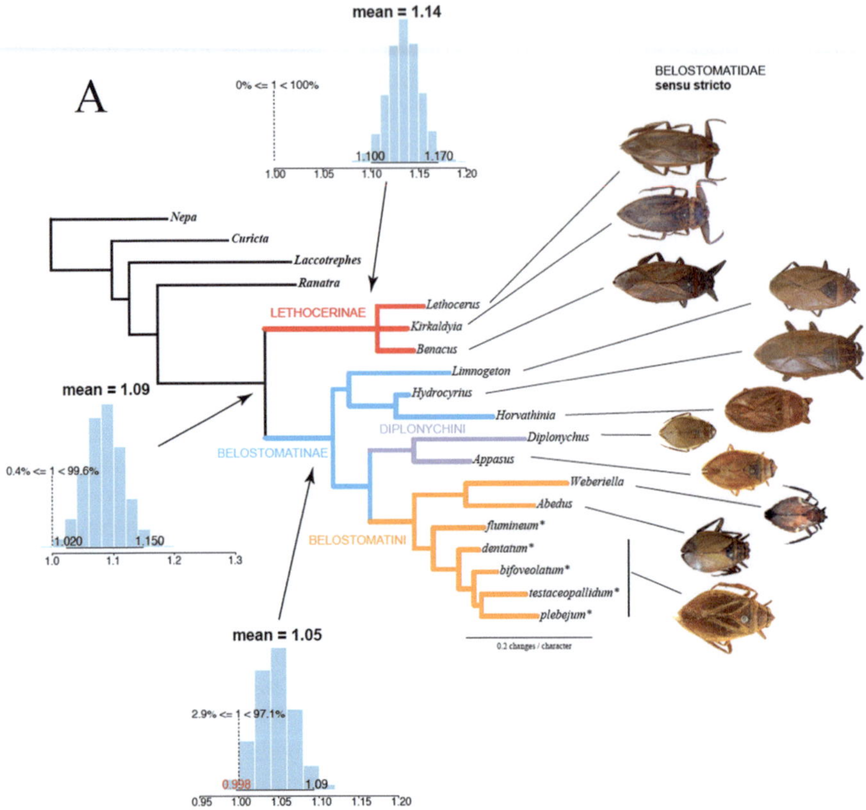


**Fig. 12.1** (**a–c**) Phylogeny of Belostomatidae sensu stricto (i.e., without fossils) and evolution of sexual size dimorphism (SSD) in total body length through water bugs. Groups of *Belostoma* species are represented by an asterisk. Outgroups (Nepidae) are in bold. (**a**) Bayesian consensus phylogram including highlighted clades estimated by maximum parsimony and maximum likelihood, with *Appasus*, *Benacus*, and *Kirkladyia* resurrected (based on Ribeiro et al. 2018). Histograms represent the posterior samples of different sexual size dimorphism in body size ancestral reconstructions by Bayesian MCMC method for estimating ancestral states. Values near to the base of histograms are the 95% probability intervals (i.e., 95% HDI). Values smaller than 1.0 are in red. Clades are colored accordingly to proposed classifications. A female-biased size dimorphism is supposed to be found earlier than that in emergent-brooders since this condition is the ancestral state in Belostomatidae. Only 2.9% of the credible values are smaller than 1.0 in the Belostomatinae node. (**b**) Posterior probabilities for evolution of the three-state threshold brooding behavior character in which the underlying model for evolution of the liability (Felsenstein 2012; Revell 2014) is an Ornstein-Uhlenbeck (OU) model with $\alpha = 0.12$ (3.1e-06 – 3.2). The results suggest that the ancestor of all Belostomatidae was a non-brooder, because it had the liability exceeding most likely the non-brooder threshold character state. Mean effective sample sizes (ESSs) for all parameters were higher than 200. Here, *Horvathinia* was considered to be a back-brooder. (**c**) Nonphylogenetic, ordinary least square (OLS) regression analysis in which the Ribeiro et al. (2018) branch lengths were dismissed (i.e., by considering a starlike phylogeny), because the fit of the OLS model was better than regression adopting the Ornstein-Uhlenbeck process (RegOU) and phylogenetic generalized least square (PGLS) models. By comparing SSD in total body length, calculated as $\log_{10}$ (F/M + 0.5), with brooding behavior degrees for each group, we evaluated how

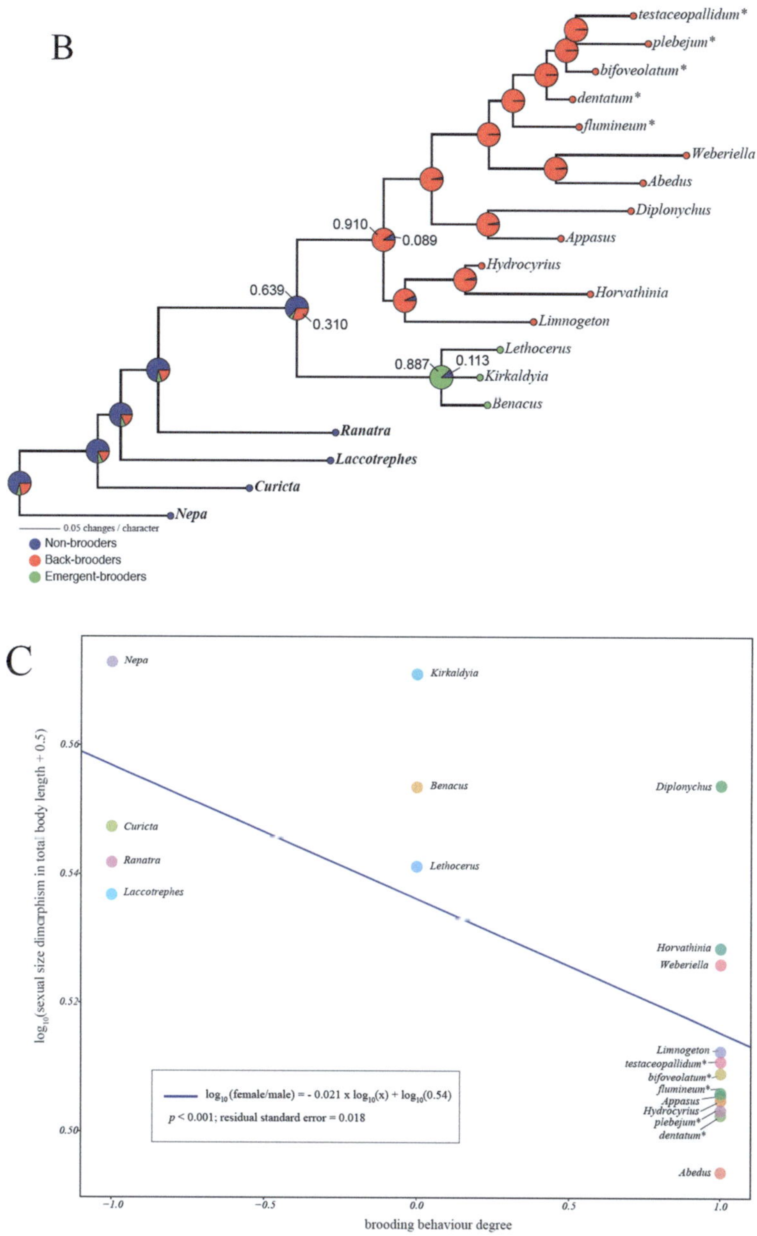


**Fig. 12.1** (continued) changes in brooding behavior conditions had contributed to changes in SSD. After controlling for the body size of males and females via multiple regression (not shown), such approach allowed to identify that the intensification of brooding behavior mediated by males is a possible target of selection (according to Davis and Roth 2008), whatever the alternative phylogenetic scenarios have been used (*Horvathinia* was considered to be a back-brooder in this plot). SSD in body length showed strong and highly statistically significant phylogenetic signal ($K = 1.407$; $P < 0.001$)

Despite being monophyletic, Ribeiro et al. (2018) failed to find clear internal relationship in the Lethocerinae clade. Based on a morphological dataset, the authors suggested that *Benacus* and *Kirkaldyia* are supported by the head equal to the anterior portion of pronotum and narrower than its largest width, a homoplastic synapomorphy. *Lethocerus* can be defined by the absence of spines or tubercles on the surface of apex of ventral diverticulum, a homoplastic condition also supported by *Horvathinia*.

Belostomatinae was also characterized based on some symplesiomorphic conditions. For example, Lauck and Menke (1961) and Lee (1991) stressed that the absence of divisions in the abdominal sternites differentiated it from the other subfamilies. However, Ribeiro et al. (2018) have based the monophyly of this subfamily on six unambiguously morphological synapomorphies. The classical male genitalic condition stated by Lauck and Menke (1961) (i.e., phallosoma fused with ventral diverticulum), the apex of second valvulae bearing spines and/or protuberances, and the foretarsi bearing both claws vestigial were some of the traits that supported such a monophyly.

*Horvathinia* has been accepted in Belostomatinae, and is supported by the abdominal sternites 3–7 divided laterally by weak, suture-like folds into median- and parasternites, a reversal according to Ribeiro et al. (2018). In fact, such recent findings are in conflict with previous hypotheses (Lauck and Menke, 1961; Lee 1991; Mahner 1993) that defined Belostomatinae as having the abdominal sternites 3–7 undivided laterally by weak folds. The resurrected genus *Appasus* (see Polhemus 1995) and *Diplonychus* have been formally recognized as a separate tribe named Diplonychini with strong support. According to Ribeiro et al. (2018), the transverse bridge of basal plate clearly jointed and entire is one of the unambiguously non-homoplastic synapomorphies. The authors still advocated that the differentiation of hemelytrum in the hemelytral membrane appeared independently in some instances in Belostomatidae, being more developed in the clade *Abedus* + *Weberiella*, moderately developed in *Limnogeton*, and completely absent in many species of *Diplonychus*. *Belostoma* was recovered as invalid by Mahner (1993) considering it very conservative and mostly distinguished by its plesiomorphic set of characteristics, such as membrane of hemelytrum distinctly larger than the greatest width of clavus, thus being wrongly considered equivalent to an ancestral Belostomatini. Ribeiro et al. (2018) recovered *Belostoma* as a monophyletic group, based on the presence of fourth segment of the antenna with only one fingerlike projection slightly more bulbous than prolongations of segments 2 and 3, an unambiguous homoplastic synapomorphy. *Abedus* was supported by the presence of straight apex of the ventral diverticulum (fused with phallosoma)—a unique unambiguous non-homoplastic synapomorphy. As with Mahner (1993), *Weberiella* has been supported by Ribeiro et al. (2018) by the lateral margins of abdomen not smooth but interrupted at the borders between the segments (a scalelike abdomen), an unambiguous non-homoplastic apomorphy.

## 12.3 Support for Smith's Hypothesis (Smith 1997)?

In aquatic environments, a fundamental constraint on the embryo brooding is related to the oxygen availability, since there can be considerable costs to making oxygen available to the whole egg mass (Fernández et al. 2000). Paternal care in Belostomatidae has probably evolved in response to reducing the risk of dehydration and/or increasing egg aeration (Munguía-Steyer et al. 2008; Requena, Munguía-Steyer & Machado 2014). Accordingly, all species studied so far exhibit paternal care for eggs, wherein the females lay on the males' back (Belostomatinae) or on emergent plant stems (Lethocerinae) (Smith 1997). Lethocerinae males (called "emergent-brooders") supply water to the egg mass laid by usually one partner female on the emergent vegetation above the water surface, whereas Belostomatinae males (called "back-brooders") show behavior such as brooding, pumping, or swimming to the water surface, providing oxygen to the egg mass on male's back laid by one or multiple partner females simultaneously (Fig. 12.2) (Smith 1997). Since all other Belostomatinae are back-brooders, we should expect *Horvathinia* to be a back-brooder as well, and describing its brooding behavior has been for many years the "holy grail" of those interested in the reproductive ecology of this group.

Smith (1997) proposed a fascinating hypothesis on the evolution of paternal care in giant water bugs. According to his hypothesis, ancestral species of Belostomatidae

**Fig. 12.2** Brooding male of Lethocerinae (*Kirkaldyia deyrolli*) (**a**), a brooding male and a female of Belostomatinae (*Appasus japonicus*) (**b**), and a male brooding two clutches on his back (**c**)

had small body size and they laid the smaller eggs under the water surface. Natural selection acting on body size (through egg size) allowed individuals to explore new food niches, such as large aquatic prey animals. As a result, he suggested that as large females produced larger eggs, the giant water bugs faced a problem: the large eggs have a low surface/volume ratio, which hampers gas exchange in water. Thus, females started laying eggs on the emergent vegetation (in Lethocerinae) or the back of conspecific males (in Belostomatinae), thus resolving the low surface/volume "problem" (Smith 1997).

According to this hypothesis, current Lethocerinae, the largest-bodied subfamily in Belostomatidae, are expected to eat larger vertebrate prey more than other aquatic insects. Lethocerinae species prefer to eat mainly vertebrate animals such as fish and anurans (Smith 1997). For example, *Lethocerus americanus* prefers to eat large fish (Schumann et al. 2012). Some researchers observed *Lethocerus* attacking birds (Hungerford 1919; Menke 1979) and water snakes (Wilson 1958). In recent years, it has also been reported that *Kirkaldya deyrolli* (belonging to the subfamily Lethocerinae) eats snakes (*Rhabdophis tigrinus*, *Amphiesma vibakari*, and *Gloydius blomhoffii*) (Mori and Ohba 2004; Ohba 2012) and turtles (*Mauremys reevesii*) (Ohba 2011) in their natural habitats in Japan. The nymphs possess a claw on the terminal segment of the raptorial foreleg that is crucial for capturing prey. Young nymphs with highly curved claws caught proportionally larger prey than older nymphs with less curved claws. When young nymphs (first to third instars) encounter prey that are larger than themselves, they first hook the claws of their raptorial legs onto the animal, and then use all their legs to pin the larger animals, fully clasping the prey body (Fig. 12.3). The young nymphs are forced to capture prey larger than themselves because smaller prey is less abundant during the hatchling season (Ohba et al. 2008). Older nymphs have an overall well-developed body size, but smaller prey is regularly utilized. Indeed, all *K. deyrolli* nymphs are able to feed on prey of a particular length (around 3 cm). They change catching behavior based on changes in raptorial characteristics to maximize prey resources acquired at each developmental stage (Ohba and Tatsuta 2016).

In contrast, prey size increases as predator nymphs grow in Belostomatinae, *Diplonychus indicus* (Cloarec 1992), *Belostoma oxyurum* (Perez-Goodwyn 2001), and *Appasus japonicus* (Ohba et al. 2008), same as a general predator's rule stating that predators catch prey that are smaller than they are (Nentwig and Wissel 1986; Warren and Lawton 1987). Moreover, Belostomatinae does not digest vertebrates well, because the saliva does not have enzymes capable of digesting protein-rich meals such as vertebrate animals (Swart et al. 2006).

Smith's (1997) hypothesis on the evolution of paternal care in water bugs has been deeply influential in behavioral ecology. Accordingly, he mentioned some predictions about an evident female-biased size dimorphism found across lethocerine lineages. Since appropriate oviposition sites on emergent vegetation may be in short supply, lethocerine females may aggressively compete for access to them, destroying non-filial eggs and opening space for their own eggs (Ichikawa 1990). Infanticide should occur in species in which females are considerably larger than males and thus able to displace the males and destroy the clutch they are caring for (Smith

**Fig. 12.3** A first instar of Lethocerinae (*Kirkaldyia deyrolli*) grasping larger vertebrate (tadpole) prey by using all six legs

1997). In species slightly or totally monomorphic in size, infanticides should be rare and multiple clutches should be the rule (Ichikawa 1990; Smith 1997). In the most recent edition of the Alcock's (2013) book, for instance, this hypothesis is explained in detail and illustrated in the chapter on parental care. According to him, in a given moment in the past, belostomatine males were able to return to the aquatic environment probably because individuals were small enough for their eggs to survive in the water, and thus the "infanticide phenomenon" became rare or practically inexistent among them.

This hypothesis on the evolution of sexual size dimorphism (SSD) in water bugs was tested and now seems to be clearly connected to the intensification of paternal care in Belostomatidae, whatever alternative phylogenetic scenario is assumed (assuming *Horvathinia* specimens as being emergent-brooders, back-brooders, or non-brooders) (Ribeiro et al. unpubl. data). Unlike Smith's scenario, a female-biased size dimorphism in total body length is supposed to be found earlier than that in lethocerines (i.e., emergent-brooders) since this condition was more likely to be the ancestral state in Belostomatidae and in other Nepidae lineages. Also, the shift to back-brooding behavior seems to be associated with evolutionary decrease in SSD after controlling for phylogenetic history (Figs. 12.1b, c). Whereas the infanticide phenomenon might occasionally explain the maintenance of the increasing in SSD values among lethocerines, such hypothesized phenomenon does not necessarily explain the origin of SSD biased toward females in Belostomatidae. Assuming the model showed (Fig. 12.1c), the results suggest that the ancestor of all

Belostomatidae was most likely a non-brooder (about 64%). As stressed by Smith (1997), however, oviposition on emergent vegetation seems to be the ancestral state in the family Belostomatidae.

The SSD values greater than one (i.e., biased to females) seem to be otherwise based on the tendency of related Nepidae species to be similar because they share a common history (according to Morand's 1997 ideas) (Table 12.1). Accordingly, this seems to be due to selection-mediating fertility in the females (Andersson 1994; Blanckenhorn 2009). In Belostomatinae, on the other hand, it is known that clutch size on the male dorsum depends both on the quantity of mature ova inside females and on the available area for egg-laying on males (Iglesias et al. 2010). So, Belostomatinae males have acquired a longer body (maybe a wider dorsum as well) than that of females presumably because they are being selected by females to offer a larger surface for egg-laying (but see Iglesias et al. 2012 and Thrasher et al. 2015 to the contrary).

The tendency for the degree of sexual dimorphism to vary with body size (see Fairbairn 1997; Dale et al. 2007; Davis and Roth 2008) refers to an allometric pattern that has been formalized into Rensch's rule (Rensch 1959). Biologists interested in documenting and quantifying this size allometry have plotted log-transformed values of female versus male body length (e.g., Davis and Roth 2008), whereby slopes below 1.0 indicate negative allometry and, in turn, an allometric pattern consistent with Rensch's rule. Accordingly, this means a greater variability in male size among lineages than in female size (Fairbairn 1997). Without phylogenetic correction or any statistical treatment, there seems not to be a tendency for small male back-brooders to be either narrower or shorter than their conspecific females (Fig. 12.4). Conversely, the relationship between female and male size—being represented by total body length—in Belostomatidae seems to be hypoallometric (i.e., slopes smaller than 1.0) (Fig. 12.4b) (Ribeiro et al. unpubl. data). Ultimately, these results provide some evidence that Belostomatidae display allometry for SSD consistent with Rensch's rule. So, this could indicate that a pattern emerges in Belostomatidae which is characterized by a greater variation in male body size than female body size (according to Fairbairn and Preziosi 1994).

## 12.4 The Role of Paternal Care and Oviposition Site Selection

### *12.4.1 Belostomatinae*

Paternal care of Belostomatinae species plays a role in maintaining oxygen and humidity conditions of their egg masses (Smith 1976a, b, 1997; Munguía-Steyer et al. 2008). Paternal behavior, i.e., brood pumping, brood stroking, and surface brooding, is essential to egg development and survival (Smith 1997). Eggs increase

**Table 12.1** Brooding behaviour categories, body size measurements of male and females of the Belostomatidae (sensu Ribeiro et al. 2018), and Nepidae genera *Curicta*, *Laccotrephes*, *Nepa*, and *Ranatra*

| Taxa | BC | *N* | Males: TBL (mean ± SD) | Females: TBL (mean ± SD) | SSD (TBL) | SDI (TBL) |
|---|---|---|---|---|---|---|
| **Lethocerinae** | | | | | | |
| *Benacus griseus* | EB | 5 | 52.5 ± 6.3 | 57.2 ± 6.4 | 0.553 | 0.131 |
| *Kirkaldyia deyrolli* | EB | 11 | 52.0 ± 2.3 | 61.6 ± 2.5 | 0.571 | 0.178 |
| *Lethocerus* | EB | 71 | 63.8 ± 9.8 | 70.4 ± 10.0 | 0.541 | 0.101 |
| **Belostomatinae** | | | | | | |
| *Abedus* | BB | 42 | 28.6 ± 4.1 | 31.0 ± 7.4 | 0.494 | −0.013 |
| *Appasus* | BB | 57 | 19.8 ± 4.2 | 21.4 ± 3.6 | 0.505 | 0.015 |
| *Belostoma* gr. *bifoveolatum* | BB | 223 | 21.2 ± 1.2 | 21.7 ± 1.2 | 0.509 | 0.024 |
| *Belostoma* gr. *dentatum* | BB | 82 | 36.6 ± 6.2 | 36.9 ± 5.4 | 0.502 | 0.009 |
| *Belostoma* gr. *flumineum* | BB | 28 | 21.1 ± 2.8 | 21.7 ± 2.4 | 0.506 | 0.015 |
| *Belostoma* gr. *plebejum* | BB | 83 | 12.3 ± 1.5 | 12.6 ± 1.5 | 0.503 | 0.010 |
| *Belostoma* gr. *testaceopallidum* | BB | 28 | 29.9 ± 2.4 | 30.9 ± 2.0 | 0.511 | 0.029 |
| *Diplonychus* | BB | 54 | 16.7 ± 3.4 | 16.4 ± 2.6 | 0.554 | 0.155 |
| *Horvathinia* | ?, unknown brooding behavior category | 26 | 25.9 ± 1.7 | 27.1 ± 1.6 | 0.528 | 0.070 |
| *Hydrocyrius* | BB | 64 | 56.0 ± 10.2 | 55.3 ± 10.3 | 0.506 | 0.015 |
| *Limnogeton* | BB | 40 | 48.7 ± 6.0 | 51.6 ± 4.7 | 0.513 | 0.032 |
| *Weberiella rhomboides* | BB | 16 | 21.0 ± 0.8 | 22.0 ± 0.8 | 0.526 | 0.063 |
| **Nepinae** | | | | | | |
| *Curicta* | NB | 54 | 20.7 ± 4.8 | 23.1 ± 5.6 | 0.547 | 0.116 |
| *Laccotrephes* | NB | 24 | 24.3 ± 11.3 | 27.3 ± 11.6 | 0.537 | 0.091 |
| *Nepa* | NB | 10 | 16.9 ± 0.9 | 20.4 ± 0.8 | 0.573 | 0.183 |
| **Ranatrinae** | | | | | | |
| *Ranatra* | NB | 230 | 32.4 ± 7.1 | 35.0 ± 7.4 | 0.542 | 0.103 |

Brooding behavior categories (BC) scored for giant water bug genera

*N* refers to the number of measurements recorded either in those references cited or based on material examined

Taxa in bold refer to the water bug subfamilies

All measurements are given in millimeters, except for SSD which was calculated as $\log_{10}(F/M) + 0.5$ and SDI which was estimated by taking the ratio of the larger (female) to the smaller sex (male) and subtracting 1

*TBL* total body length, *SSD* sexual size dimorphism, *SDI* sexual dimorphism index

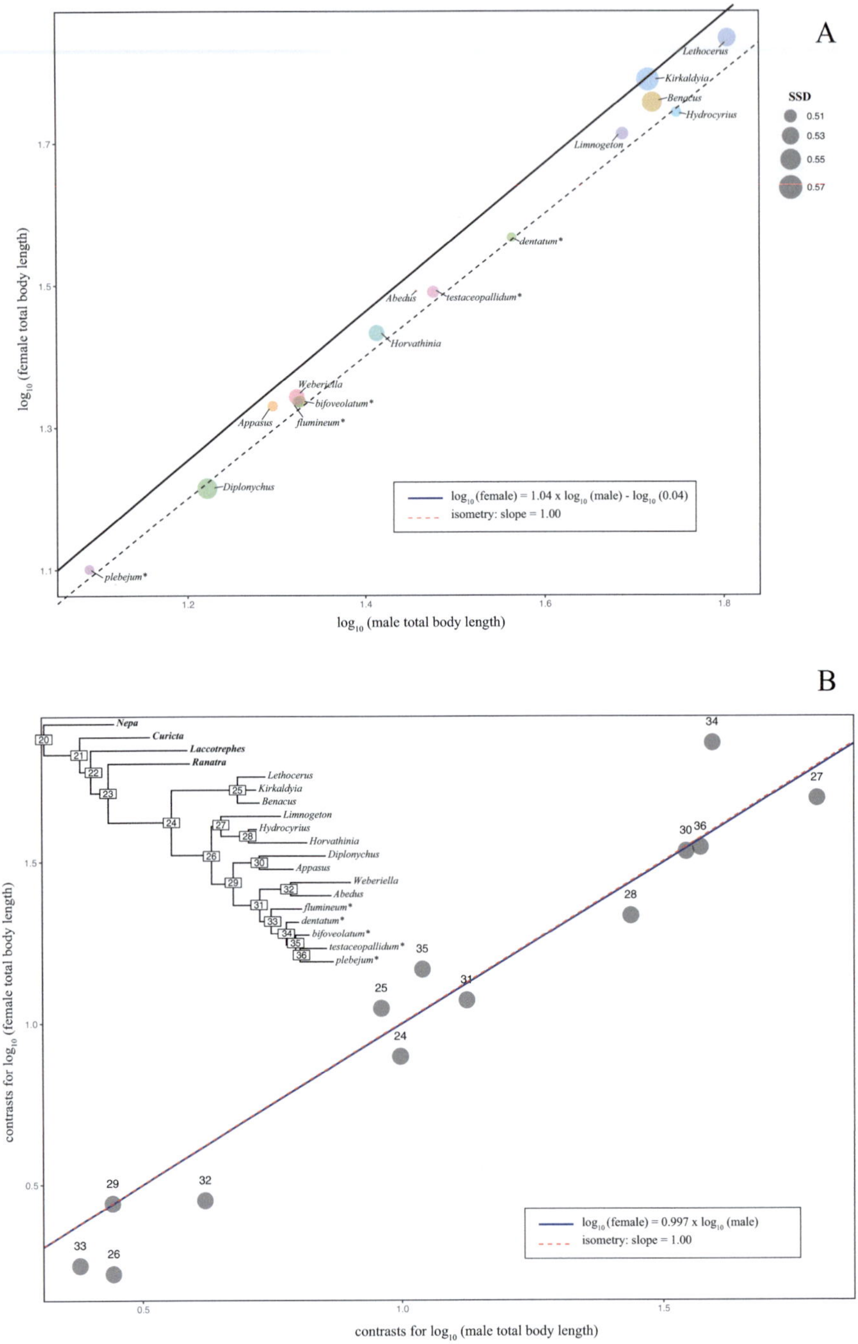


**Fig. 12.4** (**a**, **b**) Use of standardized major axis regression to assess allometry for sexual size dimorphism. Groups of *Belostoma* species are represented by an asterisk. (**a**) $\text{Log}_{10}$ (female size) versus $\log_{10}$ (male size) for each group used in the phylogeny of Belostomatidae (Ribeiro et al. 2018).

in volume by absorbing water through the hydropyle (Madhavan 1974; Venkatesan and Rao 1980). Thus, egg-caring males change brood pumping in response to the growth of the eggs, while non-brooded/submerged eggs suffer lower hatching rate in *Abedus breviceps* Stål (Munguía-Steyer et al. 2008). However, paternal care behavior incurs a cost in terms of foraging efficiency (Crowl and Alexander 1989), mobility (Kight et al. 1995), and longevity (Gilg and Kruse 2003) in *Belostoma flumineum*. Thus, males sometimes remove the egg pad in response to egg pad size (Kight and Kruse 1992) and low-temperature conditions (Kight et al. 2000) and when the males are isolated from the females (Kight et al. 2011). Males with small egg pads are more likely to discard their eggs in autumn (when breeding adults are young) than in spring (Kight et al. 2011). However, there are also studies that did not detect the cost of egg-caring in *Abedus breviceps* in the wild condition (Munguía-Steyer and Macías-Ordonez 2007) and *Belostoma lutarium* in the laboratory condition (Thrasher et al. 2015).

When the females lay egg masses, the mating pairs sometimes repeatedly copulate during a single mating sequence; this is known in many species of Belostomatinae (Smith 1979a, b; Ichikawa 1989). Frequent copulation has been interpreted as an adaptation that reduces sperm competition (Parker 1970; Simmons 2001) and risk of cuckoldry (Smith 1979a). In *A. herberti*, 98% of newly hatched nymphs were attributable to males that had brooded their eggs, indicating that the male's repeated copulation assures his paternity of the eggs (Smith 1979b). However, in *Appasus major* collected from the field, 28.4% of the eggs on the back of males were of other male origin (Inada et al. 2011). These differences between *Abedus* and *Appasus* may be derived from ecological differences between these two species, i.e., differences in cost of brooding. Where costs are higher (less food) selection on paternity assurance may be higher. These possibilities will be interesting subjects for future study.

### 12.4.2 Lethocerinae

Lethocerinae males supply water to the egg mass laid by the partner female on the emergent vegetation or branch above the water surface (Ichikawa 1988; Smith and Larsen 1993; Macías-Ordóñez 2003). Males supply eggs with water until hatching. As in Belostomatinae, submerged eggs cannot hatch (Ichikawa 1988). In addition to watering, there are two roles of the brooding behavior of *K. deyrolli* males: defense

**Fig 12.4** (continued) Such graph illustrates an isometric pattern (test of isometry, $P = 0.0736$, see Warton et al. 2006). Sexual size dimorphism (SSD) in total body length was calculated as $\log_{10}$ (F/M + 0.5). (**b**) Independent contrasts of $\log_{10}$ (female size) versus $\log_{10}$ (male size). The tree topology of Ribeiro et al. (2018) and all available Belostomatidae genera were used. Outgroups (Nepidae) are in bold and were excluded. Such graph illustrates apparently a hypoallometric pattern, which is consistent with Rensch's rule (test of isometry, $P < 0.0001$). Numbers represent the nodes in the tree against infanticidal females (Ichikawa 1990, 1991, 1995) and protection of the egg mass from predators (ants) (Ohba and Maeda 2017). Egg predation by females occurs probably due to the female-biased operational sex ratio during the reproductive season (Ichikawa 1990, 1991, 1993). The brooding males attack the infanticidal females but they may not be able to overcome them because females are larger than males. By destroying the eggs of her competitor, a female can obtain the mating partner and ensure male care of her eggs without having to search for a non-brooding mate within the limited reproductive season (Ichikawa 1990). Egg-caring males stay on the egg mass during night probably protecting the clutch against females (Ichikawa 1995), and shorten the egg period by uniform watering of all eggs, thus regulating the hatching synchronization within an egg mass (Ohba 2002) (Fig. 12.5a). The egg-caring male can protect its eggs from ants by means of physical and chemical defense. In the chemical defense, males release a chemical substance

**Fig. 12.5** Synchronized hatching between individual eggs in an egg mass (**a**), and aggregated hatchlings (small triangles) under their egg mass and the attendant male in *Kirkaldyia deyrolli* (large triangle) (**b**)

with a grassy smell; this deters ants from attacking the eggs (Ohba and Maeda 2017). After hatching, males were found situated just below the water on the natal substrate, and the first instar nymphs were aggregated around the substrate (Fig. 12.5b). The position and posture of the males were constant during daylight. When disturbed, the attendant males displayed aggressive behavior, brandishing their forelegs toward the source of disturbance. Cannibalistic behavior in the nymphs was well developed 72 h after hatching, when the nymphs had already dispersed from the natal substrate. The suppression of sibling cannibalism in younger nymphs would promote the maintenance of tight nymphal aggregations and consequently extend male care to the first instar stage in this species (Ohba et al. 2006).

### 12.4.3 Sexual Selection in Paternal Care

Once the male's care for the eggs became indispensable for egg survival, sexual selection may have played a secondary role in the maintenance of paternal care in Belostomatidae. It is hypothesized that male care in animals has evolved under the pressure of sexual selection (Tallamy 2000, 2001; Alonzo 2012; Requena et al. 2014). In addition to offspring survival, male care behavior may enhance reproductive success through increased mating opportunities, because it provides direct benefits to mates, thereby attracting them (Tallamy 2000; Alonzo 2012). Moreover, paternal care can also provide an honest signal of the male's ability to defend offspring (Tallamy 2000, 2001). Therefore, males of species with established paternal care can advertise their parental intent to females (Hoelzer 1989; Kelly and Alonzo 2009). In addition to the harvestmen (Nazareth and Machado 2010, Requena and Machado 2015) and assassin bugs (Thomas and Manica 2005), the sexual selection hypothesis is now supported by two genera of Belostomatinae (Ohba et al. 2016, 2018), belonging to Diplonychini (*Appasus* and *Diplonychus*) (Ribeiro et al. 2018). However, there is little information on the sexual selection hypothesis in other tribes in Belostomatinae and Lethocerinae. In Lethocerinae, the females of *K. deyrolli* do not give the egg-caring males their own eggs without first destroying the eggs of the other female (i.e., by performing infanticide) (Ichikawa 1990, 1991), whereas a male of *L. americanus* was found to attend multiple egg masses (Smith and Larsen 1993). The volume of literature concerning mating systems of Belostomatidae is increasing gradually, but there is not enough information on the relationship between female choice and male care. This information will be essential to understand the evolution of paternal care in Belostomatidae, as well as in other arthropods. During courtship displays, the males of Belostomatidae perform a pumping (push-up) display at the water surface. Males that displayed for longer durations secured more copulations in *Abedus indentatus* (Haldeman) (Kraus 1989). Thus, further work is needed to establish whether push-up displays indicate male parental ability, and whether females discriminate between males based on these displays (Smiseth 2014).

## 12.5 Future Directions

Studies with Belostomatidae and the evolution of paternal care through its members have raised a variety of curious unanswered questions. (1) May push-up displays used by males that seem to attract females and a variety of sophisticated brooding behaviors found among belostomatines be indeed monitored by females and provide information about male's parental ability? (2) There is no formal test of the predictions of the Smith's hypothesis (Smith 1997) concerning the evolution of body size in Belostomatidae. In Smith's scenario, mean body sizes of emergent-brooders are predicted to be larger than those of back-brooders. Were shifts to back-brooding behavior associated with a decrease in mean body size after correcting for the phylogenetic effect? If so, has the same happened in minimum and maximum values of total body length and greatest width of body? Are the mean ancestral body sizes predicted by Smith correct? Such scenario can now be explored since new phylogenies are being produced. (3) Belostomatine males make extreme efforts that seem to increase paternity. Some *Abedus* species, for example, interrupt ovipositing females frequently to insist on copulating again. So, is there a sexually antagonistic coevolution between male and female genitalia in Belostomatidae? It is possible that the presence of spurs and spines usually seen at the apexes of the second valvulae of Belostomatine females might evolve in response to such male insistence. Likewise, the parameres in the male genitalia of all Belostomatine species seem to be conveniently disassociated from the intromittent phallosoma. The hooklike shape of parameres could facilitate a good interaction with the apex of genital chamber's third valvulae, so that to force females copulate to males.

## 12.6 Conclusion

We have summarized here important advances in the understanding of paternal care in Belostomatidae. A recent phylogeny has suggested the inclusion of *H. pelocoroides* within the subfamily Belostomatinae instead of comprising the monobasic subfamily Horvathiniinae related to the subfamily Belostomatinae. So, with Belostomatidae subdivided into two subfamilies, a phylogenetic perspective was given associated with a behavioral approach to discuss the evolution of paternal care. After controlling the phylogenetic signal, a female-biased size dimorphism is supposed to be found earlier than that in emergent-brooders since this condition is the ancestral state in Belostomatidae. So, it has been suggested that the ancestor of all Belostomatidae was most likely to be a non-brooder. After controlling for the body size of males and females, we now can identify that the intensification of brooding behavior mediated by males is a possible target of selection, because it seems to be associated with evolutionary decrease in the sexual size dimorphism (SSD) in total body length, whatever the alternative phylogenetic scenarios have been used. Still, the infanticide phenomenon does not necessarily explain the origin of SSD biased toward females in Belostomatidae. Rather, it seems to be based on the tendency of related Nepidae species to be similar because they share a common history.

# References


Alcock J (2013) Animal behavior: an evolutionary approach, 10th edn. Sinauer Associates, Sunderland

Alonzo SH (2012) Sexual selection favours male parental care, when females can choose. Proc R Soc B Biol Sci 279:1784–1790

Andersson M (1994) Sexual selection. Princeton University Press, Princeton

Arnqvist G, Rowe L (2005) Sexual conflict. Princeton University Press, Princeton

Blanckenhorn WU (2009) Case studies of the differential-equilibrium hypothesis of sexual size dimorphism in two dung fly species. In: Fairbairn DJ, Blanckenhorn WU, Szekely T (eds) Sex, size and gender roles: evolutionary studies of sexual size dimorphism. Oxford University Press, Oxford, pp 106–114

Cloarec A (1992) The influence of feeding on predatory tactics in a water bug. Physiol Entomol 17:25–32

Clutton-Brock TH (1991) The evolution of parental care. Princeton University Press, Princeton

Crowl TA, Alexander JE (1989) Parental care and foraging ability in male water bugs (*Belostoma flumineum*). Can J Zool 67:513–515

Cullen M (1969) The biology of giant water bugs (Hemiptera: Belostomatidae) in Trinidad. Proc R Entomol Soc Lond A 44:123–136

Costa JT (2006) The other insect societies. The Belknap Press of Harvard University Press, Cambridge

Dale J, Dunn PO, Figuerola J, Lislevand T, Székely T, Whittingham LA (2007) Sexual selection explains Rensch's rule of allometry for sexual size dimorphism. Proc R Soc B 274:2971–2979

Davis CM, Roth VL (2008) The evolution of sexual size dimorphism in cottontail rabbits (Sylvilagus, Leporidae). Biol J Linn Soc 95:141–156

De Carlo JA (1966) Un nuevo género, nuevas especies y referencias de otras poco conocidas de la familia Belostomatidae (Hemiptera). Rev Soc Entomol Argent 28:97–109

Dupuis C (1955) Les génitalia des hémiptères hétéroptères (génitalia externes des deux sexes; voies ectodermiques femelles). Revue de la morphologie. Lexique de la nomenclature. Index bibliographique analytique. Mém Mus Natl Hist Nat A Zool 6:183–278

Estévez AL, Ribeiro JRI (2011) *Weberiella* De Carlo, 1966 (Insecta: Heteroptera: Belostomatidae) revisited: redescription with a key to the genera of Belostomatidae and considerations on back-brooding behaviour. Zoologischer Anzeiger 250:46–54

Fairbairn DJ (1997) Allometry for sexual size dimorphism: pattern and process in the coevolution of body size in males and females. Annu Rev Ecol Syst 28:659–687

Fairbairn DJ, Preziosi RF (1994) Sexual selection and the evolution of allometry for sexual size dimorphism in the water strider, *Aquarius remigis*. Am Nat 144:101–118

Felsenstein J (2012) A comparative method for both discrete and continuous characters using the threshold model. Am Nat 179:145–156

Fernández M, Bock C, Pörtner HO (2000) The cost of being a caring mother: the ignored factor in the reproduction of marine invertebrates. Ecol Lett 3:487–494

Fittkau EJ (1977) Kinal and kinon, habitat and coenosis of the surface drift as seen in Amazonian running waters. Geo-Eco-Trop 1:9–20

Gilg MR, Kruse KC (2003) Reproduction decreases life span in the giant water bug (*Belostoma flumineum*). Am Midl Nat 149:306–319

Hebsgaard MB, Andersen NM, Damgaard J (2004) Phylogeny of the true water bugs (Nepomorpha: Hemiptera-Heteroptera) based on 16S and 28S rDNA and morphology. Syst Entomol 29:488–508

Hoelzer G (1989) The good parent process of sexual selection. Anim Behav 38:1067–1078

Hua J, Li M, Dong P, Cui Y, Xie Q, Bu W (2009) Phylogenetic analysis of the true water bugs (Insecta: Hemiptera: Heteroptera: Nepomorpha): evidence from mitochondrial genomes. BMC Evol Biol 9:134–145

Hungerford HB (1919) The biology and zoology of aquatic and semi-aquatic Hemiptera. Kansas Univ Sci Bull 11:1–341
Ichikawa N (1988) Male brooding behaviour of the giant water bug *Lethocerus deyrollei* Vuillefroy (Hemiptera: Belostomatidae). J Ethol 6:121–127
Ichikawa N (1989) Breeding strategy of the male brooding water bug, *Diplonychus major* esaki (Heteroptera: Belostomatidae): is male back space limiting? J Ethol 7:133–140
Ichikawa N (1990) Egg mass destroying behaviour of the female giant water bug *Lethocerus deyrollei* Vuillefroy (Heteroptera: Belostomatidae). J Ethol 8:5–11
Ichikawa N (1991) Egg mass destroying and guarding behaviour of the giant water bug, *Lethocerus deyrollei* Vuillefroy (Heteroptera: Belostomatidae). J Ethol 9:25–29
Ichikawa N (1993) Biased operational sex ratio causes the female giant water bug *Lethocerus deyrollei* to destroy egg masses. J Ethol 11:151–152
Ichikawa N (1995) Male counterstrategy against infanticide of the female giant water bug *Lethocerus deyrollei* (Hemiptera: Belostomatidae). J Insect Behav 8:181–188
Iglesias MS, Valverde AC, Sol-Gaspe M, Crespo FA (2010) Occurrence of sexual size dimorphism in *Belostoma elegans* (Heteroptera: Belostomatidae). Entomol News 121:31–34
Iglesias MS, Crespo FA, Valverde ADC (2012) Is parental care behavior in *Belostoma* species an evolutionary cause for their common sexual size dimorphism pattern? Entomol Sci 15:155–161
Inada K, Kitade O, Morino H (2011) Paternity analysis in a egg-carrying aquatic insect *Appasus major* (Hemiptera: Belostomatidae) using microsatellite DNA markers. Entomol Sci 14:43–48
Kelly NB, Alonzo SH (2009) Will male advertisement be a reliable indicator of paternal care, if offspring survival depends on male care? Proc R Soc B Biol Sci 276:3175–3183
Kight SL, Batino M, Zhang Z (2000) Temperature-dependent parental investment in the giant water bug *Belostoma flumineum* (Heteroptera: Belostomatidae). Ann Entomol Soc Am 93:340–342
Kight SL, Kruse KC (1992) Factors affecting the allocation of paternal care in water bugs (*Belostoma flumineum* Say). Behav Ecol Sociobiol 30:409–414
Kight SL, Sprague J, Kruse KC, Johnson L (1995) Are egg-bearing male water bugs, *Belostoma flumineum* Say (Hemiptera: Belostomatidae), impaired swimmers? J Kansas Entomol Soc 68:468–470
Kight SL, Tanner AW, Coffey GL (2011) Termination of parental care in male giant waterbugs, *Belostoma flumineum* Say (Heteroptera: Belostomatidae) is associated with breeding season, egg pad size, and presence of females. Invertebr Reprod Dev 55:197–204
Kraus W (1989) Surface wave communication during courtship in the giant water bug, *Abedus indentatus* (Heteroptera: Belostomatidae). J Kansas Entomol Soc 62:316–328
Lauck DR, Menke AS (1961) The higher classification of the Belostomatidae (Hemiptera). Ann Entomol Soc Am 54:644–657
Lee C (1991) Morphological and phylogenetic studies on the true water bugs. Nat Life 21:1–183
Macías-Ordóñez R (2003) On the reproductive behavior and population ecology of *Lethocerus colossicus* Stal (Heteroptera: Belostomatidae). Folia Entomol Mex 42:161–168
Madhavan MM (1974) Structure and function of the hydropyle of the egg of the bug, Sphaerodema molestum. J Insect Physiol 20:1341–1349
Mahner M (1993) Systema Cryptoceratorum Phylogeneticum (Insecta, Heteroptera). Zoologica 48:1–302
Menke AS (1965) A new South American toe biter (Hemiptera, Belostomatidae). Contrib Sci 89:2–4
Menke AS (1979) Family Belostomatidae. In: Menke AS (ed) The semiaquatic and aquatic Hemiptera of California (Heteroptera: Hemiptera), Bulletin of the California Insect Survey. University of California Press, Berkeley, pp 76–86
Morand S (1997) Comparative analyses of continuous data: the need to be phylogenetically correct. In: Grandcolas P (ed) The origin of biodiversity in insects: phylogenetic tests of evolutionary scenarios, Mémoires du Muséum national d'Histoire naturelle, pp 73–90
Moreira FFF, Barbosa JF, Ribeiro JRI, Alecrim VP (2011) Checklist and distribution of semiaquatic and aquatic Heteroptera (Gerromorpha and Nepomorpha) occurring in Brazil. Zootaxa 2958:1–74

Mori A, Ohba S (2004) Field observations of predation on snakes by the giant water bug. Bull Herpetol Soc Jpn 2004:78–81. (in Japanese)
Mukai Y, Baba N, Ishii M (2005) The water system of traditional rice paddies as an important habitat of the giant water bug, *Lethocerus deyrollei* (Heteroptera: Belostomatidae). J Insect Conserv 9:121–129
Munguía-Steyer R, Favila ME, Macias-Ordonez R (2008) Brood pumping modulation and the benefits of paternal care in *Abedus breviceps* (Hemiptera: Belostomatidae). Ethology 114:693–700
Munguía-Steyer R, Macias-Ordonez R (2007) Is it risky to be a father? Survival assessment depending on sex and parental status in the water bug *Abedus breviceps* using multistate modelling. Can J Zool 85:49–55
Nazareth TM, Machado G (2010) Mating system and exclusive postzygotic paternal care in a Neotropical harvestman (Arachnida: Opiliones). Anim Behav 79:547–554
Nentwig W, Wissel C (1986) A comparison of prey lengths among spiders. Oecologia 68:595–600
Ohba S (2002) Synchronized mechanism and its meaning in the egg hatching of the giant water bug, *Lethocerus deyrolli* (Heteroptera: Belostomatidae). Jpn J Entomol 5:157–164. (in Japanese with English Abstract)
Ohba S (2011) Field observation of predation on a turtle by a giant water bug. Entomol Sci 14:364–365
Ohba S (2012) Field observation of predation on a Japanese mamushi, *Gloydius blomhoffii*, by a giant water bug, *Kirkaldyia deyrolli*. Jpn J Entomol 15:92–93. (in Japanese with English Abstract)
Ohba S, Hidaka K, Sasaki M (2006) Notes on paternal care and sibling cannibalism in the giant water bug, *Lethocerus deyrolli* (Heteroptera: Belostomatidae). Entomol Sci 9:1–5
Ohba S, Maeda A (2017) Paternal care behaviour of the giant water bug *Kirkaldyia deyrolli* (Heteroptera: Belostomatidae) against ants. Ecol Entomol 42:402–410
Ohba S, Matsuo S, Trang Huynh TT, Kudo S (2018) Female mate preference for egg-caring males in the giant water bug *Diplonychus rusticus* (Heteroptera Belostomatidae). Ethol Ecol Evol 30:477–484
Ohba S, Okuda N, Kudo S (2016) Sexual selection of male parental care in giant water bugs. R Soc Open Sci 3:150720
Ohba S, Tatsuta H (2016) Young giant water bug nymphs prefer larger prey: changes in foraging behaviour with nymphal growth in *Kirkaldyia deyrolli*. Biol J Linn Soc 117:601–606
Ohba S, Tatsuta H, Nakasuji F (2008) Variation in the geometry of foreleg claws in sympatric giant water bug species: an adaptive trait for catching prey? Entomol Exp Appl 129:223–227
Parker GA (1970) Sperm competition and its evolutionary consequences in the insects. Biol Rev 45:525–567
Perez-Goodwyn PJ (2006) Taxonomic revision of the subfamily Lethocerinae Lauck & Menke (Heteroptera: Belostomatidae). Stuttgarter Beitr zur Naturkd Ser A 695:1–71
Perez-Goodwyn PJ (2001) Size selective predation by *Belostoma oxyurum* (Heteroptera: Belostomatidae) on *Aedes aegypti* (Diptera: Culicidae) larvae. Rev Soc Entomol Argent 60:139–146
Polhemus JT (1995) Nomenclatural and synonymical notes on the genera *Diplonychus* Laporte and *Appasus* Amyot and Serville (Heteroptera: Belostomatidae). Proc Entomol Soc Wash 97:649–653
Rensch B (1959) Evolution above the species level. Columbia University Press, New York
Revell LJ (2014) Ancestral character estimation under the threshold model from quantitative genetics. Evolution 68:743–759
Requena GS, Machado G (2015) Effects of egg attendance on male mating success in a harvestman with exclusive paternal care. Behav Ecol 26:926–935
Requena GS, Munguía-Steyer R, Machado G (2014) Chapter 8: Paternal care and sexual selection in Arthropods. In: Macedo RH, Machado G (eds) Sexual selection. Perspectives and models from the Neotropics. Elsevier, Oxford, pp 201–233

Ribeiro JRI, Meyin-a-Ebong SE, Le-Gall P, Guilbert E (2014) A taxonomic synopsis of *Limnogeton* Mayr, 1853 (Insecta: Hemiptera: Heteroptera: Belostomatidae). Zootaxa 3779:573–584

Ribeiro JRI, Moreira FFF, Alecrim VP, Barbosa JF, Nessimian JL (2009) Espécies de heterópteros dulciaquícolas (Hemiptera, Heteroptera, Gerromorpha e Nepomorpha) registradas no Estado do Rio de Janeiro, Brasil. Arq Mus Nac Rio de Janeiro 67:303–312

Ribeiro JRI, Ohba S, Pluot-Sigwalt D, Stefanello F, Bu W, Meyin-A-Ebong SE, Guilbert E (2018) Phylogenetic analysis and revision of subfamily classification of Belostomatidae genera (Insecta: Heteroptera: Nepomorpha). Zool J Linnean Soc 182:319–359

Royle NJ, Smiseth PT, Kölliker M (2012) The evolution of parental care. Oxford University Press, Oxford

Schnack JA, Estévez AL (2005) On the taxonomic status of the genus *Horvathinia* Montandon (Hemiptera: Belostomatidae). Zootaxa 1016:21–27

Schumann DA, Cavallaro MC, Hoback WW (2012) Size selective predation of fish by *Hydrophilis triangularis* (Coleoptera: Hydrophilidae) and *Lethocerus americanus* (Hemiptera: Belostomatidae). J Kansas Entomol Soc 85:155–159

Shuster SM, Wade MJ (2003) Mating systems and strategies. Princeton University Press, Princeton

Simmons LW (2001) Sperm competition and its evolutionary consequences in the insects. Princeton University Press, Oxford

Smiseth PT (2014) Parental care. In: Shuster DM, Simmons LW (eds) The evolution of insect mating systems. Oxford University Press, Oxford, pp 221–241

Smith RL (1976a) Brooding behavior of a male water bug *Belostoma flumineum*. J Kansas Entomol Soc 49:333–343

Smith RL (1976b) Male brooding behavior of the water bug *Abedus herberti*. Ann Entomol Soc Am 69:740–747

Smith RL (1979a) Paternity assurance and altered roles in the mating behaviour of a giant water bug, *Abedus herberti* (Heteroptera: Belostomatidae). Anim Behav 27:716–725

Smith RL (1979b) Repeated copulation and sperm precedence: paternity assurance for a male brooding water bug. Science 205:1029

Smith RL (1997) Evolution of paternal care in the giant water bugs (Heteroptera: Belostomatidae). In: Choe J, Crespi B (eds) The evolution of social behavior in insects and arachnids. Cambridge University Press, Cambridge, pp 116–149

Smith RL, Larsen E (1993) Egg attendance and brooding by males of the giant water bug *Lethocerus medius* (Guerin) in the field (Heteroptera: Belostomatidae). J Insect Behav 6:93–106

Spooner D (1938) The phylogeny of the hemiptera based on a study of the head capsule. Ill Biol Monogr 16:1–102

Swart CC, Deaton LE, Felgenhauer BE (2006) The salivary gland and salivary enzymes of the giant waterbugs (Heteroptera; Belostomatidae). Comp Biochem Physiol A Mol Integr Physiol 145:114–122

Tallamy DW (2000) Sexual selection and the evolution of exclusive paternal care in arthropods. Anim Behav 60:559–567

Tallamy DW (2001) Evolution of exclusive paternal care in arthropods. Annu Rev Entomol 46:139–165

Thomas L, Manica A (2005) Intrasexual competition and mate choice in assassin bugs with uniparental male and female care. Anim Behav 69:275–281

Thrasher P, Reyes E, Klug H (2015) Parental care and mate choice in the giant water bug *Belostoma lutarium*. Ethology 121:1018–1029

Venkatesan P, Rao TR (1980) Water loss by eggs of *Diplonychus* sp. (Hemiptera: Belostomatidae). J Kansas Entomol Soc 53:587–594

Warren P, Lawton J (1987) Invertebrate predator-prey body size relationships: an explanation for upper triangular food webs and patterns in food web structure? Oecologia 74:231–235

Warton DI, Wright IJ, Falster DS, Westoby M (2006) Bivariate line-fitting methods for allometry. Biol Rev 81:259–291

Wilson CA (1958) Aquatic and semiaquatic Hemiptera of Mississippi. Tulane Stud Zool 6:116–170

# Chapter 13
# The Aquatic Lepidopterans: A Mysterious and Unknown Fauna

**Iasmim De-Freitas, Julia De Agostini, and Vanessa Stefani**

**Abstract** The order Lepidoptera is represented by the popular butterflies and moths, and is identified by the presence of scales on the wings with an extraordinary variety of colors and sizes. Although most Lepidoptera are exclusively terrestrial there are some lineages associated with the aquatic environment. Little is known about aspects of the biology and life history of its representatives leading to a gap in scientific knowledge. Thus, we can find aquatic Lepidoptera that have tracheal breathing that can remain outside the water for a limited time, being considered semiaquatic larvae, while larvae with branchial respiration are aquatic while retaining all their submerged larval development. In this chapter, possible evolutionary paths are explored which aquatic lepidopteran larvae had to live in aquatic environments and we examine the incipient studies that address life aspects of these extraordinary animals.

**Keywords** Moths · Butterflies · Parapoynx · Petrophila · Crambidae · Caterpillar

When we talk about "the unknown tropical fauna," we often mean the animals of which little is known about their role in their habitat. This is considered true for some specimens of Lepidoptera, which spend a part of their lives on aquatic ecosystems.

Lepidoptera is one of the best known groups of insect, due to its exuberance, diversity, and economic importance. The lepidopterans are commonly known as moths and butterflies and can be mostly identified by the presence of scales on their wings and for their wide variety of colors and shape patterns. It is possible to find many studies about these insects in the literature, but most of them are focused on agriculture for their potential as herbivores in plantations or on taxonomic surveys of these holometabolous insects' adult forms. However, little is known about the biology and life history of this order, which leads to a gap in the

---

I. De-Freitas · J. De Agostini · V. Stefani (✉)
Laboratório de História Natural e Reprodutiva de Artrópodes (LHINRA), Instituto de Biologia, Universidade Federal de Uberlândia, Uberlândia, MG, Brazil



K. Del-Claro, R. Guillermo (eds.), *Aquatic Insects*,
https://doi.org/10.1007/978-3-030-16327-3_13

scientific knowledge that mostly affects the species with part of their development in aquatic environments.

Currently, Lepidoptera has about 157,578 described species in the world (Zhang 2011) out of which approximately 740 species (representing 4.7% of the total) are believed to go through an aquatic larval stage (Mey and Speidel 2008). Nevertheless, the aquatic lepidopterans are relevant in aquatic ecosystems and can represent up to 32% of the macroinvertebrate community in a lotic system (Luiza-Andrade et al. 2017).

## 13.1 Evolution and Adaptation to the Aquatic Life

With the arising and radiation of the angiosperms in the Cretaceous Period, a new food resource became widely available promoting coevolutionary process and adaptive radiations in insect herbivorous including lepidopterans (Mey and Speidel 2008). During this coevolutive events, these insects were capable of exploiting all the parts of a plant and have possibly harnessed the resources of other ecosystems such as aquatic and semiaquatic environments. It is important to highlight that the transition from the aquatic to the semiaquatic life has independently occurred multiple times. Still, it has taken several adaptations to enable these forms of life to explore this new habitat, emphasizing changes in physiology and locomotion (Welch 1919).

As for terrestrial specimens, the respiratory system of the Lepidoptera larvae is formed by spiracles connecting with tracheas and a branched network of tubes that allows the air to flow throughout the body. However, in aquatic ecosystems, besides the tracheal breathing, it is also possible to find some individuals with bronchial breathing. The ones with a respiratory tracheal system are called semiaquatic and can survive out of the water for a certain period. These specimens have hydrophobic bristles throughout the body that cover the entire dorsal surface of the animal that when submerged allow the seizure of air, which is also stored around the respiratory spiracles that allows the animal to perform gas exchange. These characteristics can be observed in detail on the caterpillar of *Paracles klagesi* Rothschild 1910 (Erebidae: Arctiinae) as it has two kinds of bristles: one is thorn shaped (filiform) lying around the respiratory spiracles laterally along its body, and the other is drop shaped (pectiform), and it is located on the dorsal surface; both are hydrophobic (Fig. 13.1).

To feed, the caterpillar submerges. However, it must get back to the water surface periodically. This activity promotes the renewal of the air that is stocked between the tracheal spiracles at the base of the hydrophobic bristles. These spiracles open up in a cavity that has long chitin filaments until the lumen, directly connected to the connection trunk which leads to the trachea (Fig. 13.1). The number, anatomy, and histology of the spiracles vary among the species.

The aquatic species usually have filamentous gills (Fig. 13.2) that can show up at different larval instars varying with the species. Each gill filament has a cavity that connects directly to the hemocele. At the last instars, the gills become more ramified, and each filament will have a main trachea. At the axial position, the gill

**Fig. 13.1** Scheme of the tracheal respiratory system of *Paracles klagesi* based on Welch 1922. (**a**) The submerged larva—the arrow indicates the air bubbles between the bristles; (**b**) detail of the body segments showing the bristles; (**c**) orange drop-shaped bristles; (**d**) filiform bristles on the tracheal spiracle; (**e**) anatomic detail of the filiform bristle; (**f**) tracheal spiracles; (**g**) internal scheme of the tracheal bristle (photographs: **a**–**f** Iasmim De-Freitas)

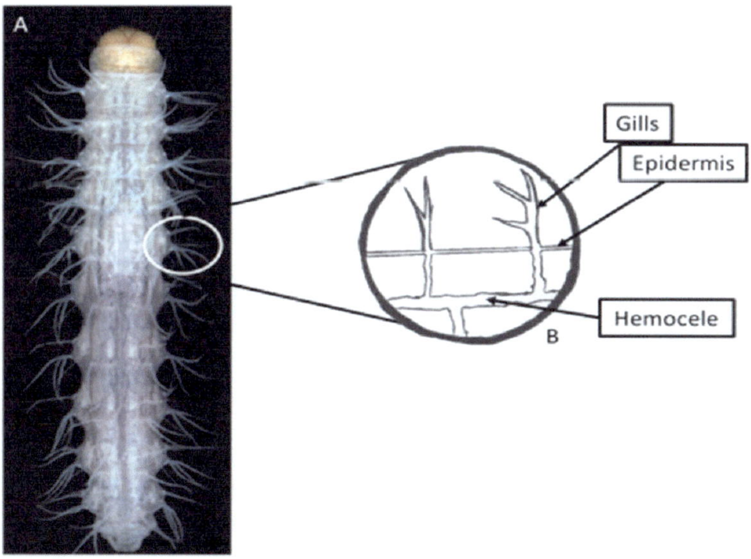


**Fig. 13.2** Scheme of the *Usingeriessa onyxalis* branchial respiration based on Welch 1922. (**a**) *Usingeriessa onyxalis* caterpillar in dorsal view; (**b**) internal scheme of a gill (photograph of the aquatic larva: courtesy of M. Alma Solis, Nathan E. Harms, Eugenie Phillips-Rodríguez, Sonja J. Scheffer, Matthew L. Lewis, Daniel H. Janzen, Winnie Hallwachs, and Mark A. Metz; 2018. Aquatic Larvae of Two Acentropines, *Usingeriessa onyxalis* (Hampson) and *Oxyelophila callista* (Forbes) (Lepidoptera: Crambidae), Entomological Society of Washington, 120(1):180–195)

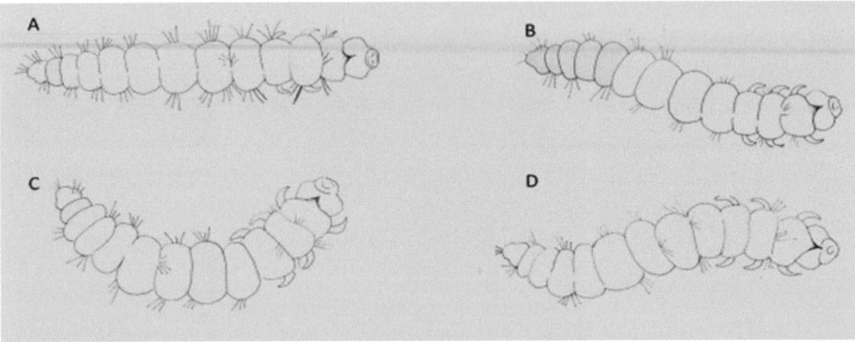


**Fig. 13.3** Locomotion scheme. (**a**) Inert caterpillar; (**b**) the beginning of the bending; (**c**) the bending; (**d**) whip effect

filaments, along with the trachea, extend together in fine branches and form the tracheoles (Fig. 13.2).

The locomotion in terrestrial species can be typified mainly as "inchworm," while the aquatic ones developed a mechanism characterized by horizontal movements along the water column, like a whip in the posterior body parts where, initially, the anterior body part is flexed to one side horizontally and with a fast movement that provides the propulsion required to swim (Fig. 13.3).

## 13.2 Biology and Life History

Erebidae, Nepticulidae, Coleophoridae, Tortricidae, Pyralidae, Cosmopterigidae, Crambidae, and Noctuidae (Hamada et al. 2014; Mey and Speidel 2008; Mccafferty and Minno 1979) are the main families of Lepidoptera that are possible to find in literature with aquatic representatives. However, even though it is known that these species have aquatic larvae, there are not enough studies about their life histories and biological aspects. Having that in mind, some known neotropical exceptions will be presented in the following paragraphs.

The Arctiinae (Erebidae) subfamily has more than 3600 species; the *Paracles* genus has two aquatic representatives: the *Paracles klagesi* Rothschil 1910 and the *Paracles laboulbeni* Bar 1873 (Meneses et al. 2013; Mey and Speidel 2008). The *P. klagesi* aquatic larvae have tracheal breathing and were found in two distinct regions of aquatic environments in Brazilian territory—in lakes near the city of Timon (Maranhão State, 43°01′52.2″W, 05°03′03.3″S) and at "Vereda" lakes of the Neotropical Savanna near the city of Uberlândia (Minas Gerais State, 48°14′02″W, 18°54′52″S). Meneses et al. (2013) observed that the caterpillar eats two types of aquatic macrophytes: *Eichhornia* sp. (Pontederiaceae) and *Cabomba caroliniana* (Cabombaceae). The pupal stage consists of a fiber cocoon which allows buoyancy and is always adhered to a plant (Fig. 13.4). The *P. laboulbeni* caterpillar has tracheal

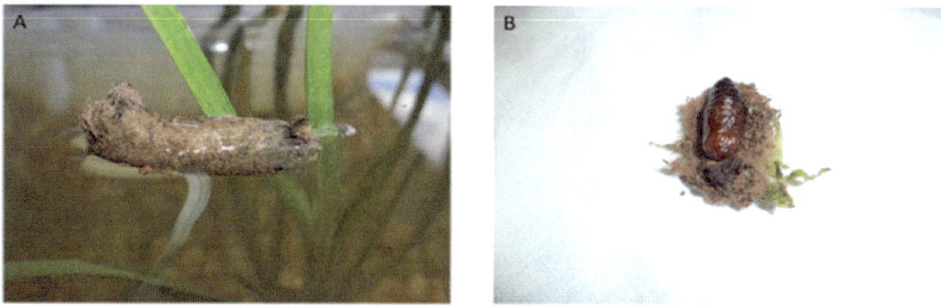


**Fig. 13.4** (**a**) The *Paracles klagesi* pupa in its natural environment, with the cocoon made of hydrophobic fibers that allow buoyancy; (**b**) the fiber wrap was taken off so the cocoon cloud be visualized (photographs: **a** and **b** Iasmim De-Freitas)

breathing and eats macrophytes like *Echinochloa spectabilis*, *Hymenachne amplexicaulis*, *Oryza grandiglunis*, and algae of the *Oedogonium* genus (Adis 1983). The larval development varies from 30 to 35 days and the period of the greatest abundance was recorded from May to June, which precedes the rise in water level. However, its abundance is reduced when the water level is low (Adis 1983).

The Crambidae family has about 600 described species (Hamada et al. 2014). Among the species that have an aquatic larval stage, there is the *Niphograpta albiguttalis* Warren 1889, distributed between Argentina and Brazil. Its total development, since the egg to the adult form, lasts about 35 days; the adult females oviposit on the leaves of *Eichhornia crassipes* (Pontederiaceae), primarily in areas where the epidermis has been damaged, which will feed during the five larval instars (Julien 2001). However, despite the *N. albiguttalis* having been considered a biological control agent against the water hyacinth, currently it is considered a plague of *E. crassipes*, because when the oldest larvae eat the leaves, the plants become waterlogged and sink (Canavan et al. 2014).

Other species that have aquatic larvae are *Donacaula forficella* Thunberg 1794 (found in China, Europe, and South Africa); *Nymphula nitidulata* Hufnagel 1767 (found in Europe, Japan, Turkey, Armenia, Russia, and China); *Schoenobius gigantella* Denis and Schiffermüller 1775 (found in Europe and China); *Cataclysta lemnata* Linnaeus 1758 (found in Europe, Morocco, and Iran); and *Elophila nymphaeata* Linnaeus 1758 (found in Europe) (Vallenduuk and Cuppen 2004); however, little is known about their life histories and biology. It is important to highlight that the geographic extension of these species can be higher than it was described.

The eccentric species *Acentria ephemerella* Denis and Schiffermüller 1775 (Crambidae) draws attention for having females with paedomorphosis (the retention of the juvenile or even larval traits into later life), which means that even though they are adults they still live in the aquatic environments and do not develop wings. The most interesting aspect about the *A. ephemerella* happens during its mating because the females need to swim to the surface and put the abdomen above the water surface so that the male can copulate (Vallenduuk and Cuppen 2004). However, this species can only be found in Central Europe. No other species of aquatic lepidoptera with paedomorphosis has been found, even in tropical regions.

The Pyralidae family has about 5000 described species (Hamada et al. 2014). The aquatic larva of the *Parapoynx rugosalis* Moschler 1890 has gills and a study carried out in Costa Rica found that *P. rugosalis* uses a tent made of leaves that protects it from predators; as its development advances this tent is changed (Mueller and Dearing 1994). Another member of the Pyralidae family is the *Samea multiplicalis* Guenee 1854 (Knopf and Habeck 1976) that also has a wide distribution that extends from the Southwest of the United States to Argentina. The average development time for each stage of life of *S. multiplicalis* is 4 days to the eggs; 15.6 days to the larvae; and 5.2 days to the pupa; also, the number of eggs laid by a female in its natural environment was 94 eggs.

Among the most interesting aspects found among the representatives of the Pyralidae family, there is the Californian species *Parargyractis confusali* Walker, which has gills and the females of which go into the water to oviposit between 220 and 296 eggs on submerged rocks. The *P. confusali* larva builds silk chambers in depressions or cracks in the rocks and feeds on resources around them. When the cocoon is packed, it has an outer casing, which has openings for the water circulation (Tuskes 1977).

It is clear that studies on aquatic Lepidoptera are very scarce, being the great majority realized for species in temperate regions. Even though the studies about the biology and natural history of these specimens are almost unknown for tropical environments, it is important to emphasize morphological works (Solis et al. 2018), molecular studies (Park et al. 2014), and studies of biological controls (Hickel et al. 2017). These aspects mentioned above could be the gateway to new research with a focus on aquatic lepidoptera, not only for scientific knowledge but also for the ecological importance they exert as bioindicators (Stoops et al. 1998) and as herbivores (Habeck 1974), and the economic importance as food resources (Ramos-Elorduy 2006; Williams and Williams 2017) and agricultural pests (Mey and Speidel 2008; McGaha 1954). It is known that tropical aquatic environments are constantly polluted by anthropic interference, such as the discharge of urban sewage, the flow of agricultural nutrients, and the waste of confined animals that alter these sensitive environments and that compromise the survival of species that we will not even know.

**Acknowledgments** We wish to acknowledge the help provided by an anonymous reviewer who made valuable improvements in our research text, to Lanna Issa dos Santos Barcelos for making the beautiful schemes, and finally to CNPq (Iasmim De-Freitas) and FAPEMIG (Júlia De Agostini) for financial support.

## References

Adis J (1983) Eco-entomological observations from the Amazon. Occurrence and feeding habits of the aquatic caterpillar *Palustra laboulbeni* Bar, 1873 (Arctiidae: Lepidoptera) in the vicinity of Manaus, Brazil. Acta Amazon 13(1):31–36

Canavan K, Coetzee JA, Hill MP, Paterson ID (2014) Effect of water trophic level on the impact of the water hyacinth moth *Niphograpta albiguttalis* on *Eichhornia crassipes*. Afr J Aquat Sci 399(2):203–208

Habeck D H (1974) Parapoynx Caterpillars In Relation With Aquatic In Florida. Department of Entomology and Nematology 32611. p 15–18

Hamada N, Nessimian JL, Querino RB (2014) Insetos aquáticos na Amazônia brasileira: taxonomia, biologia e ecologia. Editora do INPA, Manaus, p 724

Hickel E, Prando FH, Domingos SE (2017) Lagartas nas lavouras catarinenses de arroz irrigado: Ocorrência, monitoramento e manejo integrado. Epagri, Itajaí, p 182

Julien MH (2001) Biological control of water hyacinth with arthropods: a review to 2000. ACIAR Proc 102:8–20

Knopf KW, Habeck DH (1976) Life history and biology of *Samea multiplicalis*. Environ Entomol 5(3):539–542

Luiza-andrade A, Montag ALF, Juen L (2017) Functional diversity in studies of aquatic macroinvertebrates community. Scientometrics 111(3):1643–1656

Mccafferty WP, Minno MC (1979) The aquatic and semi-aquatic Lepidoptera of Indiana and adjacent areas. Entomol Great Lakes 12:179–187

Mcgaha YJ (1954) Contribution to the biology of some Lepidoptera that feed on certain plants of aquatic flowers. Trans Am Microsc Soc 73(2):167–177

Meneses AR, Bevilaqua MVO, Hamada N, Querino RB (2013) The aquatic habit and host plants of Paracles klagesi (Rothschild) (Lepidoptera, Erebidae, Arctiinae) in Brazil. Rev Bras Entomol 57(3):350–352

Mey W, Speidel W (2008) Global diversity of butterflies (Lepidoptera) in freshwater. Hydrobiologia 595(1):521–528

Mueller UG, Dearing MD (1994) Predation and avoidance of tough leaves by aquatic larvae of the moth *Parapoynx rugosalis* (Lepidoptera: Pyralidae). Ecol Entomol 19(2):155–158

Park JS, Kim MJ, Kim SS, Kim I (2014) Complete mitochondrial genome of an aquatic moth, Elophila interruptalis (Lepidoptera: Crambidae). Mitochondrial DNA 25:275–277

Ramos-Elorduy J (2006) Threatened edible insects in Hidalgo, Mexico and some measures to preserve them. J Ethnobiol Ethnomed 2(1):51

Solis MA, Harms NE, Phillips-Rodríguez E, Scheffer SJ, Lewis ML, Janzen DH, Hallwachs W, Metz MA (2018) Aquatic larvae of two Acentropines, *Usingeriessa onyxalis* (Hampson) and *Oxyelophila callista* (Forbes) (Lepidoptera: Crambidae). Ann Washington Entomol Soc 120(1):180–195

Stoops CA, Adler PH, Mccreadie JW (1998) Ecology of aquatic Lepidoptera (Crambidae: Nymphulinae) in South Carolina, USA. Hydrobiologia 379:33–40

Tuskes PM (1977) Observations on the biology of *Parargyractis confusalis*, an aquatic pyralid (Lepidoptera: Pyralidae). Can Entomol 109(5):695–699

Vallenduuk HJ, Cuppen HMJ (2004) Aquatic live caterpillars (Lepidoptera: Pyraloidea: Crambidae) from Central Europe. A key to larvae and autecology. Lauterbornia 49:1–17

Welch PS (1919) The aquatic adaptations of *Pyrausta penitalis* Grt. (Lepidoptera). Ann Entomol Soc Am 12(3):213–226

Welch PS (1922) The respiratory mechanism in certain aquatic Lepidoptera. Trans Am Microsc Soc 41(1):29–50

Williams DD, Williams SS (2017) Aquatic insects of Siân S. and its potential to contribute to the diet of the human population in global expansion. Insects 8(3):72

Zhang ZQ (2011) Animal biodiversity: an introduction to higher-level classification and taxonomic richness. Zootaxa 3148:212

# Chapter 14
# Urban Aquatic Insects

**Javier Muzón, Lía S. Ramos, and Alejandro del Palacio**

**Abstract** The increasing growth of urban areas on natural ecosystems has seriously affected wetlands. Aquatic insects, as key components of urban wetlands, are critically impacted by human environmental changes and practices. The main threats are those derived from the lost, replacement, or fragmentation of natural habitats, ecosystem homogenization, and modification of hydrological, sedimentological, and thermal wetland characteristics due to the surrounding urban matrix. Therefore, understanding the mechanisms by which urbanization processes affect biodiversity, and in particular how the biota responds to alteration of their habitats, is crucial for integrating the environment in proper urban planning. In this chapter we analyze the relationships between aquatic insects and environmental factors, including human influences and threats in urban wetlands. For this purpose, we compiled studies from around the world, especially from the neotropical region, addressing biological patterns and associated environmental processes in urban areas and endeavors. We discerned the main environmental threats and clues for the maintenance of the increase of the insect biodiversity, including the creation of new, man-made, wetlands. In addition, we focus on the importance of insect knowledge as indicators of environmental health in urban wetlands and the promotion of the citizen science to improve their conservation.

**Keywords** Urbanization · Biodiversity · Ecology · Freshwater · Colonization

## 14.1 Introduction

An urban area can be defined as a cultural landscape where people live at high densities, and extensive impervious surface areas are occupied by built structures and infrastructure, forming a dynamic patch mosaic whose structure, function, and

---

J. Muzón (✉) · L. S. Ramos · A. del Palacio
Laboratorio de Biodiversidad y Genética Ambiental (BioGeA), UNDAV, Buenos Aires, Argentina



K. Del-Claro, R. Guillermo (eds.), *Aquatic Insects*,
https://doi.org/10.1007/978-3-030-16327-3_14

dynamics are determined by coupled human-environment interactions (Pickett et al. 2011; Wu 2014). The process by which urban ecosystems are created is the urbanization (McIntyre 2000), and it has as a consequence transformed pristine natural landscapes to agricultural and suburban, and finally to urban landscapes. These impacts hinge on spatial scales from global and regional (e.g., mosquito population changes in response to global warming (Martin and Lefebvre 1995)), through landscape (e.g., electric lights, fast-moving traffic, replacement of mangroves by docks) down to highly local (e.g., water butt communities, unburied organic matter, wood stored in a warehouse). Yet the highly local is related to the regional and global; for example discarded vehicle tires encourage the spread of mosquitos and transported goods from harbor to harbor carry biota with specific habitat requirements. In addition, a worldwide increasing of infectious diseases associated with wetlands has been related, among other causes with a high human encroachment of natural environments, reductions in biodiversity, and changes in the number of vector breeding sites or in reservoir host distribution (Niemelä and MacDonnell 2011).

The Ramsar Convention's 10th meeting of the Conference of the Contracting adopted Resolution X.27 on Wetlands and urbanization (RAMSAR 2008) recognized that wetlands in urban and peri-urban areas can provide a range of important ecosystem services—benefits to people—but also that in many countries wetlands are increasingly becoming degraded because of spreading urbanization. As a complement of cities' green spaces, wetlands are also considered as blue spaces (Wu 2014). There are several papers dealing extensively with urban wetland ecology, as for example Chadwick et al. (2006), Meyer et al. (2005), Urban et al. (2006), and Walsh et al. (2005) for streams and Hassall (2014) for ponds.

Urban wetlands are ecosystems seriously threatened by the increase of the human population which, in the last decades, have been inclined to favor the growing of urban centers instead of rural areas; in addition, Latin America is one of the regions with the highest projected growth of urban population (De Sherbinin et al. 2009). This trend increases the pressure over the remaining green and blue spaces, as large parcels of land are deeply modified due to the real state pressure. At the same time the increase of the impervious surfaces which locally modify the evapotranspiration regime or increase of air and water pollution is a main driver to reduce availability, quality, and size of urban wetlands.

On the other hand, remaining wetlands are modified in order to fit human uses, practices, and preferences or new ones are made by man for different purposes, as drainage systems, garden, and ornamental and industrial ponds (Hassall 2014). The main threats are those derived from the lost or replacement of natural habitats (by filling, drainage, and excavation), ecosystem homogenization (i.e., decreasing microhabitat diversity through simplification of wetland coasts, reduction of riparian vegetation), and modification of hydrological, sedimentological, and thermal wetland characteristics due to the surrounding urban matrix and the introduction of alien species. These modifications impact negatively all levels of biodiversity and the ecosystem services they provide, implying an amelioration of life quality.

Nowadays, there are an increasing number of studies focusing on the improvement of management practices for urban wetlands and strategies for biodiversity conservation. One of the main objectives of those studies is to improve environmental quality and increase the ecosystem services they provide. Taking into account that insects represent one of the main taxa in any terrestrial or aquatic ecosystem, concurrently studies about the ecological role of the aquatic insects in these ecosystems and their value as biological indicators are ramping.

## 14.2 Aquatic Insects in Urban Wetlands

Modern cities tend to harbor different kinds of wetlands, natural, modified, and artificial, from rivers, streams, rivulets to ponds, lakes, ditches, dams, and phytotelmata (Fig. 14.1). Almost all main orders of aquatic insects can be potentially inhabitants of urban wetlands, excluding probably those extremely sensitive to environmental impacts as chemical pollution, habitat homogenization, or highly estenoic (i.e., aquatic Mecoptera). Nevertheless, as mentioned before, urbanization generates ecosystem homogenization and we could assume that a particular urban wetland exhibits less species richness than a similar one, non-anthropized, in the same region (Fig. 14.1a, b). Even though cities negatively affect natural wetlands, reducing their number, size, and heterogeneity, there is also a positive effect of human developing, the availability of new artificial wetlands which partially mitigate adverse impacts (Fig. 14.1c) (Schnack et al. 2000; Kadoya et al. 2004; Fontanarrosa et al. 2009; Holtmann et al. 2018). On the other hand, if these new artificial wetlands offer habitat for autochthonous species, we could expect an increase of more generalist or synanthropic species.

**Fig. 14.1** Urban wetland diversity. (**a**) Pergamino stream crossing Pergamino city (Buenos Aires province, Argentina), (**b**) Pergamino stream downstream the city (Buenos Aires province, Argentina), (**c**) highway ditch in Avellaneda city (Buenos Aires province, Argentina), (**d**) artificial pond, Japanese Garden (Buenos Aires city), (**e**) artificial pond in a public park (Buenos Aires city, Argentina), artificial shoreline, (**f**) artificial pond in a public park (Buenos Aires city, Argentina), artificial shoreline

**Fig. 14.1** (continued)

As aquatic insects mostly have complex amphibiotic life cycles (Fig. 14.2), urbanization impacts on aquatic and terrestrial ecosystems could affect the success of their populations. Interrelation between aquatic, semiaquatic, and terrestrial/aerial developmental stages, as for example adult emergence and maturation, mating, oviposition, and pupae survival, could be seriously affected by urbanization, leading to population losses (Smith et al. 2009). Also for those insects with a complete aquatic life cycle (e.g., several Nepomorpha as Belostomatidae, Corixidae, or Notonectidae) that eventually can fly and disperse through the terrestrial environment (Fig. 14.2a), it is necessary to consider for its conservation the structure of the surrounding terrestrial matrix.

Moreover, several management activities and/or urban infrastructure can act as ecological traps for many animals, especially for insects (Hale et al. 2015) (Fig. 14.3). Ecological traps are structures that can attract adult insects, especially those in dispersal or reproductive movements, as for example several species of Ephemeroptera which are attracted by higher polarization of reflected light of dark and shiny asphalt roads, probably because they use horizontally polarized light to identify ponds as potential oviposition sites (Smith et al. 2009; Malik et al. 2010; New 2015).

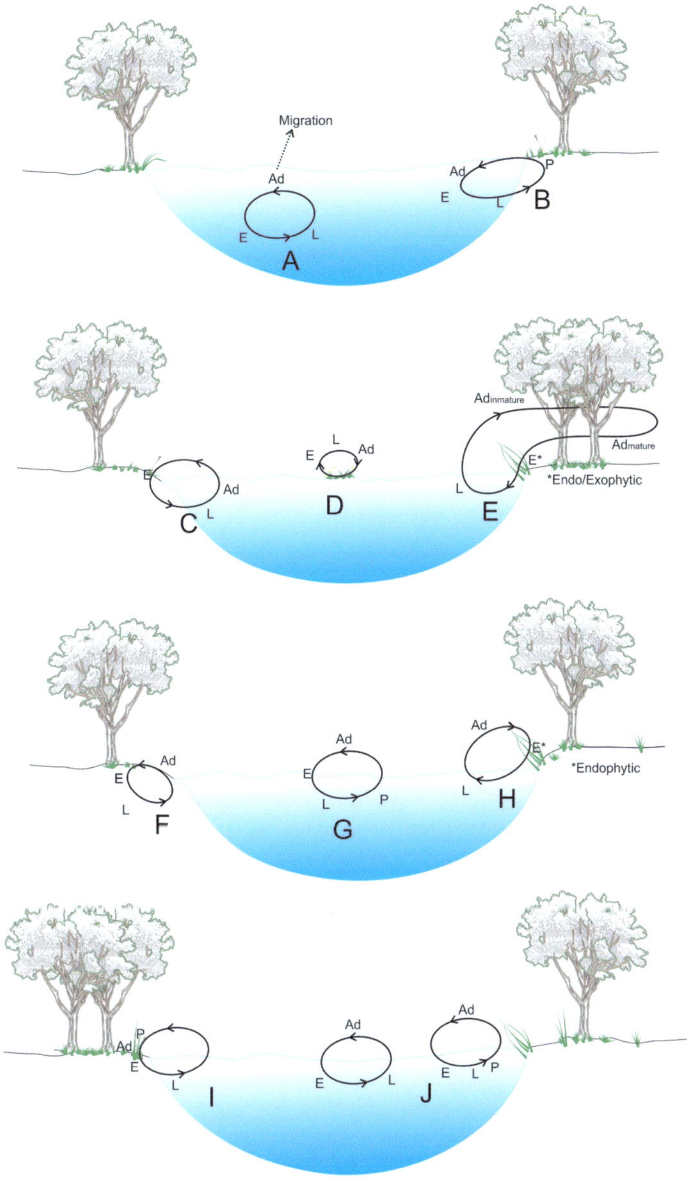


**Fig. 14.2** Diagram of general types of aquatic/semiaquatic insect life cycles. E: eggs, L: larvae, P: pupae, Ad: adults. (**a**) Fully aquatic, eventually with migration flights (e.g., Corixidae, Notonectidae, Belostomatidae), (**b**) aquatic with terrestrial/riparian pupae (e.g., Dytiscidae, Gyrinidae, Elmidae), (**c**) aquatic or semiaquatic with riparian/aerial eggs (e.g., Hydrometridae, Gerridae, Belostomatidae only genus *Lethocerus*), (**d**) semiaquatic pleustonic (e.g., Orthoptera), (**e**) aquatic with terrestrial/aerial adults with prereproductive phase flying far from wetland (e.g., several Odonata), (**f**) fully riparian (e.g., Gelastocoridae, Ochteridae, Saldidae), (**g**) aquatic larvae and pupae with eggs over the water surface and terrestrial/aerial adults (e.g., Diptera Culicidae), (**h**) aquatic larvae with endophytic eggs on riparian plants and terrestrial/aerial adults (e.g., several Odonata), (**i**) terrestrial with aquatic larvae (e.g., Megaloptera), (**j**) aquatic with terrestrial/aerial adults (e.g., Odonata Anisoptera with exophytic oviposition, Trichoptera, several Diptera)

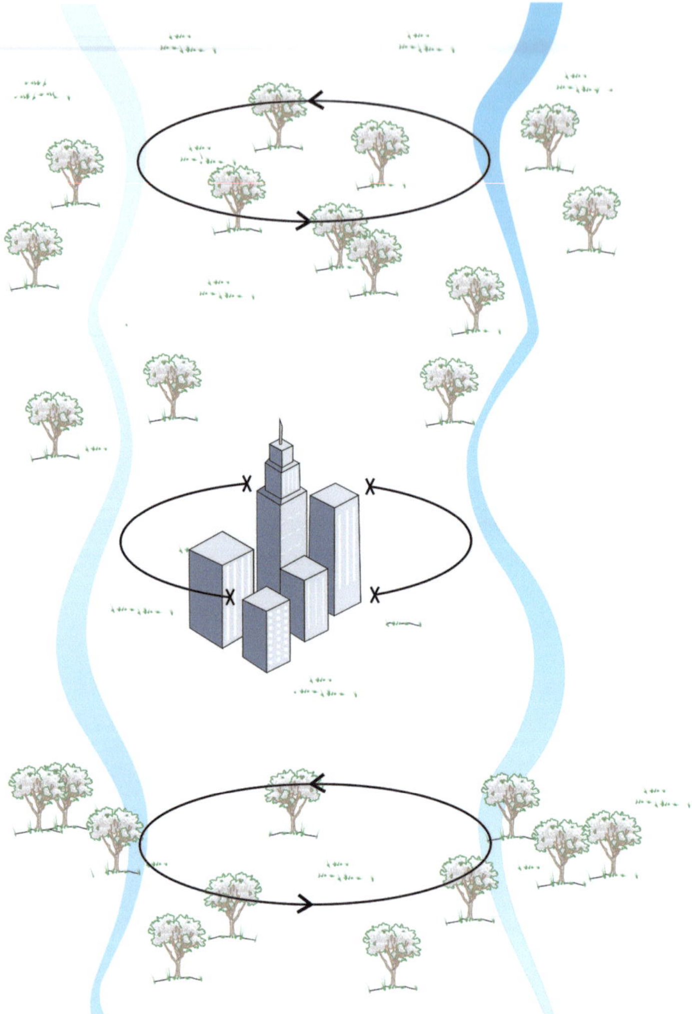

**Fig. 14.3** Cities acting as an ecological trap. Diagram showing biota relationship between two different streams and the interference of urban infrastructure. Modified from Smith et al. (2009)

## 14.3 Main Threats and Impacts

There are a series of common characteristics or processes in urban centers proposed as threats for aquatic insects:

- Habitat fragmentation: Fragmentation reduces wetland size and connectivity and could affect negatively diversity. Wetland fragmentation in cities may be due to different types of human constructions or infrastructure (e.g., roads and buildings). Lost, replacement, or size reduction can be mentioned as a first impact on wetlands and its communities. There are some infrastructure, as bridges or road culverts, that also can act as barriers, mainly for lotic insect migration (Blakely et al. 2006). In addition, loss of connectivity between aquatic and terrestrial habitats, as for example modification of shorelines, could have a critical impact on those taxa with amphibiotic cycles with eggs and/or pupas in an aquatic/terrestrial interphase (Fig. 14.1e, f) (Smith et al. 2009).

- Nevertheless, there are no conclusive evidences about negative effects of wetland fragmentation on insects (Kadoya et al. 2004; Goertzen and Suhling 2013; Talaga et al. 2017). Suitability of habitats is assumed as more relevant than size and connectivity in some studies, perhaps due to the fact that most aquatic insects have a high dispersion capacity (Goertzen and Suhling 2013; Ramos et al. 2017). Whether wetland size could or could not influence insect diversity, there are some indirect environmental characteristics that could be primarily affected by size and shape, as for example shoreline longitude and heterogeneity of riparian vegetation (Samways and Steytler 1996; Kadoya et al. 2004) (Fig. 14.1d–f). Artificial man-made wetlands, as for example dams and road ditches, can host a significant specific diversity (Schnack et al. 2000; Goertzen and Suhling 2013; Ramos et al. 2017; Holtmann et al. 2018) counterbalancing negative impacts of other human constructions (Fig. 14.1c, d). On the other hand, an increase of ß diversity was positively correlated with habitat fragmentation in urban areas because fragmentation increases habitat heterogeneity and promotes edge species (Jones and Leather 2012; Talaga et al. 2017).
- Chemical pollution: Including diffuse, extensive, and continuous emissions deposited through atmospheric pollutants (from industry, traffic-related emissions, etc.) and point sources (as industry deposits, waste deposits, landfills, sewer, etc.). In addition, wetlands can receive pollutant inputs, through stormwater runoff and treated wastewater effluents (Niemelä and MacDonnell 2011; Turpin-Nagel and Vadas 2016). Unlike soil, pollution negative impacts on wetlands are fast and harder to contain, both on running and standing waters. Aquatic insects are negatively affected not only by the chemical itself, but also by alteration produced by it (e.g., changes in oxygen demand, conductivity, pH). Other important sources of water pollution are insecticides, commonly used in cities mostly for pest and mosquitoes' control. Some special ecosystems, as phytotelmata, are greatly threatened by insecticide spraying (Talaga et al. 2017).
- Light pollution: Artificial light is a new stressor for biotic components, recently incorporated by humans in the last 150 years. The attraction potential that artificial lights have over insects is recognized albeit not well known because light does not affect uniformly species and sexes; cities' light pollution can modify photoperiod responses and reduce nightime drift of larval aquatic insects in urban streams by disrupting their circadian rhythms (Henn et al. 2014). These impacts can modify inter- and intraspecific interactions and communities' structures, implying the disruption of ecosystem functioning, for which the term ecological light pollution was coined (Longcore and Rich 2004).
- Resource disruption: Habitat fragmentation, exotic vegetation, and human water consumption can affect negatively resource availability for many insect groups.

For example, differences in the composition of the riparian vegetation can produce changes in leaf litter decomposition, thus modifying the relative abundance among functional feeding groups, or oviposition substrate availability (Frankie and Ehler 1978; Smith 2007; Talaga et al. 2017).

- Urban heat island effect: It has been observed in most cities of the world that air temperatures are significantly higher than in the surrounding rural areas; this alteration of the climate is known as heat island effect. It can be a very complex phenomenon, showing a mosaic of different temperatures according to city design, distribution, and abundance of green spaces and wetlands (New 2015). These differences are not restricted to a specific season in the year, and vary according to the characteristics of the city, such as size and type of buildings (Niemelä and MacDonnell 2011). The causes of this effect are the supplanting of vegetation by impervious surfaces, which reduces evaporation and evapotranspiration (Niemelä and MacDonnell 2011; New 2015). There are several effects of the heat island effect on general urban biodiversity including aquatic communities. Higher temperatures affect the periods of growth and flowering of plants that affect or suppress ecological successions, which modifies the structures of the communities. In addition, the heat island favors the settlement of thermophilic species and alien species from warmer climates.
- Invasive species: Urban populations are fundamental in trade and transport plants and animal alien and/or invasive species and provide opportunities for a range of different introduction pathways (Hassall 2014). There are several examples of deliberated fish introduction in cities' ponds both as ornamental and sports fishing species. These invasive species could have a direct impact on aquatic insects' diversity and abundance, modifying their taxocoenosis (Muzón et al. 2005; Buria et al. 2007). On the other hand, invasive aquatic plants have a negative impact on wetlands, modifying water courses, reducing runoff, replacing native species, etc. (Howard 1999). Nonetheless, the effect of invasive exotic plants is not always detrimental to endemic biodiversity, since many of these plants behave ecologically similar to native plants (Samways 2005). Some species of dragonflies are excluded when invasive exotic trees provide a lot of shade in the habitat, but if the indigenous species are mixed with the native ones, the species do not seem to be affected (Kinvig and Samways 2000).

## 14.4 Aquatic Insects as Target Taxa for Urban Wetland Restoration and Conservation

As urban wetlands are nowadays considered key ecosystems to improve life quality and cities' sustainability, there are increasing studies focusing on evaluating and monitoring its environmental health, and evaluating the efficiency of recovery and/or conservation practices (Kadoya et al. 2004; Blakely et al. 2006; Hassall 2014; New 2015; Nieto et al. 2017).

Systematic and ecological knowledge of insects from urban wetlands is a major input for monitoring ecosystem health and evaluating best practices to improve biodiversity levels in cities (Morley and Karr 2002). On the other hand, urban insects' knowledge helps the promotion of the citizen science which in turn improves their conservation (Dickinson et al. 2010; Buldrini et al. 2015; Code 2017; Dumakude and Graham 2017).

Aquatic insects have been proposed as a reliable bioindicator group, both in natural and anthropogenic ecosystems. A bioindicator is a species or a group of species, whether animals, plants, or microorganisms, that can live under relatively specific environmental conditions under a narrow range of tolerance to one or more environmental factors of biotic or abiotic origin (Segnini 2003). Their presence in a habitat is informative of a particular state of its environment; they have been used to evaluate the magnitude of the disturbance produced by the human being, to monitor population trends in other species, and to identify and locate areas of high regional biodiversity.

Aquatic macroinvertebrates, particularly insects, are the most used bioindicators to assess environmental health. Many countries have a long history of using macroinvertebrates to monitor the environmental health of aquatic ecosystems. Benthic macroinvertebrates are key components of aquatic food webs that link organic matter and nutrient resources (e.g., leaf litter, algae, and detritus) with higher trophic levels. With the sensitive life stage and relatively long life span, they have the ability to integrate the effects of short-term environmental variations. Besides, these assemblages are made up of many species among which there is a wide range of trophic levels and pollution tolerances, therefore providing strong information for interpreting cumulative effects. Community structure of the assemblages frequently changes in response to environmental disturbances in predictable ways, which is the basis for development of biocriteria to evaluate anthropogenic influences (Li et al. 2010). Since different taxa are found in different geographical areas, biotic indices must be regionally specific.

At present, aquatic macroinvertebrates are the most widely used organisms in the biomonitoring of wetlands impacted by human disturbance (Bonada et al. 2006). It has been shown that, among them, insects are good subrogates, reducing monitoring cost and time (Smith 2007; Simaika and Samways 2012). Furthermore, as mentioned before, several aquatic insects have complex amphibiotic life cycles that could make them good monitors simultaneously on aquatic and terrestrial ecosystems (Kietzka et al. 2016).

There are several insect groups proposed or used as good biomonitors (e.g., Ephemeroptera, Odonata, Plecoptera, Hemiptera, Coleoptera, Diptera, Trichoptera), both together and separately. There are many indexes considering several taxa at once as for example:

- Biotic index, which evaluates community structure and makes use of the indicator species concept without placing undue emphasis on species that do not appear in significant numbers (Hilsenhoff 1977).
- Family biotic index, which requires only family identifications (Hilsenhoff 1988; Pérez and Ramírez Restrepo 2008).

- EPT index, for Ephemeroptera, Plecoptera, and Trichoptera: This index works under the assumption that the higher the water quality of streams, the higher is the specific richness due to the low tolerance of these taxa (Lenat and Penrose 1996; Hepp et al. 2013).

Also, there are indexes that are not based on supra-specific taxa or multi-taxa groups, but rather consider the information at the specific level for a particular group. For example, Odonata species were successfully used to test habitat quality (Dias-Silva et al. 2010; Dutra and De Marco 2015; de Oliveira-Junior et al. 2015; Valente-Neto et al. 2016; Miguel et al. 2017). Also there are several indexes based on odonate species: the Dragonfly Biotic Index, DBI (Simaika and Samways 2009, 2012; Samways and Simaika 2016), the Dragonfly Association Index (Chovanec et al. 2015), and the Odonate River Index, ORI (Golfieri et al. 2016), and the synanthropy level of urban wetlands through the Nuorteva index (Ramos et al. 2017).

Finally, there are some indexes used to describe environmental conditions in order to evaluate taxa diversity, as for example the Habitat Integrity Index (HII), widely used in South America (Nessimian et al. 2008). This index considers a protocol of 12 items that describe several environmental conditions, assessing characteristics that include patterns of land use adjacent to the riparian vegetation, the width of the riparian forest and its state of preservation, the condition of the riparian forest within a radius of 10 m, a description of the type of sediment and presence of retaining debris, the structure and erosion of the banks, the characteristics of the substrate of the stream bed, the aquatic vegetation and debris, and the distribution of rapids, pools, and meanders. Each of these items is scored on a scale of four to six points, and these values are summed and divided by the maximum possible score to provide a final value of 0–1 for the HII, with the highest values indicating the greatest stream integrity (Nessimian et al. 2008; Miguel et al. 2017). Although originally used for birds, the index has been used for many aquatic insects such as Heteroptera (Dias-Silva et al. 2010), Odonata (Miguel et al. 2017), and Trichoptera (Pereira et al. 2012).

## 14.5 South American Background

Most studies done in urban ecosystems have to do with megacities. The landscape that these ecosystems form is more similar among the cities of the world, regardless of their location, than with the surrounding ecosystem. However, most neotropical cities do not have these characteristics, but they are smaller (Talaga et al. 2017); this can modify the relationship with the surrounding ecosystems and the response of biodiversity. Additionally, there are few studies dealing with the processes that act within urban plots over time, except in studies of environmental impact due to specific sources of contamination with a strong sociological/epidemiological content or in many cases economic.

In the Neotropics, most of the studies are developed in streams (Couceiro et al. 2007; Dias-Silva et al. 2010; Feitosa et al. 2012; Miguel et al. 2017; Hepp et al.

2013; Monteiro-Júnior et al. 2014, 2015; de Oliveira-Junior et al. 2015) and to a lesser extent temporary ponds (Fontanarrosa et al. 2009; Ramos et al. 2017) and phytotelmata (Talaga et al. 2017). These studies are mainly focused on comparison of insects' diversity between urban and rural or pristine environments. There are few studies dealing with the processes that act within urban plots over time, except in studies of environmental impact due to specific sources of contamination with a strong sociological/epidemiological content or in many cases economic.

Surprisingly, studies of urban streams from an ecological point of view are scarce. Morley and Karr (2002) surveyed the world literature published between 1991 and 2001 and found that only 30 papers were concerned with the direct estimation of human effects on the biota in urban streams.

## 14.6 Wetlands as Promoters of Urban Biodiversity

As urban green spaces, wetlands (blue spaces) are main providers of many important ecological services to local communities (Wu 2014). In fact, among others they can act as flood control, groundwater replenishment, shoreline stabilization, storm protection, sediment and nutrient retention, water purification, and reservoirs of biodiversity. In addition, wetlands are key habitats for the development of educational programs aimed to the general public on natural sciences and also to initiate several citizen science programs. Hassall (2014) proposes a socio-ecosystem approach for the environmental management of urban ponds in order to minimize the arising conflicting priorities over ecological functions and services.

A sustainable urban design (including disciplines like architecture, landscape architecture, civil engineering, and urban planning) urgently needs information inputs from basic and applied research. Nevertheless, in a global context, the knowledge of urban ecosystems is far from being balanced between terrestrial and aquatic ones, being wetlands far less known than terrestrial, as well as their biota (Pickett et al. 2011). Hassall (2014) identifies three main research needs: the roles of design and location of urban ponds in influencing biodiversity, the function of urban wetlands for stormwater and pollution management, and the public perceptions of urban ecosystems and how those perceptions are influenced by interactions with natural systems.

Regarding aquatic insects, and in order to prevent major impacts or to better manage urban wetlands, we need to know not only species inventories, but also life cycle requirements, dispersal traits and patterns, ecological relationships, how species respond to the already urbanization threats, long-term consequences of changing landscapes on population genetics and species persistence, etc. (Smith and Lamp 2008). This kind of information is very uncommon for neotropical taxa, even though there are some important works published in Brazil (e.g., Feitosa et al. 2012; Monteiro-Júnior et al. 2014; Nicasio and Juen 2015; de Oliveira-Junior et al. 2017).

Considering that urbanization impacts on wetlands cause a decrease in their biodiversity, optimizing urban design is crucial for minimizing impacts and promoting biodiversity levels. Several requirements were proposed for the construction of new artificial ponds, which were summarized by Hassall (2014). Goertzen and Suhling (2013) suggest that conservation urban planning in urban centers must focus on a general improvement of habitat structure; for example, they proposed that moderate anthropogenic disturbance can increase habitat diversity and promote species of temporary and pioneer ponds especially at ruderal sites. Increasing biological value of urban wetlands through reduction of impact level, for example restoration of degraded shorelines and banks, removal of invasive alien trees, and recovering of riparian natural vegetation, can significatively increase specific richness, specially insects (Samways and Taylor 2004; Suren and McMurtrie 2005; Samways and Sharratt 2010; Goertzen and Suhling 2013). Also, Goertzen and Suhling (2013) have suggested at the city scale that different pond types regarding water level and succession stage would be beneficial for increasing biodiversity levels. In a similar way, a study carried out by Fontanarrosa et al. (2009) in temporary urban ponds in Buenos Aires found that seasonally irregular filling pattern due to local climatic conditions (humid temperate cities of the Southern Hemisphere) and physical characteristics of ponds (pool depth and area) were most significative to community structure.

Regarding minimizing of impacts, there are several studies around the world on how some threats affect aquatic biodiversity, especially aquatic insects. Management actions or best practices proposed in those studies must be considered for sustainable urban design; the region or the city the proposal is carried out does not matter (Walsh et al. 2005). For example, rehabilitate the natural heterogeneity of riparian ecosystems (Holmes et al. 2008); prevent and manage ecological traps (Hale et al. 2015); for those impacts produced by dark asphalt roads three mitigation measures have been proposed: to keep gravel roads in susceptible areas, to cover it with small-sized white gravel, or paint the asphalt with matte white stripes (Malik et al. 2010). Blakely et al. (2006) studied the effect of culverts as barriers for stream restoration and they found that the constriction of streams by culverts and associated physical barriers should be minimized in order to promote insect oviposition rates and sustain biodiversity levels.

## References

Blakely TJ, Harding JS, Mcintosh AR, Winterbourn MJ (2006) Barriers to the recovery of aquatic insect communities in urban streams. Freshw Biol 51:1634–1645. https://doi.org/10.1111/j.1365-2427.2006.01601.x

Bonada N, Prat N, Resh VH, Statzner B (2006) Developments in aquatic insect biomonitoring: a comparative analysis of recent approaches. Annu Rev Entomol 51:495–523. https://doi.org/10.1146/annurev.ento.51.110104.151124

Buldrini F, Simoncelli A, Accordi S, Pezzi G, Dallai D (2015) Ten years of citizen science data collection of wetland plants in an urban protected area. Acta Bot Gall 162:365–373. https://doi.org/10.1080/12538078.2015.1080187

Buria L, Albariño R, Villanueva VD (2007) Impact of exotic rainbow trout on the benthic macroinvertebrate community from Andean-Patagonian headwater streams. Fundam Appl Limnol 168:145–154. https://doi.org/10.1127/1863-9135/2007/0168-0145

Chadwick MA, Dobberfuhl DR, Benke AC, Huryn AD, Suberkropp K, Thiele JE (2006) Urbanization affects stream ecosystem function by altering hydrology, chemistry, and biotic richness. Ecol Appl 16:1796–1807. https://doi.org/10.1890/1051-0761(2006)016[1796:UASEFB]2.0.CO;2

Chovanec A, Schindler M, Waringer J, Wimmer R (2015) The dragonfly association index (Insecta: Odonata)—a tool for the type-specific assessment of lowland rivers. River Res Appl 31:627–638. https://doi.org/10.1002/rra.2760

Code A (2017) How Our Gardening Choices Affect the Health of Our Waterways. WINGS: ESSAYS ON INVERTEBRATE CONSERVATION, 5–10.

Couceiro SRM, Hamada N, Luz SLB, Forsberg BR, Pimentel TP (2007) Deforestation and sewage effects on aquatic macroinvertebrates in urban streams in Manaus, Amazonas, Brazil. Hydrobiologia 575:271–284. https://doi.org/10.1007/s10750-006-0373-z

de Oliveira-Junior JMB, Shimano Y, Gardner TA, Hughes RM, De Marco L Jr, Juen L (2015) Neotropical dragonflies (Insecta: Odonata) as indicators of ecological condition of small streams in the eastern Amazon. Austral Ecol 40:733–744. https://doi.org/10.1111/aec.12242

de Oliveira-Junior JMB, De Marco P, Dias-Silva K, Leitão RP, Leal CG, Pompeu PS, Gardner TA, Hughes RM, Juen L (2017) Effects of human disturbance and riparian conditions on Odonata (Insecta) assemblages in eastern Amazon basin streams. Limnologica 66:31–39. https://doi.org/10.1016/j.limno.2017.04.007

De Sherbinin A, Rahman A, Barbieri A, Fotso J, Zhu Y (2009) Urban population-development-environment dynamics in the developing world: case studies and lessons learned. Committee for International Cooperation in National Research in Demography (CICRED), Paris

Dias-Silva K, Cabette HSR, Juen L, De Marco P Jr (2010) The influence of habitat integrity and physical-chemical water variables on the structure of aquatic and semi-aquatic Heteroptera. Zoologia 27:918–930. https://doi.org/10.1590/S1984-46702010000600013

Dickinson JL, Zuckerberg B, Bonter DN (2010) Citizen science as an ecological research tool: challenges and benefits. Annu Rev Ecol Evol Syst 41:149–172. https://doi.org/10.1146/annurev-ecolsys-102209-144636

Dumakude N, Graham M (2017) Assessing wetland health using a newly developed land cover citizen science tool for use by local people who are not wetland specialists. S Afr J Environ Educ 33:71. https://doi.org/10.4314/sajee.v.33i1.6

Dutra S, De Marco P (2015) Bionomic differences in odonates and their influence on the efficiency of indicator species of environmental quality. Ecol Indic 49:132–142. https://doi.org/10.1016/j.ecolind.2014.09.016

Feitosa MAC, Julião GM, Costa MDP, Belém B, Pessoa FAC (2012) Diversity of sand flies in domiciliary environment of Santarém, state of Pará, Brazil: species composition and abundance patterns in rural and urban areas. Acta Amaz 42:507–514. https://doi.org/10.1590/S0044-59672012000400008

Fontanarrosa MS, Collantes MB, Bachmann AO (2009) Seasonal patterns of the insect community structure in urban rain pools of temperate Argentina. J Insect Sci 9:1–17. https://doi.org/10.1673/031.009.1001

Frankie GW, Ehler LE (1978) Ecology of Insects in Urban Environments. Annu Rev Entomol 23:367–387. https://doi.org/10.1146/annurev.en.23.010178.002055

Goertzen D, Suhling F (2013) Promoting dragonfly diversity in cities: major determinants and implications for urban pond design. J Insect Conserv 17:399–409. https://doi.org/10.1007/s10841-012-9522-z

Golfieri B, Hardersen S, Maiolini B, Surian N (2016) Odonates as indicators of the ecological integrity of the river corridor: development and application of the odonate river index (ORI) in northern Italy. Ecol Indic 61:234–247. https://doi.org/10.1016/j.ecolind.2015.09.022

Hale R, Coleman R, Pettigrove V, Swearer SE (2015) Identifying, preventing and mitigating ecological traps to improve the management of urban aquatic ecosystems. J Appl Ecol 52:928–939. https://doi.org/10.1111/1365-2664.12458

Hassall C (2014) The ecology and biodiversity of urban ponds. Wiley Interdiscip Rev Water 1:187–206. https://doi.org/10.1002/wat2.1014

Henn M, Nichols H, Zhang Y, Bonner TH (2014) Effect of artificial light on the drift of aquatic insects in urban central Texas streams. J Freshw Ecol 29:307–318. https://doi.org/10.1080/02705060.2014.900654

Hepp LU, Restello RM, Milesi SV, Biasi C, Molozzi J (2013) Distribution of aquatic insects in urban headwater streams. Acta Limnol Bras 25:1–09. https://doi.org/10.1590/S2179-975X2013005000014

Hilsenhoff WL (1977) Use of arthropods to evaluate water quality of streams. Wisconsin Department of Natural Resources. Tech Bull 100:1–14

Hilsenhoff WL (1988) Rapid field assessment of organic pollution with a family-level biotic index. J North Am Benthol Soc 7:65–68. https://doi.org/10.2307/1467832

Holmes PM, Esler KJ, Richardson DM, Witkowski ETF (2008) Guidelines for improved management of riparian zones invaded by alien plants in South Africa. S Afr J Bot 74:538–552. https://doi.org/10.1016/j.sajb.2008.01.182

Holtmann L, Juchem M, Brüggeshemke J, Möhlmeyer A, Fartmann T (2018) Stormwater ponds promote dragonfly (Odonata) species richness and density in urban areas. Ecol Eng 118:1–11. https://doi.org/10.1016/j.ecoleng.2017.12.028

Howard G (1999) Especies invasoras y humedales. Ramsar COP7 DOC 24:1–11

Jones EL, Leather SR (2012) Invertebrates in urban areas: a review. Eur J Entomol 109:463–478. https://doi.org/10.14411/eje.2012.060

Kadoya T, Suda SI, Washitani I (2004) Dragonfly species richness on man-made ponds: effects of pond size and pond age on newly established assemblages. Ecol Res 19:461–467. https://doi.org/10.1111/j.1440-1703.2004.00659.x

Kietzka GJ, Pryke JS, Samways MJ (2016) Aerial adult dragonflies are highly sensitive to in-water conditions across an ancient landscape. Divers Distrib 23:14–26. https://doi.org/10.1111/ddi.12493

Kinvig RG, Samways MJ (2000) Conserving dragonflies (Odonata) along streams running through commercial forestry. Odonatologica 29:195–208

Lenat DR, Penrose DL (1996) History of the EPT taxa richness metric. Bull North Am Benthol Soc 13:557–565

Li L, Zheng B, Liu L (2010) Biomonitoring and bioindicators used for river ecosystems: definitions, approaches and trends. Procedia Environ Sci 2:1510–1524. https://doi.org/10.1016/j.proenv.2010.10.164

Longcore T, Rich C (2004) Ecological light pollution. Front Ecol Environ 2:191–198. https://doi.org/10.1890/1540-9295(2004)002[0191:ELP]2.0.CO;2

Malik P, Hegedüs R, Kriska G, Akesson S, Robertson B, Horvath G (2010) Asphalt surfaces as ecological traps for water-seeking polarotactic insects: how can the polarized light pollution of asphalt surfaces be reduced? In: Asphaltenes: characterization, properties and applications. Nova Science Publishers, Inc., Hauppauge, pp 81–119

Martin PH, Lefebvre MG (1995) Malaria and climate: sensitivity of malaria potential transmission to climate. Ambio 24:200–207

McIntyre NE (2000) Ecology of urban arthropods: a review and a call to action. Ann Entomol Soc Am 93:825–835. https://doi.org/10.1603/0013-8746(2000)093[0825:EOUAAR]2.0.CO;2

Meyer JL, Paul MJ, Taulbee WK (2005) Stream ecosystem function in urbanizing landscapes. J North Am Benthol Soc 24:602–612. https://doi.org/10.1899/04-021.1

Miguel TB, Oliveira-Junior JMB, Ligeiro R, Juen L (2017) Odonata (Insecta) as a tool for the biomonitoring of environmental quality. Ecol Indic 81:555–566. https://doi.org/10.1016/j.ecolind.2017.06.010

Monteiro-Júnior CS, Juen L, Hamada N (2014) Effects of urbanization on stream habitats and associated adult dragonfly and damselfly communities in central Brazilian Amazonia. Landsc Urban Plan 127:28–40. https://doi.org/10.1016/j.landurbplan.2014.03.006

Monteiro-Júnior CS, Juen L, Hamada N (2015) Analysis of urban impacts on aquatic habitats in the central Amazon basin: adult odonates as bioindicators of environmental quality. Ecol Indic 48:303–311. https://doi.org/10.1016/j.ecolind.2014.08.021

Morley SA, Karr JR (2002) Assessing and restoring the health of urban streams in the Puget sound basin. Conserv Biol 16:1498–1509. https://doi.org/10.1046/j.1523-1739.2002.01067.x

Muzón J, Spinelli GR, Pessacq P, von Ellenrieder N, Estevez AL, Marino PI, Perez Goodwyn PJ, Angrisano EB, Diaz F, Mazzucconi S, Rossi G, Salomón OD (2005) Insectos acuáticos de la Meseta del Somuncura, Patagonia, Argentina. Inventario preliminar. Rev Soc Entomol Argent 64:47–67

Nessimian JL, Venticinque EM, Zuanon J, De Marco P Jr, Gordo M, Fidelis L, Batista JA, Juen L (2008) Land use, habitat integrity, and aquatic insect assemblages in Central Amazonian streams. Hydrobiologia 614:117–131. https://doi.org/10.1007/s10750-008-9441-x

New TR (2015) Insect conservation and urban environments. 244p. Cham: Springer. https://doi.org/10.1007/978-3-319-21224-1

Nicasio G, Juen L (2015) Chironomids as indicators in freshwater ecosystems: an assessment of the literature. Insect Conserv Divers 8:393–40.3. https://doi.org/10.1111/icad.12123

Niemelä J, MacDonnell MJ (2011) Urban ecology: patterns, processes and applications, vol 374. OUP Oxford, Oxford. https://doi.org/10.1093/acprof:oso/9780199563562.001.0001

Nieto C, Ovando XMC, Loyola R, Izquierdo A, Romero F, Molineri C, Rodriguez J, Rueda Martin P, Fernandez H, Manzo V, Miranda MJ (2017) The role of macroinvertebrates for conservation of freshwater systems. Ecol Evol 7:5502–5513. https://doi.org/10.1002/ece3.3101

Pereira LR, Cabette HSR, Juen L (2012) Trichoptera as bioindicators of habitat integrity in the Pindaíba river basin, Mato Grosso (Central Brazil). Ann Limnol Int J Limnol 48:295–302. https://doi.org/10.1051/limn/2012018

Pérez GR, Restrepo JJR (2008) Fundamentos de limnología neotropical (Vol. 15). Universidad de Antioquia. Colombia

Pickett STA, Cadenasso ML, Grove JM, Boone CG, Groffman PM, Irwin E, Kaushal SS, Marshall V, McGrath BP, Nilon CH, Pouyat RV, Szlavecz K, Troy A, Warren P (2011) Urban ecological systems: scientific foundations and a decade of progress. J Environ Manag 92:331–362. https://doi.org/10.1016/j.jenvman.2010.08.022

Ramos L, Lozano F, Muzón J (2017) Odonata diversity and synanthropy in urban areas: a case study in Avellaneda City, Buenos Aires, Argentina. Neotrop Entomol 46:141–150. https://doi.org/10.1007/s13744-016-0443-5

RAMSAR (2008) Resolution X.27: wetlands and urbanization. 10th meeting of the conference of the parties to the convention on wetlands. http://archive.ramsar.org/pdf/res/key_res_x_27_e.pdf

Samways MJ (2005) Insect diversity conservation. 356p. Cambridge University Press. https://doi.org/10.1017/CBO9781107415324.004

Samways MJ, Sharratt NJ (2010) Recovery of endemic dragonflies after removal of invasive alien trees: contributed paper. Conserv Biol 24:267–277. https://doi.org/10.1111/j.1523-1739.2009.01427.x

Samways MJ, Simaika JP (2016) Manual of freshwater assessment for South Africa: dragonfly biotic index

Samways MJ, Steytler NS (1996) Dragonfly (Odonata) distribution patterns in urban and forest landscapes, and recommendations for riparian management. Biol Conserv 78:279–288. https://doi.org/10.1016/S0006-3207(96)00032-8

Samways MJ, Taylor S (2004) Impacts of invasive alien plants on Red-Listed South African dragonflies (Odonata). S Afr J Sci 100:78–80

Schnack JA, De Francesco FO, Colado UR, Novoa ML, Schnack EJ (2000) Humedales antrópicos: Su contribución para la concervacióm de la biodiversidad en los dominios subtropical y pampásico de la Argentina. Ecol Austral 10:63–80

Segnini S (2003) El uso de los macroinvertebrados bentónicos como indicadores de la condición ecológica de los cuerpos de agua corriente. Ecotropicos 16:45–63
Simaika JP, Samways MJ (2009) An easy-to-use index of ecological integrity for prioritizing freshwater sites and for assessing habitat quality. Biodivers Conserv 18:1171–1185. https://doi.org/10.1007/s10531-008-9484-3
Simaika JP, Samways MJ (2012) Using dragonflies to monitor and prioritize lotic systems: a South African perspective. Org Divers Evol 12:251–259. https://doi.org/10.1007/s13127-012-0104-4
Smith B (2007) Diversity of adult aquatic insects in Hamilton urban streams and seepages. J Water Atmos 4005:1–19
Smith RF, Lamp WO (2008) Comparison of insect communities between adjacent headwater and main-stem streams in urban and rural watersheds. J North Am Benthol Soc 27:161–175. https://doi.org/10.1899/07-071.1
Smith RF, Alexander LC, Lamp WO (2009) Dispersal by terrestrial stages of stream insects in urban watersheds: a synthesis of current knowledge. J North Am Benthol Soc 28:1022–1037. https://doi.org/10.1899/08-176.1
Suren AM, McMurtrie S (2005) Assessing the effectiveness of enhancement activities in urban streams: II. Responses of invertebrate communities. River Res Appl 21:439–453. https://doi.org/10.1002/rra.817
Talaga S, Dézerald O, Carteron A, Leroy C, Carrias JF, Céréghino R, Dejean A (2017) Urbanization impacts the taxonomic and functional structure of aquatic macroinvertebrate communities in a small Neotropical city. Urban Ecosyst 20:1001–1009. https://doi.org/10.1007/s11252-017-0653-6
Turpin-Nagel K, Vadas TM (2016) Controls on metal exposure to aquatic organisms in urban streams. Environ Sci Process Impacts 18:956–967. https://doi.org/10.1039/c6em00151c
Urban MC, Skelly DK, Burchsted D, Price W, Lowry S (2006) Stream communities across a rural-urban landscape gradient. Divers Distrib 12:337–350. https://doi.org/10.1111/j.1366-9516.2005.00226.x
Valente-Neto F, De Oliveira RF, Rodrigues ME, Juen L, Swan CM (2016) Toward a practical use of Neotropical odonates as bioindicators: testing congruence across taxonomic resolution and life stages. Ecol Indic 61:952–959. https://doi.org/10.1016/j.ecolind.2015.10.052
Walsh CJ, Roy AH, Feminella JW, Cottingham PD, Groffman PM, Morgan RP (2005) The urban stream syndrome: current knowledge and the search for a cure. J N Am Benthol Soc 24:706–723. https://doi.org/10.1899/04-028.1
Wu J (2014) Urban ecology and sustainability: the state-of-the-science and future directions. Landsc Urban Plan 125:209–221. https://doi.org/10.1016/j.landurbplan.2014.01.018

# Chapter 15
# The Tinbergen Shortfall: Developments on Aquatic Insect Behavior that Are Critical for Freshwater Conservation

**Fabio de Oliveira Roque, Francisco Valente-Neto, Marciel Elio Rodrigues, Francine Novais Souza, Davidson Gomes Nogueira, Ana Cláudia Piovezan-Borges, Alan P. Covich, and Michael J. Samways**

**Abstract** Humans have altered biodiversity worldwide, including accelerating species extinctions and loss of ecosystem services. Loss of many species is occurring even before they are collected for study, and researchers have recognized seven knowledge shortfalls that affect biodiversity and its use in conservation. This loss is especially true in aquatic ecosystems highly threatened by human pressures, including habitat transformation and degradation, and water extraction. Here, we propose the "Tinbergen shortfall" in honor of Nikolaas Tinbergen, related to limited knowledge of aquatic insect behavior in tropical regions, and how this shortfall highlights the need to have more knowledge on aquatic insect behavior for improving applied ecology, specifically biomonitoring, as it is the field where aquatic insect behavior information is most often used in measuring water quality, ecological integrity, and conservation. Through a systematized literature search in Web of Knowledge

---

F. de Oliveira Roque (✉)
Instituto de Biociências, Universidade Federal de Mato Grosso do Sul, Campo Grande, MS, Brazil

Centre for Tropical Environmental and Sustainability Science (TESS) and College of Science and Engineering, James Cook University, Cairns, QLD, Australia

F. Valente-Neto · D. G. Nogueira · A. C. Piovezan-Borges
Instituto de Biociências, Universidade Federal de Mato Grosso do Sul, Campo Grande, MS, Brazil

M. E. Rodrigues · F. N. Souza
Departamento de Ciências Biológicas, Universidade Estadual de Santa Cruz (UESC), Ilhéus, BA, Brazil

A. P. Covich
Odum School of Ecology, University of Georgia, Athens, GA, USA

M. J. Samways
Department of Conservation Ecology and Entomology, Stellenbosch University, Matieland, South Africa



K. Del-Claro, R. Guillermo (eds.), *Aquatic Insects*,
https://doi.org/10.1007/978-3-030-16327-3_15

database, we showed the limitation of our current knowledge on aquatic insects, behavior, and its use in conservation. We demonstrate a bias toward temperate countries, orders (e.g., Lepidoptera, Orthoptera, and Heteroptera), behavioral topic (feeding behavior), and limited use of aquatic insect behavioral knowledge in conservation science. Although aquatic insects having a long-established history in biomonitoring protocols worldwide, the use of behavior knowledge, such as oviposition and dispersal, is still limited. We emphasize the behavior that proxies can be used in biomonitoring and conservation studies in cases where information is unavailable, and how behavior information can improve recent approaches such as simulation modelling. Aquatic insect behavior can better inform conservation strategies based on scientific evidence, and we use some examples, including the creation and maintenance of protected areas, the reduction of anthropogenic impacts on animal behavior, and the development of specific agendas for animal species or groups of species that are particularly relevant for their value and changes in human behavior. The challenges ahead for biodiversity conservation will require addressing Tinbergen shortfall for aquatic insects and the increase of our understanding about the behavior of our own species in relation to the others.

**Keywords** Biomonitoring · Insect-human interactions · Anthropogenic impacts · Biodiversity loss · Bioindicators

## 15.1 Introduction

Species extinctions and imperilment in freshwater ecosystems have increased worldwide due to human impacts that are resulting in the loss of ecosystem services (Dudgeon et al. 2006; Strayer and Dudgeon 2010; Dudgeon 2014). Human activities have resulted in at least 10,000 freshwater species going globally extinct or becoming imperiled (Strayer 2006; IUCN 2017; Vörösmarty et al. 2010). Because each species has specific ecological traits and behaviors, when a species is lost, many ecological traits and behaviors that can be fundamental for ecosystem services and to community structure are also missing (Covich et al. 2004; Luck et al. 2009; Dirzo et al. 2014). Loss of many species is occurring even before they are collected for study. Because most species are not yet formally described (the Linnean shortfall) their ecological and geographical distributions are poorly understood (Lomolino 2004; Whittaker et al. 2005). Knowledge gaps related to the biology of invertebrate species and their importance in freshwater ecosystems are examples of bias, while the lack of cross-ecosystem integration requires much more consideration (Di Marco et al. 2017).

These knowledge gaps affect recommendations and prevent effective conservation actions, and are a cause for increasing concern. Here we summarize several key features. First, the knowledge gap regarding aquatic insects is a major concern as they are highly speciose, and it is a group that plays many important roles in providing material and nonmaterial ecological services. Balian et al. (2008) estimate that the total number of freshwater animal species is about 1,260,000, and aquatic insects are the dominant group (60.4%). Second, the most threatened ecosystems in tropical regions are freshwater rivers, lakes, and wetlands. These ecosystems are losing biodiversity at an alarming rate due to deforestation, habitat degradation, pollution, eutrophication, siltation, impoundments, channelization, flood control, exotic species invasions, overexploitation, fisheries, increasing salinity, and climate change (Dudgeon et al. 2006; Samways et al. 2010; Winemiller et al. 2016; Pelicice et al. 2017).

Improving inclusion of information about little studied but highly important groups is at the heart of global conservation and sustainability strategies. Improvement requires overcoming shortfalls of various types of knowledge. Seven previously proposed shortfalls (Hortal et al. 2015) pinpoint knowledge gaps for species taxonomy (Linnean), distribution (Wallacean), abundance (Prestonian), evolutionary patterns (Darwinian), abiotic tolerances among species' niches (Hutchinsonian), species traits (Raunkiæran), and biotic interactions (Eltonian). These shortfalls are especially valid for determining how aquatic insects are protected and conserved. Here, we propose a new shortfall, the "Tinbergen shortfall" in honor of Nikolaas Tinbergen (one of the founders of modern ethology, and winner of the Nobel Prize, https://www.nobelprize.org/prizes/medicine/1973/tinbergen/facts/). This shortfall relates to limited knowledge of aquatic insect behavior in general in tropical regions. This focus highlights the needs for devoting multidisciplinary studies of aquatic insects' behavior in a broad sense throughout the tropics to solve problems in conservation and biomonitoring. We explore the implication of this shortfall for biodiversity conservation of aquatic insects and propose some avenues for future progress in understanding how insects function in freshwater ecosystems.

Although the link between animal behavior and biodiversity conservation has a long history, it was only recently that these areas were merged into a field termed conservation behavior (see Sutherland 1998; Blumstein and Fernández-Juricic 2004). Since then, there have been advances using behavioral information for captive breeding, species reintroduction, relocation, reducing animal-human conflicts, protected area design, and wildlife management. However, most studies have focused on mammals and birds. Only recently have some studies focused on aquatic invertebrates, and they have shown that understanding behavior can inform better decision-making in conservation (Samways 2007; Samways et al. 2010). Aquatic insect behavioral science is a rapidly developing discipline, and the knowledge generated is highly relevant for conservation and biomonitoring (Heino and Peckarsky 2014). Most initiatives for conservation depend on behavioral data. For instance, to assess the conservation status of a species, it is important to know aspects of its movement and reproduction, because these behaviors are critical parameters for understanding the distribution and abundance of populations and communities, as well as their relationships with spatial, temporal, and environmental variables. Despite their importance, most studies related to conservation of population, communities, and metacommunities in tropical freshwaters are limited in their inferences because of the gaps in knowledge of aquatic insect behavior.

In this chapter, we evaluate trends in conservation literature considering aquatic insect behavior, including insect groups, biological behavioral traits, geography, ecological systems, and level of biological organization that are needed to better comprehend the Tinbergen shortfall. We then explore the implications of having more knowledge of aquatic insect behavior for improving applied ecology. The focus here is on biomonitoring because it is the field where aquatic insect behavior information is most often used in measuring water quality and ecological integrity. Finally, we discuss some challenges on using information on aquatic insect behavior in a conservation context, especially through connecting of animal behavior and conservation (Berger-Tal et al. 2011). We include example of direct and indirect anthropogenic impacts on animal behavior that have consequences for populations and communities, behavior-based management, and behavioral indicators to other processes that are of conservation concern (e.g., use of aquatic insects as bioindicators of water quality).

## 15.2 The Extent of the Challenge for Filling the Knowledge Gap

Monitoring biases and trends in the literature on insect conservation is crucial to tracking the progress of the discipline. We evaluated trends in the aquatic insect literature that cover aspects of behavior and conservation. We searched the literature using ISI Web of Knowledge database in December 2018, using as keywords the combination of the name of each 31 existing insect orders AND one of the following words related to behavior: "behavior," "feeding behavior," "reproductive behavior," and "breeding behavior." In the second step, we associate each insect order AND the keywords related to the behavior AND "conservation". Among the selected papers, the following information was extracted: insect group, country of the study, year, main key words that are selected in the article, type of ecosystem, and full reference.

Supporting our proposal of the Tinbergen shortfall for aquatic insects, we found bias toward geographical region, taxonomic order, environmental system, and behavioral topics in studies involving behavior and conservation. Most studies have been done in temperate countries (United States had the most undertaken), with terrestrial orders (e.g., Lepidoptera, Orthoptera, and Heteroptera), and were about behavioral traits focusing on feeding behavior (Fig. 15.1; Table 15.1—complementary material). Despite some changing trends, studies on behavior of aquatic insects are still poorly connected with conservation science and conservation priorities. Overcoming this challenge may require improving the integration among conservation and behavioral studies on aquatic insects, but more importantly this integration requires strong connection with real-world applications, as

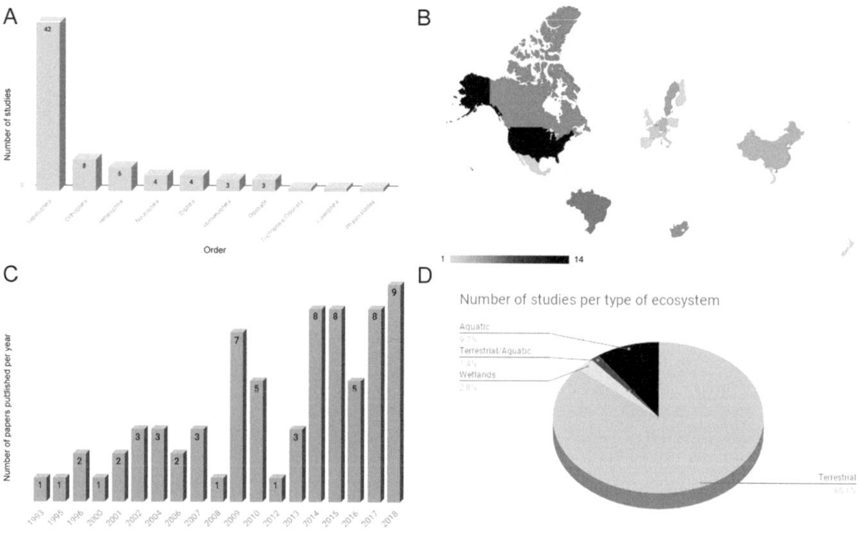


**Fig. 15.1** (**a**) Number of studies selected per insect order that assessed topics relating behavior and conservation. (**b**) Number of studies published by country; light and dark colors represent countries with few and large number of papers, respectively. (**c**) Trend in the number of papers published per year. (**d**) Number of studies by ecosystem types

already highlighted by Berger-Tal et al. (2011) for animal behavior and conservation in general.

## 15.3 Using Behavior Knowledge in Applied Issues: Biomonitoring

Aquatic insects have a long-established history in biomonitoring protocols worldwide, because high biotic diversity responds to various anthropogenic changes, including deforestation, urbanization, toxins, and organic pollution (Rosenberg and Resh 1993; Buss et al. 2015). For example, in lotic systems, anthropogenic impacts that increase water temperature and decrease dissolved oxygen cause a great reduction in species richness of Ephemeroptera, Plecoptera, and Trichoptera (EPT). The effects of anthropogenic changes on species abundance and distribution depend on species-specific tolerances to altered conditions and resources that may ultimately lead to local extinctions. However, we are far from understanding, particularly in the neotropics, how anthropogenic impacts alter species' behaviors, and whether these changes result in fitness reductions that have consequences for populations and communities (Peckarsky et al. 1997; Heino and Peckarsky 2014).

**Table 15.1** Number of articles found in the database Web of Science: In dark gray, articles selected in order that have at least one of the key words of behavior

| Order | Behavior | Behavior + conservation | Behavior stricto sensu | Feeding behavior | Feeding behavior + conservation | Feeding behavior stricto sensu | Reproductive behavior | Reproductive behavior + conservation | Reproductive behavior stricto sensu | Breeding behavior | Breeding behavior + conservation | Breeding behavior stricto sensu |
|---|---|---|---|---|---|---|---|---|---|---|---|---|
| Archaeognatha | 18 | 0 | 0 | 1 | 0 | 0 | 3 | 0 | 0 | 0 | 0 | 0 |
| Blattodea | 88 | 3 | 0 | 28 | 1 | 0 | 6 | 0 | 0 | 4 | 0 | 0 |
| Coleoptera | 5360 | 161 | 0 | 1435 | 30 | 1 | 696 | 20 | 0 | 218 | 17 | 0 |
| Collembola | 227 | 3 | 0 | 52 | **0** | 0 | 13 | 0 | 0 | 3 | 0 | 0 |
| Dermaptera | 87 | 0 | 0 | 11 | 0 | 0 | 12 | 0 | 0 | 3 | 0 | 0 |
| Diplura | 5 | 0 | 0 | 1 | 0 | 0 | 2 | 0 | 0 | 0 | 0 | 0 |
| Diptera | 6163 | 102 | 0 | 170 | 31 | 0 | 712 | 11 | 0 | 305 | 10 | 2 |
| Embioptera | 17 | 1 | 0 | 1 | 0 | 0 | 3 | 0 | 0 | 0 | 0 | 0 |
| Ephemeroptera | 193 | 7 | 0 | 77 | 0 | 0 | 11 | 0 | 0 | 5 | 0 | 0 |
| Grylloblattodea | 4 | 1 | 0 | 1 | 0 | 0 | 0 | 0 | 0 | 0 | 0 | 0 |
| Hemiptera | 2660 | 39 | 0 | 926 | 16 | 0 | 318 | 3 | 0 | 59 | 0 | 0 |
| Heteroptera | 1197 | 12 | 5 | 297 | 3 | 0 | 178 | 1 | 1 | 18 | 1 | 0 |
| Hymenoptera | 9668 | 230 | 2 | 1257 | 35 | 1 | 1351 | 31 | 0 | 234 | 3 | 0 |
| Isoptera | 769 | 9 | 0 | 188 | 1 | 0 | 59 | 1 | 0 | 20 | 2 | 0 |
| Lepidoptera | 5594 | 220 | 38 | 1576 | 46 | 4 | 629 | 18 | 0 | 129 | 10 | 0 |
| Mantodea | 68 | 0 | 0 | 14 | 0 | 0 | 11 | 0 | 0 | 1 | 0 | 0 |
| Mecoptera | 47 | 0 | 0 | 17 | 0 | 0 | 9 | 0 | 0 | 1 | 0 | 0 |
| Megaloptera | 24 | 0 | 0 | 1 | 0 | 0 | 1 | 0 | 0 | 0 | 0 | 0 |
| Neuroptera | 233 | 9 | 3 | 58 | 5 | 1 | 23 | 1 | 0 | 3 | 1 | 0 |
| Odonata | 755 | 43 | 2 | 90 | 3 | 0 | 152 | 7 | 0 | 49 | 8 | 2 |
| Orthoptera | 1286 | 26 | 8 | 242 | 3 | 0 | 195 | 4 | 0 | 41 | 6 | 0 |
| Phasmatodea | 24 | 1 | 1 | 2 | 0 | 0 | 4 | 0 | 0 | 1 | 0 | 0 |

| | | | | | | | | | | | | |
|---|---|---|---|---|---|---|---|---|---|---|---|---|
| Phthiraptera | 52 | 1 | 0 | 8 | 1 | 0 | 3 | 0 | 0 | 11 | 1 | 0 |
| Plecoptera | 158 | 0 | 0 | 54 | 0 | 0 | 9 | 0 | 0 | 0 | 0 | 0 |
| Protura | 2 | 0 | 0 | 0 | 0 | 0 | 1 | 0 | 0 | 0 | 0 | 0 |
| Psocoptera | 31 | 1 | 0 | 7 | 0 | 0 | 0 | 0 | 0 | 0 | 0 | 0 |
| Siphonaptera | 48 | 0 | 0 | 14 | 0 | 0 | 3 | 0 | 0 | 2 | 0 | 0 |
| Strepsiptera | 30 | 0 | 0 | 0 | 0 | 0 | 7 | 0 | 0 | 0 | 0 | 0 |
| Thysanoptera | 303 | 4 | 0 | 109 | 1 | 0 | 27 | 0 | 0 | 14 | 0 | 0 |
| Trichoptera | 256 | 13 | 1 | 77 | 0 | 0 | 9 | 0 | 0 | 0 | 0 | 0 |
| Zoraptera | 9 | 0 | 0 | 1 | 0 | 0 | 4 | 0 | 0 | 0 | 0 | 0 |
| Zygentoma | 13 | 0 | 0 | 2 | 0 | 0 | 1 | 0 | 0 | 0 | 0 | 0 |

In light gray, the articles selected in order that had at least one of the key words linked to the behavior plus the word conservation. Blank cells, the articles in order that actually use behavior in conservation matters

The two pathways in which aquatic insect behavioral metrics have been used in conservation and biomonitoring (see Berger-Tal et al. 2011 for animal behavior in general) are (1) behavioral indicators that provide an early warning to population decline or habitat degradation before community responses are evident (e.g., Roque et al. 2018) and (2) behavioral indicators used to monitor water conditions (Rosenberg and Resh 1993). Studies of feeding behavior are also highly relevant and have a long-established application in developing metrics for aquatic ecology and biomonitoring (Merrit et al. 2008; Rosenberg and Resh 1993). The focus on oviposition and dispersal behavior is valuable because these types of behavior affect survival and reproduction (hence recruitment), so providing essential information on population and community dynamics that can improve biomonitoring.

Basic knowledge about behavior is fundamental to better development of biomonitoring and conservation studies. Researchers use animal behavior to track a species' movement, dispersal, and feeding behaviors to decide the best time and place for collecting different species that are effective bioindicators. For example, dragonflies are best sampled on sunny days between 10:00 and 15:00 h, due to their known thermal restrictions (Corbet 2004). Blackflies include important disease vector species and are known to be associated with rapid-flowing waters, so the sampling strategy to study these animals should focus on these habitats. Some species of *Chironomus* (Chironomidae) prefer habitats with low dissolved concentrations and high organic matter, and therefore are effective bioindicators of organic matter pollution (Rosenberg and Resh 1993).

Another key information that behavioral studies can provide to biomonitoring relies on the classical relationship between biodiversity and environment, a cornerstone of biomonitoring. Any change in the environment can affect survivorship, reproduction, and growth. When these effects are scaled up to the community level, distributional changes emerge. Reproduction is a particularly interesting life history aspect that is used to better understand the biodiversity-environment relationship of aquatic insects because adult female insects can select favorable environmental conditions to lay their eggs and increase their offspring's chances of survivorship and population size (Encalada and Peckarsky 2012; Lancaster and Downes 2014; Heino and Peckarsky 2014). Some aquatic insects, such as the damselfly *Calopteryx splendens*, select more oxygenated, fast-flowing water for laying eggs, which significantly increases egg development time while decreasing their mortality, compared to slow-flowing waters (Siva-Jothy et al. 1995). Others species identify suitable structures to lay their eggs, such as protruding substrates (rocks, stems, wood). For example, the number of oviposition sites (protruding rocks) influenced the density of larvae of the mayfly *Baetis rhodani* and a caddisfly *Rhyacophila dorsalis*, whose adults oviposit exclusively on emergent rocks in riffles (Lancaster and Downes 2014). The neonate and mid-stage larva distribution pattern of mayfly *Baetis rhodani* is positively related to egg masses, suggesting a persistence of maternal effects on larval distribution patterns (Lancaster et al. 2011). Damselflies may select specific patches of a water body with certain vegetation structure to oviposit, where larvae will develop (Guillermo-Ferreira and Del-Claro 2011). In the neotropics, oviposition behavior of dragonflies was used to understand the relationship between species richness and forest loss (Rodrigues et al. 2018). The number of species of exophytic and epiphytic (species which lay eggs on the emerged surface of substrates) behavior was positive, but negatively related to riparian forest loss, suggesting that forest loss may favor species with particular oviposition behaviors (Rodrigues et al. 2018). These studies demonstrate that a strong effect of adult behavior on larva distribution and oviposition trait behaviors can be incorporated in ecology and biomonitoring to better understand biodiversity-environment relationships.

The oviposition behavior involves decisions by female insects, such as to stay at their natal sites or to disperse from their natal sites to another site with more suitable environmental conditions. In this way, oviposition behavior and dispersal are interconnected, although the former is a small-scale phenomenon and the latter is related to larger scale regional processes (dispersal between sites). Species dispersal behavior may change the classical biodiversity-environment relationship, an important connection recognized in metacommunity ecology (Leibold et al. 2014; Leibold and Chase 2018). Species may leave a high-quality patch as a result of intense competition and may colonize low-quality patches, and where mortality rates are greater than reproduction rates. The high mortality is overcome by dispersers from high-quality patches (source patches) maintaining populations in the low-quality patches (sink patches). In turn, these source-sink dynamics could mask biodiversity-environment relationship, affecting biomonitoring (Heino 2013). For example, specialist species can be present in low-quality patches not because the conditions and resource levels are suitable in those patches, but as a result of dispersal of individuals from a close high-quality patch. Despite this theoretical perspective, information on dispersal such as the maximum distances, direction, and frequency of movement is unknown for most aquatic insects (Lancaster and Downes 2014). Consequently, dispersal proxies are increasingly being used in biomonitoring and conservation studies such as distance-related measures among sites, dispersal traits, and graph-based approaches (e.g., Erős et al. 2012; Heino 2013). However, a recent study showed that dispersal traits did not predict colonization (Downes and Lancaster 2018), as usually assumed, indicating that much more information is needed regarding this type of behavior.

Although adult aquatic insects can fly and consequently disperse longer distances than larvae, drift is also important at smaller spatial scales, such as the stream reach scale (Downes and Lancaster 2018). Larvae may actively select for microhabitats with specific backgrounds or substrates to occupy (Tavares et al. 2018), which may have effects at macroscale and determine species distribution and community composition. Maternal behavior detected in early instars of *B. rhodani* and *R. dorsalis* is a result of female selection of sites to oviposit but is obscured in later instars (Lancaster and Downes 2014). This result is a consequence of density-dependent mortality of early instar instars, mainly due to predation, and dispersal (drift). These effects of mortality and drift on later instars could mask the environmental-biodiversity relationship, if it is assumed that high-quality habitats are more abundant than low-quality habitats. Drift can be even more important for community patterns than aerial dispersal. In a large-scale experiment (22 km stream length),

Downes and Lancaster (2018) demonstrated that larval drift is the main dispersal route used to reach downstream habitats and that dispersal traits frequently used to infer dispersal did not capture colonization ability. These results suggest the need to improve our current knowledge of aquatic insect dispersal to improve conservation planning, restoration, and biomonitoring of aquatic systems (Erős et al. 2012; Heino 2013; Tonkin et al. 2018).

Understanding the behavior of aquatic insects and its application in biomonitoring and conservation is still incipient, particularly in the Neotropical region. This knowledge gap leads to weak or even erroneous conclusions. This means that we need an improvement in information on insect behavior to enhance biomonitoring, with many more studies required on the large number of species despite the difficulties associated with analysis of aquatic insects (e.g., direct observation of larval behavior). For conservation planning and biomonitoring related to accelerating rates of habitat loss, fragmentation, and transformation, we could use some behavioral proxies, such as species traits (e.g., Rodrigues et al. 2018) and distance proxies for dispersal (Tonkin et al. 2018). Alternatively, we could use other approaches to improve conservation planning and to predict future human impacts such as niche modelling (Diniz-Filho et al. 2010). Simulation modelling studies is another interesting tool that can be used to inform conservation and biomonitoring actions (Economo 2011; Siqueira et al. 2014). These models require some biological characteristics to bring the simulations close to reality. Among the required information, dispersal rates and oviposition preferences are particularly important in the case of aquatic insects.

Valente-Neto et al. (2018) used a simulation model to predict biodiversity patterns in a riverine network, and showed the model-predicted observed patterns for the less mobile insects. Despite not knowing the number of eggs and dispersal rates for many aquatic insects, they used two proxies (mean of number of eggs and a dispersal kernel). The same model was also used in a biomonitoring context (Siqueira et al. 2014). We expect that in the near future, we can include more specific behavioral information in modelling approaches and biomonitoring, such as oviposition preferences and maximum dispersal distance, and improve the predictability of biodiversity patterns.

## 15.4 Using Aquatic Insect Behavior Information in Conservation

In the last decade, many studies have emphasized the key role animal behavior plays in conservation practice. Here we will give some examples of how the understanding of aquatic insect behavior can inform better conservation strategies based on scientific evidence.

The most effective conservation strategy for a target species or group is the creation and maintenance of protected areas. Aquatic insect distribution information has been used in prioritizing areas for protection through systematic planning processes (Nieto et al. 2017). For many reasons, the selection of priority areas for conservation has often used species distribution. As a result, the creation of the protected areas has been strongly based on terrestrial biota, and the threats to these species related to the terrestrial ecosystems (Azevedo-Santos et al. 2018). Because the physical organization of freshwater drainage networks influences the dispersal behavior of aquatic insects (Tonkin et al. 2018), ignoring the spatial relations of aquatic habitats can result in less effective planning of the locations of protected areas for conserving aquatic insect metapopulations and metacommunities in the long term (Erős et al. 2012). The behavior of target species is essential when designing protected areas at the landscape scale. For example, the Serra de São José Wildlife Reserve was created to protect and conserve dragonflies in the Atlantic Forest, and has had considerable success based on the locations of many streams with suitable habitats for these species (Bedê et al. 2015).

Another important strategy for aquatic insect conservation is reducing anthropogenic impacts on animal behavior. When humans alter the environment, the fitness value of existing behavioral strategies changes and can lead to local extinctions. Not only do such impacts affect population dynamics and local extinction, but they also lead to the extinction of behaviors. Anthropogenic effects on gene pools may select certain genes and exclude others that would express unique behaviors or behaviors of interest for ecological services. Moreover, in some species, aquatic insects may exhibit behavioral syndromes or personalities. In these cases, genetic diversity results in different phenotypic traits that are interpreted as a diversity of behavioral patterns among individuals. To lose such behavioral diversity can be compared to lose biodiversity and ecological processes that rely on animal behaviors (see Cordero-Rivera 2017). We may even be successful in conserving a flagship species, but at the same time conservation attempts may exclude certain genes that produce behaviors that are typical for this species. For instance, if we conserve dragonflies, but not their typical oviposition sites or habitat selection behaviors, we lose what makes a dragonfly a dragonfly. In other words, we lose the natural genetic components of entire lineages. Ultimately, such losses may impair the stability of populations and lead to long-term extinction events.

Hence, reducing and mitigating anthropogenic stress on aquatic insect behavior are critical for conservation. Studies have produced some of the information needed for a general framework for conserving biodiversity that is likely to be useful for aquatic insects. First reducing habitat loss should be a top priority for conservation planners because land conversion is still the main reason of biodiversity loss in tropical regions (see Rodrigues et al. 2016; Roque et al. 2018). Strategies that reduce anthropogenic threats to natural systems would benefit aquatic insects. Losses of species are clearly increased by deforestation, habitat degradation and transformation, overexploitation, pollution, eutrophication, siltation, impoundments, flood control, exotic species invasions, fisheries, increasing salinity, and climate change. In general, "macro" conservation initiatives depend less on specific knowledge about aquatic insects, but they are intended to implement general public policies for biodiversity conservation. However, there is no reason to wait for solving the entire

Tinbergen shortfall before conservation action. We can improve conservation strategies as more information on aquatic insects becomes available to avoid wasting effort, money, and time.

It is unrealistic to propose behavioral studies of all aquatic insect species to support conservation strategies. The first reason is the Linnean and Wallacean shortfalls. How could researchers study aquatic insect behavior when many species are still not formally described or have a huge gap in their distribution? In addition, behavioral studies demand considerable fieldwork to assess key variables of an insect species (e.g., movement, reproductive behavior, selection of habitats for oviposition) and there are relatively few insect specialists to carry out these studies (Samways 2018). It is important to develop specific agendas for animal species or groups of species of particular relevance, such as endangered species or key species. In these cases, behavioral studies could be part of specific conservation strategies for sustaining these species.

Human and aquatic insects interact in many positive and negative ways. Behavior manipulation is an important field that can inform conservation strategies to mitigate human and aquatic insect conflicts or booster beneficial interactions. For example, manipulating elements of the landscapes (e.g., adding potential perches, oviposition sites) can attract and readjust the behavior of aquatic insect predators (e.g., dragonflies), which could increase the consume of mosquitos and other insects, reducing potential conflicts between humans and these vectors of diseases (Faithpraise et al. 2014). Controlling insect populations through species-specific manipulation of behavior such as sexual communication, without adversely affecting nontarget species, has a long history in applied entomology, particularly in the context of agriculture (Foster and Harris 1997). Beyond the studies on manipulation of behavior for controlling some species of mosquitos, not much has been done on other aquatic insect groups. The use of behavior manipulation (e.g., applications of sex pheromones for attract sexual patterns) as a strategy to improve the populations of red listed aquatic insects is a completely open avenue.

According to the United Nations Food and Agriculture Organization, almost 50% of the world's economy is based directly and indirectly on the use of biological resources. The sustainable use of the biodiversity and ecosystem services for people is the core of conservation strategies (CBD 2014). People directly use aquatic insects in many different ways (Suter and Cormier 2014), such as for food, robotic inspiration, bioindicators, bio-jewels, and pets. Aquatic insects are also involved in critical ecological services, such as water quality, decomposition, and food web regulation. Some species have nonmaterial value, including art, cinema, folklore, music, literature, and religious activities. So, improving the knowledge on aquatic insect behavior has implications for all these areas. Without greater knowledge of the aquatic insect behavior, the promise of using aquatic insects in a sustainable way to improve human health and conservation will remain largely unfulfilled.

Related areas of study such as ecological restoration need to be integrated in establishing protected areas and other large projects because these activities can contribute enormously to biodiversity conservation and global stability and resilience, especially given the current and projected impacts of land-use changes and climate change (Steffen et al. 2015). There are many initiatives to restore aquatic ecosystems around the world. However, in most cases, aquatic insects are neglected, or they are used only as bioindicators of environmental conditions rather than as integral parts of the entire ecosystem. Although there are few documented cases of captive breeding and reintroductions of aquatic insects, some successful reintroduction attempts (e.g., Hemiptera, see Ahlroth et al. 2003; Odonata, see Dolny et al. 2018; Plecoptera, see Zwick et al. 2011) indicate that these activities can enhance restoration and increase the value of protected areas. Because behavior is central for the adaptation of animals to their environments, understanding the ways that aquatic insect species perceive different environments, and how their learning processes affect changes in behavior, is fundamental in planning the best reintroduction strategies. As an example, studies about spatial learning in dragonflies have shown that they remember their positive and negative experiences in different environments and adjust their future space use accordingly (Eason and Switzer 2006). Captive breeding and reintroduction programs can use this kind of behavioral information to increase the fitness of a species in changing environments. In summary, reducing the Tinbergen shortfall is critical to increasing the success of reintroduction and conservation of aquatic insects.

Human behavior is causing a major loss of biodiversity. Changes in human behavior are necessary to achieve more sustainable relationships between humans and the rest of nature. The success of a conservation program can depend on both our knowledge of behavior of aquatic insects and human actions. Aquatic insects and humans interact in complex ways, so changing the behavior of humans in relation to the insects is among the main challenges for aquatic insect conservation. According to Samways (2018), insect conservation throughout the twenty-first century calls for action and new perspectives: (1) philosophy (establishing the ethical foundation), research, policy, psychology (understanding how to engage humans in insect conservation action), practice (implementation of action), and validation (establishing how well we are doing at conserving insects). In this way, educational programs, insect watching movements, citizen science initiatives (e.g., Nerbonne and Nelson 2004; Buytaert et al. 2014), and monitoring programs of freshwater ecosystems (e.g., Buss et al. 2015) can all help humans to better understand the importance of aquatic insects. Like bird or butterfly watching, aquatic insect watching has a huge potential to reconnect people with nature, and this is currently happening. People are already engaged in observing dragonflies, such as Dragonfly Pond Watch, a volunteer-based project of the Migratory Dragonfly Partnership (MDP)—https://xerces.org/dragonfly-migration/pondwatch/; Dragonfly India Citizen Science movement. Moreover, there are many volunteer programs of aquatic monitoring using insects as bioindicators in place across the world. New communication technologies have boosted the flow of data and knowledge in such initiatives, as the RiuNet, a *Citizen Science* project dedicated to assessing the ecological status of a river reach by following the steps of a standardized protocol available in the RiuNet app (http://www.riunet.net). In this context, we have an unprecedented opportunity to create big collaborative networks around the conservation of aquatic insects.

Species extinction is just the most evident consequence of biodiversity loss. In fact, when a species goes extinct, our planet also loses genes, behaviors, interactions, ecological services, and so on. Beyond the potential consequences for humans already explored in this chapter, the loss of animal behaviors due to anthropogenic impacts led to the "extinction of experiences"—in other words, people are becoming less likely to have direct contact with nature (Soga and Gaston 2016).

New fields such as conservation psychology and personalized ecology which focus on intention-behavior exhibited by people are promising areas that link animal and human behavior toward conservation (Gaston et al. 2018; Simaika and Samways 2018). New perspectives such as "behaviorscape," which integrates landscape ecology, behavior ecology, and biodiversity ecology, provide a solid theoretical framework to understand biodiversity patterns with implication for conservation. However, it has not yet been used for aquatic insects. The challenges ahead for biodiversity conservation will require not only the improvement of the Tinbergen shortfall for aquatic insects but also an increase in our understanding of the behavior of our own species in relation to all the others.

## References

Ahlroth P, Alatalo RV, Holopainen A et al (2003) Founder population size and number of source populations enhance colonization success in water striders. Oecologia 137:617–620

Azevedo-Santos VM, Frederico RG, Fagundes CK et al (2018) Protected areas: a focus on Brazilian freshwater biodiversity. Divers Distrib 1–7. https://doi.org/10.1111/ddi.12871

Balian EV, Segers H, Leveque C et al (2008) The freshwater animal diversity assessment: an overview of the results. Hydrobiology 595:627–637

Bedê LC, Machado ABM, Piper W et al (2015) Odonata of the Serra de São José – Brazil's first wildlife reserve aimed at the conservation of dragonflies. Notulae Odonatologicae 8:117–155

Berger-Tal O, Polak T, Oron A et al (2011) Integrating animal behavior and conservation biology: a conceptual framework. Behav Ecol 22:236–239

Blumstein DT, Fernández-Juricic E (2004) The emergence of conservation behavior. Conserv Biol 18:1175–1177

Buss DF, Carlisle DM, Chon T-S et al (2015) Stream biomonitoring using macroinvertebrates around the globe: a comparison of large-scale programs. Environ Monit Assess 187:4132

Buytaert W, Zulkafli Z, Grainger S et al (2014) Citizen science in hydrology and water resources: opportunities for knowledge generation, ecosystem service management, and sustainable development. Front Earth Sci 2:1–21

CBD (2014) Global Biodiversity Outlook 4. https://www.cbd.int/gbo4/. Accessed 17 Jan 2019

Corbet PS (2004) Dragonflies: behaviour and ecology of odonata. Harley Books, Colchester, England

Cordero-Rivera A (2017) Behavioral diversity (ethodiversity): a neglected level in the study of biodiversity. Front Ecol Environ 5:7

Covich AP, Austen MC, Barlocher F et al (2004) The role of biodiversity in the functioning of freshwater and marine benthic ecosystems. Bioscience 54:767–775

Di Marco M, Chapman S, Althor G et al (2017) Changing trends and persisting biases in three decades of conservation science. Glob Ecol Conserv 10:32–42

Diniz-Filho JAF, De Marco JP, Hawkins BA (2010) Defying the curse of ignorance: perspectives in insect macroecology and conservation biogeography. Insect Conserv Divers 3:172–179

Dirzo R, Young HS, Galetti M et al (2014) Defaunation in the Anthropocene. Science 345:401–406
Dolny A, Sigutova H, Ozana S et al (2018) How difficult is it to reintroduce a dragonfly? Fifteen years monitoring *Leucorrhinia dubia* at the receiving site. Biol Conserv 218:110–117
Downes BJ, Lancaster J (2018) Itinerant, nomad or invader? A field experiment sheds light on the characteristics of successful dispersers and colonists. Freshw Biol 63:1394–1406
Dudgeon D, Arthington AH, Gessner MO et al (2006) Freshwater biodiversity: importance, threats, status and conservation challenges. Biol Rev 81:163–182
Dudgeon D (2014) Threats to freshwater biodiversity in a changing world. In: Freedman B (ed) Global environmental change. Springer, Dordrecht, pp 243–253
Eason PK, Switzer PV (2006) Spatial learning in dragonflies. Int J Comp Psychol 19:268–281
Economo EP (2011) Biodiversity conservation in metacommunity networks: linking pattern and persistence. Am Nat 177:167–180
Encalada AC, Peckarsky BL (2012) Large-scale manipulation of mayfly recruitment affects population size. Oecologia 168:967–976
Erős T, Olden JD, Schick RS et al (2012) Characterizing connectivity relationships in freshwaters using patch-based graphs. Landsc Ecol 27:303–317
Faithpraise FO, Idung J, Usibe B et al (2014) Natural control of the mosquito population via Odonata and Toxorhynchites. Int J Innov Res Sci Eng Technol 3:12898–12911
Foster SP, Harris MO (1997) Behavioral manipulation methods for insect pest-management. Annu Rev Entomol 42:123–146
Gaston KJ, Soga M, Duffy JP et al (2018) Personalised ecology. Trends Ecol Evol 33:916–925
Guillermo-Ferreira R, Del-Claro K (2011) Oviposition site selection in *Oxyagrion microstigma* Selys, 1876 (Odonata: Coenagrionidae) is related to aquatic vegetation structure. Int J Odonatol 14:275–279
Heino J (2013) The importance of metacommunity ecology for environmental assessment research in the freshwater realm. Biol Rev 88:166–178
Heino J, Peckarsky BL (2014) Integrating behavioral, population and large-scale approaches for understanding stream insect communities. Insect Sci 2:7–13
Hortal J, de Bello F, Diniz-Filho JAF et al (2015) Seven shortfalls that beset large-scale knowledge of biodiversity. Annu Rev Ecol Evol Syst 46:523–549
IUCN (2017) Redlist of threatened species, version 2017-3. http://www.iucnredlist.org. Accessed 08 Jan 2019
Lancaster J, Downes BJ, Arnold A (2011) Lasting effects of maternal behaviour on the distribution of a dispersive stream insect. J Anim Ecol 80:1061–1069
Lancaster J, Downes BJ (2014) Maternal behaviours may explain riffle-scale variations in some stream insect populations. Freshw Biol 59:502–513
Leibold MA, Holyoak M, Mouquet N (2014) The metacommunity concept: a framework for multi-scale community ecology. Ecol Lett 7:601–613
Leibold MA, Chase JM (2018) Metacommunity ecology, vol 59. Princeton University Press, Princeton
Lomolino MV (2004) Conservation biogeography. In: Lomolino MV, Heaney LR (eds) Frontiers of biogeography. New directions in the geography of nature. Sinauer Associates, Sunderland, MA, pp 293–296
Luck GW, Harrington R, Harrison PA et al (2009) Quantifying the contribution of organisms to the provision of ecosystem services. Bioscience 59:223–235
Merrit RW, Cummins KW, Berg MB (2008) An introduction to aquatic insects of North America, 4th edn. Kendall Hunt Publishers, Dubuque
Nerbonne JF, Nelson KC (2004) Volunteer macroinvertebrate monitoring in the United States: resource mobilization and comparative state structures. Soc Nat Resour 17:817–839
Nieto C, Ovando XMC, Loyola R et al (2017) The role of macroinvertebrates for conservation of freshwater systems. Ecol Evol 7:5502–5513
Peckarsky BL, Cooper SD, McIntosh AR et al (1997) Extrapolating from individual behavior to populations and communities in streams. J N Am Benthol Soc 16:375–390

Pelicice FM, Azevedo-Santos VM, Vitule JR et al (2017) Neotropical freshwater fishes imperiled by unsustainable policies. Fish Fish 18:1119–1133
Rodrigues ME, Roque FO, Quintero JMO et al (2016) Nonlinear responses in damselfly community along a gradient of habitat loss in a savanna landscape. Biol Conserv 194:113–120
Rodrigues ME, Roque FO, Guillermo-Ferreira R et al (2018) Egg-laying traits reflect shifts in dragonfly assemblages in response to different amount of tropical forest cover. Insect Conserv Divers. https://doi.org/10.1111/icad.12319
Roque FO, Menezes JFS, Northfield T et al (2018) Warning signals of biodiversity collapse across gradients of tropical forest loss. Sci Rep 8:1622
Rosenberg DM, Resh VH (1993) Freshwater biomonitoring and benthic macroinvertebrates. Springer, New York
Samways MJ (2007) Insect conservation: a synthetic management approach. Annu Rev Entomol 52:465–487
Samways MJ, McGeoch MA, New TR (2010) Insect conservation: a handbook of approaches and methods. Oxford University Press, Oxford
Samways MJ (2018) Insect conservation for the twenty-first century. In: Shah MM, Sharif U (eds) Insect science diversity, conservation and nutrition. InTechOpen, London
Simaika JP, Samways MJ (2018) Insect conservation psychology. J Insect Conserv 22:635–642
Siqueira T, Durães LD, Roque FDO (2014) Predictive modelling of insect metacommunities in biomonitoring of aquatic networks. In: Ferreira CP, Godoy WAC (eds) Ecological modelling applied to entomology. Springer, Amsterdam, pp 109–126
Siva-Jothy MT, Gibbons DW, Pain D (1995) Female oviposition site preference and egg hatching success in the damselfly *Calopteryx splendens xanthostoma*. Behav Ecol Sociobiol 37:39–44
Soga M, Gaston KJ (2016) Extinction of experience: the loss of human-nature interactions. Front Ecol Environ 14:94–101
Steffen W, Richardson K, Rockström J et al (2015) Planetary boundaries: guiding human development on a changing planet. Science 347:6223
Strayer DL (2006) Challenges for freshwater invertebrate conservation. J N Am Benthol Soc 25:271–287
Strayer DL, Dudgeon D (2010) Freshwater biodiversity conservation: recent progress and future challenges. J N Am Benthol Soc 29:344–358
Suter GW, Cormier SM (2014) Why care about aquatic insects: uses, benefits, and services. Integr Environ Asses Manag 11:188–194
Sutherland WJ (1998) The importance of behavioural studies in conservation biology. Anim Behav 56:801–809
Tavares RI, Pestana GC, Rocha AD et al (2018) Come to the dark side: habitat selection of larval odonates depends on background visual patterns. Ecol Entomol 43:640–646
Tonkin JD, Altermatt F, Finn DS et al (2018) The role of dispersal in river network metacommunities: patterns, processes, and pathways. Freshw Biol 63:141–163
Valente-Neto F, Durães L, Siqueira T et al (2018) Metacommunity detectives: confronting models based on niche and stochastic assembly scenarios with empirical data from a tropical stream network. Freshw Biol 63:86–99
Vörösmarty CJ, McIntyre PB, Gessner MO et al (2010) Global threats to human water security and river biodiversity. Nature 467:555–561
Whittaker RJ, Araújo MB, Paul J et al (2005) Conservation biogeography: assessment and prospect. Divers Distrib 11:3–23
Winemiller KO, McIntyre PB, Castello L et al (2016) Balancing hydropower and biodiversity in the Amazon, Congo, and Mekong. Science 351:128–129
Zwick P, Becker G, Wagner R et al (2011) The fauna of the Breitenbach in central European stream ecosystems. In: Wagner R, Marksen J, Zwick P et al (eds) The long term study of the Breitenbach. Weinheim, Wiley-Blackwell, pp 195–485

# Chapter 16
# Genetic Connectivity in Conservation of Freshwater Insects

**Drielly da Silveira Queiroga, Renan Fernandes Moura, and Jessica Ware**

**Abstract** Ecosystems and species are disappearing fast and the conservation of isolated and fragmented landscapes is not enough to maintain healthy populations. However, populations from fragmented and impacted landscapes may be benefited if there are pathways allowing their connection. These pathways enable the exchange of individuals, allowing species to increase their genetic diversity and resilience to stochastic events by recolonization and phenotypic adaptations. In aquatic ecosystems, climate changes and water exploration are impacting the species' capability to disperse among populations and survival. In this scenario, aquatic insects are even more threatened as most of them have terrestrial and aquatic life stages, suffering impacts from both environments. Focusing in this aspect, this chapter aims to provide an initial insight about how population connectivity can be used in conservation strategies as well as methods of measuring genetic connectivity. Here we selected studies with odonates, ephemeropterans, and other aquatic insects to exemplify how river dynamics can influence the direction of gene flow and dispersal patterns of individuals, besides showing the main approaches used in this study area. By contributing to the understanding of this necessary field, we hope to stimulate new researchers to engage in the conservation of aquatic insects.

**Keywords** Conservation genetics · Endangered arthropods · Entomology · Habitat fragmentation · Molecular markers

D. S. Queiroga (✉)
Department of Biology, University of São Paulo (USP), Ribeirão Preto, Brazil

R. F. Moura
Universidade Federal de Uberlândia – UFU – Centro de Estudos do Cerrado/Laboratório de Ecologia Comportamental e de Interações (LECI), Uberlândia, MG, Brazil

J. Ware
Biology Department, Rutgers University, Newark, NJ, USA



K. Del-Claro, R. Guillermo (eds.), *Aquatic Insects*,
https://doi.org/10.1007/978-3-030-16327-3_16

## 16.1 An Overview

Globally environments are disappearing faster than we can access their diversity. Scientists, environmentalists, and governments have issued dire warnings about biodiversity loss and the consequences of such losses for human health and environment permanence. There are many causes for biodiversity loss ranging from the exploitation of natural resources to the introduction of exotic species and global warming. According to World Wide Fund for Nature (WWF), a rich biodiversity is fundamental for the balance and stability of ecosystems, providing vital ecosystem services such as climate regulation, nutrient and waste management, flood control, coastal protection, provision of food and freshwater, fuel, medicines, building materials, fertile soils, and breathable air. Biodiversity has also a wide potential for economic use, especially for biotechnology. Extinction is a natural fate for any species, and occurs at some background rate; however, anthropogenic activities have increased extinction rates across species, causing catastrophic diversity loss. Preservation is imperative when it comes to aquatic environments given the intimate relationship between water and all living beings. Although aquatic environments are essential to human health, they are often the most exploited environments. Human reliance on freshwater water for consumption, agriculture and industrial purposes, and effluent dilution has resulted in freshwater vulnerability.

Recognizing the main components and dynamics of aquatic environments is the first step to establish conservation strategies. For instance, the spatial arrangement of freshwater systems is widely recognized as a fundamental attribute to be investigated. Freshwater systems are organized in drainage basins, which have distinct geological processes of formation, being composed of interconnected springs, streams, rivers, and lakes along the basin. The water direction and speed are determined by the region's topography, which can create rapids or standing water sites. The basin's arrangement is known as dendritic network, since the interconnection among streams and rivers resembles the design of a tree, with numerous small branches interconnecting and increasing in thickness as they approach to the trunk. When an ecological disturbance occurs, the direction of water flow and the disturbance origin can be used to predict negative impacts along the basins. This information combined with the regional topography can be useful to define conservation strategies within threatened areas.

Other aspects of river basins can help us to predict species diversity. For example, springs and small streams have small volumes of water and low sunlight incidence. As streams interconnect, becoming large rivers, their both the water volume and the sunlight incidence increase. For aquatic insects, the sunlight incidence and structural changes of rivers, as well as the properties of terrestrial surroundings (that may provide microhabitats), are fundamental to determine what species may occur at a given point of the watershed.

Many aquatic insects may increase or decrease in abundance in response to environmental disturbances; these species are known as bioindicators and may be used to evaluate the environmental quality of aquatic ecosystems. Changes in the relative abundance of bioindicator species may be enough to indicate whether a given environment is under stress. Thus, it is imperative to preserve bioindicators as they can alert us before environmental impacts become large enough to extinguish ecologically important species. Unfortunately, bioindicators cannot be used as universal conservation tools as they are an indirect and relatively subjective measuring of environmental quality. So what other approaches can be used to conserve biodiversity? Currently, there are several conservation approaches for aquatic organisms, but the most common practices have serious economical and logistical issues. For instance, protecting regions where threatened populations live is essential, but it is not possible for all populations. Research species conservation programs cost a lot of time and resources, and many species are widely distributed, making it infeasible to be applied for all known threatened populations. Since we are not able to protect all species at the same time, priorities should be established: How do we evaluate which species deserves more attention?

To best determine which species require the most attention, many scientists have turned to genetic tools to identify genetically diverse populations and their relative genetic connectivity. Such studies are fundamental for conservation genetics, that uses genetic analyses for management and conservation purposes. The International Union for Conservation of Nature (IUCN) recommends the preservation of high genetic diversity in populations. High genetic diversity may allow populations to more quickly adapt to environmental changes and reduce their levels of inbreeding—these two features are related to the increase in survival and fitness of natural populations. Thus, high genetic diversity and connectivity among populations—which increase genetic diversity—are one of the most important factors for the preservation of natural populations.

The importance of genetic diversity in the maintenance of natural populations has become more widely studied recently thanks to an increase in the number of conservation genetic studies that seek to combine genetic studies of populations with conservation. Many of these studies focus on understanding the mechanisms that lead to the persistence of a population and how populations respond under some natural or anthropogenic environmental stress. In this sense, this chapter provides some basic concepts to understand the studies that approach genetic connectivity in freshwater environments, focusing mainly on aquatic insects.

## 16.2 What Is Genetic Connectivity?

Genetic connectivity is the exchange of genetic material among populations of the same species, allowing the genetic diversity maintenance. This exchange occurs through the processes of **dispersion** and **migration**, which are movements performed by individuals or populations (Kool et al. 2013). While migration is characterized by long-distance movements, often related to seasonality—as seen in birds and fishes—dispersal is used more commonly to describe any local movements performed individuals. Dispersion can be directional when an individual goes to a "planned" place, such as its breeding site or randomly when reaching "non-planned" places, other populations, or uninhabited sites. Dispersion allows individuals to reach different populations, but genetic connectivity only occurs through mating between individuals. This reproductive connection among populations is known as **gene flow** and it may increase **genetic diversity** by introducing new alleles into a target population, changing local allele frequencies (Whitlock and Mccauley 1999).

Links among populations may be reduced or prevented due to **barriers**. Barriers are any obstacle to species dispersal. They may be structural or even environmental variations, whether naturally occurring or not. They act by separating populations although sometimes individuals may overcome barriers, resulting in the connection of two or more populations. When a barrier is strong enough, however, that no individual can transpose it, the populations become isolated, causing a genetic diversity loss (Crook et al. 2015).

The importance of gene flow for the maintenance of populations is clear, and measurements of gene flow reflect the evolutionary history of species (Ye et al. 2018). Studies on genetic connectivity have sought to understand the evolutionary processes and natural history of species. This knowledge has helped researchers to develop conservation efforts, and manage endangered species, including the mountain lion, manatee terns and harpies (Castilho et al. 2012). The pressures suffered by species vary among distinct environments, reinforcing the importance of specific studies for each species or environment: freshwater and terrestrial species greatly differ in response to environmental pressures. Within the freshwater habitat, differences occur in water flow, substrate, etc., making each environment relatively unique (Campbell Grant et al. 2007). For terrestrial species, physical structures such as vegetation, mountains, or urbanness may act as barriers to dispersal (Anderson et al. 2010). By contrast, barriers in aquatic environments are physical-chemical (i.e., temperature and water oxygen rate) more often than structural (i.e., waterfalls, dams, or rapids) (McRae 2006).

### *16.2.1 What Is the Importance of Genetic Connectivity in Aquatic Environments?*

In freshwater systems, water speed and depth, and chemistry can act as barriers to organismal dispersion, as well terrestrial environments, that ultimately limit and shape aquatic systems (Phillipsen and Lytle 2013). Barriers may limit movements of organisms among rivers, generating high genetic structuring by isolating certain populations. Connectivity in aquatic systems has four dimensions: (1) longitudinal, throughout the channel; (2) lateral, in floodplain inundation events; (3) vertical, between the water mirror and the other water layers in environments with great depth; and (4) temporal, due to seasonal and stochastic events. Given this, McGlashan and Hughes (2001) suggest three possible scenarios: (1) negligible gene flow among drains over long periods, resulting in strong genetic structuring; (2) moderate gene flow among closely connected adjacent drains; and (3) extensive genetic flow, which reflects a high dispersion potential that can overcome the isolation imposed by drainage structures.

The structure of most basins has a tree pattern, where their trunk and branches form a structure known as a **dendritic network** (Campbell Grant et al. 2007). This structure can be found in many freshwater environments—except in isolated lakes—and is organized hierarchically, from the source (upstream) to the mouth (downstream), following the direction of water flow. This dendritic hierarchy was proposed by Robin L. Vannote et al. (1980), in a theory known as the "River Continuum Concept" (RCC). In the RCC, there are three groups of fluvial gradients, classified by its number of streams confluence: headwaters (stream order 1–3), midreaches (stream order 4–6), and lower reaches (stream order > 6) (Fig. 16.1). Each stream has different characteristics—i.e., variation in shading provided by riparian forests, sediment type, and water volume—that allow us to predict which organisms are expected to occur at each stream point. Since the environmental requirements of organisms vary along a fluvial gradient, RCC theory can be used to predict potential barriers to dispersion. This idea was tested by Meffe and Vrijenhoek (1988) and Hughes et al. (2009); they proposed a model based on RCC to explain the connectivity of aquatic populations called the "Stream Hierarchy Model" (SHM).

Under the SHM, connectivity patterns among populations of riverine species are a consequence of the dendritic arrangement of streams (Hughes et al. 2013). Under this model, populations from distinct sites within the same stream would show higher connectivity than populations present in different streams, but in the same sub-catchment. Subsequently, population connectivity would decrease among distinct sub-catchments and catchments, following a hierarchical dendritic model,

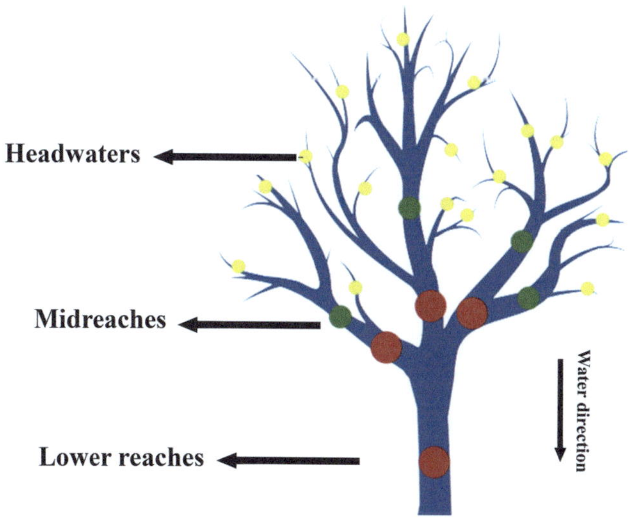


**Fig. 16.1** Representation of a watershed in its tree-like dendritic pattern. The hierarchical structure is represented by circles: yellow (headwaters), green (midreaches), and red (lower reaches)

where connectivity decreases as the scales are increased. For instance, shredders—organisms that feed on coarse particulate organic material—are expected to be found in headwaters. These animals are represented by organisms such as mayflies (Ephemeroptera) or stoneflies (Plecoptera) who will feed on fallen leaves, abundant in headwaters and small streams due to riparian vegetation. In bigger streams and rivers, food might not be accessible for these organisms since most organic matter is composed of fine particles that may be taken away by strong water streams (Horne and Goldman 1994). Therefore, as RCC predicts a tree pattern, with habitats of these mayflies and stoneflies connected by large rivers, SHM predict that structural and chemical changes which occur along stream orders will be barriers to the dispersal of aquatic species, rather than the distance or isolation of streams (Keller et al. 2012).

While in lotic systems the main barriers to species distribution are the terrestrial environment between headwaters, waterfalls, and dams; the major barrier in lentic systems is the stratification of water bodies that may occur in deep lakes. Stratification is a phenomenon where different water layers of varying temperatures occur simultaneously, and the less dense layers—those with the highest temperatures—float on the denser ones with a minimum degree of mixing between them. This occurs due to the absence of aquatic currents that mix the surface water (heated by the atmosphere) with the deep water. However, stratification is a dynamic phenomenon and the water layers behave differently every season. In temperate lakes, in the early spring, when temperatures are increasing, the ice layer breaks and the water temperature increases to 4 °C, reducing its density. As summer approaches, air temperature heats surface waters and decreases their density, but deeper water remains cold due to low light incidence. In summer, there is a water layer called **thermocline**, a band of rapid temperature changes, which occurs between **hypolimnion**, the lowest and coldest lake layer, and **epilimnion**, the upper warm layer. During the fall season, the epilimnion and hypolimnion waters start to mix as their temperature equalizes until the winter, where occurs an inverse stratification, with the high-density water on top (Horne and Goldman 1994). These changes directly affect the biota in these environments. Changes in the water temperatures across seasons, for example, limits the occurrence of some species. Changes in this cycle can cause drastic population reductions (bottlenecks), that may result in a loss of genetic diversity and extinction events.

Knowing the main limitations to the dispersion of organisms is fundamental to assess the connectivity between populations. Factors such as mobility capacity, life cycle, environmental dependences, and sensibility can influence the dispersal potential of species and it should be taken into account when planning studies to assess connectivity between populations. There are several techniques for assessing connectivity between populations. However, it is always indicated to combine two or more techniques to increase the accuracy of the results and to decide what techniques are the best will depend on the organism in question. The following is a summary of the two major categories of connectivity assessment techniques seen in most studies.

### 16.2.2 How Can One Evaluate Connectivity?

There are several methods for estimating connectivity between populations. Since each method has advantages and disadvantages, researchers should choose methods tailored to the organisms and environments to be evaluated. Such methods are grouped into two categories: **direct** and **indirect estimates** (Kool et al. 2013). In direct estimates, individuals or gametes are samples used to estimate gene flow (i.e. Villella et al. 2004). Mark-recapture techniques are common direct approaches (Fig. 16.2); however, they are relatively imprecise about the actual gene exchange between populations (Kool et al. 2013). For example, dispersing individuals may not be successful in finding a mate if they eventually arrive at inadequate habitats that reduce their chances of a successful mating. On the other hand, indirect estimates investigate the allele frequencies between populations. They are commonly used due to its instant response, without the uncertainty of recapture and the need of waiting for individuals to travel between populations. Despite these advantages, indirect measurements have disadvantages: genetic samples assess the outcomes of past connectivity, but recent changes to populations may not be yet detectable (Rosa 2009).

### 16.2.3 Molecular Markers: What Genetic Analyses Are Used to Evaluate Population Connectivity?

#### 16.2.3.1 Mitochondrial DNA (mDNA)

Mitochondrial genes are maternally inherited by the offspring. The main characteristic of mtDNA markers is their non-Mendelian pattern of inheritance, with low rate of genetic recombination. Polymerase chain reaction (PCR) can be used to

**Fig. 16.2** Mark-recapture techniques are used to assess individual's movements among populations. The picture shows an individual from *Erythemis credula* (Libellulidae) marked on its right hindwing

amplify copies of the mtDNA which can be sequenced and analyzed. mtDNA are useful markers to estimate phylogenetic relationships, estimate genetic distance, and discriminate subpopulations and overall biogeographic history.

#### 16.2.3.2 Microsatellites

Microsatellites, also known as short tandem repeats (STRs) or simple sequence repeats (SSRs), are small sequences of DNA (1–6 base pairs), organized in tandem, randomly distributed in the eukaryotic genome. Using a small sample of DNA, microsatellites can be amplified by primers—segments of nucleic acids necessary for the initiation of DNA replication—in a polymerase chain reaction (PCR) technique, where a small and specific region of the genome is amplified by DNA polymerase enzymes. Due to their high frequency, polymorphism, and distribution along the genome of many species, microsatellites are used for tests on recent genetic splits among populations as estimators of genetic distances, discriminating subpopulations, genetic diversity, gene flow, bottleneck effects, hybridism, and individual identification.

#### 16.2.3.3 Single-Nucleotide Polymorphism (SNP)

SNPs reflect the variation in DNA sequences at a single nucleotide in a specific position of the genome. They present low rates of mutation, and are commonly used as genetic markers for population-based, phylogeographic, and phylogenetic studies.

**Freshwater Environments**

Freshwater environments encompass a wide variety of formations, with different geological history, and biotic and abiotic components. They are divided into three types: lotic, lentic, and wetlands, which are a specific type of lentic system.

**Lotic** systems have flowing waters and can vary from springs, of few centimeters of deep, to large water bodies such as the Brazilian Amazon River, the biggest river on earth, with 6.992 km in length. Lotic systems can vary in speed, form, and size. Springs and rivers are examples of lotic systems, as are streams, creeks, waterfalls, and rapids. Rivers have unidirectional flow, continuous physical change, high degree of spatial and temporal heterogeneity at all scales, and high habitat variability throughout their orders and bear a specialized biota adapted to live in running water conditions.

**Lentic** systems are those where the water flows slowly or is completely stopped. They are lakes, lagoons, water puddles, wetlands, and even seasonal pools. They can range from a temporary rainwater pool with few centimeters depth to areas such as the Baikal lake of Russia, with a maximum depth of

1740 m. While lotic systems present a continuous flow that homogenizes the water temperature, lentic systems have a thermal stratification that results in water layers of different temperatures.

**Wetlands** are areas where the soil is saturated or flooded (permanently or seasonally). The main wetland types are swamps, marshes, bogs, fens, and peatlands. Wetlands occur on every continent and are important to water purification, flood control, and carbon sink. The largest wetlands are the Amazon River basin, the South American Pantanal, the West Siberian Plain, and the Sundarbans in the Ganges-Brahmaputra delta (Fig. 16.3).

**Fig. 16.3** An example of lotic system, a stream surrounded by a riparian forest in the Cerrado biome, southeastern, Brazil

## 16.3 New Approaches in Freshwater Connectivity: Riverscape Genetics

There are many studies aiming to assess how habitat heterogeneity influences genetic structure of populations in terrestrial environments by combining methods from landscape ecology, spatial statistics, and population genetics (Tischendorf and Fahrig, 2000; Alp et al. 2012; Phillipsen et al. 2015). These studies comprise a sub-discipline known as landscape genetics that, more recently, has been expanded to be used in marine environments and defined as "seascape genetics" (Davis et al. 2018). Both landscape and seascape genetics try to explain how specified habitats can increase or decrease the movements of individuals and test whether certain models of individual movements can explain connectivity better than other models.

For river environments, an approach called riverscape genetics (RG) is commonly used. RG combines methods from landscape ecology and population genetics. It seeks to understand processes and factors that interfere in population dynamics to gather information that can be used to develop conservation strategies. Davis et al. (2018) define RG as "an area of study that evaluates the effect of riverscape features on spatial genetic variation." Thus, it evaluates the effect of riverscape features on spatial genetic variation considering the riverscape as a continuum where habitats have permeable boundaries but also discontinuities among patches that can influence river organisms.

Studies in RG use continuous and discrete sampling designs to identify physical barriers and correlations between spatial genetic variation and abiotic factors. Most studies about RG, however, have been focused on discrete barriers such as waterfalls and dams (isolation by barrier hypothesis, i.e. Coleman et al. 2018) and on the physical distance among populations (isolation by distance hypothesis, i.e. Finn et al. 2006). For insects, recent studies concluded that the greatest obstacle to the dispersion of most aquatic insects is distance (Sabando et al. 2011; Short and Caterino 2009; Watanabe et al. 2008, 2010). Although their winged form can overcome most physical barriers, they often fail to travel long distances (Pfeiler and Markow 2017), with the exception of long-distance migrants like the dragonfly *Pantala flavescens* (e.g., Hobson et al. 2012; May 2013). An example of low dispersal is found in the study carried out by Chaput-Bardy et al. (2008) with the damselfly *Calopteryx splendens* (Fig. 16.4), where they found clear isolation by distance among populations. Adults cannot disperse long distances due to their limited flight capacity. By contrast, larvae can disperse greater distances than adults—despite their lower active dispersal capacity—as they are carried by the stream flow.

**Fig. 16.4** The damselfly *Calopteryx splendens* (male) defending its territory on water. Photo by Kilodk on pixabay.com

Compiling several studies on riverscape genetics, Davis et al. (2018) evaluated how they have approached riverscape features in distinct aquatic taxa and pointed a few prospects for future studies. They found that studies addressing the effects of natural and anthropogenic barriers commonly seek explanations based on isolation by distance and by barriers to explain genetic diversity, but rarely these two hypotheses alone have fully explained the patterns found. Although their importance is recognized as structuring processes of genetic diversity, there is another kind of isolation that has been used to study certain organisms such as fishes. Isolation by resistance occurs when effects of multiple landscape variables, such as precipitation and temperature, are the main limitations to gene flow among populations. In this scenario, populations can respond to physical-chemical variations as barriers to dispersion more than physical barriers or distance (McRae 2006). Although there are few studies that combine all these approaches (barriers, distance, and resistance), the use of isolation by resistance concept combined with isolation by distance and barriers provides better responses to river systems.

Besides the hypotheses that try to explain connectivity patterns based on potential dispersal limits, there are others that predict the direction of gene flow. Created by Paz-Vinas et al. (2015), the Downstream Increase in Intraspecific Genetic Diversity (DIGD) hypothesis describes the asymmetrical genetic pattern found in many groups of aquatic organisms, predicting gene flow direction and estimating the genetic variation source found in a species population. DIGD suggests that dendritic networks allow asymmetrical gene flow, with the direction of gene exchange between populations usually following the water flow direction, since many species cannot travel against it (Keller et al. 2012).

Certain phenotypes can be selected by specific pressures on populations. When this occurs, organisms usually show strict adaptations that give them advantages to explore local resources better than nonspecialized neighbors, increasing their fitness. Whitehead et al. (2011) observed that living populations of the killifish *Fundulus heteroclitus* had adaptations that allowed them to tolerate pollution. In a stressed environment, these adaptations were mediated by genes related to osmotic shock, by increasing velocity response to these conditions. The link between spatial adaptive genetic variation and riverscape heterogeneity is the focus of gene-environment association (GEA) studies, which aim to find associations between genes and environment conditions.

Although GEA studies have been reporting strong correlations between genes and environment, their main challenge is in the study of wild populations. However, there are problems in assessing adaptations in wild populations regarding the ability of isolating the environmental variables of interest. An alternative to these problems is computer programs that carry out genetic selection simulations, such as LOSITAN (Looking for Selection in a Tangled dataset) (Antao et al. 2008). These programs help to understand adaptive evolution in natural populations by evaluating how dispersal and connectivity affect adaptive genes across populations. This information can be used to elucidate how environmental disturbances (such as biological invasions, diseases) affect natural populations.

An important issue is that GEA studies do not always reflect recent conditions and events. Adaptations settle in populations only after several generations, demanding a large timescale for some species depending on the time between generations. Thus, in many cases, spatial genetic structure of populations can be attributed to ancient historical events, as the Pleistocene glacial cycles or mega-flood events that affected the genetic connectivity among populations (Wong et al. 2004). Geological processes can structure or interrupt connections between populations on an entire river network. Current and past geological processes that affect genetic variation of populations allow us to preserve different species by providing information about the resilience, adaptive capacity, and natural history of populations. To evaluate the importance of past processes, scientists have developed molecular techniques capable of detecting the effects of ancient geological processes on current genetic patterns. Ye et al. (2018) using mDNA and other molecular markers found that the current connectivity patterns in *Metrocoris sichuanensis* (Hemiptera: Gerridae) were established in the last glacial event and did not change until the present. Despite the interesting results of these techniques, most studies have been conducted using fishes, which limits the extrapolation of these results to other aquatic organisms.

Riverscape genetics is a new and specialized study area that uses landscape genetics applied in river studies, but just as every new approach the methods involved are constantly being discussed and improved. Perhaps, the greatest challenge of RG today is to find a way to incorporate the dendritic spatial arrangement of rivers and the water directional movement on its modeling. Although landscape genetics addresses spatial variables, the dendritic arrangement of rivers induces organisms to exhibit dispersion patterns and connectivity that are dependent on the conformation of water bodies, flow, and physicochemical factors, and approaches that focus on these particularities are needed (Eros and Campbell Grant 2015). These particularities have been studied in seascape genetics studies; however, studies focusing on riverine organisms and other organisms than fishes can help us to establish ecological generalizations.

## 16.4 Genetic Connectivity in the Study of Aquatic Insects

In aquatic environments, there are several ways to use the dispersive potential of a species to approach connectivity, but knowledge regarding behavior and natural history of the involved species is required to evaluate their dispersal ability (Hughes 2007). For instance, it is known that flight is the major dispersal mechanism of winged insects while larvae and species with fully aquatic adult forms apparently move very little within water streams, resulting in significant genetic differentiation. Despite their limited dispersion, aquatic insects can eventually disperse overland. Boersma and Lytle (2014) reported one of these few events in the water bug *Abedus herberti* (Hemiptera: Belostomatidae), a flightless aquatic insect. They argue that even strictly aquatic insects, can have some level of genetic connectivity, as in drought periods in which these insects are compel to disperse overland. Differently, winged insects can travel through barriers, and some species can travel for distances as long as entire countries.

Differences in flight potential are not necessarily related among taxonomic orders. Among Odonata, for instance, there is a clear difference between the flight capability of anisopterans and zygopterans, where anisopterans, with larger bodies and wings, can fly greater distances than zygopterans, overall. Anderson (2009) and Watts et al. (2004) studied two extreme examples of dispersal ability in these groups. Anderson (2009) used mark-recapture techniques to confirm earlier reports of large-scale transoceanic dispersion in *Pantala flavescens* (Anisoptera), suggesting an incredible flight capability for this dragonfly. Conversely, Watts et al. (2004) used microsatellite markers to study the UK populations of the endangered damselfly *Coenagrion mercuriale* (Zygoptera) and found a fine-scale dispersal with high genetic structure among them, where isolation by distance develops within 10 km. While *P. flavescens* is genetically homogeneous among the continental scales due to its increased flight capacity, *C. mercuriale* has a low dispersive potential compared to other Odonata, even other Zygoptera, due to its lower flight capability and vulnerability to habitat fragmentation. This vulnerability to habitat fragmentation makes its dispersive potential relatively short, not exceeding 1.5–2 km. In addition to the high isolation by distance (IBD) (within 10 km) experienced by *C. mercuriale*, fragmentation effects make its populations more isolated than in natural conditions, increasing the risk of extinction.

Fragmentation effects harm aquatic insects as they can use the terrestrial environments as habitats, mating places, and/or corridors for migration, which allows the maintenance of gene flow among populations or subpopulations. Deforestation and land use negatively affect genetic diversity by creating barriers to dispersal and excluding adult habitats. There is a variety of studies addressing these questions, mostly with headwater-specialized insects. Alexander et al. (2011) studied the mayfly *Ephemerella invaria* (Ephemeroptera) in forested and partially deforested headwater catchments in the Mid-Atlantic Region of the Eastern United States. Using molecular markers, they found a strong negative correlation between genetic diversity and deforestation on first-order catchments. As the deforestation rate increases the genetic diversity of *E. invaria* populations decreases, so intact terrestrial environments are necessary to maintain the connectivity among populations that otherwise would be extinct. Extinction events can be followed by recolonizing events performed by survivors, but even so they may cause genetic diversity reduction by excluding populations that could be a source of genetic variation.

Isolation by distance is an important hypothesis to explain the genetic structuring in many aquatic insects as black flies, damselflies, dragonflies, caddisflies, and mayflies. Studying the Japanese caddisfly *Stenopsyche marmorata* (Trichoptera), Yaegashi et al. (2014) combined microsatellite techniques with empirical field observations of flight behavior and spatial distribution and found a significant IBD in populations separated from each other for more than 33 km. This caddisfly generally disperses along the water flow, probably occurring in both larval and adult stages, but occasionally winged adults can disperse among water catchments.

Interestingly, these researchers observed that the lateral dispersion of *S. marmorata* allows the existence of two distinct lineages, separated in upland (regions upper 250 m a.s.l.) and lowland (< 250 m a.s.l.) lineages. The main difference between them is a voltinism (number of generations per year) adaptation mediated by temperature (as we saw on GEA). Upland lineages experience monthly accumulated temperature (MAT) < 90 °C and exhibit univoltine life cycle (one reproductive event per year), whereas lowland lineages experience MAT > 90 °C and bivoltine life cycle (two reproductive events per year). Thus, besides IBD, these differences in life cycles may reduce the gene flow between *S. marmorata* populations, acting as a reproductive barrier.

Strictly aquatic insects are frequently considered to exhibit high isolation among populations from different streams, considering their low capacity of terrestrial dispersion, but that does not always occur. Miller et al. (2002) studied the dispersive potential and gene flow of *Ambrysus thermarum* (Hemiptera: Naucoridae), which has an exclusive aquatic life cycle. To evaluate its dispersive potential, they positioned traps throughout the stream margins to capture adults during dispersal movements between populations, but they did not find any of them. Hence, they did not expect to find genetic any connectivity among populations, but results showed strong genetic similarity among them. Combining these two results and the fact that seasonal streams were common in the study area, the authors concluded that there are consecutive periods of extinction and recolonization in this ephemeral environment, homogenizing the allele frequencies among populations. When there are consecutive extinctions, the new populations only have as source of genetic variation the few individuals that eventually travel between these populations, explaining the low genetic differentiation observed.

Although Miller's findings concerning *A. thermarum* emerged from recent historical events of seasonal streams, there are cases where the natural factors influencing populations are depicted from much older periods. Using mDNA, Baker et al. (2003) investigated the spatial genetic structure of the caddisfly *Cheumatopsyche* sp. (Trichoptera) in Australian streams. They suggest that *Cheumatopsyche* was isolated in small populations during the Pleistocene, but over time there was an expansion of these populations that increased the dispersion and gene flow between them. The gene flow increase caused the high genetic homogenization observed nowadays in a large geographic scale despite distance between them. Another example is the study carried out by Finn et al. (2007) with the giant water bug *Abedus herberti* in Sky Island, USA. Using mDNA as well, they found that *A. herberti* populations are isolated due to the Pleistocene climatic cycles that suppressed the connections among headwater populations. In addition, the increasing rate of climatic warming in this region accelerated the disappearance of ephemeral streams in which this species lives.

Here, we exposed a few examples illustrating how most researchers approach genetic connectivity to study aquatic insects. Hypotheses as IBD have been used to explain dispersion limitation for several taxa. However, as we have seen, the genetic connectivity patterns of each taxon will depend on life history traits and historical events that affect each environment. In addition to these factors, considering historical processes and the health of adjacent habitats is an effective way to investigate connectivity of aquatic insects.

## 16.5 Perspectives for the Conservation of Aquatic Insects

The importance of aquatic insects for the maintenance of aquatic and terrestrial systems has been defended for a long time. Aquatic insects contribute to many ecological functions and can admit distinct trophic positions, being filterers, collectors, predators, and others. The disappearance of one of these categories unbalances the whole system, reducing the amount of available oxygen in water, as well as affecting the rate of energy transfer. Aquatic insects transfer energy when predated by fishes and other predators, and the energy flow can even travel from the aquatic system to the terrestrial environment benefiting terrestrial predators such as birds, mammalians, and amphibians. For the maintenance of aquatic insects and their ecological functions, several conservation approaches have been developed, aiming most of the time specific species or environments. As we have seen in this chapter, genetic strategies allied with conservation strategies are helping to clarify connectivity patterns of aquatic insects and raised hypotheses that seek to explain their distribution (Frankham et al. 2012). Some hypotheses can be applied to all aquatic insects, such as IBD, and others can vary according to the environment type and the developmental stage of individuals. Even so, it is time to use this knowledge to guide management policies of aquatic and adjacent environments.

Habitat fragmentation, whether occurring directly in the aquatic environment through dams or in the terrestrial environment through deforestation, is the greatest challenge for aquatic insect conservation (Bunn and Arthington 2002). Studies performed by Fagan (2012), Petersen et al. (2004), Alexander et al. (2011), and others showed evidences of habitat fragmentation effects on genetic structure of aquatic insect populations. Even highly dispersive species may encounter barriers to their dispersion, but undoubtedly the most vulnerable species are those with limited dispersive potential. Some of them form naturally isolated populations, which makes them more susceptible to local extinctions. *C. mercuriale*, a threatened damselfly, can disperse only 10 km, but in a continuous open agricultural land its dispersal is reduced to 1.5–2 km. This scenario is not different for *A. herberti*, a naucorid, where populations are highly isolated by the headwater shortening caused by global warming. Climate change would dry completely the mountain headwaters in which *A. herberti* inhabits, causing not only its local extinction and genetic diversity loss but also of many other aquatic species. These studies may serve as a benchmark to define conservation units and should be considered when the implementation of ecological corridors (they increase connectivity among populations) and environmental restorations are necessary.

A novel approach for conserving aquatic insects is to use certain threatened species with a wide distribution as "umbrella species." It is common for many governments and society in general to not devote much attention to insect conservation, focusing attention on more charismatic groups such as mammals and birds. However, when we focus on aquatic insects it is possible to relate the preservation and health of these organisms to the quality of water and thus to human health. The main taxa associated to this idea are Ephemeroptera, Plecoptera, and Trichoptera (known as EPT); they are used in many rapid protocols that assess the quality of aquatic environments (Elbrecht et al. 2015). They inhabit headwaters and other aquatic environments relatively undisturbed, and can be used to protect a wide variety of other taxa that depend not only on aquatic habitats, but also on the terrestrial ones.

To conclude, the conservation of aquatic insects directly depends on preserving their aquatic systems, which ultimately depicts the most valuable resource for us humans: the water. Nonetheless, all approaches depend on governmental managements and the creation of policies that minimize the human impacts on these systems. Studies that take into account genetic composition of populations for the establishment and maintenance of genetic diversity sources are necessary for biodiversity conservation. However, anthropogenic issues are beyond the directly observed impacts on aquatic systems. As some studies have shown, global temperature is still rising and wildlife is suffering with a fast environmental change that most species are not able to deal with. Freshwater systems are dependent on the water supply in wet seasons for their persistence; thus changes in rainfall regimes in addition to the exploitation of water tables can impact these environments at large scales, affecting the fauna and human needs. In this scenario, it is essential to optimize the natural resource conservation using modern tools, such as genetic markers allied to large-scale public policies that address the diversity loss problem as a real threat to the human health.

## References

Alexander LC, Hawthorne DJ, Palmer MA, Lamp WO (2011) Loss of genetic diversity in the North American mayfly *Ephemerella invaria* associated with deforestation of headwater streams. Freshw Biol 56:1456–1467. https://doi.org/10.1111/j.1365-2427.2010.02566.x

Alp M, Keller I, Westram AM, Robinson CT (2012) How river structure and biological traits influence gene flow: a population genetic study of two stream invertebrates with differing dispersal abilities. Freshw Biol 57:969–981. https://doi.org/10.1111/j.1365-2427.2012.02758.x

Anderson RC (2009) Do dragonflies migrate across the western Indian Ocean? J Trop Ecol 25:347–358

Anderson CD, Epperson BK, Fortin MJ et al (2010) Considering spatial and temporal scale in landscape-genetic studies of gene flow. Mol Ecol 19:3565–3575. https://doi.org/10.1111/j.1365-294X.2010.04757.x

Antao T, Lopes A, Lopes RJ, Beja-Pereira A, Luikart G (2008) LOSITAN: a workbench to detect molecular adaptation based on a Fst-outlier method. BMC Bioinformatics 9(323). https://doi.org/10.1186/1471-2105-9-323

Baker AM, Williams SA, Hughes JM (2003) Patterns of spatial genetic structuring in a hydropsychid caddisfly (*Cheumatopsyche* sp. AV1) from southeastern Australia. Mol Ecol 12:3313–3324. https://doi.org/10.1046/j.1365-294X.2003.02011.x

Boersma KS, Lytle DA (2014) Overland dispersal and drought-escape behavior in a flightless aquatic insect, *Abedus herberti* (Hemiptera: Belostomatidae). Southwest Nat 59:301–302. https://doi.org/10.1894/N09-FRG-07.1

Bunn SE, Arthington AH (2002) Basic principles and ecological consequences of altered flow regimes for aquatic biodiversity. Environ Manag 30:492–507. https://doi.org/10.1007/s00267-002-2737-0

Campbell Grant EH, Lowe WH, Fagan WF (2007) Living in the branches: population dynamics and ecological processes in dendritic networks. Ecol Lett 10:165–175. https://doi.org/10.1111/j.1461-0248.2006.01007.x

Castilho CS, Marins-Sá LG, Benedet RC, Freitas TRO (2012) Genetic structure and conservation of Mountain Lions in the south-Brazilian Atlantic rain forest. Genet Mol Biol 35(1):65–73

Chaput-Bardy A, Lemaire C, Picard D, Secondi J (2008) In-stream and overland dispersal across a river network influences gene flow in a freshwater insect, Calopteryx splendens. Mol Ecol 17:3496–3505. https://doi.org/10.1111/j.1365-294X.2008.03856.x

Coleman RA, Gauffre B, Pavlova A, Beheregaray LB, Kearns J, Lyon J, Sasaki M, Leblois R, Sgro C, Sunnucks P (2018) Artificial barriers prevent genetic recovery of small isolated populations of a low-mobility freshwater fish. Heredity 120(6):515–532

Crook DA, Lowe WH, Allendorf FW et al (2015) Human effects on ecological connectivity in aquatic ecosystems: integrating scientific approaches to support management and mitigation. Sci Total Environ 534:52–64. https://doi.org/10.1016/j.scitotenv.2015.04.034

Davis CD, Epps CW, Flitcroft RL, Banks MA (2018) Refining and defining riverscape genetics: how rivers influence population genetic structure. Wiley Interdiscip Rev Water e1269. https://doi.org/10.1002/wat2.1269

Elbrecht V, Feld CK, Gies M et al (2015) Genetic diversity and dispersal potential of the stone fly *Dinocras cephalotes* in a central European low mountain range. Freshw Sci 33:181–192. https://doi.org/10.1086/674536

Eros T, Campbell Grant EH (2015) Unifying research on the fragmentation of terrestrial and aquatic habitats: patches, connectivity and the matrix in riverscapes. Freshw Biol 60:1487–1501. https://doi.org/10.1111/fwb.12596

Fagan WF (2012) Connectivity, fragmentation, and extinction risk in dendritic metapopulations. Ecology 83:3243–3249

Finn DS, Theobald DM, Black WC IV, Poff NL (2006) Spatial population genetic structure and limited dispersal in a rocky mountain alpine stream insect. Mol Ecol 15:3553–3566. https://doi.org/10.1111/j.1365-294X.2006.03034.x

Finn DS, Blouin MS, Lytle DA (2007) Population genetic structure reveals terrestrial affinities for a headwater stream insect. Freshw Biol 52:1881–1897. https://doi.org/10.1111/j.1365-2427.2007.01813.x

Frankham R, Ballou JD, Briscoe DA (2012) Introduction to conservation genetics. Cambridge University Press, Cambridge, p 640

Hobson KA, Anderson RC, Soto DX, Wassenaar LI (2012) Isotopic evidence that dragonflies (Pantala flavescens) migrating through the Maldives come from the northern Indian subcontinent. PLoS One 7:e52594. https://doi.org/10.1371/journal.pone.0052594

Horne AJ, Goldman CR (1994) Limnology, vol 2. McGraw-Hill, New York

Hughes JM (2007) Constraints on recovery: using molecular methods to study connectivity of aquatic biota in rivers and streams. Freshw Biol 52:616–631. https://doi.org/10.1111/j.1365-2427.2006.01722.x

Hughes JM, Schmidt DJ, Finn DS (2009) Genes in streams: using DNA to understand the movement of freshwater fauna and their riverine habitat. Bioscience 59:573–583. https://doi.org/10.1525/bio.2009.59.7.8

Hughes JM, Huey JA, Schmidt DJ (2013) Is realised connectivity among populations of aquatic fauna predictable from potential connectivity? Freshw Biol 58:951–966. https://doi.org/10.1111/fwb.12099

Keller D, Van Strien MJ, Holderegger R (2012) Do landscape barriers affect functional connectivity of populations of an endangered damselfly? Freshw Biol 57:1373–1384. https://doi.org/10.1111/j.1365-2427.2012.02797.x

Kool JT, Moilanen A, Treml EA (2013) Population connectivity: recent advances and new perspectives. Landsc Ecol 28(2):165–185

May ML (2013) A critical overview of progress in studies of migration of dragonflies (Odonata: Anisoptera), with emphasis on North America. J Insect Conserv 17:1–15

McGlashan DJ, Hughes JM (2001) Low levels of genetic differentiation among populations of the freshwater fish *Hypseleotris compressa* (Gobiidae: Eleotridinae): implications for its biology, population connectivity and history. Heredity (Edinb) 86:222–233. https://doi.org/10.1046/j.1365-2540.2001.00824.x

McRae BH (2006) Isolation by resistance. Evolution 60:1551–1561

Meffe GK, Vrijenhoek RC (1988) Conservation genetics in the management of desert fishes. Conserv Biol 2:157–169. https://doi.org/10.1111/j.1523-1739.1988.tb00167.x

Miller MP, Blinn DW, Keim P (2002) Correlations between observed dispersal capabilities and patterns of genetic differentiation in populations of four aquatic insect species from the Arizona White Mountains, U.S.A. Freshw Biol 47:1660–1673. https://doi.org/10.1046/j.1365-2427.2002.00911.x

Paz-Vinas I, Loot G, Stevens VM, Blanchet S (2015) Evolutionary processes driving spatial patterns of intraspecific genetic diversity in river ecosystems. Mol Ecol 24:4586–4604. https://doi.org/10.1111/mec.13345

Petersen I, Masters Z, Hildrew AG, Ormerod SJ (2004) Dispersal of adult aquatic insects in catchments of different land use. J Appl Ecol 41:934–950. https://doi.org/10.1111/j.0021-8901.2004.00942.x

Pfeiler E, Markow TA (2017) Population connectivity and genetic diversity in long distance migrating insects: divergent patterns in representative butterflies and dragonflies. Biol J Linn Soc 122:479–486. https://doi.org/10.1093/biolinnean/blx074

Phillipsen IC, Lytle DA (2013) Aquatic insects in a sea of desert: population genetic structure is shaped by limited dispersal in a naturally fragmented landscape. Ecography (Cop) 36:731–743. https://doi.org/10.1111/j.1600-0587.2012.00002.x

Phillipsen IC, Kirk EH, Bogan MT et al (2015) Dispersal ability and habitat requirements determine landscape-level genetic patterns in desert aquatic insects. Mol Ecol 24:54–69. https://doi.org/10.1111/mec.13003

Rosa AJDM (2009) Marcadores moleculares e suas aplicações em estudos populacionais de espécies de interesse zootécnico. Embrapa Cerrados, Planaltina, DF

Sabando MC, Vila I, Peñaloza R, Véliz D (2011) Contrasting population genetic structure of two widespread aquatic insects in the Chilean high-slope rivers. Mar Freshw Res 62:1–10. https://doi.org/10.1071/MF10105

Short AEZ, Caterino MS (2009) On the validity of habitat as a predictor of genetic structure in aquatic systems: a comparative study using California water beetles. Mol Ecol 18:403–414. https://doi.org/10.1111/j.1365-294X.2008.04039.x

Tischendorf L, Fahrig L (2000) On the usage and measurement of landscape connectivity. Oikos 90:7–19. https://doi.org/10.1034/j.1600-0706.2000.900102.x

Vannote RL, Minshall GW, Cummins KW et al (1980) The river continuum concept. Can J Fish Aquat Sci 37:130–137. https://doi.org/10.1139/f80-017

Watanabe K, Monaghan MT, Omura T (2008) Longitudinal patterns of genetic diversity and larval density of the riverine caddisfly *Hydropsyche orientalis* (Trichoptera). Aquat Sci 70:377–387. https://doi.org/10.1007/s00027-008-8099-9

Watanabe K, Monaghan MT, Takemon Y, Omura T (2010) Dispersal ability determines the genetic effects of habitat fragmentation in three species of aquatic insect. Aquat Conserv 20:574–579. https://doi.org/10.1002/aqc.1124

Watts PC, Rouquette JR, Saccheri IJ et al (2004) Molecular and ecological evidence for small-scale isolation by distance in an endangered damselfly, *Coenagrion mercuriale*. Mol Ecol 13:2931–2945. https://doi.org/10.1111/j.1365-294X.2004.02300.x

Whitehead A, Galvez F, Zhang S et al (2011) Functional genomics of physiological plasticity and local adaptation in killifish. J Hered 102:499–511. https://doi.org/10.1093/jhered/esq077

Whitlock MC, Mccauley DE (1999) Indirect measures of gene flow and migration: F(ST) ≠ 1/(4Nm + 1). Heredity (Edinb) 82:117–125. https://doi.org/10.1038/sj.hdy.6884960

Wong BBM, Keogh JS, McGlashan DJ (2004) Current and historical patterns of drainage connectivity in eastern Australia inferred from population genetic structuring in a widespread freshwater fish *Pseudomugil signifer* (Pseudomugilidae). Mol Ecol 13:391–401. https://doi.org/10.1046/j.1365-294X.2003.02085.x

Yaegashi S, Watanabe K, Monaghan M, Omura T (2014) Fine-scale dispersal in a stream caddisfly inferred from spatial autocorrelation of microsatellite markers. Freshw Sci 33:172–180. https://doi.org/10.1086/675076

Ye Z, Yuan J, Li M et al (2018) Geological effects influence population genetic connectivity more than Pleistocene glaciations in the water strider Metrocoris sichuanensis (Insecta: Hemiptera: Gerridae). J Biogeogr 45:690–701. https://doi.org/10.1111/jbi.13148

# Chapter 17
# Aquatic Insects as a Source for Biomimetics

**Stanislav N. Gorb and Elena V. Gorb**

**Abstract** Understanding functional principles of materials, structures, sensors, actuators, locomotion, control systems, behavior of aquatic insects is of major scientific interest. On the other hand, this basic knowledge is also highly relevant for technical applications. One of the greatest challenges for today's engineering science is miniaturization. Insects have solved many problems correlated with extremely small size, during their evolution. Zoologists, entomologists, morphologists, neurobiologists have collected a huge amount of information about the structure and function of such living micromechanical systems. This information can be utilized to mimic them for industrial applications. There are following main technology areas, where aquatic insect's solutions may be applied: (1) materials science and technology, (2) surface science, (3) science of adhesives, (4) optics, (5) photonics, (6) sensorics, and (7) robotics. A few selected examples are discussed in this chapter, but having a great number of described aquatic insect species as a source for inspiration, one may expect a lot more ideas from aquatic insects science for biomimetics.

**Keywords** Cuticle · Bioinspiration · Biomimetics · Functional morphology · Materials · Surfaces · Insecta

## 17.1 Introduction

Insects have solved a variety of technical problems in their evolution and have evolved a great variety of shapes and structures. Although often intricate and fragile, they can nonetheless deal with extreme mechanical loads. Many functional solutions in insects are based on a variety of ingenious structural solutions. Elaborate cuticular structures, rich sensory equipment of insects including complex eyes, and various receptors (chemo-, hygro-, thermo-, mechanoreceptors, etc.) taken together

S. N. Gorb (✉) · E. V. Gorb
Department Functional Morphology and Biomechanics, Zoological Institute of the Kiel University, Kiel, Germany
e-mail: sgorb@zoologie.uni-kiel.de



K. Del-Claro, R. Guillermo (eds.), *Aquatic Insects*,
https://doi.org/10.1007/978-3-030-16327-3_17

with a compact brain reveal interesting motion control patterns and remarkable behavioral features (Gorb 2011; Gorb and Gorb 2016). Aquatic insects are not an exception: due to their aquatic or amphibiotic habit, they evolved many particularly interesting functional solutions, which are of special interest for biomimetics.

Understanding functional principles of insect materials, structures, sensors, actuators, locomotion patterns, control systems, and behavior is of major scientific interest, since we can learn about their functional principles and biological role. On the other hand, this knowledge is also highly relevant for technical applications. One of the greatest challenges for today's engineering science is miniaturization. Aquatic insects have solved many problems dealing with their small size and amphibiotic life, during their evolution. Zoologists, entomologists, morphologists, and neurobiologists have collected a huge amount of information about the structure and function of such living micromechanical systems. This information can be utilized to mimic them for diverse technical applications.

The following are the main technological areas, where "insect solutions" may be of interest: (1) materials science and technology, (2) surface science, (3) science of adhesives, (4) optics, (5) photonics, (6) sensorics, and (7) robotics (Fig. 17.1). Possible innovations may also appear on the boundary between aquatic entomology and the areas listed above. Only a few selected examples are discussed in this chapter, but having a huge amount of other described aquatic insect species as a source for inspiration one may expect a lot more ideas for biomimetics (Gorb 2011).

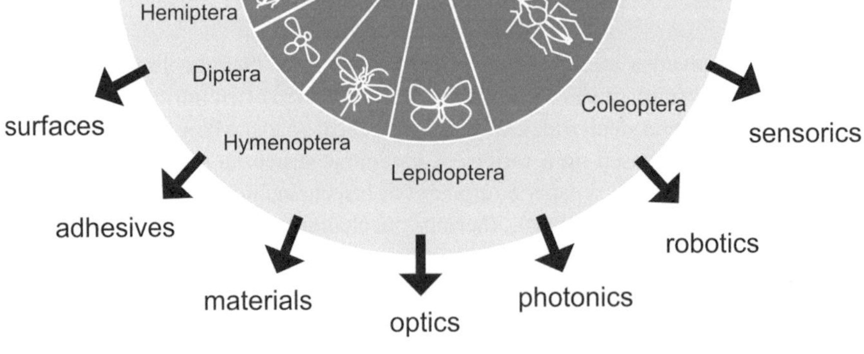


**Fig. 17.1** Diagram of insect diversity as a source for biomimetic ideas in various areas of technology (Gorb 2011)

## 17.2 Materials

The insect exoskeleton is called a cuticle. Being an interface between a living animal and environment, the cuticle of an insect serves many functions. (1) It limits the dimensions of the exoskeleton and is a basis for muscle insertion (mechanical function and locomotion). (2) It is an important element in organism defense against a variety of external factors, such as mechanical stress and dry, wet, cold, or hot environments. (3) It takes part in the transport of diverse epidermal secretions and serves as a chemical reservoir for the storage of metabolic waste products. (4) A variety of cuticular structures are parts of various receptors. (5) The cuticle, its coloration pattern, and chemical components are also important for thermoregulation and are often involved in diverse communication systems. (6) Specialized cuticular protuberances may serve a variety of functions, such as oxygen retention under water, food grinding, and body cleaning (grooming).

Cuticle is a layered composite material (Fig. 17.2a), which consists of two principal components: chitin and protein. Chitin is a structural polymer, found in insects in crystalline form. The structure of chitin molecules and their arrangement are well studied, but the structure of chitin-protein assemblages still remains unknown.

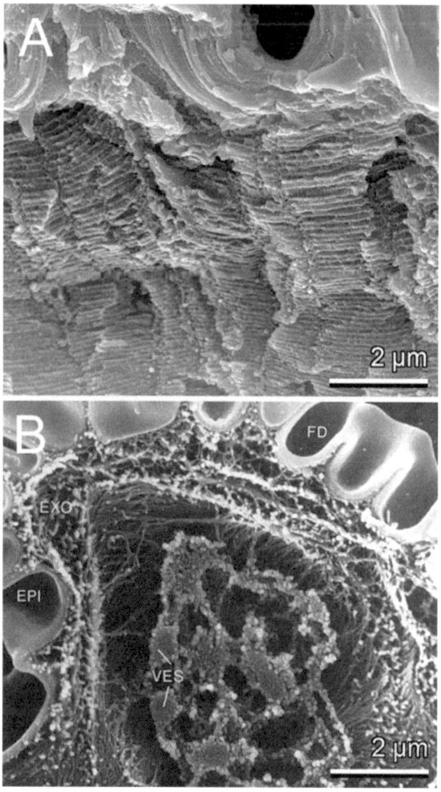


**Fig. 17.2** Dragonfly cuticle. (**a**) Fracture through the stiff tanned cuticle of the rear side of the head in the adult dragonfly *Aeshna mixta*, scanning electron microscopy (SEM) (Gorb 2011). Note the layered structure of the exocuticle. (**b**) Neck membrane of the adult damselfly *Ischnura elegans*, SEM (Gorb 2000). Semi-thin Spurr-embedded sections treated with Maxwell's solution, SEM. *EPI* epicuticle, *EXO* exocuticle, *FD* folds, *VES* vesicles

There are three morphs of chitin differing in the molecule packing and polarities of adjacent chains: α-, β-, and γ-chitin (Rudall 1963). The arrangement of molecules usually varies within the different morphs of chitin. All three morphs of chitin are classified as viscoelastic polymers (Hepburn and Chandler 1978). Chitin crystals always occur in insects as bundles of microfibrils of 25–30 Å in diameter. Microfibrils are always associated with protein in a chitin-protein complex. Chitin-bearing cuticular microfibrils have a complex pattern of orientation in the three-dimensional space of the cuticle.

Arthrodial and caterpillar-like cuticles are materials of great flexibility, great extensibility, and reasonable strength. The interaction between chitin fibers and protein matrix is very loose, as has been demonstrated in deformation and fractographic studies. Under tension, these cuticles are visco-plastic and show so-called necking, as do the steel specimens under the same conditions (Hepburn and Chandler 1976). However, there is remarkable variation regarding the extent to which such cuticles can be extended (Vincent and Wood 1972; Vincent and Wegst 2004). There are two types of membranous cuticle organizations found in insects. Highly extensible membranes can extend more than 1000% of their length (Vincent 1981). This cuticle is highly specialized at the molecular level due to its protein composition (Hackman 1975; Hackman and Goldberg 1987). The explanation for such extreme extension ability is the specific alignment of chitin fibers; one can also assume that the protein matrix is not completely bound to the chitin fraction. The second type of membranous cuticle is folding laminated cuticle with a lower degree of extension. This cuticle is found in the neck membranes of dragonflies (Fig. 17.2b) (Gorb 2000).

Resilin is a protein that has its molecular structure in the form of a three-dimensional network of chains having only a few stable cross-links (Andersen 1963, 1964, 1966). These covalent cross-links are fluorescent due to the presence of di- and trityrosine (Andersen and Weis-Fogh 1964). Amino acid composition of resilin from different species is quite similar (Hepburn 1985). Resilin was initially described in the prealar arm and wing-hinge ligaments of the locust *Schistocerca gregaria* and in the pleuro-subalar muscle tendon of the dragonfly *Aeshna grandis*. Meanwhile it has been revealed in different structures of numerous insect species including aquatic dragonfly larvae (Büsse et al. 2017). The resilin-bearing structures usually show recoverable mechanical deformability, so that the resilin-containing cuticle has been called rubberlike cuticle, in contrast to the tanned "solid" cuticle. Resilin can be reversibly extended over 200% of its initial length. Its principal physical properties are its long-range elasticity and lack of either creep or stress relaxation (Weis-Fogh 1960, 1961; Andersen and Weis-Fogh 1964), but Kovalev et al. (2018) recently showed a viscoelastic response of resilin. The resilin-containing structures work as mechanical springs under tension and compression (Rothshild et al. 1975; Frazier et al. 1999; Gorb 2004). Resilin has been revealed in different areas of insect wings (Gorb 1999; Haas et al. 2000a, b; Appel and Gorb 2014), where it presumably prevents material fatigue in folds and promotes wing profile adaptation to aerodynamic situation in flight (Appel et al. 2015; Rajabi et al. 2015, 2016, 2017).

Insect cuticle demonstrates, in different functional systems, a gradient of material properties, which may range from very stiff areas of condyle of joints to membranous areas between leg segments (Gorb 2000). These gradients depend on the fiber density, fiber orientation, polymerization degree of the matrix, and thickness of single layers.

Meanwhile, composite materials are widely spread in various areas of technology. However, the fiber size in technical systems usually remains in the micrometer range and fibers normally have random or preferable orientation within the matrix. Insect cuticle may inspire materials scientists with its helicoidal arrangement of nanofibers in successive layers as it is the case in the exocuticle and with its gradient-like materials properties.

In the last years, many efforts were undertaken to synthesize artificial rubberlike material protein of the cuticle (resilin) (Elvin et al. 2005). There are many areas of potential applications for such a material ranging from medical rubbers to technical sealing systems.

## 17.3 Surfaces

Aquatic insect surface structures may serve many different functions, such as air retention, food grinding, and body cleaning (Fig. 17.3). Some examples of such surfaces in purely terrestrial and aquatic insects seen in a scanning electron microscope are given in Fig. 17.4.

Because of the structural and chemical complexity of insect surfaces, exact functional principles have been studied for only a few systems. Due to the broad diversity of functions of insect surfaces, inspirations from entomology are currently in focus in a broad range of research topics in engineering sciences: adhesion, friction, wear, lubrication, filtering, sensorics, wetting phenomena, self-cleaning, antifouling,

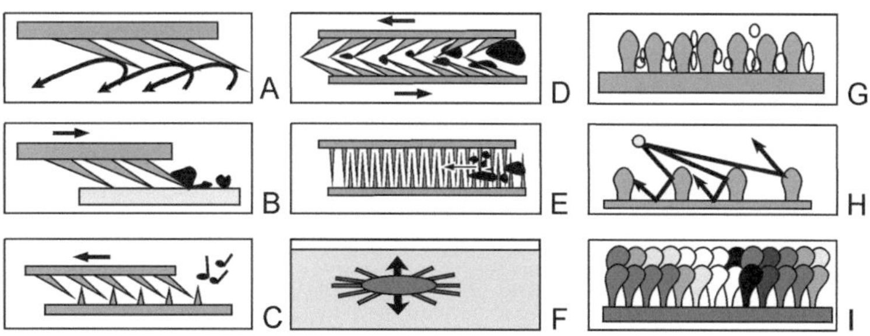


**Fig. 17.3** Diagram of functions of cuticular microstructures in aquatic insects. (**a**) Aerodynamically active surfaces in adults. (**b**) Grooming. (**c**) Sound generation. (**d**) Food grinding. (**e**) Filtration devices. (**f**) Hydrodynamically active surfaces. (**g**) Air retention. (**h**) Thermoregulation. (**i**) Body coloration pattern (Gorb 2001, 2011)

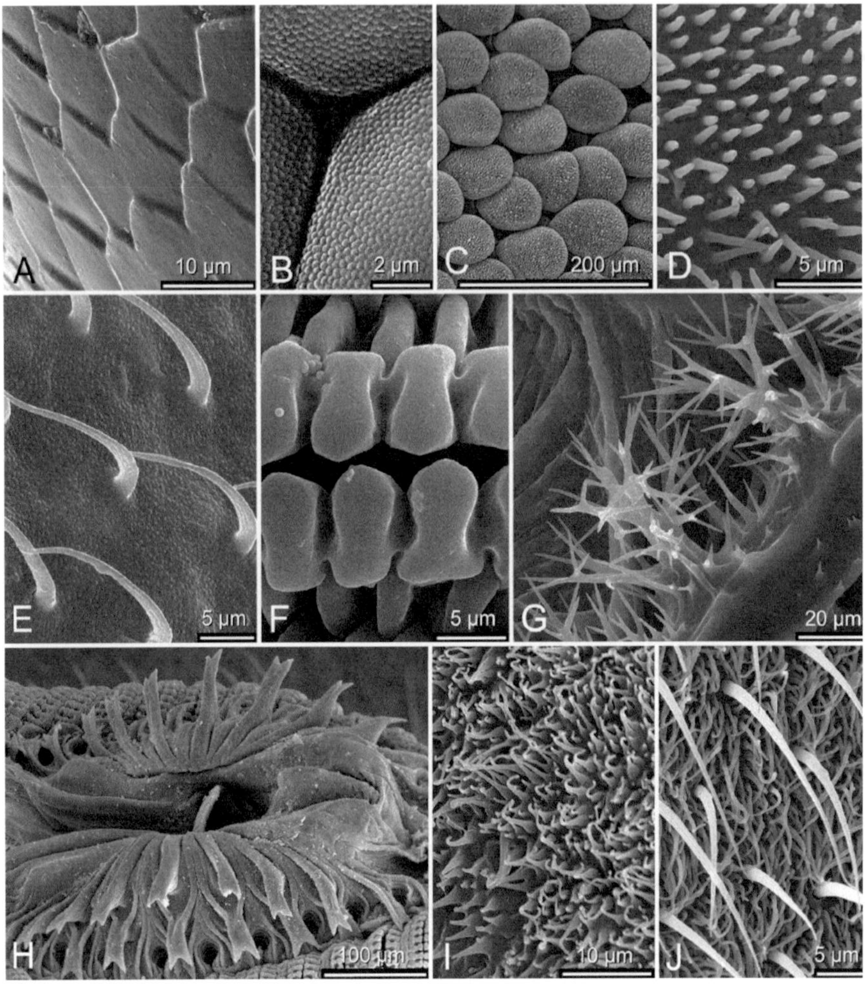


**Fig. 17.4** Functional diversity of non-innervated cuticular protuberances in insects, SEM. (**a**) Unspecialized polygonal surface on the tarsus of the scarabaeid beetle *Melolontha melolontha*. (**b**) Ommatidia surface in the calliphorid fly *Calliphora vicina*. (**c**) Scales on the dorsal surface of the elytron in the scarabaeid beetle *Hoplia sp*. (**d**) Same, the surface of a single scale. (**e**) Wing surface in the bibionid dipteran *Bibio ferruginatus*. (**f**) "Pseudotrachea" of the labellum in the *C. vicina*. (**g**) Filter system of the spiracle in the tenebrionid beetle *Tenebrio molitor*. (**h**) Prestomal teeth in the *C. vicina*. (**i**) Plastron in the nepid bug *Ranatra linearis*. (**j**) Air retaining hair coverage in the water strider *Gerris lacustris*

thermoregulation, optics, etc. Since insect surfaces are multifunctional, it makes them even more interesting from the perspective of biomimetics.

One of the challenges in designing moving parts of microelectromechanical devices (MEMS) is the fabrication of joints allowing precise motion of parts about one rotational axis or multiple axes. One problem is the high friction, stiction, and wear rate of joints (Scherge et al. 1999; Komvopoulos 2003). Wear of the interacting surfaces is a consequence of friction, affecting the material's contact points by becoming deformed or being torn away. Friction and wear are strongly correlated processes, by which the points of the surfaces in contact change their topography continuously. Capillary adhesion, due to the presence of a water layer in contact, can account for a great part of the measured friction or lead to the stiction between a contact pair (Scherge et al. 1999). These are critical issues limiting the operational lifetime and negatively influencing the technological potential of MEMS. Conventional methods of lubrication cannot always be used, especially in devices with medical applications. Friction reduction in some man-made mechanical systems is based on the different hardness of elements in contact (Miyoshi 2001; Li et al. 2004), on the use of hydrophobic surfaces, and on applying surface texture, which minimizes the real contact area between two solid surfaces. Ideas from studying surface properties of insect joints, which have stiffer outer layers located on softer ones, specialized microstructure on the contact pair (Fig. 17.5), and particular fiber orientation in the exocuticle, might represent an interesting set of principles towards a solution (Perez Goodwyn and Gorb 2004; Barbakadse et al. 2006).

Aquatic insects move on land, in the air, and in water. The resistance against motion mediated by surrounding media and by mechanical contact with various materials was an evolutionary factor, which contributed to the appearance of many surfaces adapted to reduce such resistance. On the other hand, one always needs friction to generate force, in order to move on a substrate or to overcome the drag caused by friction elsewhere. A living motion system becomes optimized when it is capable of minimizing friction at one end of the system while maximizing it at the other end (Radhakrishnan 1998). In other words, insects, like any other organism, need a combination of maximum friction required for acceleration, deceleration, and maneuvering, combined with minimum friction in joints for economic energy expenditure. Adhesion phenomena can also positively or negatively contribute to the functionality of such a system.

In insect joints working under lower loading forces, but much higher frequencies than vertebrate joints (Wootton and Newman 1979; Gronenberg 1996), the joint surfaces are usually smooth or present a combination of wavy and smooth counterparts (Fig. 17.5a–e). Underlying tissues are penetrated with channels, which are presumably responsible for delivering lubricants into the contact area. The specialized surface structures have been shown to confer friction-reducing properties in insect joints (Perez Goodwyn and Gorb 2004). Technical developments of such contact pairs are realized using laser technology and 3D printing (Vagts et al. 2017, 2018) (Fig. 17.5f, g).

Small bristles, scales, and microtrichia of the wings of flying insects (Bocharova-Messner and Dmitriev 1984) presumably may generate microturbulences around such structures in flight due to building a kind of lubricating layer of air between an airflow and the insect's surface. This could possibly decrease friction during high-speed flight.

Some insect surfaces especially in aquatic species are covered by a superhydrophobic (non-wettable) cuticle, which has an external layer consisting of both cuticle

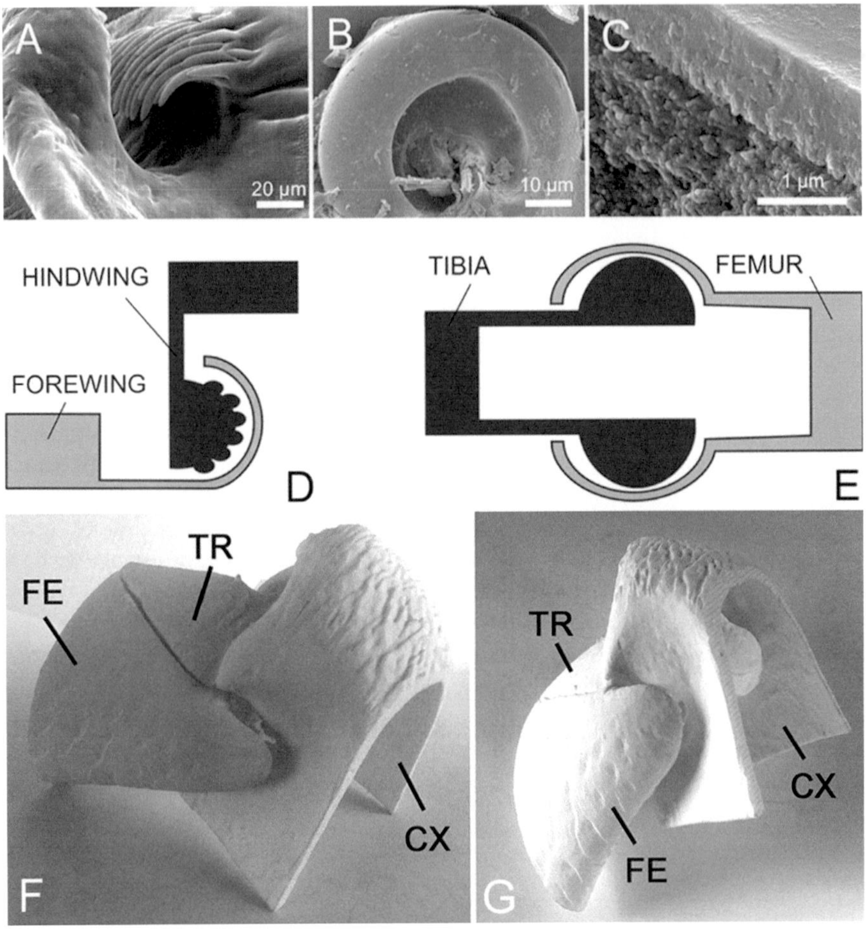


**Fig. 17.5** Examples of micro-joints in insects. (**a**) Lateral view of the wing double-wave locking mechanism in the bug *Coreus marginatus* (forewing part), SEM. This joint provides interlocking between both wings on the same body side in the anterior direction allowing them to slide in the medial and lateral directions (Perez Goodwyn and Gorb 2004). (**b**) Medial aspect of the femorotibial joint (femoral part) of the leg in the beetle *Melolontha melolontha*, SEM. (**c**) Fracture of the material of the joint in the beetle *M. melolontha*, SEM. (**d**) Diagram of the wing-locking mechanism shown in **a**. (**e**) Diagram of the femorotibial joint shown in **b** and **c**. Constructional and mechanical principles found in such joints can be used to design joints in technical actuators. (**f**, **g**) 3D printed coxo-trochanter-tibial joint of the beetle *Pachnoda marginata* based on the data obtained from the micro-CT scan (Vagts et al. 2017). *CX* coxa, *FE* femur, *TR* trochanter

microstructure and epicuticular waxes. The layer may contain wax projections, with dimensions ranging from hundreds of nanometers to micrometers (Fig. 17.6). The roughness of such surfaces together with their hydrophobic properties decreases wettability, which is reflected in a greater contact angle of water droplets on such surfaces compared to smooth surfaces of the same chemical composition. In some

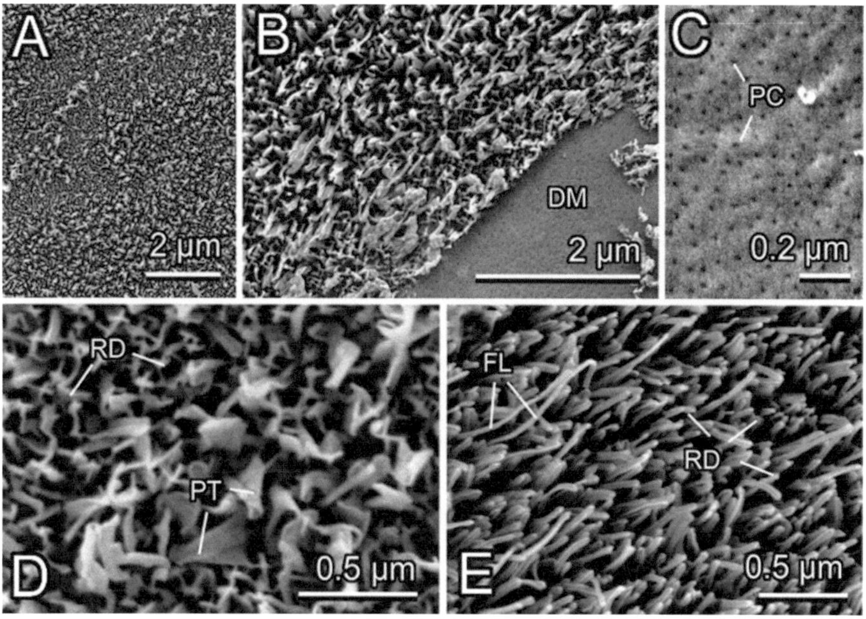


**Fig. 17.6** Dorsal wing surface of the adult dragonfly *Calopteryx splendens*, SEM: young mature male surface of wing membrane covered with 3D wax (**a**, **d**, **e**); wing membrane surface with damage (scratches) in wax crystalline layer (**b**, **c**). *DM* damage/scratch, *FL* wax filaments, *PC* porous channels of cuticle, *PT* wax platelets, *RD* wax rods, *WC* wax projections (Gorb et al. 2009)

plant surfaces, this property results in their ability to be cleaned by rolling drops of water (Barthlott and Neinhuis 1997, 1998). Similar insect structures in insects, such as those on wings of representatives of Odonata, Ephemeroptera, and Neuroptera, are extremely non-wettable and self-cleaning (Wagner et al. 1996; Gorb et al. 2009). In addition, these surfaces prevent biofouling under water.

Many aquatic and semiaquatic arthropods have sculptured surfaces involved in holding air underwater for respiration. Such surfaces called plastrons usually contain fields of microtrichia, very small cuticle protuberances (Heckmann 1983). These structures appear convergently in various arthropod taxa, as an adaptation to aquatic environments: Collembola, Lepidoptera, Coleoptera, Heteroptera, Diptera, Araneae, and Diplopoda (Thorpe and Crisp 1947; Hinton 1976; Messner 1988). Some terrestrial insects, such as Aphididae (Auchenorrhyncha), also bear similar structures in the form of bristles, mushroomlike spines, or stigmal plates, which can protect their surfaces from moisture (Heie 1987). In water striders and some spiders, the anti-wetting surfaces of legs and the ventral body side are involved in the locomotion mechanism of walking on water.

Especially interesting are the genera *Halobates* (Gerridae) and *Haloveloides* (Veliidae), the only open seawater dwellers (Perez Goodwyn 2009). Representatives from the genus *Halobates* have outstanding water protection structures. The microtrichia pile is composed of specialized prolongations. Each microtrichium has a

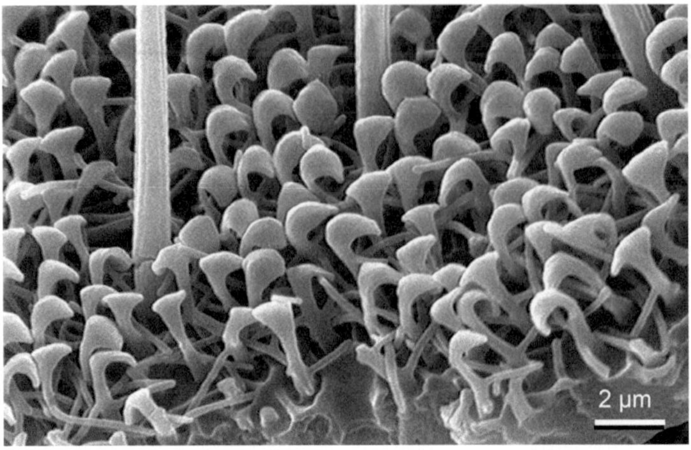


**Fig. 17.7** Thoracic microtrichia cover of *Halobates germanicus* (Gerridae), SEM. Note golf club-shaped heads and interlocking extensions between shafts (Perez Goodwyn 2009)

thickened head, which is several times wider than the shaft and usually tilted to one side like in a golf club (Fig. 17.7). On the shaft itself, there are up to four perpendicular branches interlocking the microtrichia, the shafts of which are 0.8–1.1 µm apart. This structure provides effective water protection in case of being submerged, a frequent situation in the open sea (Perez Goodwyn 2009).

Highly specialized areas of cuticle, responsible for sound generation, usually consist of patterns of cuticular plates, seldom of microtrichia fields (Hinton 1970). When these surfaces slide over each other, sound is generated. Such structures have been previously described from elytra, abdomina, and coxae in phylogenetically distantly related arthropod taxa, including bugs Cimicomorpha, Pentatomomorpha, and beetles (Gogala 1984; Palestrini et al. 1987). The sound frequency corresponds to the periodicity of these structures contacting functionally corresponding surfaces and to the speed of sliding. These systems should be, in addition, wear resistant. However, rigorous experiments supporting or rejecting this statement are absent in the literature.

Filtration systems are usually equipped with long bristles. Such systems are well known from mouthparts of aquatic insect larvae. The filtering system of insect spiracles that keeps the tracheae free of dirt particles is composed of branched acanthae. However, in some cases, completely different principles may be involved in the functioning of these systems. The labellum of many dipterans bears so-called pseudotracheae (Gracham-Smith 1930; Elzinga and Broce 1986) whose outgrowths have a complex material structure. Labellum and pseudotracheae are driven by muscles, resilin springs, and hydraulic pressure, enabling a change in the diameter of the filtration sieve, depending on the size of the particles in the food.

Functional surfaces in technical systems have many functional requirements, which can be fulfilled by using principles learned from biology. Figure 17.8 lists reported activities on the implementation of biologically inspired microstructure

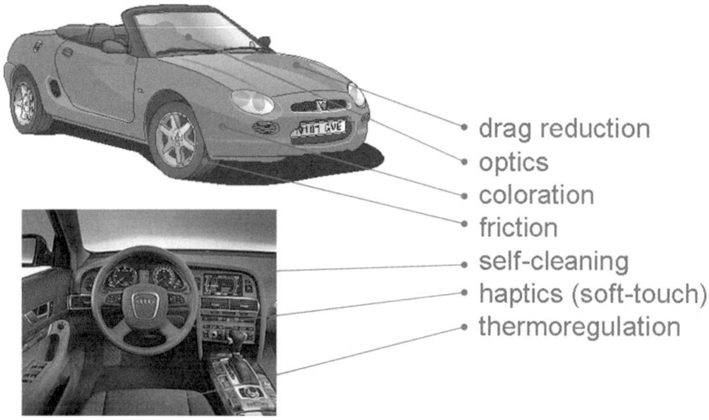


**Fig. 17.8** An example from the car industry that applies functional surfaces, where microstructure development uses ideas from biological studies on insects and some other organisms

into product design. Since insects bear a huge variety of such microstructures, many of which have not even been previously described, a systematic approach to insect surface science would be very desirable. The important step in this direction is establishing a database of insect functional surfaces. Furthermore, we need more experimental studies, targeted to answering the relationship between structure at various levels of organization and its function. We believe that enhancing the pool of new ideas from biology will provide a leap in the surface technology of tomorrow.

## 17.4 Attachment Devices

There is a variety of specialized biological systems that prevent separation of two surfaces. These systems are often called *attachment devices* (Gorb 2001). Some of them are based entirely on mechanical principles, while others additionally rely on the chemistry of polymers and colloids (Scherge and Gorb 2001; Habenicht 2002). There are at least three reasons for using adhesives in biology and in the industry as well: (1) they join dissimilar materials; (2) they improve stress distribution within the joint; and (3) they increase design flexibility (Waite 1983). These reasons are relevant both to the evolution of natural attachment systems and to the design of man-made joining materials.

In general, adhesive-bond formation consists of two phases: contact formation and generation of intrinsic adhesion forces across the joint (Naldrett 1992). The action of the adhesive can be supported by mechanical interlocking between irregularities of the surfaces in contact. Increased surface roughness usually results in an increased strength of the adhesive joint in the glue-based systems due to the increased contact area between the contacting surfaces and the solidified adhesive substance. Strong adhesion is also possible between two ideally smooth surfaces. Adhesive organs, which may be used for attachment to substrates as well as for catching prey, demonstrate a huge diversity among insects due to their structural and chemical properties. Biological adhesion underlies the organization of all living tissues. There is no doubt that many functional solutions have evolved independently in different insect lineages.

Many species of insects are supplied with diverse attachment devices, the morphology depending on the species' biology and particular function, in which the attachment device is involved. The evolutionary background and animal behavior influence the specific composition of attachment systems in each particular species. There are eight fundamental classes of attachment principles found in insects: (1) hooks, (2) lock or snap, (3) clamp, (4) spacer, (5) suction, (6) expansion anchor, (7) adhesive secretions (glue), and (8) friction (Gorb 2001). However, different combinations of these principles also occur in existing attachment structures. Three types of adhesion at the organism level are known: (1) temporary adhesion allowing an organism to attach strongly to the substrate and detach quickly when necessary (see the subsection below about locomotory attachment devices); (2) transitory adhesion permitting simultaneous attachment and movement along the substrate; (3) permanent adhesion involving the secretion of cement. These three types of adhesion do not have the same purpose and use different adhesive systems.

Walking machines usually use suckers to hold onto vertical and overhanging surfaces. A primary disadvantage of this attachment principle is that a very smooth substrate surface is required. A future goal is walking robots capable of walking on a variety of surfaces. Insects can walk rather well on smooth and structured substrata, on inclines, vertical surfaces, and some of them even on the ceiling. One such example is the hairy surface of legs in flies and beetles. This system uses a secretion enabling hairs to attach and detach very quickly to diverse substrata. The hair structure includes a mechanism that delivers the secretion, in extremely small amounts, directly to the contact area (Ishii 1987), and only then, when contact to the substrate is achieved.

Generally, in their evolution, insects have developed two distinctly different mechanisms to attach themselves to a variety of substrates: with smooth pads or with setose/hairy surfaces. Due to the flexibility of the material of the attachment structures, both mechanisms can maximize the possible contact area with the substrate, regardless of their microsculpture. Tenant setae are relatively soft structures (Niederegger et al. 2002). In *Calliphora* flies, their tips are usually compressed, widened, and bent at an angle of about 60° to the hair shaft (Bauchhenss and Renner 1977). Some male beetles possess mushroomlike terminal tips of setae (Fig. 17.9a).

Different forces may contribute to the resulting attachment force: capillary adhesion and intermolecular van der Waals forces. A contribution of intermolecular interaction to the overall adhesion has been shown in experiments on the adherence of beetles (Stork 1980) and beetle setae (Stork 1983) on a glass surface. The presence of claws, decrease of air pressure, decrease of relative humidity, or electrostatic forces do not influence beetle attachment on smooth substrata. In the terrestrial chrysomelid beetles, the resulting attachment force directly depends on the number

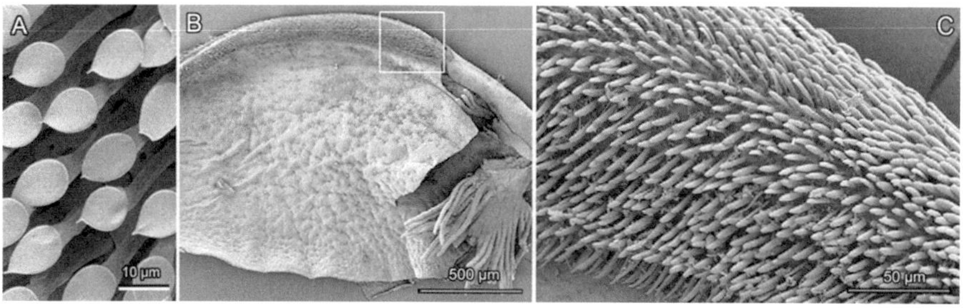


**Fig. 17.9** Insect attachment devices, SEM. (**a**) Ventral surface of the tarsus in the male of the terrestrial beetle *Gastrophysa viridula* (Gorb et al. 2007). (**b**, **c**) Underwater attachment structures on the ventral surface of the gills of the mayfly larva *Epeorus assimillis* (Ditsche-Kuru et al. 2010): overview of a whole-gill lamella (**b**); antero-proximal part of the pad (**c**)

of single hairs contacting the surface. The contribution of intermolecular interaction and capillary force has been demonstrated for the fly *Calliphora vicina* in a nanoscale experiment using an atomic force microscope (Langer et al. 2004). Attachment forces increase when the contacting surfaces slide against each other. This may explain why flies placed on a smooth undersurface always move their legs in a lateral-medial direction (Wigglesworth 1987; Niederegger and Gorb 2003). During these movements, setae slide over the surface in order to obtain optimal contact.

The size of single points in hairy attachment devices gets smaller and their density higher as the body mass of the animal group increases (Scherge and Gorb 2001; Arzt et al. 2003). The fundamental importance of contact splitting for adhesion on smooth and rough substrata has been explained by a very small effective elastic modulus of the fiber array (Persson 2003). Adhesion enhancement by division of the contact has also been demonstrated experimentally (Peressadko and Gorb 2004; Gorb et al. 2007). An additional advantage of patterned surfaces is the reliability of contact on various surface profiles and the increased defect tolerance of individual contacts.

In aquatic insects, hairy attachment surface has been observed on the ventral surface of the abdominal gills of the mayfly larvae (Ditsche-Kuru et al. 2010) (Fig. 17.9b, c). These microstructures evolved independently from the leg adhesive pads mentioned above and adhere not only to the smooth stone surfaces, but also to those covered with biofilms.

Hairy leg attachment pads are promising candidates for biomimetics of robot soles adapted for locomotion. Similar principles can be applied to the design of microgripper mechanisms with an ability to adapt to a variety of surface profiles (see paragraph *Robotics* in this chapter). There are very interesting insect adhesive systems combining microstructure and glue, such as the mouthpart apparatus used by rove beetles (genus *Stenus*) to trap the animals they feed on (Betz et al. 2009). The beetles' extended, rod-shaped lower lips have two sticky pads and shoot out like a catapult to capture potential prey (Fig. 17.10a, b). It seems that the adhesive function is based not only on the subdivision of the contact area into many smaller

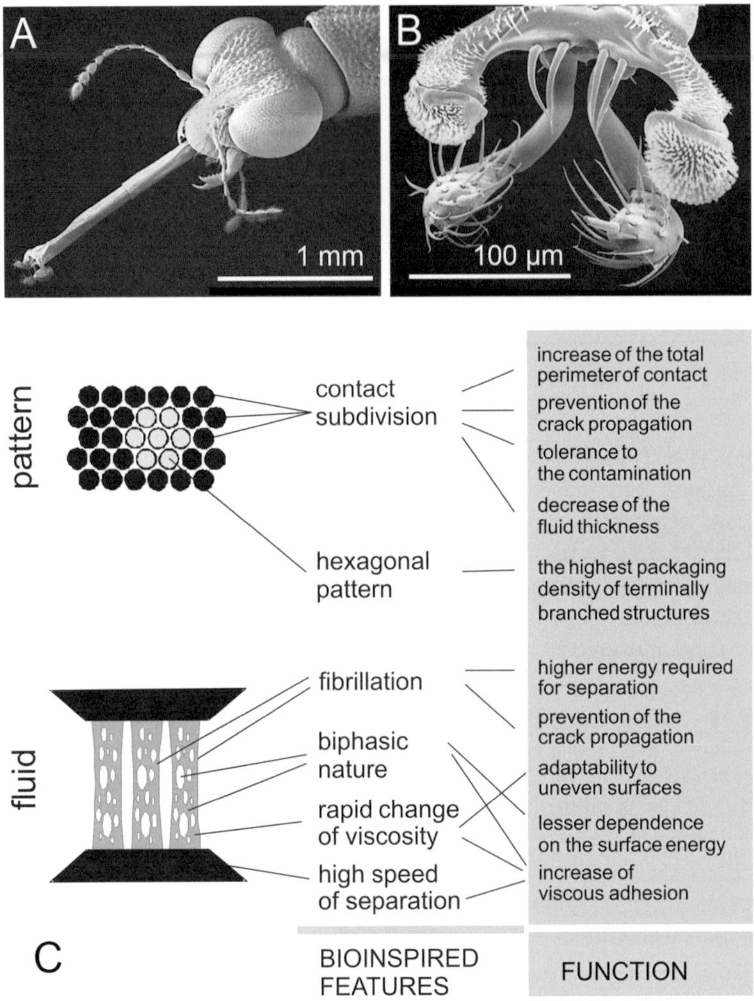


**Fig. 17.10** The adhesive capture apparatus of the rove beetle *Stenus comma*. (**a**) SEM image of the head with the protruding lower lip (labium). (**b**) SEM image of the tip of the labium with adhesive pads. (**c**) Summary of the functional principles with biomimetic potential in the system under investigation (Betz et al. 2009)

ones, but also on very special features of the beetle glue itself: (a) fibrillation at the surfaces' separation (energy dissipation mechanism and crack propagation-preventing mechanism), (b) biphasic nature of the glue (universal wettability mechanism), and (c) rapid change of its viscosity (viscose force adhesion mechanism).

The industry of adhesives is presently following three main goals (Hennemann 2000): (1) an increase in the reliability of glued contact; (2) mimicking of natural, environment-friendly glues; and (3) development of mechanisms for application of a minute amount of glue to the surface. An additional challenge is the use of substances and/or mechanisms that allow multiple attachments and detachments and enable attachment to a variety of surfaces. This section clearly demonstrates the strong potential of insect adhesives for inspiring development of technical adhesives that fulfill the listed requirements.

## 17.5 Optics

Ommatidial gratings are antireflective structures on the eyes of insects, especially those which are nocturnally active (Fig. 17.11). These protuberances are very small (200 nm in diameter) and increase visual efficiency through decreased surface reflection in their density and increased photon capture for a given stimulus condition (Parker et al. 1998; Vukusic and Sambles 2003).

Such a grating is particularly useful on a curved corneal surface, since it increases the transmission of incident light through the cornea compared with a smooth surface. For an increase in transmission and reduced reflection, a continuous matching of the refractive indexes at the boundary of both adjacent materials is very critical. If the periodicity of the grating is smaller than the wavelength of transmitting light, only light of zero order can be reflected or transmitted. For a constant ratio of both materials (cuticle-air) at the boundary between media, the electromagnetic field strength of incoming light is nearly constant (Bernhard et al. 1965). This region, therefore, can be considered as homogenous and an effective refractive index can be given.

Grooming is a very important function for insects that sometimes live in extremely dirty or dusty environments. Their rich sensory equipment of eyes and antennae has to be kept clean in order to respond adequately to external signals. Many insects bear specialized cleaning structures (Schönitzer and Lawitzky 1987;

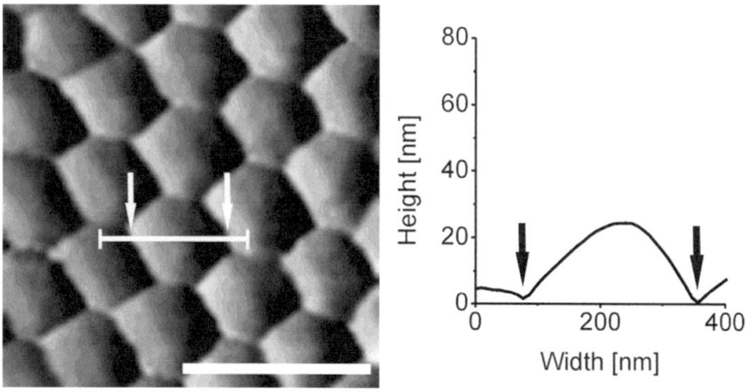


**Fig. 17.11** Atomic force microscopic (AFM) error channel image (left) and profile (right) of ommatidia of the moth *Laothoe populi*. White bar marks the region used for measurements of surface profile. The grating borders indicated by arrows in the left picture correspond to the arrows shown in the surface profile diagrams. Scale bar = 500 nm (Peisker and Gorb 2010)

Francouer and Loiselle 1988). Others rely on the micro- and nanostructured surfaces with an antiadhesive function. Ommatidia gratings are a multifunctional surface that employs self-cleaning due to the real contact surface reduction mechanism (Peisker and Gorb 2010).

## 17.6 Structural Coloration

Structural coloration due to the presence of scales and bristles is well known in insects, such as butterflies (Grodnicky 1988; Ghiradella 1989) and beetles (Schultz and Hadley 1987). For example, scales of some scarabaeid beetles bear additional microtrichia on their surfaces responsible for lusterless appearance of the elytra surface. Scales of some curculionid beetles contain photonic crystals inside (Fig. 17.12). The coloration pattern serves for species and sex recognition, and also for camouflage and mimicry. The most interesting type of structural coloration is called iridescence, which is well known in many insect species (Ghiradella et al. 1972; Huxley 1975).

The iridescence is a result of optical interference within multilayer structures (Ghiradella 1991) that are rather complex in their architecture, and may be incorporated into systems that can produce several different optical effects. Such effects include diffraction-assisted reflection angle broadening (Vukusic et al. 1999, 2000a), all-structural color mixing, and strong polarization effects (Vukusic et al. 2000b).

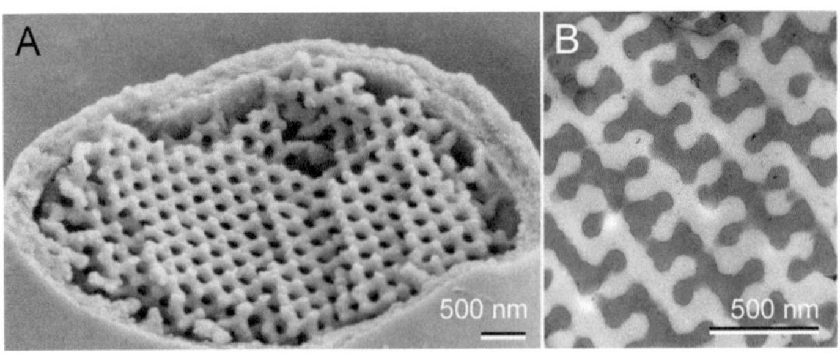


**Fig. 17.12** Photonic crystals of the dorsal scalelike hairs in the curculionid beetle *Phyllobius argentatus*. (**a**) SEM micrograph of the cross fracture through the scale. (**b**) Transmission electron microscopy (TEM) micrograph of the oblique ultrathin section of the scale

## 17.7 Sensorics

Insects carry out flow sensing in various contexts using mechanosensory hairs (Gnatzy and Heusslein 1986) (Fig. 17.13a). The wide variety in hair length and number in insects can be partly explained by the physics of the medium and its interactions with the hair (Humphrey et al. 1993, 2003). For example, filiform hairs provide crickets with a highly sensitive receptor system capable of differentiating air-particle movements in the low-frequency range due to their diverse combinations of directional selectivity, sensitivity, and frequency response.

Crickets also possess composite mechanoreceptors. For example, the coupling of filiform hairs with campaniform sensilla creates a mechanoreceptor with an extended working range (Heusslein et al. 2009). In addition, the composite mechanoreceptor serves as a tactile receptor system superior in its spatial detection range.

The arrangement of hairs also has important sensory effects (Casas and Dangles 2010). The spatially heterogeneous information provided by a transverse flow around a cylinder implies that hairs should be placed all around a cylinder, maximizing the chances to perceive a source coming from any angle (Steinmann et al. 2006). Cricket airflow sensors have recently been a source of inspiration for building artificial airflow sensors (Dijkstra et al. 2005). Design guidelines for building flow-sensing MEMS arrays were also based on biomimetic ideas from studies on the cricket's cerci. A spatial arrangement of MEMS hairs with a large range of angles relative to flow direction on a dedicated platform increases the sensitivity of such sensors. Such design could represent a major advance to the actual mounting of MEMS hairs on a horizontal plate. This is, however, not a trivial task in MEMS fabrication.

In aquatic bugs *Notonecta* (Fig. 17.13b), the superhydrophobic hair coverage keeps stable air layer between the hairs and by this generates a water-layer interface at the body. Vibrations caused by predators or prey can be captured by this interface and effect in deformations in the regions of hair sockets and be further transmitted to the mechanosensory neurons (Mail et al. 2018). Recently, this mechanosensor inspired development of biomimetic sensory device for pressure and flow sensing in underwater applications.

## 17.8 Robotics

Insect-inspired robotics has strongly developed in the last two decades. Because insects are structurally and functionally complex, a complete technical copy of an animal in hardware and software is not possible (Delcomyn 2004, 2007). There are two different opinions on this matter. Some researchers suggest that as many features of an animal as possible should be incorporated into a robot (Ritzmann et al. 2000, 2004). Other researchers argue that including too many animal-like features into a robot can even impair performance (Yoneda and Ota 2003). Which features of

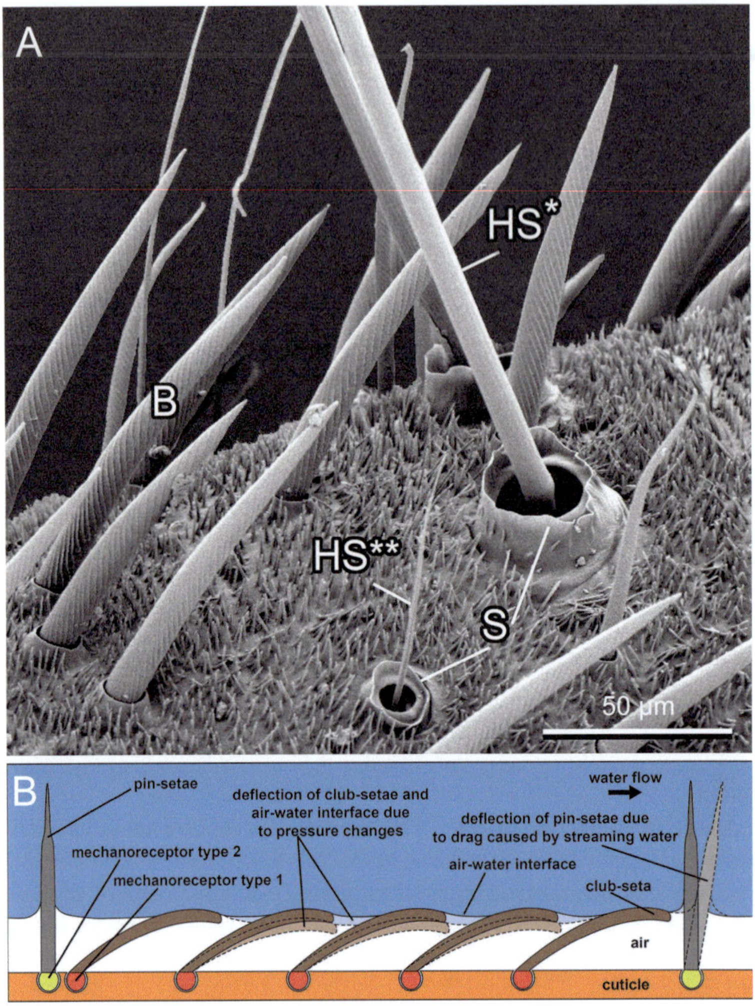


**Fig. 17.13** Insect mechanosensors. (**a**) SEM of cercal filiform hairs and campaniform sensilla of the cricket *Gryllus bimaculatus*. Cercus surface showing a filiform hair with a long hair shaft (*HS**) inserted in a large cuticular socket (*S*) and a filiform hair with a very short hair shaft (*HS***) inserted in a small socket (*S*) (Heusslein et al. 2009). (**b**) Proposed *Notonecta* forewing surface function (Mail et al. 2018). An air layer is kept in between the setae. The club setae (dark brown) support the air–water interface. If pressure increases, the air is compressed and the air–water interface is deformed. This deflects the seta (light brown, dashed outline). Mechanoreceptors (red) connected with the cuticle (orange) at the base of each seta should enable monitoring of the setal deflection. The pins (dark gray) most likely penetrate the air–water interface. If so they should be deflected (light gray, dashed outline) by water flow. This deflection most likely is sensed by cuticular mechanoreceptors (yellow). *B* bristles

insects and which aspects of their locomotion are really important for implementation into robot design?

One significant contribution to insect-inspired robotics can be search for nonconventional ways to move parts of the body, for example due to specialized joints, springs, and latches. Work is already in progress on a variety of novel actuators, such as electro-elastomers (Pei et al. 2003) and ionic polymeric-conductor composites (Shahinpoor 2003). Another contribution is the understanding and technical implementation of the concept of the dynamic stability allowing insects to move on land, in the air, or underwater with high speed and to be able to perform strong maneuverability. Especially challenging is locomotion on the wall and ceiling (Daltorio et al. 2005a, b, 2007, 2009). Generally, an incorporation of biomechanical principles into robots has recently contributed to the better performance of these robots (Altendorfer et al. 2001; Witte et al. 2004). The sensory feedback is critical to a fully functional, agile walking robot (Schmitz et al. 2001). This is the least developed area of bioinspired robotics. The reason is that it is difficult to make small and light but effective artificial sensors. Finally, central control of early robots must be changed to more distributed control that is typical for biological systems including insects (Delcomyn 1999; Thor et al. 2018).

Figure 17.14 shows some implementations of insect-based research into robotic systems. Many labs worldwide have designed hexapod robots, but probably the most elaborate ones originate from the Quinn Lab at Case Western Reserve University (Cleveland, USA) (Fig. 17.14a, b). The Prolero, RHex, and Whegs™ series robots utilize a method of locomotion that combines the advantages of wheels and legs (wheel-legs). Wheels are relatively simple and allow a vehicle to move over terrain quickly. Legs allow robots to run on a plain surface, to climb obstacles that are higher than what a wheeled vehicle would be able to climb over, and even to swim. These robots can be equipped with adhesive materials that make them capable of climbing up a glass wall (Fig. 17.14c, d). Structural data on cockroach legs (Frazier et al. 1999) were used to machine insect legs from aluminum blanks, in order to mimic chain-like tarsi capable of optimal grip with various kinds of terrain (Fig. 17.14f, g). In the future, mechanical design of such machines will rely on 3D printing and other additive manufacturing methods (Fig. 17.5f, g). This in turn will allow use of (1) exoskeletons, (2) complex lightweight constructions, and (3) composite materials consisting of different materials and representing sandwiches or lattices at different scales (from macro- and micro- down to the nanoscale). Moreover, using this kind of manufacturing technology may enable direct mimicking of insect constructions obtained through the microcomputed tomography from real insects.

Caterpillar-like locomotion that basically consists of two grippers in the anterior and posterior parts of the body and a set of muscles in between allowing body bending has inspired design of the InspiRatnik robot (Fig. 17.14e). It was recently tested by the company TETRA GmbH (Ilmenau, Germany) for the applications in the 3D environment, such as set of cables or pipes.

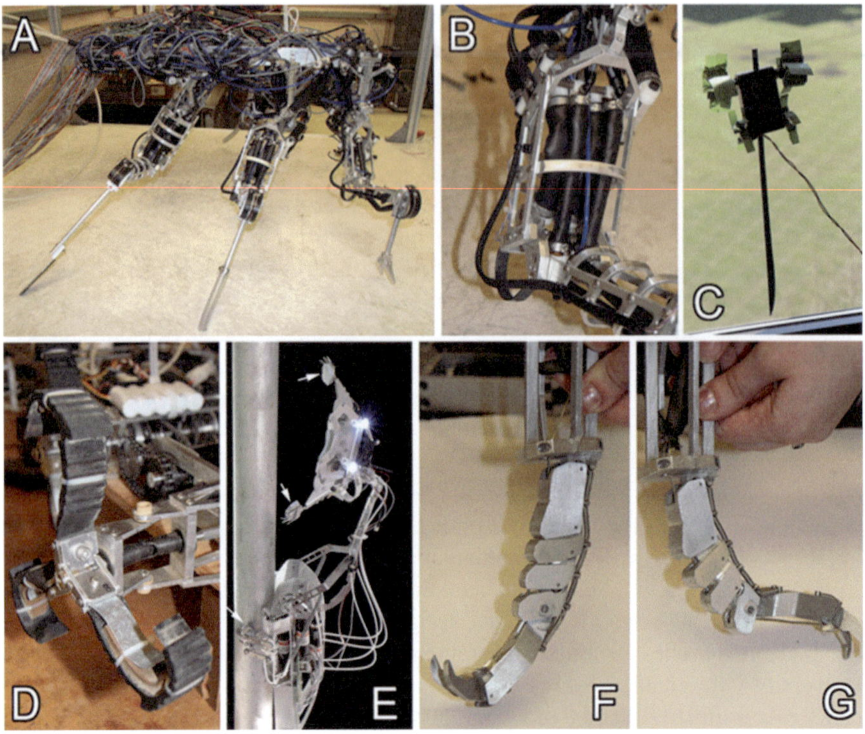


**Fig. 17.14** Some examples of insect-inspired robots. (**a**, **b**) Robot with a hexapod kinematics based on studies of the cockroach performed in the Ritzmann Lab in the Biology Department at Case Western Reserve University (CWRU, Cleveland, USA). It has a total of 24° of freedom with five for each front leg, four for the middle legs, and three for the rear legs. (**c**) Climbing Mini-Whegs™ scales vertical glass surfaces using compliant, adhesive feet (Daltorio et al. 2005a, b, 2007, 2009). Like all Mini-Whegs™ robots, it has one drive motor that turns all four wheel-legs in an alternating diagonal gait. (**d**) Legs of the typical Whegs™ robot. (**e**) InspiRatnik robot based on the principle of locomotion of geometrid caterpillars. (**f**, **g**) Gripping leg of the robot based on the inspiration from an insect tarsus. (**a–d**, **f**, **g**) Robots built in the Quinn Lab (CWRU, Cleveland, USA). (**e**) Robot built by TETRA GmbH (Ilmenau, Germany)

## 17.9 Future Perspectives

What can be done to advance the field of insect-inspired biomimetics? First and obviously, additional research on insects will help, not just in the application of biological knowledge to engineering problems, but on the biological systems themselves. Incorporating more biological knowledge into the design of artificial systems will improve their performance, but the fact of the matter is that biologists still do not have a complete understanding of how insect materials are constructed, what their performance is in different situations, how insect locomotion is generated, controlled, regulated, etc. Hence, many technological areas will benefit from additional entomological research.

Additionally, a huge variety of insects and their systems have not been previously studied at all. That is why screening for new interesting systems seems to remain an extremely important research field in the nearest future.

Only few examples of bioinspired systems were tested in order to compare their performance with that of conventional systems or their biological prototypes. The field of biomimetics in general would benefit greatly from the development of a set of standard tests that can be used to evaluate the performance of bioinspired materials and systems. The ultimate objective of biomimetics will be to improve the performance of biologically inspired systems in the real world. Then, it appears to be essential that these systems have to be subjected to real tests of performance. Clearly, the kind of tests will vary from system to system and be different for systems that have different performance objectives.

## References


Altendorfer R, Moore N, Komsuolu H, Buehler M, Brown HB, Mcmordie D, Saranli U, Full R, Koditschek DE (2001) RHex: a biologically inspired hexapod runner. Auton Robot 11:207–213

Andersen SO (1963) Characterization of a new type of cross-linkage in resilin, a rubber-like protein. Biochim Biophys Acta 69:249–262

Andersen SO (1964) The cross-links in resilin identified as dityrosine and trityrosine. Biochim Biophys Acta 93:213–215

Andersen SO (1966) Covalent cross-links in a structural protein, resilin. Acta Physiol Scand 66(Suppl. 263):1–83

Andersen SO, Weis-Fogh T (1964) Resilin, a rubber-like protein in arthropod cuticle. Adv Insect Physiol 2:1–65

Appel E, Gorb SN (2014) Comparative functional morphology of vein joints in Odonata. In: Zoologica. Schweizerbart Science Publishers, Stuttgart, pp 1–104

Appel E, Heepe L, Lin C-P, Gorb SN (2015) Ultrastructure of dragonfly wing veins: composite structure of fibrous material supplemented by resilin. J Anat 227(4):561–582

Arzt E, Gorb S, Spolenak R (2003) From micro to nano contacts in biological attachment devices. Proc Natl Acad Sci USA 100:10603–10606

Barbakadse N, Enders S, Gorb S, Arzt E (2006) Local mechanical properties of the head articulation cuticle in the beetle *Pachnoda marginata* (Coleoptera, Scarabaeidae). J Exp Biol 209:722–730

Barthlott W, Neinhuis C (1997) Purity of the sacred lotus or escape from contamination in biological surfaces. Planta 202:1–8

Barthlott W, Neinhuis C (1998) Lotusblumen und Autolacke: Ultrastruktur pflanzlicher Grenzflächen und biomimetische unverschmutzbare Werkstoffe. In: Nachtigall W, Wisser A (eds) 4. Bionik - Kongress, München 1998. Gustav Fischer Verlag, Stuttgart, Jena, Lübeck, Ulm, pp 281–293

Bauchhenss E, Renner M (1977) Pulvillus of *Calliphora erythrocephala* Meig. (Diptera; Calliphoridae). Int J Insect Morphol Embryol 6(3/4):225–227

Bernhard CG, Miller WH, Moller AR (1965) Insect corneal nipple array - a biological broad-band impedance transformer that acts as an antireflection coating. Acta Physiol Scand S 63(Suppl. 243):5

Betz O, Körner L, Gorb SN (2009) An insect's tongue as the model for two-phase viscous adhesives? Adhesion 6(1):32–35

Bocharova-Messner OM, Dmitriev AZ (1984) Morphological and functional analysis of the wing venation in Odonata according to the data of the scanning electron microscopy. In: IX Congress of All-Union Entomological Society. Naukova Dumka, Abstr. Theses. Kiev, p 65. (In Russian)

Büsse S, Hörnschemeyer T, Gorb SN (2017) The head morphology of *Pyrrhosoma nymphula* larvae (Odonata: Zygoptera) focusing on functional aspects of the mouthparts. Front Zool 14(25):1–13

Casas J, Dangles O (2010) Physical ecology of fluid flow sensing in arthropods. Annu Rev Entomol 55:505–520

Daltorio KA, Gorb SN, Peressadko A, Horchler AD, Ritzmann RE, Quinn RD (2005a) A robot that climbs walls using micro-structured polymer feet. International Conference on Climbing and Walking Robots, pp 131–138

Daltorio KA, Horchler AD, Gorb SN, Ritzmann RE, Quinn RD (2005b) A small wall-walking robot with compliant, adhesive feet. IEEE/RSJ International Conference on Intelligent Robots and Systems, pp 4018–4023

Daltorio KA, Wei TE, Gorb SN, Ritzmann RE, Quinn RD (2007) Passive foot design and contact area analysis for climbing mini-whegs (TM). The 2007 IEEE International Conference on Robotics and Automation (ICRA ´07), pp 1274–1279

Daltorio KA, Wei TE, Horchler AD, Southard L, Wile GD, Quinn RD, Gorb SN, Ritzmann RE (2009) Mini-whegs TM climbs steep surfaces using insect-inspired attachment mechanisms. Int J Robot Res 28:285–302

Delcomyn F (1999) Walking robots and the central and peripheral control of locomotion in insects. Auton Robot 7:259–270

Delcomyn F (2004) Insect walking and robotics. Annu Rev Entomol 49:51–70

Delcomyn F (2007) Biologically inspired robots. In: Habib MK (ed) Bioinspiration and robotics: walking and climbing robots. I-Tech, Vienna, pp 279–300

Dijkstra M, van Baar JJ, Wiegerink RJ, Lammerink TSJ, de Boer JH, Krijnen GJM (2005) Artificial sensory hairs based on the flow sensitive receptor hairs of crickets. J Micromech Microeng 15:132–138

Ditsche-Kuru P, Koop JHE, Gorb SN (2010) Underwater attachment in current: the role of setose attachment structures on the gills of the mayfly larvae *Epeorus assimilis* (Ephemeroptera, Heptageniidae). J Exp Biol 213:1950–1959

Elvin CM, Carr AG, Huson MG, Maxwell JM, Pearson RD, Vuocolo T, Liyou NE, Wong DCC, Merritt DJ, Dixon NE (2005) Synthesis and properties of cross-linked recombinant pro-resilin. Nature 437:999–1002

Elzinga RJ, Broce AB (1986) Labellar modifications of Muscomorpha flies (Diptera). Ann Entomol Soc Am 79:150–209

Francouer A, Loiselle R (1988) Evolution du strigile chez les formicides (Hymenoptera). Nat Can 115:333–335

Frazier SF, Larsen GS, Neff D, Quimby L, Carney M, DiCaprio RA, Zill SN (1999) Elasticity and movements of the cockroach tarsus in walking. J Comp Physiol A 185:157–172

Ghiradella H (1989) Structure and development of iridescent butterfly scales: lattices and laminae. J Morphol 202:69–88

Ghiradella H (1991) Light and colour on the wing: structural colours in butterflies and moths. Appl Opt 30:3492–3500

Ghiradella H, Aneshansley D, Eisner T, Silbergleid RE, Hinton HE (1972) Ultra-violet reflection of a male butterfly: interference colour caused by thin layer elaboration of wing scales. Science 178:1214–1217

Gnatzy W, Heusslein R (1986) Digger wasp against crickets. I. Receptors involved in the anti-predator strategies of the prey. Naturwissenschaften 73:212–215

Gogala M (1984) Vibration producing structures and songs of terrestrial Heteroptera as systematic character. Biol Vestn 32:19–36

Gorb SN (1999) Serial elastic elements in the damselfly wing: mobile vein joints contain resilin. Naturwissenschaften 86:552–555

Gorb SN (2000) Ultrastructure of the neck membrane in dragonflies (Insecta, Odonata). J Zool Lond 250:479–494

Gorb SN (2001) Attachment devices of insect cuticle. Kluwer Academic Publishers, Dordrecht, Boston, London

Gorb SN (2004) The jumping mechanism of cicada *Cercopis vulnerata* (Auchenorrhyncha, Cercopidae): skeleton-muscle organisation, frictional surfaces, and inverse-kinematic model of leg movements. Arthr Str Dev 33:201–220

Gorb SN (2011) Insect-inspired technologies: insects as a source for biomimetics, chapter 13. In: Vilcinskas A (ed) Insect biotechnology, biologically-inspired systems 2. Springer Science+Business Media B.V, Berlin, pp 241–264. https://doi.org/10.1007/978-90-481-9641-8_13

Gorb SN, Gorb EV (2016) Insect-inspired architecture: insects and other arthropods as a source for creative design in architecture. In: Knippers J et al (eds) Biomimetic research for architecture and building construction, biologically-inspired systems 9. Springer International Publishing, Switzerland, pp 57–83. https://doi.org/10.1007/978-3-319-46374-2_4

Gorb SN, Sinha M, Peressadko A, Daltorio KA, Quinn RD (2007) Insects did it first: a micropatterned adhesive tape for robotic applications. Bioinspir Biomim 2:S117–S125

Gorb SN, Tynkkynen K, Kotiaho JS (2009) Crystalline wax coverage of the imaginal cuticle in *Calopteryx splendens* (Odonata: Calopterygidae). Int J Odonatol 12(2):205–221

Gracham-Smith GS (1930) Further observations on the anatomy and function of the proboscis of the blow-fly, *Calliphora erythrocephala* L. Parasitology 22:47–115

Grodnicky DL (1988) Structure and function of the scale coverage of the wings in butterflies (Lepidoptera/Hesperioidea, Papilionoidea). Entomol Rev 67:251–256

Gronenberg W (1996) Fast actions in small animals: springs and click mechanisms. J Comp Physiol A 178:727–734

Haas F, Gorb S, Wootton RJ (2000a) Elastic joints in dermapteran hind wings: materials and wing folding. Arthr Struct Dev 29:137–146

Haas F, Gorb SN, Blickhan R (2000b) The function of resilin in beetle wings. Proc Roy Soc London B 267:1375–1381

Habenicht G (2002) Kleben: Grundlagen, Technologien, Anwendung. Springer Verlag, Berlin

Hackman RH (1975) Expanding abdominal cuticle in the bug *Rhodnius* and in the tick *Boophilus*. J Insect Physiol 21:1613–1623

Hackman RH, Goldberg M (1987) Comparative study of some expanding arthropod cuticles: the relation between composition, structure and function. J Insect Physiol 33:39–50

Heckmann CW (1983) Comparative morphology of arthropod exterior surfaces with capability of binding a film of air underwater. Int Rev Ges Hydrobiol 68:715–736

Heie OE (1987) Morphological structure and adaptations. In: Aphids: biology, natural enemies and control. Elsevier, Amsterdam, pp 393–400

Hennemann O-D (2000) Kleben von Kunststoffen. Anwendung, Ausbildung, Trend. Kunststoffe 90:184–188

Hepburn HR (1985) Structure of the integument. In: Kerkut GA, Gilbert LI (eds) Comprehensive insect physiology biochemistry and pharmacology. Pergamon Press, Oxford, pp 1–58

Hepburn HR, Chandler HD (1976) Material properties of arthropod cuticles: the arthrodial membranes. J Comp Physiol A 109:177–198

Hepburn HR, Chandler HD (1978) Tensile mechanical properties and transconformational changes of chitins. In: Muzzarelli RA, Parisier ER (eds) Proceedings of the first international conference on chitin/chitosan. Massachusetts Institute of Technology, Cambridge, Mass, pp 124–143

Heusslein R, Gras H, Gnatzy W (2009) Functional coupling of cercal filiform hairs and campaniform sensilla in crickets. In: Gorb SN (ed) Functional surfaces in biology, vol 1. Springer-Verlag, Dordrecht, pp 203–233

Hinton HE (1970) Some little-known surface structures. In: Neville AC (ed) Insect ultrastructure. Blackwell Scientific Publishers, Oxford and Edinburgh, pp 41–58

Hinton HE (1976) The fine structure of the pupal plastron of simulid flies. J Insect Physiol 22:1061–1070

Humphrey JAC, Devarakonda R, Iglesias I, Barth FG (1993) Dynamics of arthropod filiform hairs. I. Mathematical modelling of the hair and air motions. Phil Trans R Soc Lond B 340:423–440

Humphrey JAC, Barth FG, Voss K (2003) The motion-sensing hairs of arthropods: using physics to understand sensory ecology and adaptive evolution. In: Barth FG, Schmid A (eds) Ecology of sensing. Springer-Verlag, Berlin, Heidelberg, New York, pp 105–115

Huxley J (1975) The basis of structural colour variation in two species of *Papilio*. J Entomol A 50:9–22

Ishii S (1987) Adhesion of a leaf feeding ladybird *Epilachna vigintioctomaculata* (Coleoptera: Coccinellidae) on a vertically smooth surface. Appl Ent Zool 22:222–228

Komvopoulos K (2003) Adhesion and friction forces in microelectromechanical systems: mechanisms measurement, surface modification techniques and adhesion theory. J Adhes Sci Technol 17:477–517

Kovalev A, Filippov A, Gorb SN (2018) Slow viscoelastic response of resilin. J Comp Physiol A 204(4):409–417. https://doi.org/10.1007/s00359-018-1248-2

Langer M, Ruppersberg P, Gorb SN (2004) Adhesion forces measured at the level of a terminal plate of the fly's seta. Proc R Soc Lond B 271:2209–2215

Li X, Madan I, Bozkurt H, Birkhofer H (2004) Lifetime of solid lubricated roller bearings. In: 14th Int. Coll. Tribology, Jan. 2004, Esslingen, Germany 3, pp 1361–1364

Mail M, Klein A, Bleckmann H, Schmitz A, Scherer T, Rühr PT, Lovric G, Fröhlingsdorf R, Gorb SN, Barthlott W (2018) A new bioinspired method for pressure and flow sensing based on the underwater air-retaining surface of the backswimmer *Notonecta*. Beilstein J Nanotechnol 9:3039–3047

Messner B (1988) Funktionelle Morphologie der Insektenkutikula am Beispiel der Plastronatmer. Wiss Z E M Arndt-Univ Greifswald Math Naturwiss R 37:27–30

Miyoshi K (2001) Solid lubrication: fundamentals and applications. Marcel Decker Inc, New York, Basel

Naldrett MJ (1992) Adhesives. In: Vincent JFV (ed) Biomechanics - materials: a practical approach. IRL Press, Oxford, pp 219–240

Niederegger S, Gorb S (2003) Tarsal movements in flies during leg attachment and detachment on a smooth substrate. J Insect Physiol 49:611–620

Niederegger S, Gorb SN, Jiao Y (2002) Contact behaviour of tenant setae in attachment pads of the blowfly *Calliphora vicina* (Diptera, Calliphoridae). J Comp Physiol A 187:961–970

Palestrini C, Piazza R, Zunino M (1987) Segnali sonori in the species di Geotrupini (Coleoptera, Scarabaeoidea, Geotrupidae). Boll Soc Entomol Ital 119:139–151

Parker AR, Hegedus Z, Watts RA (1998) Solar-absorber antireflector on the eye of an Eocene fly. Proc Roy Soc London B 265:811–815

Pei Q, Pelrine R, Stanford S, Kornbluh R, Rosenthal M (2003) Electroelastomer rolls and their application for biomimetic walking robots. Synth Met 135:129–131

Peisker H, Gorb SN (2010) Always on the bright side of life: anti-adhesive properties of insect ommatidia grating. J Exp Biol 213(Pt 20):3457–3462

Peressadko A, Gorb SN (2004) When less is more: experimental evidence for tenacity enhancement by division of contact area. J Adhes 80:247–261

Perez Goodwyn P (2009) Anti-wetting surfaces in Heteroptera (Insecta): hairy solutions to any problem. In: Gorb SN (ed) Functional surfaces in biology, vol 1. Springer-Verlag, Dordrecht, pp 55–76

Perez Goodwyn PJ, Gorb SN (2004) Frictional properties of contacting surfaces in the hemelytra-hindwing locking mechanism in the bug *Coreus marginatus* (Heteroptera, Coreidae). J Comp Physiol A 190:575–580

Persson BNJ (2003) On the mechanism of adhesion in biological systems. J Chem Phys 118:7614–7621

Radhakrishnan V (1998) Locomotion: dealing with friction. Proc Natl Acad Sci USA 95:5448–5455

Rajabi H, Ghoroubi N, Darvizeh A, Dirks J-H, Appel E, Gorb SN (2015) A comparative study of the effects of vein-joints on the mechanical behaviour of insect wings: I. Single joints. Bioinspir Biomim 10(5):056003

Rajabi H, Ghoroubi N, Darvizeh A, Appel E, Gorb SN (2016) Effects of multiple vein microjoints on the mechanical behaviour of dragonfly wings: numerical modelling. Roy Soc Open Sci 3:150610

Rajabi H, Ghoroubi N, Stamm K, Appel E, Gorb SN (2017) Dragonfly wing nodus: a one-way hinge contributing to the asymmetric wing deformation. Acta Biomater 60:330–338
Ritzmann RE, Quinn RD, Watson JT, Zill SN (2000) Insect walking and biorobotics: a relationship with mutual benefits. Bioscience 50:23–33
Ritzmann RE, Quinn RD, Fischer MS (2004) Convergent evolution and locomotion through complex terrain by insects, vertebrates and robots. Arthr Str Dev 33:361–379
Rothshild M, Schlein J, Parker K, Neville C, Sternberg S (1975) The jumping mechanism of *Xenopsylla cheopis*. III. Execution of the jump and activity. Phil Trans Roy Soc Lond B 271:499–515
Rudall KM (1963) The chitin/protein complexes of insect cuticles. In: Beament JWL, Treherne JE, Wigglesworth VB (eds) Advances in insect physiology, vol. 1. Academic Press, London, pp 257–311
Scherge M, Gorb SN (2001) Biological micro- and nanotribology. Springer-Verlag, Berlin
Scherge M, Li X, Schaefer A (1999) The effect of water on friction of MEMS. Tribol Lett 6:215–220
Schmitz J, Dean J, Kindermann T, Schumm M, Cruse H (2001) A biologically inspired controller for hexapod walking: simple solutions by exploiting physical properties. Biol Bull 200:195–200
Schönitzer K, Lawitzky G (1987) A phylogenetic study of the antenna cleaner in Formicidae, Mutillidae and Tiphiidae (Insecta, Hymenoptera). Zoomorphology 107:273–285
Schultz TD, Hadley NF (1987) Structural colors of tiger beetles and their role in heat transfer through the integument. Physiol Zool 60:737–745
Shahinpoor M (2003) Ionic polymer-conductor composites as biomimetic sensors, robotic actuators and artificial muscles - a review. Electrochim Acta 48:2343–2353
Steinmann T, Casas J, Krijnen G, Dangles O (2006) Air-flow sensitive hairs: boundary layers in oscillatory flows around arthropod appendages. J Exp Biol 209:4398–4408
Stork NE (1980) Experimental analysis of adhesion of *Chrysolina polita* (Chrysomelidae, Coleoptera) on a variety of surfaces. J Exp Biol 88:91–107
Stork NE (1983) The adherence of beetle tarsal setae to glass. J Nat Hist 17:583–597
Thor M, Strøm-Hansen T, Larsen LB, Kovalev A, Gorb SN, Baird E, Manoonpong P (2018) A dung beetle-inspired robotic model and its distributed sensor-driven control for walking and ball rolling. Artif Life Robotics. https://doi.org/10.1007/s10015-018-0456-8
Thorpe WH, Crisp DJ (1947) Studies on plastron respiration. I. The biology of *Aphelocheirus* and the mechanism of plastron retention. J Exp Biol 24:227–269
Vagts S, Schlattmann J, Busshardt P, Kleinteich T, Gorb SN (2017) The application of multi-body simulation approach in the kinematic analysis of beetle leg joints. Artif Life Robotics. https://doi.org/10.1007/s10015-017-0386-x
Vagts S, Schlattmann J, Kovalev A, Gorb SN (2018) The topology of the leg joints of the beetle *Pachnoda marginata* (Scarabaeidae, Cetoniinae) and its implication for the tribological properties. Biomimetics 3:12. https://doi.org/10.3390/biomimetics3020012
Vincent JFV (1981) Morphology and design of the extensible intersegmental membrane of the female migratory locust. Tiss Cell 13:18–31
Vincent JFV, Wegst U (2004) Design and mechanical properties of insect cuticle. Arthr Str Dev 33:187–199
Vincent JFV, Wood SDE (1972) Mechanism of abdominal extension during oviposition in *Locusta*. Nature 235:167–168
Vukusic P, Sambles JR (2003) Photonic structures in biology. Nature 424:852–855
Vukusic P, Sambles JR, Lawrence CR, Wootton RJ (1999) Quantified interference and diffraction in single *Morpho* butterfly scales. Proc R Soc Lond B 266:1403–1411
Vukusic P, Sambles JR, Ghiradella H (2000a) Optical classification of microstructure in butterfly wing-scales. Photonics Sci News 6:61–66
Vukusic P, Sambles JR, Lawrence CR (2000b) Structural colour: colour mixing in wing scales of a butterfly. Nature 404:457

Wagner T, Neinhuis C, Barthlott W (1996) Wettability and contaminability of insect wings as a function of their surface sculpture. Acta Zool 77:213–225
Waite JH (1983) Adhesion in byssally attached bivalves. Biol Rev 58:209–231
Weis-Fogh T (1960) A rubber-like protein in insect cuticle. J Exp Biol 37:889–907
Weis-Fogh T (1961) Molecular interpretation of the elasticity of resilin, a rubber-like protein. J Mol Biol 3:648–667
Wigglesworth VB (1987) How does a fly cling to the under surface of a glass sheet? J Exp Biol 129:363–367
Witte H, Hoffmann H, Hackert R, Schilling C, Fischer MS, Preuschoft H (2004) Biomimetic robotics should be based on functional morphology. J Anat 204:331–342
Wootton RJ, Newman DJS (1979) Whitefly have the highest contraction frequencies yet recorded in non-fibrillar flight muscles. Nature 280:402–403
Yoneda K, Ota Y (2003) Non-bio-mimetic walkers. Int J Robot Res 22:241–249

# Index

**A**
Abdominal-substrate stridulation, 119
*Abedus herberti*, 39, 40
Adaptations, 11, 12, 24
  dispersal (*see* Dispersal)
Adephaga, 218
Adhesive capture apparatus, 414
Adhesive organs, 412
Adipokinetic hormone (AkH), 51
Adult male mayfly, 268
*Aedes albopictus*, 25
Aepophilidae, 210
Aeshnidae, 145
Agonistic character displacement (ACD), 179–181
Agonistic interactions, 180
Alderflies, 170
Altitude, 89
Amino acid composition, 404
Amphibiotic life cycles, 352
Amphizoidae, 218
*Anax junius*, 55, 56
Anholt's model, 41
Animal behavior, 5
Anisoptera, 201
Anisozygoptera, 201
Antarctoperlaria, 118, 127, 130
Antennae
  aglomerular antennal lobe, 155
  Baetidae, 157
  chemosensory role, 150
  *D. melanogaster*, 156
  damselflies larval, 145
  intra- and interspecific communication, 140
  morpho-functional hypothesis, 154
  in Odonata, 145, 152
  Paleoptera, 146
  Plecoptera, 142, 154
  porous and digitated pegs, 153
  re-evaluation, 152
  stoneflies nymph *Paragnetina media*, 157
Anthropogenic effects, 375
Anthropogenic impacts, 368, 369
Anti-predatory behavior, 98, 99
Aphelocheiridae, 213
Aporous, 140, 141, 159
Aquatic adult life
  Coleoptera, 278, 279
  Hemiptera, 279, 280
Aquatic animals, 109
Aquatic bugs, 210
Aquatic cockroaches, 265
Aquatic environments, 382
Aquatic insects, 355, 357
  abiotic and biotic factors, 6
  adhesive-bond formation, 411
  amphibiotic habit, 8
  arthropods, 409
  attachment force, 412
  behavior, 5–7
  bioindicators, 8
  cuticle, 403
  dispersal, 6
  distribution information, 374
  ecological engineers, 3
  ecological services, 2
  essentials, 2
  exoskeleton, 403
  factors, 12
  filiform hairs, 417
  functional principles, 8
  functional solutions, 401


K. Del-Claro, R. Guillermo (eds.), *Aquatic Insects*,
https://doi.org/10.1007/978-3-030-16327-3

Aquatic insects (*cont.*)
functional surfaces, 410
genetic connectivity, 8
global distribution, 6
grating, 415
hairy attachment surface, 413
immature stages, 2
invertebrates, 2
keystone species, 2
mammals, 4
material properties, 405
MEMS, 407
microfibrils, 404
micro-joints, 408
microtrichium, 409
nanofibers, 405
parental care, 7
plant surfaces, 409
principles, 402
resilin, 404
sensory and communication abilities, 6
sexual conflict, 7
sexual selection, 7
structural and chemical complexity, 405
structural coloration, 416
surface structures, 405
technical developments, 407
technological areas, 402
tropical systems, 4
urban wetlands, 7
Vereda, 5
visual perception, 6
visual signals, 5
young biologists, 6
Aquatic macroinvertebrates, 357
Arctiinae (Erebidae) subfamily, 344
Arctoperlaria, 118, 130
*Argia angelae*, 4
Armaments, 281, 283
Arthropods, 1
Artificial man-made wetlands, 355
Aspidytidae, 219
Atomic force microscopic (AFM), 415
Attachment devices, 411

**B**
Bacteria, 107, 108
Barriers, 384
Basiconic sensilla, 140, 142
Beetles, 170
Behavioral ecology, 5
Belostomatidae, 211
distribution and phylogenetic relationships, 322, 323, 326
general biology, 322
paternal care and oviposition site selection, 330, 333
Benthic macroinvertebrates, 12
Bergmann's rule, 78
Biodiversity, 13, 15, 20, 351, 382
Biodiversity loss, 375, 378
Biogeographic hypotheses, 16
Bioindicators, 372, 376, 377, 382, 383
Bioinspired systems, 421
Biological engineers, 15
Biology, 2, 6–8
Biomimetics, 402
Biomonitoring, 368
animal behavior, 367, 369, 374
aquatic invertebrates, 367
behavioral indicators, 372
biodiversity conservation, 367
conservation planning, 374
dispersal proxies, 373
ecological restoration, 376
ecosystem, 367, 368
freshwater ecosystems, 366
human and aquatic insect, 376
knowledge gap, 366
oviposition behavior, 373
protocols, 369
reproduction, 372
simulation modelling studies, 374
Biotic environment
bacteria, 107, 108
competition, 104, 106
fungi, 108
harmonic interactions, 106
herbivory, 109
parasitism, 100–104
phylogenetic lineages, 109
predation, 95–99
predatory aquatic insects, 99, 100
Biotic Index, 357
Blaberid female, 265
Blaberidae, 206
Black flies, 170
Blattodea, 264, 265
*Boreonectes aequinoctialis*, 39, 40
Butterflies, 141, 341

**C**
Caddisflies, 158, 170, 171
Caddisfly larvae, 99, 106

Campaniform sensilla, 141, 143, 146
Cannibalism, 105
Carabidae, 236
Caterpillar, 344
Ceratopogonidae, 242
Chaetic sensilla, 140
Character displacement
ACD, 179, 181, 182
Chemical defense, 192, 198, 201, 205, 239, 247, 248
Chemical pollution, 355
Chironomids, 171
Chrysomelidae, 239
Climate warming, 38, 62, 63
Cneoglossidae, 236
Coarse particulate organic matter (CPOM), 18
Coeloconic sensilla, 140, 142, 150–151, 153, 157
Coenagrionidae, 145
Coleoptera, 216, 278, 279
Collembola, 193, 198
Colonization cycle, 40
Commensalism, 96
Communication, 6
Competition, 104, 106
Compilation aquatic insect orders, 192
Conservation behavior, 367
Contests
territorial, 174, 175, 177, 183
Convenience polyandry, 300, 302
Corixidae, 54
Corixoidea, 212
Cosmopterigidae, 246
Crambidae, 246, 344, 345
Cricket airflow sensors, 417
Cryptic female choice
description, 297
odonates, 310
postcopulatory sexual selection, 307
sexual conflict, 308
sexually antagonistic co-evolution, 312
and sperm competition, 297, 305, 306, 311, 314
Curculionidae, 239
Cuticle, 403

**D**

Damselflies, 372
and dragonflies, 170
Defense mechanisms
active and passive defenses, 192
Adephaga, 218, 247
adults and developmental stages, 192
Aepophilidae, 210
Amphizoidae, 218
anal glands and secretions, 237
Aphelocheiroidea, 213–214
aquatic entomology, 248
aquatic larval stages, 248
aromatic pygidial gland constituents, 228
arthropods, 247
Aspidytidae, 219
Belostomatidae, 211
Blaberidae, 206
Carabidae, 236
Ceratopogonidae, 242
Chrysomelidae, 239
Cneoglossidae, 236
Coleoptera, 216
Collembola, 193, 198, 199
Corixoidea, 212
Curculionidae, 239
defensive glands, 243
Diptera, 241, 242
Dryopidae, 235
dytiscid beetles, 193
Dytiscidae, 218–223, 226, 248
Elmidae, 235
Ephemeroptera, 198, 201
Epimetopidae, 234
Erirhinidae, 239
Eulichadidae, 236
Gelastocoridae, 212
Georissidae, 240
Gerridae, 208
Gerromorpha, 207
Gyrinidae, 226, 229
Haliplidae, 229, 230
Hebridae, 207
Helophoridae, 234
Helotrephidae, 215
Hermatobatidae, 208
Heteroceridae, 240
Heteroptera, 206, 208–209
Hydradephaga, 217, 219
Hydraenidae, 232–234
Hydrochidae, 234
Hydrometridae, 207
Hydrophilidae, 232–233
Hydroporinae, 222
Hygrobiidae, 230
Hymenoptera, 240
Lampyridae, 232–233, 238
Leiodidae, 238
Lepiceridae, 216

Defense mechanisms (*cont.*)
  Lepidoptera, 246
  Leptopodidae, 210
  Leptopodomorpha, 208
  Limnichidae, 240
  Lonchopteridae, 244
  Lutrochidae, 236
  Macroveliidae, 207
  Madeoveliidae, 208
  Mecoptera, 247
  Megaloptera, 215
  Meruidae, 231
  Mesoveliidae, 207
  morphological defense, 224–225, 243
  Myxophaga, 216
  Nanophyidae, 239
  Naucoroidea, 213
  Nepidae, 210, 211
  Nepoidea, 210
  Nepomorpha, 210
  Neuroptera, 215, 216
  Nitidulidae, 236
  Noteridae, 231
  Notonectoidea, 213
  Ochteridae, 213
  Ochteroidea, 212
  Odonata, 201–204
  Omaniidae, 210
  Orthoptera, 205, 206
  Paraphrynoveliidae, 207
  peritrichic ciliates, 249
  phylogenetic trees, 193
  physicochemical properties, 249
  Plecoptera, 204
  Pleidae, 214
  predator-prey interactions, 192
  Psephenidae, 236
  Ptilodactylidae, 238
  pyroglutamic acid, 243
  Saldidae, 210
  Scarabaeidae, 238
  Scirtidae, 235
  Spercheidae, 234
  Sphaeriusidae, 216
  Staphylinidae, 237, 238
  steroid deoxycorticosterone, 226
  Stratiomyidae, 244
  stridulatory organ, 243
  Syrphidae, 244
  Tabanidae, 244
  terrestrial and aquatic habitats, 192
  Torridincolidae, 216
  Trachypachidae, 240
  Trichoptera, 244, 245
  Veliidae, 207
Dendritic networks, 20
Detritus, 15
Diptera, 241, 242, 266
Dispersal
  aquatic and semiaquatic insects, 36
  and climate warming, 62, 63
  colonization
    aquatic beetles, 47
    aquatic insects, 45
    chemical and physical cues, 45
    compression, 45
    fish-associated patches, 46
    hydrophylids, 47
    mesocosms, 45, 46
    mosquitoes, 47
    polarizing surfaces, 45
  and community dynamics, 60, 61
  diapause and reproduction, 53
  evolution, 36
  and genetic change, 59, 60
  habitat template
    allozyme frequency analysis, 39
    aquatic environments, 37
    aquatic insects, 37
    climate warming, 38
    diurnal species, 37
    DNA sequence analysis, 39
    drift, 40, 41
    drought, 37
    $F_{ST}$ values, 39
    genetic exchange, 39
    hydrophilids, 38
    hyporheic zone, 39, 40
    lentic and lotic waters, 38
    lotic species, 39
    mark-recapture methods, 37
    niche breadth, 38
    passive interception traps, 41
    patchy recruitment hypothesis, 41
    physical/biological processes, 36
    upstream crawling, 41
  habitats, 36
  initiation
    accelerate development, 43
    *Baetis*, 44
    beetles and corixid, 42
    Hemiptera and Coleoptera, 43
    life cycle, 44
    *Notonecta*, 43
    photoperiod, 44
    temporary seepage pools, 42

long-distance migration
*A. junius*, 56
aquatic/semiaquatic insects, 54
butterflies, locusts and dragonflies, 54
defined, 54
ITCZ, 55, 57
Monarch butterflies, 56
Odonata, 56, 57
*P. flavescens*, 55
photoperiod, 57
*Schistocerca gregaria*, 54
metapopulations, 58, 59
and nutrient subsidies, 61, 62
passive and active, 49
polymorphism
adaptive morphology, 50
chironomids, 52
corixids and notonectids, 50
dimorphism, 50
Gerridae, 50
JH, 51
libellulid Anisoptera, 51
mesocosm, 52
non-diapause generation, 50
photoperiod, 50
requirements, 51
water striders, 50
timing, 48
Dispersion, 384
Dissolved oxygen (DO), 83
Diving beetles
arms races, 315
case of, 304
Dytiscidae, 311
female reproductive tract, 309
mating behaviour, 307
sperm heteromorphism, 311
water striders and odonates, 298
Dobsonflies, 170, 172
Dominant-subordinate interactions, 168
Dorsal wing surface, 409
Dorsoventral flattening, 20
Downstream Increase in Intraspecific Genetic Diversity (DIGD), 391
Dragonflies, 273
and damselflies, 170, 299
and damselflies larval antennae, 145, 150, 152
hygro-thermoreceptors, 157
"sunny insects", 158
Dragonfly Biotic Index, 358
Dragonfly cuticle, 403
Drift compensation, 40
Drift paradox, 40
Drought, 88
Dryopidae, 235
Dytiscidae, 54, 219–221, 226

## E

Ecogeographic rules, 76
Ecogeographical model, 90
Ecological interactions, 109
Ecological networks, 2, 3
Ecological traps, 352
Eco-morphology, 78
Ecophysiological hypothesis (EH), 79
Ectothermic species, 79
Elaborate cuticular structures, 401
Elmidae, 235
Encounter sites
adult behavior data, 127
*Capnia lacustra*, 127
*Dinocras cephalotes*, 127
emergence substrates, 126
postemergence maturation, 126
*Protonemura gevi*, 127
resonant substrates, 125
stoneflies, 125
stoneflies from Southern Hemisphere, 127
Environments inhabited
human-made environments, 24
intermittent streams, 21
lakes and ponds, 21, 22
madicolous habitats, 23
marine environments, 26
mosquito invasions, 24
phytotelma, 24, 26
rivers and streams, 18–20
temporary pools, 23, 24
water-filled tree holes, 24
wetlands, 22
Ephemeroptera, 16, 83, 140–142, 144–147, 149–152, 157–159, 170, 198, 200, 266, 269, 352
Ephemeropteran, 281, 282
Epilimnion, 386
Epimetopidae, 234
Epiophlebiidae, 145
EPT index, 358
Erebidae, 246
Erirhinidae, 239
Eulichadidae, 236

## F

Facilitation, 96, 106
Family Biotic Index, 357

Filiform hairs, 141, 143–146, 149
Filter feeders, 87
Filter-feeding blackflies, 171
Filtration systems, 410
Fine particulate organic matter (FPOM), 18
Freshwater, 322, 323
Freshwater animals, 109
Freshwater aquatic systems, 2
Freshwater environment, 76, 104, 107
Friction reduction, 407
Functional diversity, 406

**G**
Ganges-Brahmaputra delta, 389
Gelastocoridae, 212
Gene Environments Associations (GEA), 391
Genetic connectivity, 383
  advantages and disadvantages, 387
  aquatic systems, 384
  barriers, 384
  dendritic hierarchy, 385
  dispersion, 384
  epilimnion and hypolimnion, 386
  freshwater environments, 383
  freshwater habitat, 384
  freshwater systems, 384
  genetic flow, 384
  lotic systems, 386
  migration, 383
  mtDNA, 387
  RCC theory, 385
  SHM, 385
  SNPs, 388
Genetic diversity, 383
Geographic ranges, 38
Georissidae, 240
Gerridae, 208
Gerromorpha, 207
Giant water bugs
  large eggs, 328
  paternal care systems, 322
Gill filament, 342
Globally environments, 382
Gomphidae, 145
Gordiid horsehair worm, 102
Graph theory, 3
Grooming, 415
Gyrinidae, 226, 229

**H**
Habitat architecture, 86
Habitat complexity, 76, 84
Habitat fragmentation, 354, 395
Habitat Integrity Index (HII), 83, 358
Hairy leg attachment pads, 413
Haliplidae, 229, 230
*Halobates*, 26
Headwater streams, 18
Heat Island effect, 356
Hebridae, 207
Helophoridae, 234
Helotrephidae, 215
Hemiptera, 279, 280
Hermatobatidae, 208
Heteroceridae, 240
Heteroptera, 173, 174, 206
Human behavior, 377
Human-environment interactions, 350
Hydraenidae, 234
Hydraulics, 84, 88
Hydrochidae, 234
Hydrodynamics, 76, 84, 85
Hydrometridae, 207
Hydrophilidae, 54, 232, 234
Hydrophobic surfaces, 407
Hydroscaphidae, 216
Hygrobiidae, 230, 231
Hymenoptera, 240, 241, 269, 270
Hypolimnion, 386

**I**
Imagos, 267, 269
Individual habitat selection (IHS), 59
Inland aquatic system, 2
Insect attachment devices, 413
Insect-inspired robotics, 417, 419
Insect mechanosensors, 418
The International Union for Conservation of Nature (IUCN), 383
Intertropical Convergence Zone (ITCZ), 55
Intraguild predation (IGP), 105, 106
Invasive species, 356
Ionotropic receptors (IRs), 147, 153, 156
Iridescence, 416
Isolation by distance (IBD), 393
Isotomidae, 198

**J**
Juvenile hormone (JH), 51

**K**
Kleptoparasitism, 302

## L

La Crosse virus (LACV), 103
Lampyridae, 238, 239
Larval development, 345
Larval odonates, 172
Leiodidae, 238
Lentic systems, 388
Lepiceridae, 216
Lepidoptera, 246, 270, 271, 341, 342, 346
  aquatic, 342
  caterpillar submerges, 342
  hydrophobic bristles, 342
  semiaquatic, 342
  species, 342
  terrestrial specimens, 342
Leptopodidae, 210
Leptopodomorpha, 208
Lestidae, 145
Lethocerinae, 323, 333, 335
  brooding male, 327
  first instar, 329
*Leucorrhina intacta*, 51
Libellulidae, 145, 157
Light environment, 80
Light pollution, 355
Limnephilidae, 24
Limnichidae, 240
Lonchopteridae, 244
Lotic and lentic environments, 15, 17, 18, 21
Lotic systems, 388
Lutrochidae, 236

## M

Macrohabitat, 85
Macroscales, 76, 85
Macroveliidae, 207
Madeoveliidae, 208
Malaise/sticky traps, 40
Mark-recapture techniques, 387
Mate guarding
  in colubrids and anurans, 132
  emergence period of males and females, 129
  intersexual vibrational communication, 128
  *Isoperla morenica*, 130
  mating position, 131
  mechanism, 131
  Perloidea species, 128
  in Plecoptera, 131
  positive protandry value, 129
  pre-copulatory, 122
  *Pteronarcella badia*, 131
  re-mate/answer male drumming calls, 131
  sperm transmission, 129
  spermatophores, 130
  stoneflies, 128, 129, 132
  tactile contacts, 130
Mating systems, 264
Maturation, 352
Mayflies, 81, 170, 171
  and stoneflies, 158
  winged stages, 152
Mechanosensory, 417
*Mecistogaster* sp., 27
Mecoptera, 247
Megaloptera, 215, 275
Meruidae, 231
*Mesocapnia arizonensis*, 39, 40
Mesohabitats, 85, 86
Mesoscale, 76, 86, 88
Mesoveliidae, 207
Metapopulations, 58, 59
Microelectromechanical devices (MEMS), 406
Microhabitat, 12, 86, 87
Microsatellites, 388
Microscale, 76, 85, 86, 88
Mitochondrial genes, 387
Molecular markers
  microsatellites, 388
  mtDNA, 388
  SNPs, 388
Momphidae, 246
Monitoring biases and trends, 368
Monthly accumulated temperature (MAT), 394
Moths, 170, 341
Multiporous, 140, 141, 147, 150–151
Mutualisms, 96
Myxophaga, 216

## N

Nanophyidae, 239
Naucoroidea, 213
Necromones, 104
Neotropical region, 374
Nepidae, 210, 211
Nepoidea, 210, 323
Nepomorpha, 210
Networks
  harbors terrestrial, 95
  trophic, 99, 107
Neuroptera, 215, 271
Neutral theory of biodiversity, 60

Niche theory, 86
Nitidulidae, 236
Noctuidae, 246
Noteridae, 231
Notonectoidea, 213
Nuorteva index, 358

**O**

Ochteridae, 213
Ochteroidea, 212
Odonata, 16, 170, 172, 178, 179, 272–274
Odonates
  female, 300, 302, 303
  male, 303, 314
  territorial species, 306
  water striders, 312
Odontoceridae, 278
Omaniidae, 210
Ommatidial gratings, 415, 416
Oogenesis-flight syndrome, 53
Optimum thermal regime, 16
Orthoptera, 205, 206, 274
Oviposition, 104–106, 109, 173–175, 264, 266, 269–272, 279, 283, 352, 373
Ovoviviparity, 265, 269
Oxidation reduction potential (ORP), 83

**P**

*Pachydiplax longipennis*, 51
*Palaeagapetus*, 13
*Pantala flavescens*, 55
Paraphrynoveliidae, 207
Parasitism, 100, 101, 103, 104
  aquatic hymenoptera, 103
  aquatic insects, 104
  competition, 100
  female *A. aegypti*, 104
  female *Forcipomyia*, 100, 101
  gordiid horsehair worm, 102
  horsehair worms, 102
  LACV, 103
  *Leishmania*, 103
  nematodes, 102
  odonates, 102
  parasite transmission, 102
  parasites and parasitoids, 100
  pathogens and parasites, 104
  trematodes, 100
  underwater petioles, 103
  water mites, 101
Paternal care
  *Abedus* species, 336
  Belostomatidae, 322, 323, 326
  Lethocerinae, 333, 335
  and oviposition site selection, 330, 333
  sexual selection, 335
  Smith's hypothesis, 327, 328, 330, 336
  SSD, 336
Periphyton depletion, 171
Petaluridae, 145
*Philanisus plebeius*, 26
Philopatry, 59
Photonic crystals, 416
Photoperiod, 44, 81
Phototropism, 81
Phytoplankton and zooplankton production, 19
Phytotelmata, 105
Placoid sensilla, 140, 142
Plasticity, 284, 285
Plecoptera, 117–119, 122, 125, 140–142, 145, 147, 150–151, 153, 154, 157–159, 170, 171, 204, 275, 276
Pleidae, 214
Polymerase Chain Reaction (PCR) technique, 388
Polymorphism, 284
*Pontomyia*, 26
*Pontomyia oceana*, 28
Predation, 96, 99
  anti-predatory behaviors, 98, 99
  aquatic and semiaquatic organisms, 95
  aquatic insects, 97, 99
  biotic interactions, 96
  caddisfly larvae case, 99
  female *Culex* mosquitoes, 98
  habitat interferes, 96
  inducible defense, 98
  morphological defenses, 98
  nonlethal effects, 96
  odonate larvae, 98
  and parasitism, 96
  pisaurid spider preying, 97
  predator-prey interactions, 98
  predatory aquatic insects, 99, 100
  predatory signals, 98
  stonefly predators, 96
Predatory aquatic insects, 99–100, 107
Prothoracic exocrine glands, 222
Psephenidae, 236
Pseudocells, 198
Pseudostigmatidae, 26

Ptilodactylidae, 238
Pygidial gland secretion, 225
Pyralidae, 246, 346

**R**
Reproductive behavior, 271
Reproductive character displacement (RCD), 181
Reproductive strategies, 125, 129
Reproductive trade-off, 282, 283
Resilin, 404
Resource disruption, 355
Resource holding power (RHP), 169
Reynolds number, 87
Riparian shading, 80
River continuum concept (RCC), 18–20, 88
Riverscape genetics (RG)
  anthropogenic issues, 396
  aquatic environments, 392
  aquatic insects, 394
  definition, 390
  deforestation and land-use, 393
  dendritic arrangement, 392
  DIGD, 391
  features, 391
  GEA, 391
  IBD, 394
  mDNA, 392
  odonata, 393
  phenotypes, 391
  physical barriers, 390
  river environments, 390
  structuring processes, 391

**S**
Saldidae, 210
Salinity, 82
Salinization, 82
Scales
  large
    climate cycles, 13
    geological processes, 13
    global warming, 13
    *Palaeagapetus*, 13
    temperature and precipitation, 13
  local
    abiotic and environmental variables, 16, 17
    aquatic vegetation, 15
    land usage, 14, 15
    spatial factors, 14
    substrate heterogeneity, 14
Scarabaeidae, 238
Scirtidae, 235
Semi-aquatic bugs, 170, 172, 173, 207
Sensilla
  adult Paleoptera, 160
  auxiliary cells, 140
  chemoreception
    adult stage, 150, 152, 154, 155
    aquatic stage, 147, 150
    double-walled sensilla, 147, 148
    gustatory sensilla, 147
    in insects, 147
    single-walled sensilla, 147, 148
    taste, 147
  chemoreceptors
    adult stage, 150–151
    aquatic stage, 149
  cuticular parts, 140, 142
  Ephemeroptera and Plecoptera, 142
  external and internal cues, 140
  freshwater, 140
  hemimetabolous insects, 160
  intra and interspecific communication, 140
  mechanoreception
    adult stage, 146
    aquatic stage, 143, 145, 146
    campaniform, 143
    elastic "joint membrane", 143
    filiform hairs and bristles, 143
    tubular body, 143
  mechanoreceptors
    aquatic stage, 144–145
    external and internal morphology, 141
  metamorphosis, 140
  morphology of antennae, 141
  sensory neurons, *Libellula depressa*, 153
  signals, 140
  thermo-hygroreception
    adult stage, 158–160
    aglomerular antennal lobe, 156
    aquatic insects, 155
    aquatic stage, 156, 157
    coeloconic, 156
    hygro- and thermoreceptive sensillum, 156
    hygro- and thermoreceptors, 155
    mechanosensor model, 156
    molecular and neuroanatomical investigations, 156
    sensory neurons, 156
    small terrestrial animals, 155
    thermo- and hygro-sensitivity, 155

Sexual conflict
animal behaviour, 296
Brazilian caves, 314
concept, 296
after copulation, 309, 310, 315
copulation duration, 306, 314
description, 296
dytiscid beetles, 304
Dytiscidae beetle family, 313
ecology
effects, 298
evolutionary, 297
and genitalia, 311, 312
mating duration, 305–307
mating rate, 299, 300, 302, 305
odonate mating behaviour, 299
postcopulatory sexual selection, 299
reproductive behaviour, 299
water striders, 298
gomphid behaviour, 313
harassment rate of male *Gerris buenoi*, 301
intromittent organ, 314
mating frequency, 301
mating rates, 303
odonates, 300
parental care, 297
postcopulatory sexual selection, 297, 307, 308, 310, 311, 314
sex-dependent selection, 297
study, 315
volume of sperm, 313
water striders, 298, 301
whirling beetles, 314
Sexual ornaments, 264, 281, 283
Sexual selection
evolutionary pressure, 280–282
paternal care, 335
plasticity, 284, 285
polymorphism, 284
reproductive trade-off, 282, 284
Sexual size dimorphism (SSD), 329, 336
Short Tandem Repeats (STRs), 388
Sminthuridae, 198
Smith's hypothesis, 327, 328, 330
Solar irradiance, 80
Species extinction, 378
Spercheidae, 234
Sperm competition
cryptic female choice, 297, 305, 306, 311, 312, 314
definition, 296
postcopulatory behaviours, 308
Spermatheca traits, 323
Sphingidae, 246
Spongeflies, 170
Staphylinidae, 237
Stoneflies, 170, 171
adult biology, 117
antennae, 157
duets, 120
encounter sites, 125, 127, 128
*Hemimelaena flaviventris*, 119
interbeat intervals, Iberian *Capnioneura* species, 124
intersexual communication, 132
*Isoperla morenica*, 119
male of *Protonemura gevi*, 128
male vibrational signals, 121
mating and mate guarding, 128–132
mating balls, 132
and mayflies, 158
*Perla* sp. mating, 130
reproduction and sexes, 154
Southern Iberian Peninsula, 126
vibrational call complexity, 123
vibrational communication, 118–120, 122, 125
Strahler stream order, 18
Stratiomyidae, 244
Stream Hierarchy Model (SHM), 385
Stream invertebrates, 88
Stridulation, 232
Styloconic sensilla, 140, 142
Subimago, 267
Substrate heterogeneity, 84–86
Substrate roughness, 85
Surface irregularities, 86
Syrphidae, 244

**T**

Tabanidae, 244
Temperature, 79, 80
Temperature-size rule (TSR), 78
Terrestrial adult life
Blattodea, 264, 265
Diptera, 266
Ephemeroptera, 266, 269
Hymenoptera, 269, 270
Lepidoptera, 270, 271
Megaloptera, 275
Neuroptera, 271
Odonata, 272, 274
Orthoptera, 274
Plecoptera, 275, 276
Trichoptera, 277

Terrestrial insects, 409
Terrestrial matrix, 352
Territorial behavior
  aquatic insects, 170
  male color polymorphism, 181
  non-territorial dominance interactions, 172
  non-territorial tactics, 169
Territoriality
  in adult aquatic insects, 172
  agonistic character displacement, 179, 180, 182
  in aquatic insect larvae, 170–172
  coexistence and evolution, 183
  costs and benefits, 169
  damselfly *Hetaerina americana*, 176
  definition, 168
  in insects, 169, 170
  interference and interspecific aggression in *Hetaerina*, 180
  interspecific, 179
  non-territorial tactics, 169
  signals and cues, 177, 178
  social dominance, 168
  spacing, 168
  territorial animals, 168
  territorial behavior, 182, 183
  territorial interactions, 168
  territorial signals, 178
  territorial status, 174, 175, 177
  territorial tactic, 169
  types, 172–174
  wing color morph ratios of *Mnais* damselflies, 181
Territory signals
  arbitrary identity badge model, 179
  handicap model, 178
  priority model, 178
  uncorrelated asymmetry model, 178
Thermal environment, 80
Thermal melanism hypothesis (TMH), 77, 78
Thermal tolerance range, 16
Thicopteran, 282
Thoracic microtrichia, 410
Thorp and Covich's Freshwater Invertebrates, 192
*Tigriagrion aurantinigrum*, 5
Topography, 84
Torridincolidae, 216
Tortricidae, 246
Tracheal respiratory system, 343
Trachypachidae, 240
Trichoid sensilla, 140, 142–146, 150–151
Trichoptera, 140–142, 146, 147, 150, 154, 157, 158, 170, 244, 245, 277
Trophic webs, 99, 107
Turbidity, 82

**U**
Uniporous, 140, 141, 147, 150–151
Urban heat island effect, 356
Urban wetland diversity, 351
Urbanization, 24
  aquatic insects, 352, 357, 359
  bioindicator, 357
  biomonitors, 357
  chemical pollution, 355
  definition, 349
  ephemeroptera, 352
  habitat fragmentation, 354
  human-environment interactions, 350
  light pollution, 355
  neotropics, 358
  oviposition, 352
  resource disruption, 355
  riparian ecosystems, 360
  sustainable urban design, 359
  systematic and ecological knowledge, 356
  threats, 354
  urban green spaces, 359
  urban wetland, 350
  vector breeding sites, 350
  wetland, 350

**V**
Veliidae, 207
Vibrational communication
  abdominal-substrate stridulation, 119
  adult stoneflies, 119
  aquatic insects, 118
  *Diura bicaudata*, 124
  experimental studies, 124
  intersequence intervals, 122
  *Isoperla*, 122, 125
  male and female signals, 120
  Plecoptera, 118, 122
  *Pteronarcella badia*, 125
  rubbing signals, 119
  in Southern Iberian Perlodidae, 122
  stoneflies, 118, 120
  two-beat signal, 120
Vibrational signals, 6

**W**
Walking machines, 412
Water accelerates/decelerates, 84
Water beetle, 279

Water bug, 280
Water chemical composition, 82–84
Water mites, 101, 102
Water striders
copulation duration, 305
mating rate, 301
mating system, 300
odonates and diving beetles, 298
sexual conflict studies, 298
water surface, 298, 307, 314
Water temperature, 16, 79
Wetland, 389
Whegs™, 419
World Wide Fund for Nature (WWF), 382

**Z**
Zygoptera, 201

MIX
Papier aus verantwortungsvollen Quellen
Paper from responsible sources
FSC® C105338

If you have any concerns about our products,
you can contact us on
**ProductSafety@springernature.com**

In case Publisher is established outside the EU,
the EU authorized representative is:
**Springer Nature Customer Service Center GmbH**
**Europaplatz 3, 69115 Heidelberg, Germany**

Printed by Libri Plureos GmbH
in Hamburg, Germany